Microscopic Structure and Dynamics of Liquids

NATO ADVANCED STUDY INSTITUTES SERIES

A series of edited volumes comprising multifaceted studies of contemporary scientific issues by some of the best scientific minds in the world, assembled in cooperation with NATO Scientific Affairs Division.

Series B: Physics

RECENT VOLUMES IN THIS SERIES

Volume 25 – Chemistry and Physics of One-Dimensional Metals
edited by Heimo J. Keller

Volume 26 – New Developments in Quantum Field Theory and Statistical Mechanics–Cargèse 1976
edited by Maurice Lévy and Pronob Mitter

Volume 27 – Topics in Theoretical and Experimental Gravitation Physics
edited by V. De Sabbata and J. Weber

Volume 28 – Material Characterization Using Ion Beams
edited by J. P. Thomas and A. Cachard

Volume 29 – Electron–Phonon Interactions and Phase Transitions
edited by Tormod Riste

Volume 30 – Many Degrees of Freedom in Field Theory
edited by L. Streit

Volume 31 – Many Degrees of Freedom in Particle Theory
edited by H. Satz

Volume 32 – Electron and Ion Spectroscopy of Solids
edited by L. Fiermans, J. Vennik, and W. Dekeyser

Volume 33 – Microscopic Structure and Dynamics of Liquids
edited by J. Dupuy and A. J. Dianoux

Volume 34 – Path Integrals and Their Applications in Quantum, Statistical, and Solid State Physics
edited by George Papadopoulos and J. T. Devreese

Volume 35 – Correlation Functions and Quasiparticle Interactions in Condensed Matter
edited by J. Woods Halley

The series is published by an international board of publishers in conjunction with NATO Scientific Affairs Division

A	**Life Sciences**	Plenum Publishing Corporation
B	**Physics**	New York and London
C	**Mathematical and Physical Sciences**	D. Reidel Publishing Company Dordrecht and Boston
D	**Behavioral and Social Sciences**	Sijthoff International Publishing Company Leiden
E	**Applied Sciences**	Noordhoff International Publishing Leiden

Microscopic Structure and Dynamics of Liquids

Edited by

J. Dupuy
Université Claude Bernard
Villeurbanne, France

and

A. J. Dianoux
Institut Laue–Langevin
Grenoble, France

PLENUM PRESS • NEW YORK AND LONDON
Published in cooperation with NATO Scientific Affairs Division

Library of Congress Cataloging in Publication Data

Nato Advanced Study Institute, Corsica, France, 1977.
Microscopic structure and dynamics of liquids.

(Nato advanced study institute series: Series B, Physics; 33)
"Proceedings of a NATO Advanced Study Institute held in Corsica, France, September, 1977."
Includes index.
1. Liquids–Congresses. 2. Fluid dynamics–Congresses. 3. Microstructure–Congresses. I. Dupuy, Josette. II. Dianoux, A. J. III. Title. IV. Series.
QC138.N18 1977 530.4'2 78-4197
ISBN-13: 978-1-4684-0861-4 e-ISBN-13: 978-1-4684-0859-1
DOI: 10.1007/978-1-4684-0859-1

Proceedings of a NATO Advanced Study Institute held in Corsica, France, September, 1977

Softcover reprint of the hardcover 1st edition 1978
A Division of Plenum Publishing Corporation
227 West 17th Street, New York, N.Y. 10011

Preface

What did we have in mind when in May, 1976, we (Professor de Gennes, Dr. Tourand and ourselves) thought of a Summer School in the field of liquids? First, we wanted to present and discuss the new results that have been obtained recently, in particular at the high flux reactor of the Institut Laue-Langevin in Grenoble since it became operational in 1972. In order to achieve this goal, the major part of this Summer School was devoted to an extensive presentation of the general concepts and methods of studying this state of matter (time-dependent correlation functions, molecular dynamics, intermolecular forces, spectroscopic techniques...) and concentrated on a few specific systems which have seen significant development in the last few years, both theoretically and experimentally. These systems are the different classes of simple liquids: metallic liquids, ionic liquids, simple molecular liquids and the new field of superionic conductors (solid electrolytes). Furthermore, we wanted to put some emphasis on a particular research area in the field of liquids, namely critical phenomena in fluids. This was chosen both because of our personal interest in this field and the major theoretical advances which have occurred in the last ten years.

We also wished that some new powerful techniques or new theoretical approaches be presented at this School. Thus, picosecond laser techniques, theoretical calculations on dipolar fluids, and angular correlations in molecular liquids were the subject of specific seminars.

However, such a Summer School could not be fruitful if we did not hear of new results and promote scientific exchange between the participants. This was achieved through a free poster session and five Round Table discussions animated by specialists. We were thus able to exchange ideas and knowledge in the various domains of collective motions in liquids, light and neutron scattering, critical phenomena in fluids, supercooled liquids and solid electrolytes.

We thank the lecturers, the scientific animators and the specialists for the success they brought to this School, success which was clearly evident through the enthusiasm of all participants during these two weeks.

This Summer School is the third in the field of liquids which has been organized in France. The first two, which were organized in Menton by Professor Bratos, dealt essentially with the techniques for studying liquids.

This Summer School is the fifth of the kind promoted by Professor C. H. S. Dupuy (Advanced Study Institute of Material Science). Corsica is well adapted for the particular style we wanted to attain: a common life of two weeks at a congenial location, far from laboratories or cities.

We very much appreciate the help of those who understood the advantages this region could offer to our School: Mr. V. Carlotti, "Conseiller Général," Mr. X. Carlotti, Mayor of Aleria, and Mr. M. Angelini, Mayor of Tallone, who enabled all the participants to feel the warm hospitality of Corsica, hospitality that we met every day amongst the staff of the holiday village of Casabianda. Our presence at this resort was made possible through the kindness of Mr. H. P. Bonaldi and Mr. J. Talbert.

We wish to acknowledge the financial support of NATO and the comprehension of Dr. Kester from its Scientific Affairs Division, of the Region Corse, of the Institut Laue-Langevin and of the IRF-CEA.

We thank the members of the Scientific Committee, Professors Alder, Bratos, de Gennes, Powles, Springer, and Tosi, who had the difficult task of deciding orientations and choices. Finally, we thank our everyday collaborators, Dr. Chassagne, Mr. Guiraud, Mrs. Parisot and Mrs. Giraud, for all the work they have done to assure success of this School.

J. Dupuy

A. J. Dianoux

Contents

CLOSING ADDRESS

Basic Lectures

CORRELATION FUNCTIONS AND THEIR RELATIONSHIP WITH EXPERIMENTS

Jean-Pierre Hansen

Laboratoire de Physique Théorique des Liquides*
Université Paris VI, Place Jussieu, Paris (France)

*Equipe associée au C.N.R.S.

LECTURE 1

TIME-DEPENDENT CORRELATION FUNCTIONS : DEFINITIONS AND GENERAL PROPERTIES

1.1 Introduction

Time correlation functions (TCF) are powerful tools in the study of microscopic dynamic processes in gases, liquids, solids or plasmas, in or near thermodynamic equilibrium. For the statistical mechanics theorist they are of primary importance, since they yield a measure of intrinsic microscopic fluctuations, and a quantitative description of "molecular dynamics" in many-body systems. According to the choice of microscopic quantities which are correlated, TCF's probe both single-particle motions and collective modes resulting from the cooperative motions of large numbers of particles. TCF's are also a convenient bridge between microscopic and macroscopic descriptions of many-body systems; in particular they yield exact microscopic expressions for the phenomenological transport coefficients of hydrodynamics.

To the experimentalist, TCF's are of great value, since the frequency spectra from inelastic scattering (e.g. of light or neutrons), absorption (e.g. of infrared radiation) or relaxation (e.g. dielectric) experiments can be conveniently expressed in terms of a few simple TCF's. In fact certain very different experimental techniques measure spectra which can be related to the same TCF; this means that different techniques may probe identical microscopic processes (e.g. local density fluctuations) on generally different time (or frequency) scales.

The aim of the present lectures is to introduce time correlation functions in classical and quantum statistical mechanics, to give their general properties and to show how they are related to a number of experimental frequency spectra. The first lectures will be more formal and devoted to definitions, theoretical background (linear response theory and fluctuation-dissipation theorem) and techniques which allow simple phenomenological analysis of TCF's (memory function formalism). Contact with selected experimental techniques will illustrate the usefulness of theoretical concepts. This theoretical part will be kept on a rather elementary level. Because of lack of time, kinetic theories and their applications to discontinuous hard core systems will be excluded from these lectures, despite their intrinsic theoretical interest.

The last lectures will deal with the applications of TCF's to the analysis of "molecular dynamics" and their experimental probes in simple liquids, including translational motions in rare gases and liquid metals, rotational motions in simple molecular liquids and charge fluctuations in ionic liquids (essentially molten salts). Some of these dynamical processes will be analyzed in much more detail by other lecturers of this summer school, in particular by Dr. Volino (molecular liquids). There will also be much contact (and probably some overlap) with the lectures of Dr. Mc Donald who will introduce the powerful technique of computer "experiments" which allow, in particular, to measure correlation functions which are inaccessible by laboratory experiments.

TCF's have been more thoroughly reviewed by several authors and these lectures owe much to reviews by Kubo [1], Martin [2], Gordon [3], Berne [4] and Schofield [5] and to the recent books by Forster [6], Berne and Pecora [7] and Hansen and Mc Donald [8].

1.2 Definitions

Consider a system of N particles having each ν degrees of freedom. A dynamical variable associated with such a system is any scalar, vectorial or tensorial function of the instantaneous values of some, or all, of the νN coordinates q_i and νN momenta p_i :

$$A(t) = A(\{q_i(t)\}, \{p_i(t)\}).$$

In quantum mechanical systems, A is an operator (generally hermitian) which may also depend on the spins of the particles. According to whether A is an even or odd function of the momenta, it has the signature $\varepsilon_A = \pm 1$ under time reversal. This means, that for the quantum-mechanical operator, $A^* = \varepsilon_A A$ (A^* is the complex conjugate of A). Generally the dynamical variables also have a given parity under spatial inversion ($(\{q_i\}, \{p_i\} \rightarrow (\{-q_i\}, \{-p_i\})$.

The time evolution of A is given by :

$$\dot{A} = \frac{\partial A}{\partial t} = i\mathcal{L} A \tag{1.1}$$

where the Liouville operator $\mathcal{L}$ is :

$$\mathcal{L} = i\ \{H,\ \ \} = i \sum_{i=1}^{\nu N} \left(\frac{\partial H}{\partial q_i} \frac{\partial}{\partial p_i} - \frac{\partial H}{\partial p_i} \frac{\partial}{\partial q_i} \right) \quad \text{classically,}$$

$$\mathcal{L} = \frac{1}{\hbar} [H,\ \ ] \qquad \text{in quantum mechanics.} \tag{1.2}$$

H is the hamiltonian of the isolated system, { , } denotes the classical Poisson bracket, and [,] the commutator. A formal solution of (1.1) is

$$A(t) = \exp\{ i\mathcal{L} t \} A \tag{1.3}$$

where $A \equiv A(0)$ is the value of the dynamical variable at $t = 0$. In quantum mechanics (1.3) is the usual Heisenberg representation :

$$A(t) = e^{iHt/\hbar} A\, e^{-iHt/\hbar} \tag{1.4}$$

It is frequently useful to consider local dynamical variables of the general form :

$$A(\vec{r},t) = \sum_{i=1}^{N} a_i(t)\, \delta(\vec{r} - \vec{r}_i(t)) \tag{1.5}$$

where a_i is any physical quantity (e.g. the mass, the linear or angular momentum of a molecule) associated with particle i, and $\vec{r}_i(t)$ is the center of mass (CM) position of the particle. In quantum mechanical systems, a_i and $\vec{r}_i$ do not, in general, commute so that (1.5) must be properly symmetrized. We shall also deal with the spatial Fourier components of $A(\vec{r},t)$:

$$A(\vec{k},t) = \int A(r,t) e^{-i\vec{k}.\vec{r}} = \sum_{i=1}^{N} a_i(t)\, e^{-i\vec{k}.\vec{r}_i(t)} \tag{1.6}$$

A dynamical variable is said to be <u>conserved</u> if it satifies a continuity equation :

$$\frac{\partial A(\vec{r},t)}{\partial t} + \vec{\nabla} \cdot \vec{J}_A(\vec{r},t) = 0 \tag{1.7}$$

where $\vec{j}_A$ is the "current" associated with the "density" A. (1.7) is a local expression of the fact that $A^{tot} = \sum_{i=1}^{N} a_i$ is constant. A classic example is the number density ($a_i = 1$) :

$$\rho(\vec{r},t) = \sum_{i=1}^{N} \delta(\vec{r} - \vec{r}_i(t)) \tag{1.8}$$

its associated current is simply :

$$\vec{j}_\rho(\vec{r},t) = \sum_{i=1} \frac{\vec{p}_i(t)}{m_i} \delta(\vec{r} - \vec{r}_i(t)) \tag{1.9}$$

where $\vec{p}_i$ is the linear momentum of CM motion. The continuity equation for the Fourier components of a conserved variable reads simply :

$$\frac{\partial A}{\partial t}(\vec{k},t) + ik.\vec{j}_A(\vec{k},t) = 0 \tag{1.10}$$

In classical statistical mechanics the equilibrium time correlation function of two dynamical variables A and B is defined as :

$$C_{AB}(t',t'') = < A(t') \; B(t'') > \tag{1.11}$$

where the angular brackets denote either an ensemble average over initial conditions :

$$<A(t') \; B(t'')> = \int d^{\nu N}q(t'') \int d^{\nu N}p(t'') A(\{q_i(t')\}, \{p_i(t')\}) \times B(\{q_i(t'')\},\{p_i(t'')\}) \; f_o^{(N)}(\{q_i(t'')\},\{p_i(t'')\}) \tag{1.12}$$

or the average over time :

$$<A(t') \; B(t'')> = \lim_{T \to \infty} \frac{1}{T} \int_o^T A(t'+ s) \; B(t'' + s)ds \tag{1.13}$$

In (1.12), $f_o^{(N)}$ is the equilibrium distribution function in $2\nu N$-dimensional phase space. For a canonical ensemble, $f_o^{(N)} = e^{-\beta H}/Q_N$, $\beta = 1/k_BT$ and Q_N is the partition function. Both averages yield the same result in the thermodynamic limit, provided the system is ergodic, which will always be assumed henceforth.

If the dynamical variables vary in space as in (1.5), the corresponding TCF is non-local both in time and in space :

$$C_{AB}(\vec{r}', \vec{r}''; t', t'') = \langle A(\vec{r}', t')\, B(\vec{r}'', t'') \rangle \tag{1.14}$$

Since the Fourier components of such local variables are complex quantities, their TCF's are generally defined as :

$$\begin{aligned} C_{AB}(\vec{k}', \vec{k}''; t', t'') &= \langle A(\vec{k}', t')\, B^{*}(\vec{k}'', t'') \rangle \\ &= \langle A\, (\vec{k}', t')\, B(-\vec{k}'', t'') \rangle \end{aligned} \tag{1.15}$$

since the a_i are real (hermitian).

In quantum mechanics we define the one-sided TCF :

$$C_{AB}(t', t'') = \langle A(t')\, B(t'') \rangle = \mathrm{Tr}\,\{\rho\, A(t')\, B(t'')\} \tag{1.16}$$

where $\rho = e^{-\beta H}/\mathrm{Tr}\, e^{-\beta H}$ is the equilibrium density matrix which is diagonal in the basis of the eigenstates $|n\rangle$ of H; consequently :

$$\begin{aligned} \langle A(t')\, B(t'') \rangle &= \sum_n \rho_n \langle n| A(t')\, B(t'') |n\rangle \\ &= \sum_n \sum_m \rho_n \langle n\, |A|\, m \rangle\langle m\, |B|\, n \rangle\, e^{i\omega_{nm}(t'-t'')} \end{aligned} \tag{1.17}$$

where we have used (1.4) and the fact that the states $|n\rangle$ form a complete set; $\omega_{nm} = (E_n - E_m)/\hbar$, with E_n the energy eigenvalue in the state $|n\rangle$. In the absence of any degeneracy with respect to the total angular momentum (fluid in a rigid container), the energy eigenstates $|n\rangle$ can always be chosen <u>real</u>. Taking into account the time reversal signatures of A and B, it then follows from (1.17) that :

$$C^{*}_{AB}(t', t'') = \varepsilon_A\, \varepsilon_B\, C_{AB}(t'', t') \tag{1.18}$$

In general A(t') and B(t") do not commute and it is often convenient to introduce the <u>symmetrized</u> TCF :

$$C^{S}_{AB}(t', t'') = \langle \tfrac{1}{2}\, (A(t'), B(t'')) \rangle \tag{1.19}$$

where (,) denotes the anticommutator. If A and B are hermitian, C^S_{AB} is equal to the <u>real</u> part of C_{AB}, and from (1.18) :

$$C^{S}_{AB}(t', t'') = C^{S\,*}_{AB}(t', t'') = \varepsilon_A \varepsilon_B\, C^{S}_{AB}(t'', t') \tag{1.20}$$

Kubo has introduced a "canonical" TCF which plays a central role in linear response theory :

$$C^{C}_{AB}(t',t'') = \frac{1}{\beta}\int_{0}^{\beta} d\lambda \; \langle A(t' - i\hbar\lambda)\; B(t'')\rangle \qquad (1.21)$$

where :

$$A(t' - i\hbar\lambda) = e^{\lambda H} \; A(t')e^{-\lambda H}$$

Note that all three definitions (1.16), (1.19) and (1.21) reduce to the same TCF in the classical ($h \to 0$) limit. The canonical TCF has properties similar to those of the symmetrized TCF [1].

In the special case where a dynamical variable is correlated with itself (or its complex or hermitian conjugate), the TCF is called an <u>autocorrelation</u> function (ACF) :

$$C_{AA}(t',t'') = \langle A(t')\; A(t'')\rangle \qquad (1.22)$$

1.3 General Properties and Orders of Magnitude

a) The fundamental property of all equilibrium TCF's is their <u>stationarity</u>. Since quilibirum averages are stationary, i.e. independent of the time origin, the TCF's are invariant under time translations, and hence depend only on $t = t' - t''$:

$$C_{AB}(t',t'') = \langle A(t')\; B(t'')\rangle = \langle A(t' - t'')\; B(0)\rangle$$

$$= C_{AB}(t' - t'') \qquad (1.23)$$

The stationary property is in fact immediately apparent from (1.17) and allows us to write :

$$\frac{d}{ds}\langle A(t+s)\; B(s)\rangle = \langle \dot{A}(t+s)\; B(s)\rangle + \langle A(t+s)\; \dot{B}(s)\rangle = 0.$$

Thus :

$$\langle \dot{A}(t)B\rangle = -\langle A(t)\dot{B}\rangle \qquad (1.24)$$

and similarly :

$$\frac{d^2}{dt^2}\langle A(t)B\rangle = \langle \ddot{A}(t)B\rangle = -\langle \dot{A}(t)\dot{B}\rangle \qquad (1.25)$$

More generally it follows from (1.20) and (1.23) that :

$$C^{S}_{AB}(t) = \varepsilon_A\, \varepsilon_B\, C^{S}_{AB}(-t) \qquad (1.26)$$

In particular, symmetrized (or classical) autocorrelation functions are real, even functions of time, whereas one-sided ACF's have an even real part and an odd imaginary part.

b) It is clear that :

$$\lim_{t \to \infty} C_{AB}(t) = \langle AB \rangle \tag{1.27}$$

where < AB > is the static correlation function. It follows immediately from Schwarz's inequality that :

$$|C^S_{AB}(t)| \leqslant [\langle A A^+ \rangle \langle B B^+ \rangle]^{1/2} \tag{1.28}$$

In particular for an ACF :

$$|C^S_{AA}(t)| \leqslant C_{AA}(0) = \langle A A^+ \rangle \tag{1.29}$$

as one would intuitively expect.

c) In the limit $t \to \infty$, the two dynamical variables become uncorrelated, i.e. :

$$\lim_{t \to \infty} C_{AB}(t) = \langle A \rangle \langle B \rangle \tag{1.30}$$

It is often convenient to define the variable in such a way as to exclude the invariant part and to calculate the time correlation of the fluctuating parts, i.e. :

$$C_{AB}(t) = \langle [A(t) - \langle A \rangle][B - \langle B \rangle] \rangle \tag{1.31}$$

This has the advantage that $\lim_{t \to \infty} C_{AB}(t) = 0$, corresponding physically to the complete loss of correlation between the fluctuating parts of the variables.

d) If $C_{AB}(t)$ is defined as in (1.31), it is possible to define its Fourier transform $\hat{C}_{AB}(\omega)$, called the spectral function (or power spectrum), and its Fourier-Laplace transform $\tilde{C}_{AB}(z)$:

$$\hat{C}_{AB}(\omega) = \frac{1}{2\pi} \int_{-\infty}^{+\infty} e^{i\omega t} C_{AB}(t) dt \tag{1.32}$$

$$\tilde{C}_{AB}(z) = \int_{0}^{\infty} e^{izt} C_{AB}(t) dt \tag{1.33}$$

Since $C_{AB}(t)$ is bounded (cf. (1.28)), it is clear that $\tilde{C}_{AB}(z)$ is analytic in the upper half of the complex z plane. One easily verifies that $\tilde{C}_{AB}(z)$ is the Hilbert transform of $\hat{C}_{AB}(\omega)$:

$$\tilde{C}_{AB}(z) = -i \int_{-\infty}^{+\infty} d\omega \frac{\hat{C}_{AB}(\omega)}{\omega - z} \tag{1.34}$$

We see from (1.18) that the spectrum of an autocorrelation function is necessarily always <u>real</u> and satisfies :

$$\hat{C}_{AA}(\omega) = \lim_{\varepsilon \to 0} \frac{1}{\pi} R\, \tilde{C}_{AA}(\omega + i\varepsilon) \tag{1.35}$$

Explicitly :

$$\hat{C}_{AA}(\omega) = \sum_{n}\sum_{m} \rho_n \,|< m|A|n >|^2\, \delta(\omega - \omega_{nm}) \tag{1.36}$$

Since $\rho_n/\rho_m = \exp\{-\beta(E_n - E_m)\}$, it then follows, for hermitian A, that :

$$\hat{C}_{AA}(-\omega) = e^{-\beta\hbar\omega}\, \hat{C}_{AA}(\omega) \tag{1.37}$$

Since $C^S_{AA}(t)$ is real and even in t, $\hat{C}^S_{AA}(\omega)$ is real and even in ω.

e) The spectral function of an ACF is non-negative. Consider for example the classical ACF $C_{AA}(t-t') = < A(t)A^*(t') >$ and define the auxiliary variable :

$$\mathcal{A} = \frac{1}{\sqrt{2T}} \int_{-T}^{T} dt\, A(t)\, a(t)$$

where a(t) is an arbitrary function; clearly the "scalar product" $<\mathcal{A}\mathcal{A}^*>$ is non-negative :

$$<\mathcal{A}\mathcal{A}^*> = \frac{1}{2T} \int_{-T}^{T} dt \int_{-T}^{T} dt'\, a(t') < A(t)A^*(t') > a^*(t') \geqslant 0$$

Now choose $a(t) = \exp\{i\omega t\}$, go from the variables t,t' to t, $\tau = t-t'$ and take the limit $T\to\infty$; this yields :

$$\lim_{T\to\infty} <\mathcal{A}\mathcal{A}^*> = \lim_{T\to\infty} \int_{-T}^{T} e^{i\omega\tau}\, C_{AA}(\tau)\, d\tau$$

$$= \hat{C}_{AA}(\omega) \geqslant 0 \tag{1.38}$$

f) If the liquid is homogeneous, translational invariance in space implies that non-local TCF's defined by (1.14) depend only on the relative position $\vec{r} = \vec{r}' - \vec{r}''$:

$$C_{AB}(\vec{r}', \vec{r}''; t', t'') = C_{AB}(\vec{r}' - \vec{r}'', t' - t'') \tag{1.39}$$

Translational invariance also implies that correlations between Fourier components $A(\vec{k}', t')$ and $B(\vec{k}'', t'')$ are only non-zero if $\vec{k}' = -\vec{k}''$:

$$\begin{aligned} C_{AB}(\vec{k}', \vec{k}''; t'-t'') &= \langle A(\vec{k}',t')\, B(-\vec{k}', t'')\rangle \delta_{\vec{k}',\vec{k}''} \\ &= C_{AB}(\vec{k}', t'-t'') \end{aligned} \tag{1.40}$$

Clearly $C_{AB}(\vec{k},t)$ is the spatial Fourier transform of $C_{AB}(\vec{r},t)$:

$$C_{AB}(\vec{k},t) = \int e^{-i\vec{k}.\vec{r}}\, C_{AB}(\vec{r}, t) d^3r \tag{1.41}$$

g) The correlation function $C_{AB}(t)$ of two dynamical variables A and B of opposite spatial parities is zero at all times, provided the hamiltonian, and hence the Liouville operator, has even parity which is quite generally the case. For instance the local number density (1.8) and its associated current (1.9) are uncorrelated at all times. The obviousness of the given example renders a formal proof unnecessary !

Similarly, if the dynamical variables have a well-defined signature under the reflection of all coordinates and momenta through a plane, the TCF of two variables of opposite signature is identically zero.

h) The ACF's of conserved variables have a very important property in the long wavelength ($k \to 0$) limit :

$$\begin{aligned} \lim_{k \to 0} \hat{C}_{AA}(\vec{k},\omega) &= \int d^3r \langle A(\vec{r},t)\, A(0,0) \rangle \\ &= \langle [\int d^3r\, A(\vec{r},t)]\, A(0,0)\rangle \\ &= \langle A^{tot}\, A(0,0) \rangle \end{aligned}$$

which is independent of time, since A^{tot} is constant. Hence

$$\lim_{k \to 0} \hat{C}_{AA}(k,\omega) = \text{Constant} \times \delta(\omega) \tag{1.42}$$

The "correlation time" of the long wave-length limit of the ACF of a conserved variable is thus infinite and its spectrum reduces to a δ-function. As the wave-length decreases, we expect the correlation time to decrease, and to become of the order of magnitude of characteristic microscopic times when the wave-length becomes of the order of magnitude of the inter-molecular spacing. A simple dimensional argument suggests that a typical microscopic relaxation for translational motions in liquids will be of the order of :

$$\tau \simeq a(m/\varepsilon)^{1/2}$$

where ε is the strength of the intermolecular potential, m is the mass of the particles and a the intermolecular spacing ($a \gtrsim \sigma$, where σ is the range of the repulsive potential in neutral liquids). For all simple liquids, $\tau \simeq 10^{-12}$ sec. Note that the order of magnitude of a typical "collision time" :

$$\tau_c \simeq (a - \sigma) / < v^2 >^{1/2} \simeq (a - \sigma)(m/k_B T)^{1/2} \simeq 0.1\ \tau$$

yields $\tau_c \simeq 10^{-13}$ sec, since $k_B T \simeq \varepsilon$ in the liquid range, and $(a - \sigma) / \sigma \simeq 0.1$.

From the classical mean-square angular velocity $< \omega^2 > = k_B T/I$, where I is the moment of inertia of a molecule, we find $\tau_R \simeq 10^{-12}$sec for the rotational relaxation time in liquid N_2. Vibrational frequencies on the other hand are considerably higher, so that the associated vibrational times are much shorter ($\tau_V \simeq 10^{-14}$ sec).

1.4 Frequency Moments and Sum Rules

For non-singular inter-molecular potentials, the TCF $C_{AB}(t)$ admits a Taylor expansion around t = 0 :

$$\begin{aligned} C_{AB}(t) &= \sum_{n=0}^{\infty} \frac{t^n}{n!} C_{AB}^{(n)}(0) \\ &= \sum_{n=0}^{\infty} \frac{t^n}{n!} < A^{(n)}(0)\ B(0) > \\ &= \sum_{n=0}^{\infty} \frac{t^n}{n!} (-1)^n < A(0)\ B^{(n)}(0) > \qquad (1.43) \\ &= \sum_{n=0}^{\infty} \frac{t^n}{n!} (-1)^n < A(0)\ (i\mathcal{L})^n\ B(0) > \end{aligned}$$

where the stationary condition (1.23), and equ. (1.1) have been used. $C_{AB}^{(n)}(0)$ is the nth derivative at $t = 0$. According to (1.25) the expansion of a symmetrized (or classical) TCF contains only even or odd powers in t. In particular a symmetrized ACF contains only even powers :

$$C_{AA}^{S}(t) = \sum_{n=0}^{\infty} \frac{t^{2n}}{(2n)!} \; C_{AA}^{S(2n)}$$

$$= \sum_{n=0}^{\infty} \frac{t^{2n}}{(2n)!} < (A^{(2n)}(0), A(0)) >$$

$$= \sum_{n=0}^{\infty} \frac{t^{2n}}{(2n)!} (-1)^n < A^{(n)}(0) \; A^{(n)}(0) >$$

$$= \sum_{n=0}^{\infty} \frac{t^{2n}}{(2n)!} (-1)^n <[(i\mathcal{L})^n A(0)]^2 > \qquad (1.44)$$

Differentiating the inverse Fourier transform of (1.32) 2n times with respect to t yields :

$$< \omega^{2n} >_{AA} \equiv \int_{-\infty}^{+\infty} \omega^{2n} \hat{C}_{AA}^{S}(\omega) d\omega$$

$$= (-1)^n \hat{C}_{AA}^{S(2n)}(0) \qquad (1.45)$$

Thus the frequency moments of the spectral function are directly related to derivatives of the ACF taken at $t = 0$. The latter are static correlation functions which can generally be expressed as integrals over static distribution functions, and can often be explicitly calculated for low values of n. Note that for non-local ACF's, the frequency moments are functions of k.

The continuity equation for conserved dynamical variables leads to simple expressions for the second moment sum rules, called f-sum rules. For simplicity we restrict ourselves to classical systems. From (1.45) and (1.25) :

$$< \omega^2>_{AA} = - \ddot{C}_{AA}(\vec{k},0) = < \dot{A}(\vec{k},0) \; \dot{A}(-\vec{k},0) >$$

$$= < -i\vec{k} \cdot \vec{j}_A(\vec{k},0) \; i\vec{k} \cdot \vec{j}_A(-k,0) >$$

$$= k^2 < |j_A^x(\vec{k},0)|^2 > \qquad (1.46)$$

where the last equality holds for an isotropic fluid (k chosen along Ox). For the special case where A is the local number density (1.8), the average in (1.46) is easily calculated, yielding for N identical particles of mass m :

$$< \omega^2 >_{\rho\rho} = \frac{N k_B T\, k^2}{m} = N \omega_o^2 \qquad (1.47)$$

The characteristic frequency $\omega_o \to 0$ with k, in agreement with (1.42) !

Finally we note that the spectral representation (1.34) leads directly to a "high frequency" expansion of $\widetilde{C}_{AB}(z)$ in powers of $1/z$, which is intimately related to the short time expansion (1.43). An example of such an expansion will be used in section 2.5.

1.5 An Illustration : The Velocity ACF and the Langevin Theory of Brownian Motion

To illustrate and lighten up this rather tedious lecture devoted to definitions and formalism, we briefly consider here one of the simplest TCF's, the classical velocity ACF and its relation to the theory of Brownian motion. This will allow us to introduce in an intuitive fashion some of the fundamental concepts encountered in later lectures.

Consider a Brownian particle in solution, or a tagged molecule among N identical molecules; let $\vec{v}(t)$ be the particle velocity at time t. The velocity ACF is then defined as :

$$Z(t) = \frac{1}{3} < \vec{v}(t).\vec{v}(0) > = <v_x(t)\ v_x(0)> \qquad (1.48)$$

where we use the notation Z(t) rather than $C_{\vec{v}\vec{v}}(t)$, and the statistical average is over the full phase space of the particle and the solvent (or "bath"). We shall see later that Z(t) is in principle measurable from the long wave-length limit of incoherent inelastic neutron scattering. Z(t) has also been calculated with high accuracy for a number of liquids by computer "experiments". These calculations show that Z(t) relaxes from its initial value $Z(0) = <v^2> = 3\ k_B T/m$ on a time scale of 10^{-12}sec and shows interesting features at triple point densities. Here we shall show that Z(t) is directly related to the macroscopic self-diffusion coefficient, before considering briefly the problem of Brownian motion. Consider a set of particles which, from an initial position $\vec{r}(0)$, diffuse, in a time t, to positions $\vec{r}(t)$. The self-diffusion coefficient is given by the well-known Einstein relation (which will be "derived" in lecture 3):

$$D = \lim_{t \to \infty} \frac{1}{6t} \langle |\vec{r}(t) - \vec{r}(0)|^2 \rangle \tag{1.49}$$

a relation typical of stochastic "random walk" processes. (1.49) can be rewritten in another useful form. We note first that :

$$\vec{r}(t) - \vec{r}(0) = \int_0^t \vec{v}(t')\, dt'$$

Hence :

$$\langle |\vec{r}(t) - \vec{r}(0)|^2 \rangle = \int_0^t dt' \int_0^t dt'' \langle \vec{v}(t').\vec{v}(t'') \rangle$$

$$= 2\int_0^t dt' \int_0^{t'} dt'' \langle \vec{v}(t'-t'').\vec{v}(0) \rangle$$

where we have used translational invariance; changing from the variables t',t" to t', s = t' - t" , using the definition (1.48) and integrating by parts, we obtain :

$$\langle |\vec{r}(t) - \vec{r}(0)|^2 \rangle = 6\int_0^t dt' \int_0^{t'} ds\, Z(s)$$

$$= 6t\int_0^t (1 - s/t)\, Z(s) ds$$

Upon replacing in (1.49) we obtain the important relation :

$$D = \int_0^{\infty} Z(s) ds \tag{1.50}$$

expressing the self-diffusion coefficient as the time integral of the velocity ACF. This is a first example of a Green-Kubo formula which relates a macroscopic, phenomenological transport coefficient to the time-integral of a microscopic TCF. A crude estimate of D in liquid Argon ($T \simeq 10^2$ °K), based on (1.50) is :

$$D \simeq \frac{1}{2} Z(0) \times \tau = \frac{k_B T \times \tau}{2m} \simeq 10^{-4} \text{ cm}^2/\text{sec.}$$

in qualitative agreement with experimental values.

In the case of a large Brownian particle diffusing in a solvent of light molecules, Z(t) is easily deduced from the phenomenological stochastic theory of Langevin [9], who assumes that the force acting on a Brownian particle consists of two parts : a systematic frictional force proportional to the velocity $\vec{v}(t)$, and a randomly fluctuating force, $\vec{R}(t)$, which arises from collisions with the surrounding molecules :

$$m\,\dot{\vec{v}}(t) = -\,m\,\xi\,\vec{v}(t) + \vec{R}(t) \tag{1.51}$$

where ξ is the <u>friction coefficient</u>. The random force is assumed

a) to vanish in the mean :

$$<\vec{R}(t)> = 0 \tag{1.52}$$

b) to be uncorrelated with the velocity :

$$<\vec{R}(t)\cdot\vec{v}(0)> = 0 \tag{1.53}$$

c) to have a correlation time which is infinitely short :

$$<\vec{R}(t)\cdot\vec{R}(0)> = 2\pi\,R_o\,\delta(t) \tag{1.54}$$

The meaning of (1.54) is that the spectral function of the random force is a constant, R_o ("white" spectrum):

$$R_o = \frac{1}{2\pi}\int_{-\infty}^{+\infty} <\vec{R}(t)\cdot\vec{R}(0)>\,e^{i\omega t}\,dt$$

These assumptions are reasonable ones when the Brownian particle is much larger than its neighbours, because even on a short time-scale the motion of the Brownian particle will be determined by a very large number of essentially uncorrelated collisions.

The two terms on the r.h.s. of equ. (1.51) are not independent. To see the relation between them we first write the solution to (1.51) in the form :

$$m\vec{v}(t) = m\vec{v}(0)\,\exp\{-\xi t\} + \exp\{-\xi t\}\int_o^t \exp\{\xi s\}\,\vec{R}(s)\,ds \tag{1.55}$$

Squaring and taking the mean we find, using (1.53) and (1.54) :

$$m^2 <|v(t)|^2> = m^2<|\vec{v}(0)|^2> \exp\{-2\xi t\} + \frac{\pi R_o}{\xi}\,[1-\exp\{-2\xi t\}] \qquad 1.56)$$

We now take the limit $t\to\infty$. Under equilibrium conditions :

$$<|\vec{v}(\infty)|^2> = <|\vec{v}(0)|^2> = 3\,k_BT/m$$

from which it follows that :

$$\xi = \frac{\pi\beta R_o}{3m} = \frac{\beta}{3m}\int_o^{\infty} <\vec{R}(t)\cdot\vec{R}(0)>dt \qquad (1.57)$$

From a physical point of view it is not unexpected to find a link between the frictional and random forces. If the Brownian particle were to be drawn through the medium by an external force, it would experience a systematic retarding force, proportional to the velocity, which arises from random collisions with the solvent molecules. (1.57) is in fact a special case of a very general result, known as the "fluctuation-dissipation theorem", which we shall derive in the next lecture.

The friction coefficient is also related to the diffusion coefficient. Multiplying (1.51) by $\vec{v}(0)$, taking the statistical average and using (1.53) we obtain :

$$\dot{Z}(t) = \frac{1}{3}<\dot{\vec{v}}(t)\cdot\vec{v}(0)> = -\xi Z(t)$$

which can be directly integrated to :

$$Z(t) = Z(0)\,e^{-\xi t} \qquad (1.58)$$

i.e. the Langevin theory predicts an exponential decay of Z(t). Replacing in (1.50) we immediately recover the well-known Einstein relation :

$$D = \frac{k_BT}{\xi m} \qquad (1.59)$$

Note that the velocity ACF (1.58) is very poor at short times, since it is not an even function of t as it should. This deficiency stems from the breakdown of the Langevin theory and, in particular, of the assumptions on the random force at very short times. Furthermore these assumptions are very hard to justify for particles of the same size. The question will be reconsidered in lecture 3, where the Langevin theory will be generalized and given a sound statistical mechanics basis. In particular retardation effects will be introduced through a generalized friction coefficient which will be non-local in time ("memory function") and the concept of "random force" will be given a precise statistical mechanical meaning.

LECTURE 2

LINEAR RESPONSE THEORY

In this lecture we study the behaviour of a system perturbed by an external field or probe, to which it is weakly coupled. The main result will be that the response of the system to a weak perturbation can be entirely described in terms of time correlation functions of the system at equilibrium, i.e. in the absence of the external field. Since under the influence of an external perturbation the system normally heats up, which corresponds to an energy dissipation process, the important result is generally referred to as the "fluctuation-dissipation theorem".

The hamiltonian in the presence of the external field is :

$$H = H_o + H' \tag{2.1}$$

where H_o characterizes the unperturbed system and H' is the perturbation which is generally time-dependent :

$$H'(t) = -\int d^3r' \; A(\vec{r}') \; F(\vec{r}',t) \tag{2.2}$$

where $F(\vec{r},t)$ is the time-dependent external field (scalar, vectorial or tensorial) and $A(\vec{r})$ is the dynamical variable of the system which couples to it; the minus sign is conventional.

2.1 Inelastic Scattering Cross Sections

We first consider a system interacting with an incident monochromatic beam of particles represented by a plane wave. These particles may be either neutrons, which interact with the system's nuclei, or electrons, which interact with the system's charges. The case of an incident photon beam (light scattering) is slightly more complicated and will be considered later. The neutrons probe the microscopic nuclear density $\rho_n(\vec{r})$, while electrons probe the microscopic charge density $\rho_z(\vec{r})$. We shall now show that the inelastic scattering cross section is directly related to the spectrum of the ACF's of these densities. If we consider a single incident particle located at $\vec{r}$, the perturbation is of the general form (2.2) with $A(\vec{r}') = \rho_n(\vec{r}')$ or $\rho_z(\vec{r}')$ and $F(\vec{r},t) = -V(\vec{r} - \vec{r}')$, where V is the interaction potential of the particle with a nucleus or a charge of the system :

$$H' = \int d^3r' \; \rho(\vec{r}') \; V(\vec{r} - \vec{r}') \tag{2.3}$$

For the sake of simplicity we shall assume that, in the case of neutron scattering, all nuclei in the target are identical and have zero spin, so that they all interact in the same way with the incident neutron. If the interaction is represented by a Fermi potential of the type :

$$V(r) = \frac{2\pi a\hbar^2}{m} \delta(\vec{r}) \tag{2.4}$$

this means that the scattering length a is identical for all nuclei. The corresponding scattering is then purely <u>coherent</u>. If a varies (due to the presence of different nuclear species or nuclear spins), the scattering is partly <u>incoherent</u>; but we shall at present ignore this complication.

The situation is simpler for electrons, since in that case V(r) is simply the Coulomb potential - e/r.

Let $\vec{k}_i$ and $\vec{k}_f$ be the incident and final wave vectors of the probe. In the Born approximation the initial (incident) and final (scattered) states of the probe are plane waves

$$|\vec{k}_i > \sim e^{i\vec{k}_i\cdot\vec{r}} \qquad |\vec{k}_f > \sim e^{i\vec{k}_f\cdot\vec{r}} \tag{2.5}$$

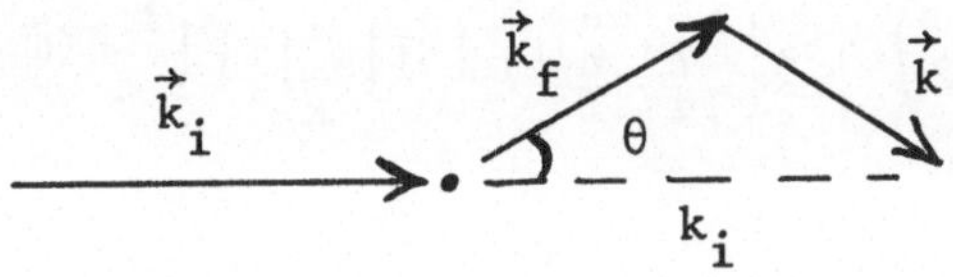

θ is the scattering angle and $\vec{k} = \vec{k}_i - \vec{k}_f$ the scattering wave vector. If $|i>$ and $|f>$ are the initial and final states of the scattering system and E_i and E_f the corresponding energy eigenvalues of H_o, i.e. the energy of the target before and after the scattering event, energy conservation requires that :

$$\hbar\omega_{fi} = E_f - E_i = \hbar\omega \tag{2.6}$$

where $\hbar\omega = \frac{\hbar^2}{2m}(k_i^2 - k_f^2)$ is the energy <u>gain</u> (if $E_f > E_i$) or <u>loss</u> (if $E_f < E_i$) of the scattered particle of mass m.
According to the "golden rule" the probability per unit time of a transition from the initial state $|\vec{k}_i>|i>$ of the target + probe to the final state $|\vec{k}_f>|f>$ is given by :

$$W_{i \to f} = \frac{2\pi}{\hbar^2} |<f| <\vec{k}_f|H'|\vec{k}_i| i>|^2 \delta(\omega - \omega_{fi}) \tag{2.7}$$

where the δ-function simply expresses energy conservation.

Using (2.3), (2.5) and the factorization property of convolution products with respect to Fourier transformation, (2.7) is easily cast in the form :

$$W_{i \to f} = \frac{2\pi}{\hbar^2} |\langle f|\rho^{+}_{\vec{k}} \, | \, i \rangle|^2 \; |V_{\vec{k}}|^2 \; \delta(\omega - \omega_{fi}) \tag{2.8}$$

where

$$V_{\vec{k}} = \int d^3 r \, e^{-i\vec{k}\cdot\vec{r}} \; V(\vec{r}) = \begin{cases} \dfrac{2\pi a\hbar^2}{m} & \text{(neutrons)} \\ -\dfrac{4\pi e}{k^2} & \text{(electrons)} \end{cases} \tag{2.9}$$

and $\rho^{+}_{\vec{k}} \equiv \rho_{-\vec{k}}$ is the kth Fourier component of $\rho_n(\vec{r})$ or $\rho_z(\vec{r})$.

To obtain the differential inelastic cross-section for scattering into the solid angle $d\Omega$ in the energy range $\hbar d\omega$, we must sum (2.8) over final states of the target, multiply by the differential element $d^3k_f \,/(2\pi)^3 = k_f^2 dk_f d\Omega/(2\pi)^3$ $= (m/\hbar^2)k_f\hbar \; d\omega d\Omega/(2\pi)^3$, divide by the incident flux $\hbar k_i/m$, and finally take the thermal average over initial states of the target :

$$\frac{d^2\sigma}{d\omega d\Omega} = \frac{k_f}{k_i}\left(\frac{m}{2\pi\hbar^2}\right)^2 |V_{\vec{k}}|^2 \sum_i \sum_f \rho_i \, |\langle f|\rho^{+}_{\vec{k}}|i\rangle|^2 \; \delta(\omega - \omega_{fi})$$

$$= \frac{k_f}{k_i}\left(\frac{m}{\hbar^2}\right)^2 \frac{1}{(2\pi)^3} |V_{\vec{k}}|^2 \sum_i \sum_f \rho_i \int_{-\infty}^{+\infty} dt \; e^{i(\omega-\omega_{fi})t} \, |\langle f|\rho^{+}_{\vec{k}}|i\rangle|^2 \tag{2.10}$$

where use was made of a well-known representation of the δ-function. Using (2.6), (1.4) and the fact that the $|i\rangle$ and $|f\rangle$ are eigenstates of H_o, we obtain :

$$e^{-i\omega_{fi}t} \; |\langle f| \; \rho^{+}_{k} |i\rangle|^2 = \langle i|\rho_{\vec{k}}(t)| \; f \rangle \langle \, f|\rho^{+}_{\vec{k}}| \; i \rangle$$

Replacing this into (2.10) and using the completeness condition, we arrive at the final result :

$$\frac{d^2\sigma}{d\omega d\Omega} = \frac{k_f}{k_i}\left(\frac{m}{2\pi\hbar^2}\right)^2 |V_{\vec{k}}|^2 \, NS(\vec{k},\omega) \tag{2.11}$$

where N is the number of scatterers (nuclei or charges) in the system and $S(\vec{k},\omega)$ is the spectral function of the nuclear or charge density autocorrelation functions :

$$S(\vec{k},\omega) = \frac{1}{2\pi}\int_{-\infty}^{+\infty} e^{i\omega t}\,\frac{1}{N}\,<\rho_{\vec{k}}(t)\;\rho_{-\vec{k}}(0)>dt \qquad (2.12)$$

In the neutron scattering literature, $S(\vec{k},\omega)$ is often referred to as the "dynamic structure factor", and the ACF of the Fourier components of the density ρ_n is called the "intermediate scattering function" :

$$F(\vec{k},t) = \frac{1}{N} < \rho_{\vec{k}}(t)\rho_{-\vec{k}}(0) > \qquad (2.13)$$

In the simple case of a monatomic (rare gas or liquid metal) fluid, $\rho_n(\vec{r})$ coincides with the local number density $\rho(\vec{r})$ defined in (1.8), and :

$$\rho_{\vec{k}} = \sum_{i=1}^{N} e^{-i\vec{k}.\vec{r}_i} \qquad (2.14)$$

We shall return to these correlation functions in our study of density fluctuations in simple liquids (lecture 4).

The important results (2.11) is an example of the relation between inelastic scattering cross sections and TCF's. Note that, according to the general properties of the power spectra of ACF's, $S(\vec{k},\omega)$, and hence the differential cross section, is real and positive, as it should !

Note finally that the integrated <u>intensity</u> of the particle beam scattered in a given direction, i.e. the differential cross section integrated over all frequencies ω, for fixed Ω , is proportional,according to (1.45), to the value of the intermediate scattering function at t = 0, which is generally referred to as the <u>static</u> structure factor :

$$\frac{d\sigma}{d\Omega} \# \int_{-\infty}^{+\infty} S(\vec{k},\omega)\,d\omega = F(\vec{k},0) \equiv S(\vec{k})$$

2.2 Linear Response Functions

We now generalize the considerations of the previous section by studying the response of the system to a weak time-dependent external field. The perturbation is of the general form (2.2); the external field can always be considered as a superposition of monochromatic plane waves. Since we are interested in the linear response of the system, it is sufficient to consider a single plane wave of wave number $\vec{k}$ and frequency ω :

$$F(\vec{r},t) = F(\vec{k})\ e^{i(\vec{k}\cdot\vec{r}-\omega t)} \tag{2.15}$$

Replacing (2.15) in (2.2) we obtain :

$$H' = -A^{+}(\vec{k})\ F(\vec{k})e^{-i\omega t} \tag{2.16}$$

To simplify notations we shall ignore the $\vec{k}$-dependence in the following calculation, i.e. assume the external field to be spatially homogeneous; the $\vec{k}$-dependence is trivially re-introduced at the end. Moreover the perturbation is assumed to vanish in the infinite past ($t = -\infty$), at which time the unperturbed system is assumed to be in thermal equilibrium. Consequently we write H' simply as :

$$H' = -A\ F(t) = -A\ [F_o\ e^{-i\omega t}\ e^{\eta t} + c.c.] \tag{2.17}$$

where $\eta > 0$ ensures that $\lim_{t\to-\infty} F(t) = 0$, and the complex conjugate is added to have a real field. The limit $\eta \to 0^+$ will be taken at the end of the calculation.

We propose to calculate the change produced in the average of the dynamical variable B by the applied field F(t) conjugate to the variable A. To simplify notations we assume that the mean value of B in the equilibrium state is 0 :

$$< B >_o = 0.$$

To calculate $< B(t) >$ to first order in the applied field, we follow the elegant derivation of Kubo [10, 1]. The time evolution of the density matrix is given by Liouville's equation :

$$\frac{\partial\rho(t)}{\partial t} = -i\mathcal{L}\rho \tag{2.18}$$

The Liouville operator $\mathcal{L}$ is linear in the hamiltonian and is hence the sum of an unperturbed part, $\mathcal{L}_o$, and an external part, $\mathcal{L}'$. We would like to solve (2.18) subject to the initial condition $\rho(-\infty) = \rho_o = e^{-\beta H_o}/\mathrm{Tr}\rho_o$. We look for a solution of the type :

$$\rho(t) = \rho_o + \Delta\rho(t) \tag{2.19}$$

which we introduce into (2.18), retaining only first order terms :

$$\begin{aligned}\frac{\partial}{\partial t}\Delta\rho(t) &= -i\mathcal{L}_o\Delta\rho(t) - i\mathcal{L}'(t)\rho_o \\ &= -\frac{i}{\hbar}[H_o, \Delta\rho(t)] + \frac{i}{\hbar}[A, \rho_o]F(t)\end{aligned} \tag{2.20}$$

This can be formally integrated to :

$$\Delta\rho(t) = -\int_{-\infty}^{t} dt'\, e^{-i(t-t')\mathcal{L}_o}\, i\mathcal{L}'(t')\rho_o \tag{2.21}$$

Using this result we calculate $\langle B(t)\rangle$ to first order in the perturbation :

$$\begin{aligned}\langle B(t)\rangle &= \mathrm{Tr}\{\Delta\rho(t)B\} \\ &= \frac{i}{\hbar}\int_{-\infty}^{t} dt'\, F(t')\mathrm{Tr}\{e^{-i(t-t')\mathcal{L}_o}[A,\rho_o]\,B\}\end{aligned} \tag{2.22}$$

Using the cyclic properties of the trace and the definition of the Liouville operator, it is easily seen that for any two dynamical variables C and D :

$$\mathrm{Tr}\{\mathcal{L}_o C\, D\} = -\mathrm{Tr}\{C\mathcal{L}_o D\} \tag{2.23}$$

Using (2.23) in (2.22), with $C \equiv [A,\rho_o]$ and $D \equiv B$, and remembering (1.3), we obtain :

$$\begin{aligned}\langle B(t)\rangle &= \frac{i}{\hbar}\int_{-\infty}^{t} dt'\, F(t')\mathrm{Tr}\{[A,\rho_o]B(t-t')\} \\ &= \frac{i}{\hbar}\int_{-\infty}^{t} dt'\, F(t')\mathrm{Tr}\{[\rho_o,B(t-t')]A\} \\ &= \frac{i}{\hbar}\int_{-\infty}^{t} dt'F(t')\mathrm{Tr}\{\rho_o[B(t-t'), A]\}\end{aligned} \tag{2.24}$$

which can be rewritten in an intuitively appealing form by introducing the after-effect function :

$$\theta_{BA}(t) = \frac{i}{\hbar} < [B(t), A]>_o \tag{2.25}$$

$$< B(t) > = \int_{-\infty}^{t} dt' \; \theta_{BA}(t - t') \, F(t') \tag{2.26}$$

Note that frequently in the literature, $\theta_{BA}(t)$ is set equal to zero for $t < 0$ (causality), allowing the integration in (2.26) to be extended to $+\infty$. We prefer to define θ_{BA} by (2.25) for all t. $< B(t) >$ is the superposition of retarded effects of the applied field at all preceding times.

Replacing (2.17) into (2.26) and using the fact that $\theta_{BA}(t)$ is real, since it is the thermal average of $i/\hbar$ times the commutator of hermitian operators, leads to :

$$< B(t) > = R \, F_o \, e^{-i(\omega+i\eta)t} \int_{-\infty}^{t} \theta_{BA}(t-t') e^{-i(\omega+i\eta)(t'-t)} dt'$$

$$= R \, F_o \, e^{-i(\omega+i\eta)t} \int_{o}^{\infty} \theta_{BA}(s) \; e^{i(\omega+i\eta)s} \, ds \tag{2.27}$$

We now define the dynamic response function (or dynamic susceptibility) :

$$\chi_{BA}(\omega) = \chi'_{BA}(\omega) + i \, \chi''_{BA}(\omega) = \lim_{\eta \to 0+} \tilde{\theta}_{BA}(\omega + i\eta) \tag{2.28}$$

which allows (2.27) to be rewritten as :

$$< B(t) > = R \, \chi_{BA}(\omega) \, F_o \, e^{-i\omega t} \tag{2.29}$$

The generalization to a spatially variable applied field coupled to a local dynamical variable is straightforward. If the system is translationally invariant in equilibrium, (2.26) becomes :

$$< B(\vec{r},t) > = \int_{\infty}^{t} dt' \int d^3r' \; \theta_{BA}(\vec{r}-\vec{r}', t-t') \; F(\vec{r}',t') \tag{2.30}$$

and the response function, or generalized susceptibility, is related to the non-local after-effect function by :

$$\chi_{BA}(\vec{k},\omega) = \lim_{\eta \to 0+} \int_{o}^{\infty} dt \int d^3r \; \theta_{BA}(\vec{r},t) \; e^{i(\omega+i\eta)-i\vec{k}\cdot\vec{r}} \tag{2.31}$$

and

$$< B(\vec{k},t) > = R\ \chi_{BA}(\vec{k},\omega)\ F(\vec{k})\ e^{i(\vec{k}\cdot\vec{r}-\ \omega t)} \tag{2.32}$$

Typical examples of linear response theory will be given in sections 2.6 (dielectric susceptibility) and 5.2 (electrical conductivity).

2.3 The Fluctuation-Dissipation Theorem

As is already evident from equations (2.24) and (2.25), the linear response of a system to an external perturbation is entirely describable in terms of equilibrium averages. In this section we make the connection more explicit by relating θ_{BA} to an equilibrium TCF. From the relation :

$$\dot{A} = \frac{i}{\hbar}[H_o, A]$$

it is easy to derive the Kubo identity :

$$-\frac{i}{\hbar}[\rho_o, A] = \int_o^\beta d\lambda\ \rho_o\ e^{\lambda H_o}\ \dot{A}\ e^{-\lambda H_o}$$

$$= \int_o^\beta d\lambda\ \rho_o\ \dot{A}\ (-i\hbar\lambda) \tag{2.33}$$

Replacing in (2.24) leads to

$$\theta_{BA}(t) = -\int_o^\beta d\lambda <\dot{B}(t - i\hbar\lambda)A>_o$$

$$= -\ \beta C^c_{\dot{B}A}(t) = -\ \beta\frac{d}{dt}\ C^c_{BA}(t) \tag{2.34}$$

where C^C_{BA} is the canonical TCF of the variables B and A, defined in (1.21). In the classical limit ($\hbar \to 0$), (2.34) reduces to :

$$\theta_{BA}(t) = -\ \beta< \dot{B}(t)A>_o\ = \beta<B(t)\ \dot{A}> \tag{2.35}$$

The response of the system to a weak external perturbation is entirely determined by the equilibrium correlation of the fluctuating dynamical variables B and A. (2.34) is the most general formulation of the "fluctuation-disspitation theorem" which was first discovered by Nyquist [11], in the case of electric networks ; the admittance of a network is directly related to the spectrum of the random noise in the circuit, due to the thermal motion of the charge carriers.

The response function is then given by :

$$\chi_{BA}(\omega) = -\beta\int_0^\infty C^c_{\dot{B}A}(t)e^{+i\omega t}\,dt$$

$$= -\beta\,\tilde{C}^c_{\dot{B}A}(\omega)$$

$$= \beta\,[C^c_{BA}(0) + i\omega\,\tilde{C}^c_{BA}(\omega)] \qquad (2.36)$$

In the important special case where B = A (or A^+ for the Fourier components of a local dynamical variable) we conclude immediately from (1.35) that :

$$\chi''_{AA}(\omega) = \pi\beta\omega\,\hat{C}^c_{AA}(\omega) \qquad (2.37)$$

This is the most concise version of the fluctuation-dissipation theorem; moreover χ'' will be shown to be associated with the dissipation of energy by the perturbed system which is the origin of theorem's designation.

There exists finally a simple relation between the Fourier-Laplace transforms of the canonical and one-sided TCF's. A straightforward calculation leads to :

$$\tilde{C}^c_{BA}(\omega) = \frac{1}{\beta\hbar\omega}\,[1 - e^{-\beta\hbar\omega}]\,\tilde{C}_{BA}(\omega) \qquad (2.38)$$

so that, when B = A, (2.37) leads to :

$$\chi''_{AA}(\omega) = \frac{\pi}{\hbar}\,[1 - e^{-\beta\hbar\omega}]\,\hat{C}_{AA}(\omega) \qquad (2.39)$$

2.4 General Properties of Response Functions

a) The properties of the after-effect function follow directly from its definition (2.25) or (2.34) and the corresponding properties of TCF's reviewed in lecture 1. In particular θ_{BA} is a stationary and real function of t; if ε_A and $\varepsilon_B = -\varepsilon_{\dot{B}}$ are the signatures of the operators A and B under time reversal, it follows from (1.20) that :

$$\theta_{BA}(-t) = -\varepsilon_A\varepsilon_B\theta_{BA}(t) \qquad (2.40)$$

b) From the definition of the response function, (2.28) and the reality of θ_{BA}, it follows immediately that on the real axis :

$$\chi_{BA}(-\omega) = \chi^{*}_{BA}(\omega) = \chi'_{BA}(\omega) - i\,\chi''_{BA}(\omega) \tag{2.41}$$

and hence χ' is an <u>even</u> function of ω, while χ'' is <u>odd.</u>

c) We introduce the Fourier transform of the after-effect function, $\theta_{BA}(\omega)$, and the convenient auxiliary function :

$$\xi_{BA}(\omega) = -\,i\,\pi\,\hat{\theta}_{BA}(\omega) \tag{2.42}$$

$\chi_{BA}(z)$ is then the Hilbert transform of ξ_{BA} :

$$\chi_{BA}(z) = \int_{-\infty}^{+\infty} \frac{d\omega}{\pi}\,\frac{\xi_{BA}(\omega)}{\omega - z} \tag{2.43}$$

$$\chi_{BA}(\omega) = \lim_{\eta\to 0+} \chi_{BA}(\omega + i\eta) = P\int_{-\infty}^{+\infty} \frac{d\omega'}{\pi}\,\frac{\xi_{BA}(\omega')}{\omega' - \omega} + i\,\xi_{BA}(\omega) \tag{2.44}$$

d) We shall henceforth restrict ourselves to the case where $B = A$; this includes in particular the important special case of the density response function ($B = A = \rho$). From $\theta_{AA}(-t) = -\,\theta_{AA}(t)$, it is then easy to check that :

$$\xi_{AA}(\omega) = \chi''_{AA}(\omega) \tag{2.45}$$

and hence from (2.44) :

$$\chi'_{AA}(\omega) = P\int_{-\infty}^{+\infty} \frac{\chi''_{AA}(\omega')}{\omega' - \omega}\,\frac{d\omega'}{\pi} \tag{2.46}$$

This last equality, which relates the real and imaginary parts of the response function, is called a <u>Kramers-Kronig relation</u>. The reciprocal relation is :

$$\chi''_{AA}(\omega) = -\,P\int_{-\infty}^{+\infty} \frac{\chi'_{AA}(\omega')}{\omega' - \omega}\,\frac{d\omega'}{\pi} \tag{2.47}$$

The "dissipative" and "dispersive" parts (cf. the analogy with the dielectric constant) of the response are hence not independent. According to (2.39) the dissipative part is directly accessible to measurement (cf. section 2.1); then the dispersive part can be determined from (2.46).

e) It follows immediately from (2.39), and the positivity of the power spectra of ACF's, that

$$\chi''_{AA}(\omega) \geqslant 0 \tag{2.48}$$

It is instructive to give a more physical proof of this relation, which shows that χ'' is linked to energy dissipation[2,6]. The rate of change of the system's energy under the influence of the perturbation H' is :

$$\frac{dE}{dt} = \frac{d}{dt} \mathrm{Tr} \; \{ \rho(t) \; H\} \quad = - \frac{dF(t)}{dt} \quad \langle A(t) \rangle$$

If one then calculates the total energy variation :

$$\Delta E = \int_{-T}^{T} \frac{dE}{dt} \, dt$$

under the influence of a monochromatic external field applied over the time interval 2T, one easily obtains, using (2.29), the condition :

$$T F_o^2 \; \omega \; \chi''_{AA}(\omega) \geqslant 0 \tag{2.49}$$

χ''_{AA} is thus directly related to energy dissipation, as anticipated.

f) The response function has an asymptotic high frequency expansion which corresponds to the short time expansion of the associated TCF. Considering the general k-dependent case, we have from (2.43) and (2.45) :

$$\chi_{AA}(k,z) = - \sum_{n=1}^{\infty} \frac{a_{2n}}{z^{2n}} \tag{2.50}$$

where :

$$a_{2n} = \int_{-\infty}^{+\infty} \frac{d\omega}{\pi} \; \chi''_{AA}(k,\omega)\,\omega^{2n-1} \tag{2.51}$$

Only even terms contribute, since χ''_{AA} is odd in ω. The expansion is asymptotic in the sense that it is only valid if $|z|$ is large compared to all characteristic frequencies in the system.

In the classical limit ($\hbar \to 0$) we have from (2.39) :

$$\chi''_A(k,\omega) = \pi\, \beta\omega\, \hat{C}_{AA}(k,\omega) \tag{2.52}$$

and hence, using (1.45) :

$$a_{2n} = \beta < \omega^{2n} >_{AA} \tag{2.53}$$

which clearly shows the link with the short-time expansion.

g) From (2.43) we also deduce the zero frequency limit of $\chi(k,z)$, called the static susceptibility $\chi_{AA}(k)$:

$$\chi_{AA}(k) = \chi_{AA}(k, z = 0) = \int_{-\infty}^{+\infty} \frac{d\omega}{\pi} \frac{\chi''(k,\omega)}{\omega}$$

$$= \beta \int_{-\infty}^{+\infty} \hat{C}_{AA}(k,\omega)\, d\omega$$

$$= \beta C_{AA}(0) \tag{2.54}$$

It is instructive to consider the important case of the density response function. Using the notations of section 2.1, we have from (2.54) and (2.13) :

$$\chi_{\rho\rho}(k) = \beta N\, S(k) \tag{2.55}$$

where S(k) is the familiar static structure factor. Now it is well known that :

$$\lim_{k \to 0} S(k) = \chi_T / \chi_T^{\circ} \tag{2.56}$$

where χ_T is the isothermal compressibility of a system of interacting particles, and $\chi_T^{\circ} = \rho k_B T$ is that of an ideal gas at the same density and temperature. χ_T can be considered as the macroscopic susceptibility of a fluid with respect to an outside pressure change. Since

$$\lim_{k \to 0} \chi(k, z = 0) = N\rho\, \chi_T \tag{2.57}$$

the dynamic susceptibility appears as a natural extension of the macroscopic susceptibility to finite wave-lengths and frequencies, and is often referred to as a "generalized" susceptibility.

This point of view has been systematically extended by Kadanoff and Martin [12] who recognized that for a system perturbed by an external field the description of the linear response in terms of TCF's must coincide with that given by the linearized equations of hydrodynamics in the limit when all relevant physical quantities vary slowly in space and time. We shall return to this point later on.

2.5 An Application : Dielectric Susceptibility

In order to illustrate linear response theory we briefly consider the dielectric response to an external periodic electric field. If the liquid contains N polar molecules, having a permanent dipole $\vec{\mu}$, the total dipole moment is :

$$\vec{M} = \sum_{i=1}^{N} \vec{\mu}_i$$

and the coupling to the external field $\vec{E}_e(t)$ is :

$$H' = - \vec{M} \cdot \vec{E}_e(t)$$

At equilibrium, in the absence of the external field, the mean total dipole moment vanishes of course.

The general relation (2.30) yields :

$$< \vec{M}(t) > = R\, \chi_e(\omega)\, \vec{E}_e(t) \tag{2.58}$$

where the external susceptibility is, for classical dynamics, using (2.29) and (2.35) :

$$\chi_e(\omega) = - \frac{\beta}{3} \int_0^{\infty} e^{i\omega t} < \dot{\vec{M}}(t) \cdot \vec{M}(0) >_0 dt \tag{2.59}$$

In the special case of a dilute solution of non-polarizable molecules, of dipole moment μ, (2.59) yields, upon neglecting the correlations between orientations of different dipoles :

$$\chi_e(\omega) = - \frac{\beta\mu^2 N}{3} \int_0^{\infty} e^{i\omega t} < \dot{\vec{u}}(t) \cdot \vec{u}(0) >$$

where $\vec{u}$ is the unit vector in the direction of the dipole. Formula (2.59) is the basis of dielectric relaxation theory.

LECTURE 3

PROJECTION OPERATORS AND MEMORY FUNCTIONS

Until now we have introduced,TCF's proved some general properties and shown their relationship with the system's linear response to an external probe. We have made no effort yet to try and calculate TCF's theoretically starting from their Statistical Mechanics definition. This is in fact a very difficult many-body problem, which can almost never be solved exactly, except for some extremely simplified models. One can however calculate TCF's or their spectra in certain limiting regimes. Remembering (1.42) one can try to calculate TCF's involving conserved variables in the limit of very long wave-lengths and low frequencies, where one expects the macroscopic phenomenological equations of linearized hydrodynamics to hold. Examples of such calculations will be given in subsequent lectures. If we are interested in high frequency behaviour, we can use the short-time expansions (1.43) or (1.44) of TCF's. If we are interested in correlation functions in gases, we can use kinetic equations [13], describing essentially binary collision processes, of which the most famous is of course the Boltzmann equation. Finally if we are interested in dilute solutions, as is the case for Brownian particles, it is reasonable to apply stochastic theories of which we have already studied the oldest and best-known example : the Langevin theory.

In this chapter we show how the Langevin equation can be generalized and rigorously derived from Statistical Mechanics. The corresponding memory function formalism does not allow an explicit calculation of TCF's but allows to cast them in an appropriate form which incorporates high frequency information and may suggest useful approximations. The formalism allows thus at least in principle to bridge the gap between high frequency and hydrodynamic behaviour starting from the high frequency side. This approach has many features in common with "generalized hydrodynamics" theories, which attempt to bridge the gap from the low frequency, long wave-length side, by introducing, very roughly speaking,k and ω-dependent transport coefficients. A classic example of such an approach is the viscoelastic theory of high-frequency shear waves (cf. lecture 4).

3.1 Projection Operators

Among the microscopic dynamic processes in a liquid, it is sometimes possible to distinguish between "slow" and "fast" motions. Consider for example the Brownian problem : the heavy particle's velocity will change slowly in time, under the influence of a rapidly varying force due to frequent collisions with solvent particles. As another example, in a pure liquid, for sufficiently small k,we expect the Fourier composents of conserved dynamical variables to vary slowly with time, while the velocity of individual molecules vary rapidly in time.

It is thus tempting to seek a description which separates slow from fast motions. This can be formally achieved by introducing the projection operators of Zwanzig [14] and Mori [15]. For the sake of clarity we shall restrict ourselves to classical Statistical Mechanics. Consider a dynamical variable A(t) the time evolution of which is given by (1.3) in terms of $A \equiv A(0)$. We may represent the phase function A by a vector ("bra") $|A>$ in a Hilbert space of dynamical variables; more generally we may use a vector in such a space to represent a set of dynamical properties of the system, but for the present we shall restrict ourselves to a single-variable description. We associate with this Hilbert space an inner (or scalar) product of two vectors, defined as the equilibrium -averaged correlation function of two dynamical variables, which satisfies the usual requirements of a scalar product :

$$< A|B > = < A^{*}B > \equiv < B\ A^{*} >$$

The Liouville operator is hermitian in the Hilbert space of dynamical variables, as shown easily from an integration by parts over phase space :

$$< (\mathcal{L} B)\ A^{*} > = < B(\mathcal{L} A)^{*} >$$

or

$$<\mathcal{L}A|B > = < A\ |\mathcal{L}B > \tag{3.1}$$

from which it follows that :

$$\begin{aligned} C_{BA}(t) = \ & < B(t)A^{*} > \equiv <A|B(t)> \\ & = <A|e^{i\mathcal{L}t}\ B> \\ & = <e^{-i\mathcal{L}t}\ A|B> \\ & = < A(-t)|B> \end{aligned} \tag{3.2}$$

in agreement with time translation invariance.

The projection of the variable B unto A is written formally in terms of a projection operator P :

$$P|B > = (<A|A>)^{-1}\ |A > < A|B > \tag{3.3}$$

The operator $Q = 1 - P$ projects into the Hilbert subspace orthogonal to the vector $|A>$. Clearly P and Q satisfy the fundamental property of projection operators :

$$P^2 = P, \qquad Q^2 = Q \tag{3.4}$$

$$PQ = QP = 0. \tag{3.5}$$

3.2 Generalized Langevin Equation

We seek an equation of motion for the dynamical variable $|A(t)>$ which explicitly separates the evolution "parallel" and "orthogonal" to $|A>$. Let Y(t) be the normalized ACF of the variable $|A>$:

$$Y(t) = <A|A(t)> (<A|A>)^{-1} \tag{3.6}$$

Clearly :

$$P|A(t)> = Y(t)|A> \tag{3.7}$$

We now look for separate equations of motion for the projected part of $|A(t)>$, i.e. for Y(t), and for the orthogonal part :

$$|A'(t)> = Q|A(t)> \tag{3.8}$$

Note that the two projected parts are not dynamical variables by themselves, in the sense that their time evolution is not simply given by Liouville's equation !

Accordingly we divide the time derivative of $|A>$ into projected and orthogonal parts :

$$|\dot{A}> = P|\dot{A}> + Q|\dot{A}>$$

$$= i\,\Omega\,|\dot{A}> + |K> \tag{3.9}$$

In analogy with the Langevin equation (cf. section 1.5), $|K>$ is called the random force at time $t = 0$. $i\Omega$ is the number :

$$i\,\Omega = <A|\dot{A}> (<A|A>)^{-1} = \frac{d}{dt}\,Y(0) \tag{3.10}$$

which vanishes, since ACF's are even functions of time. However in the more general case of a multivariable description, $i\Omega$ is a non-zero matrix, and in order to simplify the generalization, we keep the term here too. We are now in a position to write down equations of motion for Y(t) and $|A'(t)>$. A few simple manipulations [15,8] lead to :

$$|A'(t)> \;=\; \int_0^t Y(s)\,|R(t-s)>ds \qquad (3.11)$$

where :

$$|R(t)> \;=\; \exp\{iQ\mathcal{L}t\}\;|K> \qquad (3.12)$$

$|R(t)>$ is the random force at time t. $|R> \equiv |K>$, but the time evolution of $|R(t)>$ is governed by the modified Liouville operator $Q\mathcal{L}$, while $|K(t)>$ evolves according to the full Liouville operator $\mathcal{L}$. While $|K(t)>$ is at each time orthogonal to $|A(t)>$, $|R(t)>$ remains in the subspace orthogonal to $|A>$:

$$<A|R(t)> = 0 \qquad (3.13)$$

The equation of motion for Y(t) follows similarly from a few straightforward manipulations :

$$\dot{Y}(t) - i\,\Omega\,Y(t) + \int_0^t M(t-s)\,Y(s)ds = 0 \qquad (3.14)$$

where M(t) is the memory function, which is proportional to the ACF of the random force :

$$M(t) \;=\; <R|R(t)>\,(<A|A>)^{-1} \qquad (3.15)$$

Combining (3.11) and (3.14) leads to :

$$|\dot{A}'(t)> \;-\; i\Omega|A'(t)> + \int_0^t M(t-s)\;|A'(s)>ds \;=\; |R(t)> \qquad (3.16)$$

and, finally to the equation of motion for the full dynamical variable $|A(t)>$ itself :

$$|\dot{A}(t)> \;-\; i\Omega|A(t)> + \int_0^t M(t-s)\,|A(s)>\,ds \;=\; |R(t)> \qquad (3.17)$$

This is the so-called generalized Langevin equation. The classic Langevin equation of section 1.5 is recovered, for $|A> = \vec{v}$ (velocity of the Brownian particle), if $|R(t)>$ has an infinitely short correlation time, i.e. its ACF M(t) reduces to a δ-function. If the Brownian particle has a mass comparable to that of the solvent molecules (for instance if it is a tagged particle among identical molecules), the random force has a finite correlation time, and the friction coefficient is replaced by a memory function which is non-local in time.

Equation (3.14) is called the "memory function equation" for the ACF. Note that the random force does not appear explicitly, due to (3.13). The equation can be solved by taking the Fourier transform and remembering that Y(0) = 1 :

$$\tilde{Y}(\omega) \;=\; [-\,i\omega - i\Omega + \tilde{M}(\omega)]^{-1} \qquad (3.18)$$

M(t), which is an ACF, has properties similar to those listed in sections 1.3 and 1.4 for all ACF's. In particular its spectral function is non-negative. M(t) has a short-time expansion which is closely related to that of Y(t). Differentiating (3.14) with respect to time, and exploiting the fact that Y(0) = 1, and $\Omega = 0$, we obtain :

$$M(0) = -\ddot{Y}(0) = \langle \dot{A}|\dot{A}\rangle \quad (\langle A|A\rangle)^{-1} \tag{3.19}$$

Repeated differentiation yields relation between the initial time derivatives of Y and M, and hence to relations between the frequency moments μ_{2n} of $\hat{M}(\omega)$ and $y_{2n} = \langle \omega^{2n}\rangle_{AA} / \langle A|A\rangle$ of $\hat{Y}(\omega)$. In fact the μ_{2n} and y_{2n} are related as are the moments in cumulant expansions :

$$\begin{aligned} \mu_o &= y_2 \\ \mu_2 y_2 &= y_4 - y_2^2 \\ \mu_4 y_2 &= y_6 - 2y_2 y_4 + y_2^3 \\ &\ldots\ldots \end{aligned} \tag{3.20}$$

Our analysis can of course be extended to spatially non-local, or $\vec{k}$-dependent, correlation functions. If $C_{AA}(k,t)$ is the unnormalized ACF, and we drop $\Omega = 0$, the memory function equation (3.18) takes on the general form :

$$\widehat{C}_{AA}(\vec{k},\omega) = \frac{C_{AA}(\vec{k}, t = 0)}{- i\omega + \widehat{M}_{AA}(\vec{k},\omega)} \tag{3.21}$$

If $|A(\vec{k},t)\rangle$ is a <u>conserved</u> variable, we conclude from (3.15), (3.14), (3.9) and the continuity equation (1.10), that $\widehat{M}(k,\omega)$ is necessarily of the form :

$$\widehat{M}_{AA}(k,\omega) = k^2 \widetilde{D}_{AA}(k,\omega)$$

Hence in the limit $k \to 0$, $\widehat{C}_{AA}(k,\omega)$ has a simple imaginary frequency pole at the origin, corresponding to an infinite correlation (or decay) time. The interpretation of the memory function equation is now clear. If $|A\rangle$ is the only conserved variable, or if it is completely decoupled from all other conserved variables, due to symmetry (cf. section 4.1), then the random force contains all rapidly varying processes, since it remains in the subspace orthogonal to the initial value of the conserved variable which varies slowly in time for sufficiently small k. The slowly varying variable has been "projected" out, since the evolution operator of R(t) contains $Q\mathcal{L}$ rather than $\mathcal{L}$. It is frequently possible to give a fair description of the dynamics by characterizing the fast microscopic processes by a single relaxation time. Approximations of this type will be studied later.

3.3. Extensions

There are two important ways in which the formalism can be extended. In the first place, equ. (3.14) may be regarded as one in the hierarchy of similar memory function equations [15]. If we apply the previous methods to the case when $|R>$ is treated as the dynamical variable, we obtain an equ. (3.14), but describing the evolution of the projection of $|R(t)>$ along $|R>$. The kernel of the integral equation is now the ACF of the second-order random force which is orthogonal at all times both to $|R>$ and to $|A>$. As an obvious generalization of this procedure, we can write that :

$$\dot{M}_n(t) - i\Omega_n M_n(t) + \int_0^t M_{n+1}(t-s)\, \Delta^2_{n+1}\, M_n(s)\, ds = 0 \qquad (3.22)$$

where

$$M_n(t) = <R_n|R_n(t)> \; < R\,|R_n >^{-1}$$

$$R_n(t) = \exp\{i\, Q_{n+1}\mathcal{L}t\}\, Q_{n+1}|\dot{R}_n >$$

$$\Delta^2_n = < R_n|R_n >$$

The operator P_n projects an arbitrary dynamical variable along $|R_{n-1}>$, according to the definition (3.3). The operator $Q_n = (1-P_n)$, by construction, projects onto a subspace which is orthogonal to all $|R_j >$ for $j < n$. Thus the nth-order random force $|R_n(t)>$ is uncorrelated at all times with random forces of lower order.
Equ. (3.14) is a special case of (3.22) with $P \equiv P_1$, $Q \equiv Q_1$, $|A> \equiv |R_o$, $|R \equiv |R_1>$, $\Omega \equiv \Omega_o$, $Y \equiv M_o$ and $M \equiv M_1$. Repeated application of the Fourier-Laplace transform to equations of the hierarchy leads to an expression for $Y(\omega)$ in the form of a continued fraction :

$$\tilde{Y}(\omega) = \cfrac{1}{-i\omega - i\omega_o + \Delta_1^2 \cfrac{1}{-i\omega - i\Omega_1 + \cfrac{\Delta_2^2}{-i\omega - i\Omega_2 + \cfrac{\Delta_3^2}{-i\omega \;\ldots\ldots}}}} \qquad (3.23)$$

A second extension of the method which has proved particularly useful in the study of collective modes in liquids, is the generalization to the multi-variable case, where the dynamical variable of interest is not a single fluctuating property, but a set of independent variables $\{A_1, A_2, \ldots A_n\}$. We represent this set by the column vector :

$$|\underline{A}> = \begin{pmatrix} A_1 \\ A_2 \\ \cdot \\ \cdot \\ \cdot \\ A_n \end{pmatrix}$$

and $<\underline{A}|$ by the hermitian conjugate :

$$<\underline{A}| = (A_1^*, A_2^* \ldots . A_n^*)$$

The derivation of the generalized Langevin equ. for $|\underline{A}>$ follows the lines already laid down, due account being taken of the fact that the quantities involved are no longer scalars. The results may be written in matrix form :

$$|\dot{\underline{A}}> - i\underline{\Omega} \cdot |\underline{A}> + \int_o^t \underline{M}(t-s) \cdot |\underline{A}(s)> = |\underline{R}(t)> \qquad (3.24)$$

and the correlation function matrix $\underline{Y} = [Y_{ij}]$ satisfies :

$$\dot{\underline{Y}}(t) - i\underline{\Omega} \cdot \underline{Y}(t) + \int_o^t \underline{M}(t-s) \cdot \underline{Y}(s) ds = 0$$

which is solved in terms of Fourier-Laplace transforms as :

$$\widetilde{\underline{Y}}(\omega) = [-i\omega \, \underline{I} - i\underline{\Omega} + \widetilde{\underline{M}}(\omega)]^{-1} \qquad (3.25)$$

where $\underline{I}$ is the identity matrix. Note that the diagonal elements of $\underline{Y}$ are autocorrelation functions, while the off-diagonal elements are cross correlation functions.

In an ideal situation, the vector $|\underline{A}>$ should contain all slowly varying variables of the system, so that the random force $|\underline{R}>$ which is orthogonal to $|\underline{A}>$ would describe only fast processes, the slow ones having been projected out; in that case the memory function decays much faster in time than the correlation function of interest, and a Markovian approximation on $\underline{M}$ can be expected to be reasonable.

3.4 Illustration

As a rather lengthy illustration of memory function techniques, we now study single-particle translational motion in simple liquids like rare gases or liquid metals. First we must give a certain number of definitions. We have already introduced the velocity ACF (1.48) and the density ACF (2.13). From the latter we extract the "self-part" as follows ;

$$\begin{aligned} NF(k,t) &= < \sum_i \sum_j e^{-i\vec{k}\cdot(\vec{r}_i(t) - \vec{r}_j)} > \\ &= < \sum_i \sum_j e^{-i\vec{k}\cdot(\vec{r}_i(t) - \vec{r}_j)} > + < \sum_i e^{-i\vec{k}\cdot(\vec{r}_i(t) - \vec{r}_i)} > \\ &= N F_d(k,t) + NF_s(k,t) \qquad (3.26) \end{aligned}$$

The fluid is assumed isotropic, so that the correlation functions depend only on $k = |\vec{k}|$. F_d and F_s are the space Fourier transforms of the van Hove functions $G_d(r,t)$ and $G_s(r,t)$ [17]. $G_d(r,t)$ is the generalization in time of the familiar pair distribution function $g(r)$ $(G_d(r,o) \equiv g(r))$. $G_s(r,t)$ is simply the probability of finding a tagged particle, say i, a distance r away from the origin at time t, if the same particle was initially at the origin.

The microcopic density of particle i at a point $\vec{r}$ and time t is :

$$\rho_i(\vec{r},t) = \delta(\vec{r}-\vec{r}_i(t))$$

and $G_s(r,t) = V < \rho_i(\vec{r},t)\ \rho_i(0,0) >$

so that

$$G_s(r,0) = \delta(\vec{r}) \qquad (3.27)$$

For macroscopically large r, and t, $G_s(r,t)$ must satisfy the phenomenological diffusion equ. :

$$\frac{\partial G_s(\vec{r},t)}{\partial t} = D\ \nabla^2\ G_s(\vec{r},t) \qquad (3.28)$$

which is easily solved, subject to the initial condition (3.27) :

$$G_s(r,t) = \frac{1}{(4\pi Dt)^{3/2}} \exp\{-\ r^2/4\ Dt\} \qquad (3.29)$$

from which we derive immediately the familar Einstein relation (1.49) :

$$< [\vec{r}(t) - \vec{r}(0)]^2 > = 4\pi \int_0^\infty r^2 \; G_s(r,t) r^2 dr$$

$$= 6 \; Dt$$

The corresponding expression for $F_s(k,t)$ is :

$$F_s(k,t) = \exp\{-Dk^2t\}$$

which has the following lorentzian spectrum :

$$S_s(k,\omega) = \hat{F}_s(k,\omega) = \frac{1}{\pi} \frac{Dk^2}{\omega^2 + (Dk^2)^2} \quad (3.30)$$

This result is of course only valid in the limit $k \to 0$, $\omega \to 0$. $S_s(k,\omega)$ is directly related to the incoherent cross-section for neutron inelastic scattering. (3.30) indicates that for sufficiently small momentum transfers k and energy transfers ω, the spectrum should be lorentzian and its width could be used to determine the diffusion coefficient D. A cumbersome method to determine this constant indeed !

Another ACF which is useful in the study of single particle motion is the self longitudinal current ACF :

$$C_s(k,t) = <\vec{k}\cdot\vec{j}_{\vec{k}i}(t) \; \vec{k}\cdot\vec{j}_{-\vec{k}i}> \quad (3.31)$$

where the self current

$$\vec{j}_{\vec{k}i} = \vec{v}_i(t) \; e^{i\vec{k}\cdot\vec{r}_i(t)}$$

is related to the Fourier component of the self density by the continuity equation

$$\frac{\partial}{\partial t} \rho_{\vec{k}i}(t) + i\vec{k} \cdot \vec{j}_{\vec{k}i}(t) = 0$$

which immediately implies the following simple relation between F_s and C_s :

$$C_s(k,t) = -\frac{d^2}{dt^2} F_s(k,t) \quad (3.32)$$

Finally the velocity ACF is trivially linked to C_s and F_s by :

$$Z(t) = \lim_{k\to 0} \frac{1}{k^2} C_s(k,t) = -\lim_{k\to 0} \frac{1}{k^2} \frac{d^2}{dt^2} F_s(k,t) \quad (3.33)$$

In the following we shall need the short time expansion of $F_s(k,t)$, which we write, using conventional notations [8] as :

$$F_s(k,t) = 1 - \omega_o^2 \frac{t^2}{2!} + \omega_o^2 \omega_{1s}^2 \frac{t^4}{4!} \;\ldots\ldots \tag{3.34}$$

with

$$\omega_o^2 = \frac{k_B T k^2}{m}$$

$$\omega_o^2 \omega_{1s}^2 = < \ddot{\rho}_{\vec{k}i} \, \ddot{\rho}_{-\vec{k}i} >$$

$$= 3\omega_o^4 + \Omega_o^2 \omega_o^2 \tag{3.35}$$

where Ω_o^2 is proportional to the mean square force acting on the tagged particle :

$$\Omega_o^2 = \frac{k^2}{m^2 \omega_o^2} <F_{iz}^2> = \frac{\rho}{3m} \int \nabla^2 v(r) \; g(r) d^3 r \tag{3.36}$$

v(r) is the intermolecular pair potential (we restrict ourselves to pairwise additive forces). (3.33) yields then immediately the short time expansion of Z(t) :

$$Z(t) = \frac{k_B T}{m} \left[1 - \Omega_o^2 \frac{t^2}{2!} + \ldots \right] \tag{3.37}$$

We now have all the elements needed for a memory-function analysis of single particle motion. This can be done either by using the continued fraction expansion (3.22), or perhaps in a physically more transparent way, by the multivariable description. We choose the following set of dynamical wariables :

$$| \underline{A} > = \begin{pmatrix} \rho_{\vec{k}i} \\ \dot{\rho}_{\vec{k}i} \\ \sigma_{\vec{k}i} \end{pmatrix} = \begin{pmatrix} \rho_{\vec{k}i} \\ -i\vec{k}\cdot\vec{j}_{\vec{k}i} \\ \sigma_{\vec{k}i} \end{pmatrix} \tag{3.38}$$

where the variable $\sigma_{\vec{k}i}$ is the part of $\ddot{\rho}_{\vec{k}i}$ orthogonal to the two other variables :

$$\sigma_{\vec{k}i} = \ddot{\rho}_{\vec{k}i} - < \rho_{\vec{k}i} \mid \ddot{\rho}_{\vec{k}i} > \; < \rho_{\vec{k}i} \mid \rho_{\vec{k}i} >^{-1} \rho_{\vec{k}i} \tag{3.39}$$

The corresponding static correlation matrix is consequently diagonal:

$$\langle \underline{A} | \underline{A} \rangle = \begin{bmatrix} 1 & 0 & 0 \\ 0 & \omega_o^2 & 0 \\ 0 & 0 & \omega_o^2(2\omega_o^2 + \Omega_o^2) \end{bmatrix} \tag{3.40}$$

The frequency matrix is purely off-diagonal, and given by :

$$i\underline{\Omega} = \langle \underline{A} | \dot{\underline{A}} \rangle \quad (\langle \underline{A} | \underline{A} \rangle)^{-1} = \begin{bmatrix} 0 & 1 & 0 \\ -\omega_o^2 & 0 & 1 \\ 0 & -2\omega_o^2 - \Omega_o^2 & 0 \end{bmatrix} \tag{3.41}$$

The memory function matrix is given by the ACF of the random force $|\underline{R}(t)\rangle$, which is in the subspace orthogonal to $|\underline{A}\rangle$. However $\dot{A}_1 = \rho_{\vec{k}i}$ and $\dot{A}_2 = \dot{\rho}_{\vec{k}i}$ form part of the space spanned by $|\underline{A}\rangle$, since $\dot{A}_1 \equiv A_2$ and $\dot{A}_2$ can be decomposed in a component parallel to A_1 and a component orthogonal to A_1, which, according to (3.39) is identical with $A_3 = \sigma_{\vec{k}i}$. Consequently the matrix M(t) has only one non-zero entry :

$$\underline{M}(t) = \begin{bmatrix} 0 & 0 & 0 \\ 0 & 0 & 0 \\ 0 & 0 & M_s(k,t) \end{bmatrix} \tag{3.42}$$

$$\text{with :} \quad M_s(k,t) = \langle Q \dot{A}_3 | \, e^{iQ\mathcal{L}t} \, Q \dot{A}_3 \rangle \quad \langle A_3 | A_3 \rangle^{-1}$$

$$= \langle (e^{iQ\mathcal{L}t} \, Q\sigma_{\vec{k}i}) \quad Q\dot{\sigma}_{\vec{k}i} \rangle \, [\, \omega_o^2(2\omega_o^2 + \Omega_o^2)\,]^{-1} \tag{3.43}$$

Collecting the various results and inserting them in equ. (3.25), we find that the correlation function matrix takes the form :

$$\tilde{Y}(k,\omega) = \begin{bmatrix} -i\omega & -1 & 0 \\ \omega_o^2 & -i\omega & -1 \\ 0 & 2\omega_o^2 + \Omega_o^2 & -i\omega + \tilde{M}_s(k,\omega) \end{bmatrix}^{-1} \tag{3.44}$$

In particular, $\tilde{F}_s(k,\omega)$ is given by

$$\tilde{F}_s(k,\omega) = \tilde{Y}_{11}(k,\omega) = \cfrac{1}{-\,i\omega + \cfrac{\omega_o^2}{-i\omega + \cfrac{(2\,\omega_o^2 + \Omega_o^2)}{-i\omega + \tilde{M}_s(k,\omega)}}} \tag{3.45}$$

The same result can be recovered by the continued fraction expansion. Using (3.33) the correponding velocity ACF is :

$$\tilde{Z}(\omega) = \cfrac{k_B T/m}{-i\omega + \cfrac{\Omega_o^2}{-i\omega + \tilde{M}_s(0,\omega)}} \tag{3.46}$$

It is of course impossible to calculate the memory function (3.42) from Statistical Mechanics. However if our choice of dynamical variable is at all reasonable, we may expect $M_s(k,t)$ to decay on a faster time scale than F_s or Z. The simplest approximation then is the Markovian approximation :

$$M_s(k,t) = \frac{2}{\tau_s(k)}\,\delta(t) \rightarrow \tilde{M}_s(k,\omega) = \frac{1}{\tau_s(k)} \tag{3.47}$$

$\tilde{M}_s$ is frequency-independent (white spectrum), as was the random force spectrum in the simple Langevin theory of Brownian motion. $\tau(k)$ has the dimension of time. The approximation (3.47) leads to a simple analytic form for the spectrum $S_s(k,\omega)$, satisfying automatically the exact zeroth, second and fourth moment sum rules :

$$S_s(k,\ \omega) = \frac{1}{\pi}\,\frac{\tau_s(k)\ \omega_o^2(2\omega_o^2 + \Omega_o^2)}{\omega^2\tau_s^2(\omega^2 - 3\omega_o^2 - \Omega_o^2) + (\omega^2 - \omega_o^2)^2} \tag{3.48}$$

$\tau_s(k)$ is of course unknown, but can be used as an adjustable parameter in the study of incoherent neutron scattering data, or the results of computer "experiments".

The corresponding expression for $\tilde{Z}(\omega)$ is :

$$\tilde{Z}(\omega) = \frac{k_B T}{m}\ \cfrac{1}{-i\omega + \cfrac{\Omega_o^2}{-i\omega + 1/\tau_s(0)}} \tag{3.49}$$

From (1.50) we obtain immediately a relation between $\tau_s(0)$ and the macroscopic diffusion coefficient :

$$D = \tilde{Z}(0) = \frac{k_B T}{m\Omega_o^2\ \tau_s(0)} \tag{3.50}$$

(3.50) can be inverted to give [18] :

$$Z(t) = \frac{k_B T}{m}\ \left(\cos\ \delta t + \frac{1}{2\delta\tau_s(0)}\ \sin\ \delta\ t\right)\ \exp\left\{-\frac{1}{2}\,\frac{t}{\tau_s(0)}\right\} \tag{3.51}$$

Equ. (3.51) predicts that Z(t) will always stay positive if $\beta\Omega_o\, mD > 2$, but will have a negative region whenever $\beta\Omega_o\, mD < 2$. For Argon near the triple point, $\beta\Omega_o\, mD \simeq 0.9$, and the necessary inequality is easily satisfied. The behaviour is in qualitative agreement with a negative plateau observed in computer-generated velocity ACF's [16].

To conclude this section we would like to stress that the results obtained for $F_s(k,t)$ and Z(t) are most reliable at short times, since the short-time expansion has explicitly been built in. At long times, the single particle motion couples to the hydrodynamic collective modes, and this results in the so-called long-time tail of $Z(t) \simeq t^{-3/2}$ (in 3 dimensions) which was first observed by Alder and Wainwright [19]. An extensive literature exists on the subject, which is well described in a recent review by Pomeau and Résibois [20].

LECTURE 4

LONGITUDINAL AND TRANSVERSE COLLECTIVE MODES

In this lecture we study the ACF's of conserved dynamical variables in simple, one-component liquids. We shall cast these correlation functions in a form which will enable us to study collective motions, or modes, along the lines laid down in the previous lectures. There are 5 conserved variables satisfying continuity equations of the general form (1.7), and there is a hydrodynamic process associated with each of them. These variables are of course the total number of particles, the total linear momentum and the total energy. The associated local variables satisfy the 5 continuity equations :

$$\frac{\partial}{\partial t}\,\rho(\vec{r},t) + \frac{1}{m}\,\vec{\nabla}\cdot\vec{p}(\vec{r},t) = 0 \qquad \text{(number conservation)}$$

$$\frac{\partial}{\partial t}\,\vec{p}(\vec{r},t) + \vec{\nabla}\cdot\overleftrightarrow{\sigma}(\vec{r},t) = 0 \qquad \text{(momentum conservation)}$$

$$\frac{\partial}{\partial t}\,\varepsilon(\vec{r},t) + \vec{\nabla}\cdot\vec{j}_{\varepsilon}(\vec{r},t) = 0 \qquad \text{(energy conservation)}$$

$$(4.1)$$

$\vec{p} = m\,\vec{j}(\vec{r},t)$ is the momentum density, $\overleftrightarrow{\sigma}$ the stress tensor and $\vec{j}_{\varepsilon}$ the energy current. We are interested in the density ACF or its Fourier transform, the intermediate scattering function :

$$F(k,t) = \frac{1}{N} < \rho_{\vec{k}}(t)\ \rho_{-\vec{k}}(0) > \qquad (4.2)$$

In addition we shall consider the momentum ACF, or the related particle-current-correlation matrix which, because of rotational invariance, has only two independent elements : the longitudinal current ACF :

$$C_{\ell}(k,t) = \frac{1}{N} < \vec{k}\cdot\vec{j}_{\vec{k}}(t)\quad \vec{k}\cdot\vec{j}_{-\vec{k}}(0) > \qquad (4.3)$$

and the transverse current ACF :

$$C_t(k,t) = \frac{1}{2N}\ \mathrm{Tr} < (\vec{k}\wedge\vec{j}_k(t))\,(\vec{k}\wedge\vec{j}_{-\vec{k}}(0)) > \qquad (4.4)$$

From the number continuity equation it follows immediately that

$$C_{\ell}(k,t) = -\frac{d^2}{dt^2}\,F(k,t) \qquad (4.5)$$

and hence C is not an independent ACF. C_t however is independent; in fact the transverse current is completely decoupled from the longitudinal modes; this is a direct consequence of the fact that the transverse current $\vec{k} \wedge \vec{j}_{\vec{k}}(t)$ has the signature +1 under the reflection of all coordinates and momenta through a plane orthogonal to $\vec{k}$, while all longitudinal variables have signature -1. Hence, according to property g in section 1.3, all cross correlation functions between the transverse current and any longitudinal variable are zero at all times. This means that we can treat $C_t(k,t)$ separately.

4.1 Transverse Currents and Shear Modes

The transverse current has two independent components in the plane orthogonal to $\vec{k}$. Because of rotational invariance, we can choose $\vec{k}$ in the z-direction and consider the x-component of the transverse current.

First we shall derive the limiting behaviour of $\hat{C}_t(k,\omega)$ in the long wavelength, low frequency limit on the one hand, and in the high frequency limit on the other hand. The $k,\omega \rightarrow 0$ behaviour can be derived from a simple calculation based on linearized hydrodynamics. This procedure assumes that spontaneous fluctuations in microscopic quantities decay, in the mean, according to the hydrodynamic laws governing the corresponding macroscopic variables; this reasonable hypothesis due to Onsager [21] has been justified by Kadanoff and Martin [12, 6] on the basis of linear-response theory.

The momentum conservation law (4.1) must be complemented by the well-known macroscopic definition of the stress tensor [22]

$$\sigma^{\alpha\beta}(\vec{r},t) = \delta_{\alpha\beta}\, P(\vec{r},t) - \eta\left[\frac{\partial u_\alpha(\vec{r},t)}{\partial r_\beta} + \frac{\partial u_\beta(\vec{r},t)}{\partial r_\alpha}\right] + \delta_{\alpha\beta}\left(\frac{2}{3}\eta - \zeta\right) \vec{\nabla}\cdot\vec{u}(\vec{r},t) \qquad (4.6)$$

where $\vec{u}(\vec{r},t)$ is the local fluid velocity, $P(\vec{r},t)$ is the local pressure, and η and ζ are phenomenological transport coefficients called the shear and the bulk viscosities. The momentum density and the fluid velocity are related by :

$$\vec{p}(\vec{r},t) = \rho(\vec{r},t)\, m\, \vec{u}(\vec{r},t) \qquad (4.7)$$

If we assume that the deviations of the local density from its mean, $\rho = N/V$, are small, and if we choose a frame of reference in which the mean velocity of the fluid is zero, we can linearize (4.7) to yield :

$$\vec{p}(\vec{r},t) = \rho m\, \vec{u}(\vec{r},t) = m\, \vec{j}(\vec{r},t) \qquad (4.8)$$

Replacing (4.6) and (4.8) in the momentum conservation equation (4.1) we obtain the linearized Navier-Stokes equation :

$$m\left[\frac{\partial}{\partial t} - \frac{\eta}{\rho}\nabla^2 - \frac{(1/3\,\eta+\zeta)}{\rho}\vec{\nabla}\,\vec{\nabla}\cdot\right]\vec{j}(\vec{r},t) + \vec{\nabla}P(\vec{r},t) = 0 \tag{4.9}$$

We now take a spatial Fourier transform with k in the z-direction; the x-component of the resulting expression gives us an equation of motion for the transverse component of the current :

$$\frac{\partial}{\partial t} j^x_{\vec{k}}(t) + \frac{\eta}{\rho m} k^2 j^x_{\vec{k}}(t) = 0 \tag{4.10}$$

which has the form of a diffusion equation ($k^2 \rightarrow -\nabla^2$!). Multiplying through equ. (4.10) by $j^x_{\vec{k}}$ and taking a thermal average, we obtain an equation for the transverse current ACF :

$$\dot{C}_t(k,t) + \frac{\eta}{\rho m} k^2 C_t(k,t) = 0 \tag{4.11}$$

which is easily solved by taking the Fourier Laplace transform :

$$\tilde{C}'_t(k,\omega) = \frac{C_t(k,0)}{-i\omega + \frac{\eta}{\rho m}k^2} = \frac{\omega_0^2}{-i\omega + \frac{\eta}{\rho m}k^2} \tag{4.12}$$

$\tilde{C}_t(k,\omega)$ has a purely imaginary pole which corresponds to a diffusive (i.e. non-propagating) behaviour of the transverse current fluctuations in the hydrodynamic limit; $C_t(k,t)$ decays exponentially.

We now turn to the high-frequency behaviour of $\tilde{C}_t(k,\omega)$. From the memory function equation (3.21) for a conserved variable we can write :

$$\tilde{C}_t(k,\omega) = \frac{\omega_0^2}{-i\omega + k^2\tilde{D}_t(k,\omega)} \tag{4.13}$$

At sufficiently large ω, we can use a high-frequency expansion for $\tilde{D}_t$, similar to that written down for the response function in section (2.5). Remembering (3.20) we obtain ($z = \omega + i\varepsilon$) :

$$\tilde{D}_t(k,z) = -i\int_{-\infty}^{+\infty} d\omega\, \frac{\hat{D}_t(k,\omega)}{\omega - z}$$

$$\simeq \frac{i}{z}\frac{\mu_o}{k^2} = \frac{i < \omega^2 >_t}{\omega_o^2 k^2 z} \tag{4.14}$$

where :

$$< \omega^2 >_t = \int_{-\infty}^{+\infty} \omega^2 \, \hat{C}_t(k,\omega)\, d\omega$$

$$= - \ddot{C}_t(k, t = 0)$$

$$= \omega_o^2 \, \omega_{1t}^2 \tag{4.15}$$

with :

$$\omega_{1t}^2 = \omega_o^2 + \frac{k^2\rho}{m} \int g(r) \left[1 - \cos kz \right] \frac{\partial^2 v(r)}{\partial x^2} \, d^3r \tag{4.16}$$

Gathering results, $\tilde{C}_t(k, \omega)$ reads in the high frequency limit :

$$\tilde{C}_t(k, \omega) = \frac{i\omega\omega_o^2}{\omega^2 - \omega_{1t}^2} \tag{4.17}$$

Clearly $\tilde{C}_t$ has two symmetrical real poles in the high frequency limit, corresponding to propagating "shear" modes, reminiscent of transverse phonons in solids. This means that the response of the fluid to a high frequency shear will be reactive (or "elastic") rather than diffusive. In particular, in the limit of small k, (4.16) reduces to :

$$\lim_{k \to 0} \frac{\omega_{1t}^2}{\omega_o^2} = 1 + \frac{2\pi\rho\beta}{15} \int_o^\infty \frac{d}{dr} \left[r^4 \frac{dv}{dr} \right] g(r) dr \tag{4.18}$$
$$= \beta \, G_\infty / \rho$$

where G_∞ is the so-called infinite frequency shear modulus [23] which is the analog of an elastic constant in a solid, corresponding to a long wave-length dispersion relation for the shear mode :

$$\omega \simeq \sqrt{\frac{G_\infty}{\rho m}} \; k \tag{4.19}$$

The actual behaviour of C_t is expected to be "visco-elastic", i.e. somehow intermediate between the two extreme limits we have just investigated.

In order to obtain a statistical mechanical description of shear waves, we turn again to the memory function formalism. The chosen set of dynamical variables is :

$$|\underline{A} > = \begin{pmatrix} m \, j_{\vec{k}}^{x} \\ \sigma_{\vec{k}}^{xz} \end{pmatrix} \tag{4.20}$$

The static correlation function is diagonal :

$$<A|A> = V\, k_B T \begin{bmatrix} \rho m & 0 \\ 0 & G_\infty(k) \end{bmatrix} \tag{4.21}$$

where $G_\infty(k)$ is the finite wave-length generalization of G :

$$G_\infty(k) = \rho k_B T \; \frac{\omega_{1t}^2}{\omega_o^2} \tag{4.22}$$

The frequency matrix is calculated to be :

$$i\,\underline{\Omega} = \begin{bmatrix} 0 & -ik \\ \dfrac{-ikG_\infty(k)}{\rho m} & 0 \end{bmatrix} \tag{4.23}$$

Since $\dot{A}_1$ is proportional to A_2, the memory function matrix has only one non-zero element :

$$\underline{M}(k,t) = \begin{bmatrix} 0 & 0 \\ 0 & M_t(k,t) \end{bmatrix} \tag{4.24}$$

Gathering results and replacing into (3.25) we immediately obtain the Fourier Laplace transform of the correlation function matrix $\widetilde{Y}(k,\omega)$. In particular :

$$\widetilde{C}_t(k,\omega) = \omega_o^2\, \widetilde{Y}_{11}(k,\omega) = \cfrac{\omega_o^2}{-\,i\omega + \cfrac{\omega_{1t}^2}{-\,i\omega + \widetilde{M}_t(k,\omega)}} \tag{4.25}$$

This must coincide with the hydrodynamic result (4.12) in the limit $k,\omega \to 0$, allowing us to make the identification :

$$\widetilde{M}_t(0,0) = G_\infty(0)/\eta \tag{4.26}$$

For a given k, we make again the Markovian approximation :

$$M_t(k,t) = \frac{2}{\tau_s(k)}\; \delta(t) \to \widetilde{M}_t(k,\omega) = \frac{1}{\tau_s(k)} \tag{4.27}$$

Comparing with (4.13) we see that this amounts to assuming that the memory function $D_t(k,t)$ decays exponentially, with a relaxation time equal to $\tau_t(k)$. $\tau_t(k)$ can also be identified with a wave-number-dependent shear viscosity :

$$\eta(k) = G_\infty(k)\; \tau_t(k) \tag{4.28}$$

which reduces to the macroscopic shear viscosity in the long wave-length limit, according to (4.26). Analysis of data from computer experiments shows that $\eta(k)$ decreases very rapidly with k [24]. The spectrum $\hat{C}_t(k,\omega)$ now takes the form :

$$\hat{C}_t(k,\omega) = \frac{1}{\pi} R\, \tilde{C}_t(k,\omega)$$

$$= \frac{1}{\pi} \frac{\omega_o^2\, \omega_{1t}^2\, \tau_t(k)}{\omega^2 + \tau_t^2(k)\, (\omega_{1t}^2 - \omega^2)^2} \tag{4.29}$$

Our analysis does not determine the viscoelastic relaxation time $\tau(k)$ (except in the limit $k \to 0$), and it must be determined by fitting the form (4.29) of the spectrum to results of computer experiments. An alternative procedure, using a gaussian, rather than exponential memory function allows the relaxation time to be determined from the next higher-order moment sum rule, yielding a parameter - free theory [25] .

It is easy to establish the criterion for the existence of propagating transverse modes within the context of the single relaxation time approximation. The condition for $\hat{C}_t(k,\omega)$ to have a peak at non-zero frequency at a fixed value of k is simply :

$$\omega_{1t}^2\, \tau_t^2(k) > 1/2$$

It follows from this inequality that shear waves will appear for values of k greater than the critical wave vector k_c given by :

$$k_c = \tau_t^{-1}(k_c) \left[\frac{\rho m}{2\, G_\infty(k_c)} \right]^{1/2}$$

An estimate of k_c can be obtained by taking the $k \to 0$ limit, which yields the condition :

$$k_c = \frac{1}{\eta} \left[\frac{1}{2}\, \rho m\, G_\infty(0) \right]^{1/2}$$

For a Lennard-Jones fluid near the triple point, this predicts $k\sigma \simeq 0.8$ ($\sigma = 3.4$ Å in Ar), in good agreement with computer data [26]. Note that for very large k, the peak vanishes, because the molecules in the fluid behave as free particles.

To conclude this section, we give a rapid derivation of the Green-Kubo relation for the shear viscosity η. According to (4.12) we have for small k :

$$\tilde{C}_t(k,\omega) \simeq \frac{\omega_o^2}{-i\omega} \left[1 + \frac{\eta}{\rho m} \frac{k^2}{i\omega} \right] \tag{4.30}$$

so that

$$\eta = \frac{\rho m}{\omega_o^2}\left[\frac{\omega^2 \tilde{C}_t(k,\omega) - i\omega\omega_o^2}{k^2}\right]$$

On substituting for ω_o^2 from (3.35), we find that the shear viscosity which must be real, is related to the low-frequency, long-wavelength behaviour of $\tilde{C}_t(k,\omega)$ by :

$$\eta = \beta\rho m^2 \lim_{\omega\to 0} \lim_{k\to 0} \frac{\omega^2}{k^4} R\,\tilde{C}_t(k,\omega) \qquad (4.31)$$

Note the order of the limits. This expression can be cast in a more convenient form.

From :

$$\int_0^\infty \frac{k^2}{N} < \dot{j}^x_{\vec{k}}\, \dot{j}^x_{-\vec{k}} > e^{i\omega t}\, dt$$

$$= -\int_0^\infty \frac{d^2}{dt^2} C_t(k,t) e^{i\omega t}$$

$$= \omega^2 \tilde{C}_t(k,\omega) - i\omega\omega_o$$

we can rewrite (4.31) as :

$$\eta = \beta\rho m^2 \lim_{\omega\to 0} \int_0^\infty \lim_{k\to 0} \frac{1}{Nk^2} < \dot{j}^x_{\vec{k}}\, \dot{j}^x_{-\vec{k}} > e^{i\omega t}\, dt$$

And, using the continuity equation :

$$\eta = \frac{\beta}{V}\int_0^\infty < \sigma_o^{xz}(t)\, \sigma_o^{xz}(0) > dt$$

The microscopic expression for the stress tensor is also derived from the continuity equation. A straightforward calculation yields for a system of particles interacting through an isotropic pair potential :

$$\sigma^{\alpha\beta}_{\vec{k}} = \sum_{i=1}^{N}\left[m\, v_{i\alpha} v_{i\beta} + \frac{1}{2}\sum_{j\neq i}^{N} \frac{r_{ij\alpha} r_{ij\beta}}{r_{ij}^2}\, \Phi_{\vec{k}}(r_{ij})\right] e^{-i\vec{k}\cdot\vec{r}_i}$$

with

$$\Phi_{\vec{k}}(r) = r\,\frac{dv(r)}{dr}\;\frac{e^{-i\vec{k}\cdot\vec{r}} - 1}{i\vec{k}\cdot\vec{r}} \qquad (4.32)$$

4.2 Longitudinal Collective Modes and Light Scattering

The two transverse modes having been dealt with, we now turn to the study of the three remaining hydrodynamic modes, which are longitudinal. Proceeding as in the previous section, we shall first calculate the contributions of these modes to the density fluctuations, using linearized hydrodynamics and the Onsager hypothesis, before considering the extension to finite wave-numbers and frequencies.

The energy conservation equation (4.1) must be complemented by the macroscopic phenomenological expression for the energy current [22] :

$$\vec{j}_\varepsilon(\vec{r},t) = h\,\vec{u}(\vec{r},t) - \lambda\vec{\nabla}T(\vec{r},t) \tag{4.33}$$

where $h = e + P$ is the equilibrium enthalpy density (e is the internal energy per unit volume), and λ is the thermal conductivity. Using the particle number conservation equation (4.1) to eliminate $\vec{u}$, and introducing the density of heat energy :

$$Q(\vec{r},t) = e(\vec{r},t) - \frac{(e + P)}{\rho}\,\rho(\vec{r},t) \tag{4.34}$$

the energy conservation equation (4.1) is rewritten :

$$\frac{\partial}{\partial t}\,Q(\vec{r},t) - \lambda\nabla^2\,T(\vec{r},t) = 0 \tag{4.35}$$

Choosing ρ and T as independent variables, we express $\delta Q(\vec{r},t)$ in terms of the deviations $\delta\rho(\vec{r},t)$ and $\delta T(\vec{r},t)$ of the local quantities from their mean values ρ and T :

$$\delta Q(\vec{r},t) = \rho T\left(\frac{\partial s}{\partial \rho}\right)_T \delta\rho(\vec{r},t) + \rho c_v\,\delta T(\vec{r},t) \tag{4.36}$$

where s is the entropy density and c_v is the specific heat at constant volume. Replacing (4.36) in (4.35) yields :

$$\left(\frac{\partial}{\partial t} - a\,\nabla^2\right)\delta T(\vec{r},t) + \frac{T}{c_v}\left(\frac{\partial s}{\partial \rho}\right)_T \delta\dot{\rho}(\vec{r},t) = 0 \tag{4.37}$$

with :

$$a = \lambda/\rho c_v \tag{4.38}$$

Taking the Fourier transform with respect to $\vec{r}$ and the Fourier-Laplace transform with respect to time leads to :

$$- i\omega \frac{T}{c_v}\left(\frac{\partial s}{\partial \rho}\right)_T \tilde{\rho}_{\vec{k}}(\omega) + (-i\omega + ak^2)\,\tilde{T}_{\vec{k}}(\omega) = \frac{T}{c_v}\left(\frac{\partial s}{\partial \rho}\right)_T \rho_{\vec{k}} + T_{\vec{k}} \tag{4.39}$$

The Navier-Stokes equation (4.9) is transformed in a similar way to :

$$(-\omega^2 - i\omega bk^2 + \frac{c_s^2}{\gamma} k^2)\,\tilde{\rho}_{\vec{k}}(\omega) + \frac{k^2}{m}\left(\frac{\partial P}{\partial T}\right)_\rho \tilde{T}_{\vec{k}}(\omega) \tag{4.40}$$

$$= \dot{\rho}_{\vec{k}} + (-i\omega + bk^2)\rho_{\vec{k}}$$

where :

$$b = \left(\frac{4}{3}\eta + \zeta\right)/\rho m \;;\quad \gamma = c_p/c_v \;;\quad c_s^2 = \frac{\gamma}{m}\left(\frac{\partial P}{\partial \rho}\right)_T \tag{4.41}$$

From the two simultaneous linear equations (4.39) and (4.40) we can eliminate $\tilde{T}_{\vec{k}}(\omega)$ to obtain $\tilde{\rho}_{\vec{k}}(\omega)$ in the form :

$$\tilde{\rho}_{\vec{k}}(\omega) = \rho_{\vec{k}} \frac{(-i\omega+ak^2)(-i\omega+bk^2) + (\gamma-1)c_s^2k^2/\gamma}{i\omega^3 - (a+b)k^2\omega^2 - i(c_s^2+abk^2)k^2\omega + ac_s^2k^4/\gamma} \tag{4.42}$$

where we have omitted a term proportional to $T_{\vec{k}}$ which will not contribute to the final expression for $S(k,\omega)$, since temperature and density are instantaneously uncorrelated : $< T_{\vec{k}}\, \rho_{-\vec{k}} > = 0$. The processes associated with the decay of a density fluctuation can be identified by studying the poles of $\tilde{\rho}_{\vec{k}}(\omega)$ in the complex plane, i.e. the roots of the denominator in (4.42). To order k^2 we obtain the approximate solution :

$$\omega = -i D_T k^2$$
$$\omega = \pm c_s k - i\Gamma k^2 \tag{4.43}$$

where $D_T = a/\gamma$ is the thermal diffusivity, and $\Gamma = 1/2\,[a(\gamma-1)/\gamma + b]$ is the sound attenuation coefficient. These roots correspond to a purely diffusive mode, linked to heat diffusion, and two conjugate propagating sound modes, which are damped by viscosity and thermal conduction.

On multiplying through equ. (4.42) by $\rho_{-\vec{k}}$, dividing by N and taking a thermal average, we obtain an expression for $\tilde{F}(k,\omega)$. On using the approximate roots (4.43) and taking the real part of $\tilde{F}(k,\omega)$, we finally arrive at the dynamical structure factor :

$$S(k,\omega) = \frac{1}{2\pi} S(k) \left\{ \frac{\gamma-1}{\gamma} \frac{2D_T k^2}{\omega^2+(D_T k^2)^2} + \frac{1}{\gamma}\left[\frac{\Gamma k^2}{(\omega+c_s k)^2 + (\Gamma k^2)^2} + \frac{\Gamma k^2}{(\omega-c_s k)^2 + (\Gamma k^2)^2}\right]\right\} \tag{4.44}$$

The spectrum consists of three Lorentzian lines corresponding to each of the three hydrodynamic modes (4.43) : the central Rayleigh line and two Brillouin side peaks at $\omega = \pm c_s k$. In the case of a binary mixture there is an additional hydrodynamic mode, since the numbers of particles of both species are conserved separately; the additional mode is linked to concentration fluctuations and interdiffusion of both species [27].

The total integrated intensities of the Rayleigh line and of each of the Brillouin lines are, respectively

$$I_R = \frac{\gamma-1}{\gamma} S(k) \; ; \qquad I_B = \frac{1}{2\gamma} S(k)$$

$$I_R + 2I_B = S(k) \underset{k\to 0}{\simeq} \chi_T/\chi_T^{\circ}$$

$$\frac{I_R}{2I_B} = \frac{\gamma-1}{\gamma} \tag{4.45}$$

The latter ratio is the so-called Landau-Placzek ratio.

The spectrum (4.44) is in fact directly measurable by Rayleigh-Brillouin light scattering experiments. Indeed, since the wave-length λ of light is several thousand Å, the number of molecules contained in a volume λ^3 is very large (of the order of $(\lambda/\sigma)^3 \simeq 10^9$!), so that the hydrodynamic calculation should be quite accurate in describing the experimentally observed spectra. It is easily shown that the spectrum $I(k,\omega)$ of the scattered light is directly related to the spectrum of the ACF of the local deviations of the dielectric constant, $\delta\varepsilon(k,\omega)$ from its macroscopic value ε [7]. Upon expanding :

$$\delta\hat{\varepsilon}(k,\omega) = \left(\frac{\partial\varepsilon}{\partial\rho}\right)_T \delta\hat{\rho}_{\vec{k}}(\omega)$$

this correlation function is immediately related to the density ACF, if the variation of ε with temperature is neglected, which is practically always justified. Hence a quasielastic light scattering experiment measures directly $S(k,\omega)$. Note that Rayleigh-Brillouin scattering corresponds to a frequency range of 0-10 cm^{-1}, i.e. energies typical of thermal translational motions in liquids. Inelastic, or Raman scattering experiments correspond to a much higher frequency range (typically several hundred cm^{-1}) and probe the vibrational-rotational motions in molecular liquids. It is also worth mentioning at this point that even for a liquid of rare gas atoms (spherically symmetric polarizability tensor), there is a small de-polarized contribution to the Rayleigh spectrum, due to dipole-induced dipole effects, linked with local anisotropies of the atomic distribution, and to the electronic distortion of atoms during collisions [28, 29].

A high resolution Rayleigh scattering experiment allows a direct measurement of macroscopic properties : thermal diffusivity D_T, sound attenuation Γ, sound velocity c_s; integrated intensity measurements (which are difficult) lead to χ_T and γ. A rich harvest indeed !

4.3 Longitudinal Collective Modes and Inelastic Neutron Scattering

Neutron scattering experiments probe wave numbers > 0.1 $\mathring{A}^{-1}$, which are much larger than those reached by light scattering. The simple hydrodynamic description of the previous section must be extended to cover wave-lengths of the order of inter-atomic spacings. We proceed by making once more use of the memory function formalism.

A natural choice for the components of the longitudinal dynamical vector $|\underline{A}>$ is the set of conserved variables $\rho_{\vec{k}}$, $j_{\vec{k}}$ and $\varepsilon_{\vec{k}}$. In place of $\varepsilon_{\vec{k}}$, it is more convenient to choose $T_{\vec{k}}$, i.e. that part of $\varepsilon_{\vec{k}}$ which is orthogonal to $\rho_{\vec{k}}$ [5]

$$T_{\vec{k}} \# \varepsilon_{\vec{k}} - \frac{<\rho_{\vec{k}}|\varepsilon_{\vec{k}}>}{<\rho_{\vec{k}}|\rho_{\vec{k}}>} \rho_{\vec{k}} \tag{4.46}$$

This definition reduces to the temperature in the long wave-length limit; we know from standard thermodynamic fluctuation theory that temperature and density fluctuations are de-coupled in that limit, so that (4.46) is a natural generalization of the concept of temperature.

With this choice of $|\underline{A}>$ the static correlation matrix $<\underline{A}|\underline{A}>$ is diagonal, whereas the frequency matrix is purely off-diagonal, and the memory function matrix reduces to :

$$M(k,t) = \begin{bmatrix} 0 & 0 & 0 \\ 0 & M_{22}(k,t) & M_{23}(k,t) \\ 0 & M_{32}(k,t) & M_{33}(k,t) \end{bmatrix} \tag{4.47}$$

The correlation function matrix is then obtained, using (3.25). In particular we find for the longitudinal current ACF :

$$\tilde{C}_\ell(k,\omega) = \omega_o^2 \, \tilde{Y}_{22}(k,\omega)$$

$$= \frac{\omega_o^2}{- i\omega \left[1 - \frac{\omega_o^2}{S(k)}\right] + \tilde{N}_\ell(k,\omega)} \tag{4.48}$$

where the memory function $\tilde{N}_\ell(k,\omega)$ is some combination of the four non-zero partial memory functions $M_{ij}(k,\omega)$ in (4.47) [8]. It turns out that the memory functions M_{23} and M_{32} which describe the coupling between the momentum current and the heat flux, are instantaneously uncorrelated ($M_{23}(k,t=0) = M_{32}(k,t=0) = 0$) and also vanish in the hydrodynamic limit at all times. It seems hence reasonable to discard them in first approximation. In order to satisfy the 0th and 2nd order sum rules of $\hat{C}_\ell(k,\omega)$, and hence the first three moment sum rules for $S(k,\omega)$, it suffices to require that :

$$N_\ell(k,t=0) = \omega_{1\ell}^2 - \frac{\omega_o^2}{S(k)} \tag{4.49}$$

where [30] :

$$\omega_{1\ell}^2 = \frac{\langle\omega^4\rangle_{\rho\rho}}{\langle\omega^2\rangle_{\rho\rho}} = \frac{\int_{-\infty}^{+\infty} \omega^4 S(k,\omega)\, d\omega}{\omega_o^2}$$

$$= 3\omega_o^2 + \frac{k^2\rho}{m} \int g(r) \left[1 - \cos kz\right] \frac{\partial^2}{\partial z^2} v(r)\, d^3r \tag{4.50}$$

If we assume that the effect of thermal fluctuations is negligible, which amounts to setting $M_{33} = 0$, $N_\ell(k,t)$ reduces to the single memory function $M_{22}(k,t)$. We now make a single relaxation time approximation similar to that used in the transverse case, and satisfying (4.49) :

$$N_\ell(k,t) = \left[\omega_{1\ell}^2 - \frac{\omega_o^2}{S(k)}\right] \exp\left\{-t/\tau_\ell(k)\right\} \tag{4.51}$$

The corresponding viscoelastic expression for the dynamic

structure factor reads :

$$S(k,\omega) = \frac{1}{\pi} \frac{\tau_\ell(k)\omega_o^2\left\{\omega_{1\ell}^2 - \left[\omega_o^2/S(k)\right]\right\}}{\left[\omega\,\tau_\ell(k)\,(\omega^2-\omega_{1\ell}^2)\right]^2 + \left\{\omega^2\left[\omega_o^2/S(k)\right]\right\}^2} \tag{4.52}$$

Note that (4.52) does not go over into the hydrodynamic result (4.44) in the limit $k \to 0$; this is due to the neglect of thermal fluctuations. The expression (4.52) has been used to analyse neutron scattering and computer simulation data with some success, in particular in the case of liquid Rb [31]. A more sophisticated analysis, allowing to thermal fluctuations, has been made by Levesque et al. for the Lennard-Jones fluid [26] ; their expression for $S(k,\omega)$ reduces correctly to the hydrodynamic expression (4.44) in the long wavelength limit.

A general qualitative feature of experimental and molecular dynamics computer results is the persistance of a propagating "sound" mode up to considerably larger wave-numbers (up to $k \simeq 1$ Å^{-1}) in the case of liquid alkali as compared to rare gas liquids. It is tempting to attribute this behaviour to the relative "softness" of the inter-particle repulsion in the case of liquid metals.

4.4 A Remark on Simple Molecular Liquids

We briefly discuss here the dynamical structure factor for homonuclear diatomics, specifically the case of liquid N_2. The incident neutrons are scattered by the two nuclei, which are separated by a bond-length 2d (∼ 1 Å). The Fourier components of the nuclear density are :

$$\rho_{\vec{k}} = \sum_{i=1}^{N}\left[e^{-i\vec{k}\cdot\vec{r}_i^{(1)}} + e^{-i\vec{k}\cdot\vec{r}_i^{(2)}}\right] = 2\sum_{i=1}^{N} e^{-i\vec{k}\cdot\vec{R}_i}\cos(\vec{k}\cdot\vec{u}_i d) \tag{4.53}$$

where :

$$\vec{r}_i^{\,(1)} = \vec{R}_i + \vec{u}_i d$$

$$\vec{r}_i^{\,(2)} = \vec{R}_i - \vec{u}_i d$$

are the positions of the two nuclei in molecule i, $\vec{u}_i$ is the unit vector along the molecular axis, and $\vec{R}_i$ is the CM position. The intermediate scattering function is now defined as :

$$F(k,t) = \frac{1}{4N} < \rho_{\vec{k}}(t)\ \rho_{-\vec{k}}(0) > \tag{4.54}$$

If F(k,t) is expanded in spherical harmonics [32], the function can be separated into isotropic and anisotropic parts :

$$F(k,t) = j_o^2(kd)\ F_{CM}(k,t) + F_a(k,t) \tag{4.55}$$

where F_a denotes the anisotropic part, $j_o(x)$ is the 0th order spherical Bessel function (sin x/x) and $F_{CM}(k,t)$ is :

$$F_{CM}(k,t) = \frac{1}{N} < \sum_{i=1}^{N} \sum_{j=1}^{N} e^{-i\vec{k}\cdot\vec{R}_i(t)}\ e^{i\vec{k}\cdot\vec{R}_j} > \tag{4.56}$$

The influence of angular correlations on the decay of F(k,t) has been studied by Weis and Levesque [33] by a "molecular dynamics" simulation of a simple model for N_2. They conclude that, except at very large k ($k \simeq 10\ \text{Å}^{-1}$), $F_a(k,t)$ is very small, in fact zero within statistical errors. Moreover $F_{CM}(k,t)$ agrees perfectly with F(k,t) obtained for an "equivalent" Lennard-Jones fluid, with a core diameter determined by the law of corresponding states. The insensitivity of F(k,t) to anisotropy agrees with the fact that the "collective" orientational correlation function :

$$P(t) = \frac{1}{N} \sum_i \sum_j < P_2(\vec{u}_i\cdot(t) \cdot \vec{u}_j(0)) >$$

where P_2 is the second Legendre polynomial differs little from its self part $P_s(t)$ [34] . It means that little information on angular correlations in liquids of linear molecules is likely to come from neutron scattering experiments, unless the bondlength is relatively much larger than in N_2. Experimental studies in that direction, e.g. on CO_2, should be encouraged.

LECTURE 5

IONIC LIQUIDS

5.1 Classes of Simple Ionic Liquids

Because of over-all charge neutrality, an ionic system is made up of at least two components of opposite charge. The simplest examples in the liquid state are :

a) Liquid metals, made up of positive ions and conduction electrons; the dynamics of the latter are of course on a much faster time scale, which justifies a Born-Oppenheimer -type approximation leading to a pseudo-atom description which assimilates liquid metals with neutral liquids (alkali metals being treated like rare gas); the screened effective inter-ionic potential is short-ranged, compared to the bare Coulomb potential. These liquids will be studied by Dr. Evans in his lectures.

b) Electrolyte solutions are very complicated because of the presence of polar solvent molecules. An over-simplification is the "primitive model" in which the solvent only intervenes through a macroscopic dielectric function ε, which reduces the Coulomb interaction between ions. Some of our conclusions on the dynamic of molten salts may be applicable to electrolyte solutions in the long wavelength limit. Aqueous solutions are considered in Professor Enderby's lectures.

c) Molten salts (ionic melts) are made up of ions of opposite charge and comparable sizes and masses; we shall restrict our study to the simplest case : alkali-halides (i.e. spherical ions). The strong Coulomb interaction between ions manifests itself in the very large cohesive energy, which explains the high melting temperature ($T_m \gtrsim 10^3$ K, compared to $T_m \simeq 10^2$ K for Ar). Inter-ionic forces are made up of :

a) A Coulombic part (repulsive or attractive);

b) a short-range repulsion; in the Pauling model the repulsive core is described by a $1/r^n$ repulsive potential, whereas in the Born-Huggins-Mayer model the repulsive potential is of the exponential form; potential parameters are usually chosen equal to those determined by Tosi and Fumi [35] from solid state data;

c) dispersion forces which contribute relatively little, to the thermodynamic properties ($\sim$ 1 % to the cohesive energy);

d) induction forces due to the polarization of ion cores by the strong local electric fields; this is generally described by a shell model, familiar from solid state physics.

Although ion polarization has quantitative effects on structural (e.g. g(r)) and dynamical properties of molten salts [36], it does not modify their qualitative behaviour, and will be ignored to simplify our analysis.

A crude potential model incorporating the essential features is :

$$v(r_{ij}) = \frac{e^2}{\lambda} \left[\frac{1}{n} \left(\frac{\lambda}{r_{ij}} \right)^n + Z_i Z_j \frac{\lambda}{r} \right] \tag{5.1}$$

This has been used in an extensive molecular dynamics simulation [37] of dynamical properties, with $n = 9$ and $Z_+ = -Z_- = 1$; typically for NaCℓ, $\lambda \simeq 2.34$ Å. The single ion size $\lambda \simeq (\sigma_+ + \sigma_-)/2$ is in fact almost sufficient to characterize the melt, since the Coulomb repulsion between like ions keeps them apart anyway, so that the precise values of σ_+ and σ_- **separately** are not very important. Note that in the above-**mentioned** computer "experiment" the masses of both ionic species were also assumed to be equal ($m_+ = m_-$).

5.2 Self Diffusion and Electrical Conductivity

A melt containing N_+ cations of charge $\mathcal{Z}_+ e$ and N_- anions of charge $\mathcal{Z}_- e$ ($N_+ + N_- = N$) is characterized by two self-diffusion coefficients :

$$D_{+(-)} = \int_0^\infty Z_{+(-)}(t)\,dt \tag{5.2}$$

where $Z_+(t)$ and $Z_-(t)$ are the velocity ACF's of the two species. Note that in molten alkali halides, the diffusion coefficients are of the order of 10^{-4} cm^2 sec^{-1}, i.e. about five times larger than in the rare gases. We shall show that there exists a rather simple approximate relationship between D_+, D_- and the electrical conductivity σ, known as the Nerst-Einstein relation. To do this we first derive from general linear response theory the well-known Kubo formula for electrical conductivity.

In the presence of a weak homogeneous external electric field, of angular frequency ω , applied in the x-direction, the perturbation hamiltonian reads :

$$H' = - M_x E_o \cos \omega t \tag{5.3}$$

where M_x is the total dipole moment of the melt along Ox :

$$M_x = e \sum_{i=1}^{N} \mathcal{Z}_i\, x_i \tag{5.4}$$

The response of an isotropic melt is characterized by the induced electrical current along Ox :

$$e \langle j_x'(t) \rangle = e \langle \sum_i \mathcal{Z}_i\, v_{ix}(t) \rangle \tag{5.5}$$

where the prime is used to distinguish the electrical current from the particle current. According to the general formula (2.29) we have :

$$e < j'_x(t) > = \Re\, \tilde{\sigma}(\omega)\, E_o\, e^{-i\omega t} \tag{5.6}$$

For a classical system the frequency-dependent electrical conductivity is given by :

$$\tilde{\sigma}(\omega) = \beta \int_o^\infty e^{i\omega t} < e\, j'_x(t)\, \dot{M}_x(0) > dt$$

$$= \beta e^2 \int_o^\infty e^{i\omega t}\, J'(t)\, dt$$

where J'(t) is the electrical current ACF :

$$J'(t) = \frac{1}{3} < \vec{j}'(t) \cdot \vec{j}'(0) > \tag{5.7}$$

In particular the static conductivity per unit volume is given by :

$$\sigma = \frac{\beta e^2}{V} \int_o^\infty J'(t) dt \tag{5.8}$$

J'(t) splits naturally into a "self" and a "distinct" part :

$$J'(t) = J'_s(t) + J'_d(t) \tag{5.9}$$

$$J'_s(t) = \frac{N_+ \mathcal{Z}_+^2}{3} < \vec{v}_+(t) \cdot \vec{v}_+(0)> + \frac{N_- \mathcal{Z}_-^2}{3} <\vec{v}_-(t)\ \vec{v}_-(0)>$$

$$= N_+ \mathcal{Z}_+^2\, Z_+(t) + N_- \mathcal{Z}_-^2\, Z_-(t) \tag{5.10}$$

$$J'_d(t) = \frac{1}{3} \sum_i \sum_{j \neq i} < \mathcal{Z}_i\, \vec{v}_i(t) \cdot \mathcal{Z}_j\, \vec{v}_j(0) > \tag{5.11}$$

Clearly from (5.2) and (5.8) :

$$\sigma = \frac{\beta e^2}{V}\left[\mathcal{Z}_+^2 N_+ D_+ + \mathcal{Z}_-^2 N_- D_- + \int_o^\infty J'_d(t) dt\right] \tag{5.12}$$

Now $J'_d(0) = 0$, since the velocities of different particles are instantaneously uncorrelated. Hence, making the hypothesis that $J'_d(t) = 0$ at all times, we recover the so-called Nernst-Einstein relation :

$$\sigma = \frac{\beta e^2}{V}\left[\mathcal{Z}_+^2 N_+ D_+ + \mathcal{Z}_-^2 N_- D_-\right] \tag{5.13}$$

In liquid alkali halides, the deviations from this formula are about 20 %. These stem from the cross-correlation terms which we have neglected. The Nernst-Einstein relation is generalized in the form :

$$\sigma = \frac{\beta e^2}{V}\left[Z_+^2 N_+ D_+ + Z_-^2 N_- D_-\right]\left[1 - \Delta\right] \tag{5.14}$$

Experimentally Δ is always positive in alkali halides ($\Delta \simeq 0.18$ for NaCℓ). In the computer "experiment" [37] $J'(t)$, $Z_+(t)$, $Z_-(t)$ and $J'_d(t)$ were all measured simultaneously. $J'_d(t)$ turns out to be relatively small and short-lived (about 10^{-13} sec.), lending no support to a frequently proposed hypothesis of relatively long-lived ion pairs.

Before proceeding, we draw attention to a difficulty associated with non-local electrical conductivity, which is due to the long range of the Coulomb potential [38]. In the presence of a non-homogeneous electric field, we have a spatially non-local after-effect function (cf. (2.30)), and hence a $\vec{k}$-dependent conductivity tensor $\overleftrightarrow{\tilde{\sigma}}(\vec{k},\omega)$. By an obvious extension of (5.7), the longitudinal and transverse conductivities are related to the following longitudinal and transverse current correlation functions :

$$C'_\ell(\vec{k},t) = \frac{1}{N} < \vec{k} \cdot \vec{j}'_{\vec{k}}(t) \quad \vec{k} \cdot \vec{j}'_{-\vec{k}}(0) > \tag{5.15}$$

$$C'_t(\vec{k},t) = \frac{1}{2N} \mathrm{Tr} < \vec{k} \wedge \vec{j}_{\vec{k}}(t) \ \vec{k} \wedge \vec{j}'_{-\vec{k}}(0) > \tag{5.16}$$

with $\vec{j}'_{\vec{k}}(t) = \sum_{i=1}^{N} Z_i \ \vec{v}_i(t) \ e^{-i\vec{k}.\vec{r}_i(t)}$

One would expect that in the long wavelength limit :

$$\lim_{k\to 0} \frac{1}{k^2} C'_\ell(k,t) = \lim_{k\to 0} \frac{1}{k^2} C'_t(k,t) = J'(t) \tag{?}$$

so that the transverse and longitudinal conductivities would coincide with the homogeneous ($k = 0$) conductivity σ in that limit. This is not true however, because of the $1/k^2$ divergence of the Coulomb potential, and the fact that the electrostatic field is purely longitudinal. The difference is intimately related to the difference $\omega_p = (4\pi\rho e^2/\mu)^{1/2}$ (with μ the reduced mass of an anion-cation pair) between the longitudinal and transverse "optic" modes in ionic crystals. Only the transverse conductivity tends towards σ as $k \to 0$.

5.3 Longitudinal Collective Modes

Since we are dealing with a two-component system, we define partial densities :

$$\rho_{\vec{k}}^{(a)} = \sum_{i=1}^{N_a} e^{i\vec{k}\cdot\vec{r}_i} \quad , \quad a = +,-$$

which verify two separate continuity equations; hence we have 6 conserved hydrodynamic variables instead of 5. For a neutral binary mixture, an additional hydrodynamic mode is associated with concentration fluctuations and is governed by the interdiffusion coefficient [27]. Here we shall briefly study longitudinal collective modes in a binary ionic melt.

The "natural" linear combinations of the $\rho_{\vec{k}}^{(a)}$ are the charge and the mass (or number) densities; in the simple symmetrical model which we consider, mass and number densities are proportional, and we use :

$$\rho_{\vec{k}} = \rho_{\vec{k}}^{(+)} + \rho_{\vec{k}}^{(-)}$$

$$\rho'_{\vec{k}} = Z_+ \rho_{\vec{k}}^{(+)} + Z_- \rho_{\vec{k}}^{(-)} = \rho_{\vec{k}}^{(+)} - \rho_{\vec{k}}^{(-)}$$

We define the corresponding static structure factors :

$$S(k) = \frac{1}{N} < \rho_{\vec{k}} \quad \rho_{-\vec{k}} >$$

$$S'(k) = \frac{1}{N} < \rho'_{\vec{k}} \quad \rho'_{-\vec{k}} >$$

A simple fluctuation calculation [39, 40] shows that S(0) is finite, as in the neutral case, and related to a thermodynamic derivative; the same calculation shows that $S'(k) \sim k^2$ as $k \to 0$. Since this behaviour is crucial in our understanding of charge fluctuations, we prove it by a simple "perfect screening" argument [41] based on linear response theory. Consider an external charge distribution :

$$q(\vec{r}) = q^{ext} e^{i\vec{k}\cdot\vec{r}}$$

The corresponding potential

$$\varphi(\vec{r}) = \varphi^{ext} e^{i\vec{k}\cdot\vec{r}}$$

is related to $q(\vec{r})$ by Poisson's equation :

$$\varphi^{ext} = \frac{4\pi}{k^2} q^{ext}$$

The external charge couples of course to the charge density of the system and the perturbation reads :

$$H' = \frac{4\pi q^{ext} e}{k^2} \rho'_{-\vec{k}}$$

The $\vec{k}^{th}$ Fourier component of the charge <u>induced</u> in the system per unit volume is :

$$q^{ind}_{\vec{k}} = \frac{e}{V} < \rho'_{\vec{k}} >$$
$$= \frac{e}{V} \frac{\int d^{3N} r \, \rho'_{\vec{k}} \exp\left\{ - \beta H_o - \frac{4\pi\beta q^{ext} e}{k^2} \rho'_{-\vec{k}} \right\}}{\int d^{3N} r \exp \{ \quad \}}$$

which, upon linearization, becomes :

$$q^{ind}_{\vec{k}} \simeq - \frac{4\pi \beta e^2 q^{ext}}{Vk^2} < \rho'_{\vec{k}} \rho'_{-\vec{k}} >_o$$
$$= - \frac{4\pi \beta e^2 q^{ext} \rho}{k^2} S'(k)$$
$$= - \frac{K^2}{k^2} S'(k) q^{ext}$$

with $$K^2 = 4\pi \beta\rho e^2$$

The "perfect screening" of the external charge by the conducting melt now requires that :

$$q^{ext} + q^{ind}_{\vec{k}} = 0 \qquad \text{as } k \to 0$$

which leads to the limiting behaviour :

$$S'(k) \simeq \frac{k^2}{K^2} , \quad k \to 0 \tag{5.17}$$

The number (or mass) and charge density fluctuations are described in terms of the number (or mass) and charge dynamical structure factors :

$$S(k,\omega) = \frac{1}{2\pi}\int_{-\infty}^{+\infty} \frac{1}{N} < \rho_{\vec{k}}(t)\ \rho_{-\vec{k}}(0) > e^{i\omega t}\ dt \tag{5.18}$$

$$S'(k,\omega) = \frac{1}{2\pi}\int_{-\infty}^{+\infty} \frac{1}{N} < \rho'_{\vec{k}}(t)\ \rho'_{-\vec{k}}(0) > e^{i\omega t}\ dt \tag{5.19}$$

and the spectrum of the cross correlation function :

$$S''(k,\omega) = \frac{1}{2\pi}\int_{-\infty}^{+\infty} e^{i\omega t}\ \frac{1}{N} < \rho'_{\vec{k}}(t)\ \rho_{-\vec{k}}(0) > dt \tag{5.20}$$

Note that for the simple symmetrical model, number and charge density fluctuations are decoupled at all times, due to the symmetry of the hamiltonian. For "real" salts, the cross correlations can be expected to be small if the masses and sizes of the ions are not too different; this is borne out by computer "experiments" on molten NaCℓ [42].

Since number (mass) and charge density fluctuations are decoupled (or only weakly coupled), we can treat them separately. The mass density fluctuations turn out to be in all respects similar to those in neutral mixtures, and so we shall not consider them further. The remainder of the lecture will be devoted to the investigation of the possible existence of a longitudinal "optic" mode in ionic melts, similar to that observed in ionic crystals.

We begin by giving a rough argument to show that we expect a <u>high frequency</u> mode from charge fluctuations. The characteristic frequency of charge fluctuations is given approximately by the second moment sum rule of the normalized spectrum :

$$\omega_{\ell o}^2 \simeq \frac{<\omega^2>_{\ell o}}{<\omega^o>_{\ell o}} = \frac{\int_{-\infty}^{+\infty} S'(k,\omega)\omega^2\, d\omega}{S'(k)}$$

$$= \frac{k^2 k_B T/m}{k^2/K^2} \equiv \omega_p^2$$

where we have assumed equal masses ($m_+ = m_- = m$); hence we may expect a <u>finite</u> frequency mode in the long wavelength limit. Note that the same argument predicts correctly that the "longitudinal acoustic" mode has a frequency vanishing as k, since S(k) stays finite. Because we look for a <u>finite</u> frequency mode in the long wavelength limit, we do not, a priori, expect a hydrodynamic calculation, of the type sketched in lecture 4 for neutral fluids, to be very useful. However such hydrodynamic calculations have been suprisingly successful in the study of "**plasmon**" modes in strongly coupled plasmas [43, 44], so that it is worth mentioning the results obtained by Giaquinta et al. [39] and by Vieillefosse [40]. The linearized equations of hydrodynamics given earlier must be complemented by the continuity equation for the charge density, and contributions of the local electric field must be added in the definition of the energy density and the stress tensor (where they actually vanish upon linearization). The dissipative part of the macroscopic **charge current requires the introduction of two further** transport coefficients : the inter-diffusion coefficient α, which is proportional to the conductivity σ, and the thermo-electric coefficient β :

$$\vec{j}'_{diss} = - \alpha \vec{\nabla}(\mu' + e\varphi) - \beta \vec{\nabla} T$$

where φ is the electrostatic potential (determined from the charge density by Poisson's equation) and μ' is the chemical potential conjugate to the charge density. The calculation then proceeds along the lines laid down in lecture 4, and leads to the prediction of four longitudinal modes : two symmetric propagating sound modes and a thermal diffusion mode, all three of which have a damping $\sim k^2$, and a charge diffusion mode (or inter-diffusion mode) of purely imaginary frequency :

$$\omega = - i\, 4\pi\, e^2\, \alpha$$

which is independent of k ! In the long wavelength limit, this mode dominates the charge fluctuation spectrum which reduces to a Lorentzian of finite width as $k \to 0$. So hydrodynamics does not predict a propagating longitudinal "optic" mode, contrarily **to the case of dense** ionic plasmas, where the plasmon mode is correctly predicted [43, 44]. The predictions of this simple hydrodynamic calculation for ionic melts have been confirmed by a rigorous Kinetic Theory approach [45].

The computer "experiment" on the symmetric model lead however to a rather well-defined optic mode at frequencies slightly above the "plasma" frequency ω_p [37]. This is a "resonant" mode, in the sense that it is sharpest around $k\lambda \simeq 2$ (corresponding to a wavelength of roughly twice the inter-ionic spacing), and broadens both towards high k (as expected) and low k. The dispersion of the optic mode is negative ($\frac{d\omega(k)}{dk} < 0$). The optic mode has since then also been observed in computer "experiments" based on more "realistic" models including better ion pair potentials and different masses.

The question then naturally arises whether the optic mode could be observed in a coherent inelastic neutron scattering experiment. This would be most easy if both ions had nuclei with essentially coherent scattering lengths of opposite sign. In particular if $a^{+}_{coh} = - a^{-}_{coh}$, the observed spectrum would be directly proportional to $S'(k,\omega)$ (neutrons would then behave as "charged" test particles). A first experiment was done by Price and Copley [46] on **RbBr** which is unfortunately a very unfavourable case for observing charge fluctuations, since $a_{Rb} \simeq a_{Br}$. Another experiment on KBr is currently under way at I.L.L. [47] . Although the situation is more favourable in this case (since $a_{Br} \simeq 2a_{K}$), the contributions to the observed inelastic scattering cross sections of incoherent and coherent mass fluctuation components make the observation of the optic peak very difficult.

REFERENCES

[1] R. Kubo : Rep. Progr. Phys. 29, 255 (1966).
[2] P.C. Martin in "Many-Body Physics", edited by C. De Witt and R. Balian (Gordon and Breach, New York (1968)).
[3] R.G. Gordon : Advances in Magnetic Resonance 3, 1 (1968).
[4] B.J. Berne in "Physical Chemistry, volume VIII B, edited by D. Henderson (Academic Press, New York (1971)).
[5] P. Schofield in "Statistical Mechanics", volume 2 (Specialist Periodical Reports), edited by K. Singer (The Chemical Society, London (1976)).
[6] D. Forster : "Hydrodynamics, Fluctuations, Broken Symmetry and Correlation Functions" (Benjamin, Reading, Ms. (1975)).
[7] B.J. Berne and R. Pecora : "Dynamic Light Scattering" (Wiley, New York (1976)).
[8] J.P. Hansen and I.R. McDonald : "Theory of Simple Liquids" (Academic Press, London (1976)).
[9] P. Langevin : C.R. Acad. Sci. (Paris) 146, 530 (1908).
[10] R. Kubo : J. Phys. Soc. Japan 12, 570 (1957).
[11] H. Nyquist : Phys. Rev. 32, 110 (1928).
[12] L.P. Kadanoff and P. Martin : Ann. Phys. 24, 419 (1963).
[13] S. Yip and M. Nelkin : Phys. Rev. 135, A1241 (1964).
[14] R. Zwanzig in "Lectures in Theoretical Physics", vol. 3, edited by W.F. Britton, W.B. Downs and J. Downs (Interscience, New York, 1961).
[15] H. Mori : Progr. Theor. Phys. 33, 423 and 34, 399 (1965).
[16] D. Levesque and L. Verlet : Phys. Rev. A2, 2514 (1970).
[17] L. Van Hove : Phys. Rev. 95, 249 (1954).
[18] B.J. Berne, J.P. Boon and S.A. Rice : J. Chem. Phys. 45, 1086 (1966).
[19] B.J. Alder and T.E. Wainwright : Phys. Rev. A1, 18 (1970).
[20] Y. Pomeau and P. Résibois : Physics Reports 19C, N° 2 (1975).
[21] L. Onsager : Phys. Rev. 37, 405 and 38, 2265 (1931).
[22] L.D. Landau and E.M. Lifshitz : "Fluid Mechanics" (Pergamon Press, London, 1963).
[23] R. Zwanzig and R.D. Mountain : J. Chem. Phys. 43, 4464 (1965)
[24] N.K. Ailawadi, A. Rahman and R. Zwanzig : Phys. Rev. A4, 1616 (1971).
[25] D. Forster, P.C. Martin and S. Yip : Phys. Rev. 170, 155 and 160 (1968).
[26] D. Levesque, L. Verlet and J. Kürkijarvi : Phys. Rev. A7, 1690 (1973).
[27] R. Mountain and J. Deutch : J. Chem. Phys. 50, 1103 (1969).
[28] B.J. Alder, J. Strauss and J.J. Weis : J. Chem. Phys. 59, 1002 (1973).
[29] D. Oxtoby and W.M. Gelbart : Mol. Phys. 29, 1569 (1975).
[30] P.G. De Gennes : Physica 25, 825 (1958).
[31] J.R.D. Copley and S.W. Lovesey : Rep. Progr. Phys. 38, 461 (1975).

[32] V.F. Sears : Can. J. Phys. 44, 1299 (1966).
[33] J.J. Weis and D. Levesque : Phys. Rev. A13, 450 (1976).
[34] D. Levesque and J.J. Weis : Phys. Rev. A12, 2584 (1975).
[35] M.P. Tosi and F.G. Fumi : J. Phys. Chem. Solids 25, 31 and 45 (1964).
[36] M.J.L. Sangster and M. Dixon : Adv. in Phys. 25, 247 (1976).
[37] J.P. Hansen and I.R. McDonald : Phys. Rev. A11, 2111 (1975).
[38] P.C. Martin : Phys. Rev. 161, 143 (1967).
[39] P.V. Giaquinta, M. Parrinello and M.P. Tosi : Phys. Chem. Liqu. 5, 305 (1976).
[40] P. Vieillefosse : J. de Phys. 38, L-43 (1977).
[41] F.H. Stillinger and R. Lovett : J. Chem. Phys. 49, 1991 (1968).
[42] G. Ciccotti, G. Jacucci and I.R. McDonald : Phys. Rev. A13, 426 (1976).
[43] P. Vieillefosse and J.P. Hansen : Phys. Rev. A12, 1106 (1975).
[44] I.R. McDonald, P. Vieillefosse and J.P. Hansen : Phys. Rev. Letters 39, 271 (1977).
[45] M. Baus : Physica.
[46] D.L. Price and J.R.D. Copley : Phys. Rev. A11, 2124 (1975).
[47] J.R.D. Copley : private communication.

MOLECULAR DYNAMICS: PRINCIPLES AND APPLICATIONS

Ian R. McDonald

Department of Chemistry
Royal Holloway College
Egham, Surrey, England

LECTURE 1: INTRODUCTION

The purpose of this series of lectures is to describe the way in which fast computers can be used in simulating the microscopic behaviour of liquids. We shall be concerned in particular with the method of molecular dynamics, a technique developed and first exploited by Alder and Wainwright[1]. Such calculations are often referred to as computer "experiments", a usage of the term which true experimentalists sometimes deplore! The discussion will cover a variety of aspects of the method: its underlying philosophy, if that is not too pretentious a phrase; the technical problems connected with its realization; and some examples of its usefulness in liquid-state research. Naturally the selection of topics is somewhat arbitrary, but I hope that it will give an idea of the wide range of problems which have been, or could be studied.

The basic principle of the method is easily explained. A typical "experiment" consists simply in studying the time evolution of the dynamical state of a system of interacting particles by numerical integration of the classical equations of motion. The computations therefore require the repeated execution of two basic operations:

a) the calculation of the total force acting on each particle at a given time, t say ;

b) the advancement of the coordinates of all particles to their predicted values at a later time, t+h say, h being the time step in the numerical integration.

Two points can be added here. First, the total force is almost invariably calculated as the sum of pair terms, since the explicit inclusion of many-body forces is very time-consuming. Second, step (b) is carried out somewhat differently for systems in which the potential has a hard core, hard spheres being the classic example.

By proceeding in the way we have described it is possible to follow the trajectories of the individual particles in the system and extract whatever microscopic information we desire. In a typical experiment on, say, a model simulating liquid argon, the time step h would be of order 10^{-14} sec and the integration might extend over a total of several thousand time steps. The "length" of the experiment would then be of order 10^{-11} to 10^{-12} sec. This is a short time in macroscopic terms, but it is long enough to yield the data required to calculate important quantities such as the thermodynamic properties of the system and a variety of functions which measure correlations in space and time, in addition to allowing the study of all manner of other phenomena which are less easily quantifiable. The "particles" of the system may be atoms, molecules (rigid or flexible) or ions; the basic input information is the potential of interaction between particles, from which the forces acting are obtained by differentiation.

Before embarking on a discussion of the technical problems which are involved, it is worthwhile looking at some of the reasons for undertaking what are, inevitably, rather expensive calculations, requiring access to a fast computer and, of course, a large computing budget. Much the most important is the fact that computer "experiments" yield results which are essentially exact, subject to certain qualifications which we shall discuss later. They therefore represent a source of accurate, quasi-experimental data on completely well-defined models. This is an ideal situation from the theoretician's point of view, since theoretical predictions can be tested against such data with little risk of ambiguity. This is rarely true when the experimental data are obtained from real experiments on real liquids, since in that case the information one has about the interaction potentials is invariably incomplete.

This is not the end of the story so far as the theoretician is concerned. Suppose, for example, that we have developed a theory which enables us to calculate the equation of state of a fluid if the intermolecular potential is known. Clearly we can check the value of our theory by comparing its predictions with the results of a computer "experiment" based on the same potential. But a comparison of this sort is not always conclusive, since good results can be obtained as a consequence of fortuitous cancellation of errors. Frequently the approximations introduced at intermediate stages in the working-out of a theory may themselves be tested against data coming from a computer "experiment". This makes it possible to identify the specific points at which a theory needs improvement.

Going one stage further, it is easy to imagine a situation in which a theory is under development but for the present is incomplete insofar as it assumes the knowledge of certain quantities which are in fact unknown. If these quantities can be determined in the course of a computer "experiment", the consequences of the theory can be worked out, the question of how to calculate the unknown quantities within a purely theoretical framework being left as a separate problem. This type of situation is frequently encountered in the case of perturbation theories, the as-yet unknown information being the properties of a suitable reference system.

So far we have emphasized the role which molecular dynamics can play in theoretical work. But the method has a dual character insofar as it can act as a bridge between, on the one side, statistical mechanical theory and, on the other, real experiments carried out in the laboratory. Computer "experiments" can be used to test the extent to which a given intermolecular potential is able to account for the observed properties of some real liquid, be it argon, nitrogen, water or sodium chloride, to mention just a few of the systems to which this approach has been applied. Simulation can thereby be used to improve our understanding of the intermolecular forces in real liquids, to refine existing empirical potentials, and to throw light on the way in which different macroscopic properties are affected by changes in the potential.

In this brief discussion we have identified two important areas in which computer "experiments" can usefully be put to work:

a) in the development and testing of theories;

b) in checking the adequacy of potential models applicable to real liquids;

to which may be added a third -

c) in extracting information which is experimentally inaccessible or nearly so.

A good example of c) is to be found in the propagation of shear waves in argon-like liquids, an interesting phenomenon on which molecular dynamics is so far our only source of information, but many others could be quoted.

Just how much importance one attaches to these different facets of the method is largely a matter of taste. Up till now molecular dynamics has proved most useful on the theoretical side, though it has also contributed a good deal to our understanding of the behaviour of a variety of real liquids, notably nitrogen, water and a number of simple ionic melts. The problem which one faces when attempting to simulate real systems is the fact that the intermolecular potentials are very complicated functions even in the simplest liquids. Moreover, effects such as mutual polarization and other essentially many-body interactions play a significant role in many cases, but are not easy to incorporate into a molecular dynamics calculation.

The method has other limitations, of course, but these will be more readily appreciated when we have discussed some of the practical questions involved in its implementation.

The first problem which arises is that of sample size. On any moderately large computer, a CDC6600 for example, it is relatively easy to handle systems of up to 1000 particles. Unfortunately, a sample of 1000 particles is not only very small on any macroscopic scale, but if studied in isolation surface effects would play much too important a role. To minimize such effects and thereby simulate more closely the behaviour of a macroscopic system, a periodic boundary condition is almost always imposed. The N particles of our system (the system of interest) are assumed to lie in a central cell, this basic unit being surrounded on all sides by infinitely repeating periodic images of itself. In other words, each image cell contains N particles occupying the same relative positions and having the same momenta as particles in the basic cell. The equations of motion to be solved are those of particles in the central cell. When the motion of a particle carries it out through a face of the latter, it is replaced by an image particle moving in through the opposite face, and the number of particles in each cell remains equal to N.

It is customary to use a cubic cell, though there is something to be said for using a more complicated form, a truncated dodecahedron for example. The only limitation is that the periodic array of cells must fill all space. It is also frequently convenient to select the number N and the cell shape in such a way that the periodic boundary condition generates a perfect lattice appropriate to the physical system under study when the particles in the basic cell are arranged in a suitably ordered manner. Argon, for example, crystallizes in a face-centred cubic arrangement. This makes it natural to choose a cubic cell with $N = 4n^3$, where n is an integer, which explains the widespread use of samples containing N = ..., 108, 256, 500, 864, ... particles. The alkali metals, on the other hand, crystallize in body-centred cubic form, which makes it more appropriate to take $N = 2n^3$. These remarks should not be taken as implying that the molecular dynamics method can be used in a straightforward way in the study of the mechanism underlying the liquid-solid phase transition. This is certainly not the case.

Each particle in the central cell, which we shall take to be a cube, is assumed to interact with all others in the periodic array. However, to speed up the computations, it is normal to calculate forces only between particles separated by distances smaller than some value r_t, the radius of a so-called truncation sphere, r_t being typically three to four molecular diameters. More distant particles are assumed to act as a uniform background, making a contribution to the total potential energy but not to the net force acting on a given particle, i say. (Of course when calculating thermodynamic properties such as energy and pressure, account must be taken of the background.) Thus in computing the force acting on i, all interactions between i and others inside the truncation sphere centred on i are calculated explicitly, the total force being determined partly by particles inside the basic cell and partly by others in neighbouring in cells. Provided that $r_t < \frac{1}{2}L$, where L is length of the cube, it is impossible for the truncation sphere around i to include more than one image of any other particle, j say, which lies in the basic cell. This serves to attenuate the effects of the periodic boundary condition, since they appear only implicitly. When the potential is of very long range, however, truncation can lead to systematic errors and other means of summing the forces must be adopted. This is true, in particular, of ionic and polar systems.

A fundamental problem which must still be faced is that of how far the properties of a small ($N \sim 10^3$) periodic system are truly representative of the macroscopic system which the model is designed to simulate. There is no easy answer to this question; it depends very much on the form of the potential and on the phenomena under investigation. If in doubt, it is wisest to make preliminary calculations for a number of values of N and look for signs of a systematic trend in the results. Broadly speaking it is true to say that for short-ranged potentials the bulk properties are only weakly dependent on sample size for $N > 100$ and that such errors as exist are for the most part no larger than the inevitable statistical uncertainties. However, the restriction on sample size does have certain clear-cut disadvantages. First, the study of critical phenomena is largely beyond the scope of computer "experiments", because the large-scale density fluctuations characteristic of the critical point are unavoidably suppressed by use of a periodic boundary condition. Second, irrespective of the thermodynamic state, it is impossible to study spatial fluctuations having a wavelength greater than L (the length of the cube). For an 864 particle system representative of argon near its triple point it turns out that L is approximately 30 Å, so that the minimum accessible wavenumber is approximately 0.2 $Å^{-1}$. This limitation is a particularly regrettable one since it is just in this range of wavenumber that collective effects become important, notably in the appearance of propagating sound waves, and it also marks the current limit of the best inelastic neutron scattering experiments. Third, the periodic boundary condition has an effect on long-lived time correlations, since a local disturbance can move through the system and reappear, albeit in modified form, after a recurrence time $\tau_{rec} \simeq L/c$, where c is the speed of sound. For the same example of 864 argon atoms near the triple point, this estimate yields $\tau_{rec} \simeq 5 \times 10^{-12}$sec, or 500 typical integration steps. Correlations in time cannot usefully be studied over intervals longer than this.

The second major computational problem is the choice and implementation of a suitable algorithm for the integration of the equations of motion. Given the importance of this question, it is remarkable how little effort has been directed at the systematic study of the relative merits of different schemes. In making a choice there are two main considerations to be borne in mind. One is accuracy, the other is economy. Not surprisingly, these two factors are often in conflict. Algebraic simplicity is a third consideration, less important but not insignificant, since the development of a correctly working program can be a major undertaking.

In summary: the essential input information, apart from the density and temperature at which the experiment is to be carried out, is the intermolecular pair potential. The primary output consists of the trajectories of the particles, i.e. their coordinates and momenta as functions of time. The advantages of the method have already been described. Its main limitations are, first, those imposed by the small size of the system which can be studied; second, the restriction to pairwise additive forces; and, third, the fact that the calculations, at least as we have described them here, are made entirely within the framework of classical mechanics, thereby ruling out, for example, any realistic treatment of molecular vibrations.

It would not be right to end this introductory talk without making at least a brief reference to the Monte Carlo method, a technique of simulation which was first applied to statistical mechanical problems by Metropolis et al.[2]. The method involves the generation of a series of configurations of the particles of the system in a way which ensures that the configurations are distributed in phase space according to some prescribed probability density. The mean value of any configurational property determined from a sufficiently large number ($\sim 10^6$) of configurations provides an estimate of the ensemble-averaged value of that quantity, the character of the ensemble average being dependent on the particular sampling procedure which is used. The method differs from molecular dynamics insofar as no time scale is involved and the order in which the configurations are generated is of no particular significance. Thus the Monte Carlo method cannot be used in the study of time-dependent phenomena. For the calculation of thermodynamic properties, however, the Monte Carlo method is in some ways more suitable than molecular dynamics. Lack of time prevents us from discussing the method in any greater detail, but excellent accounts of its basis and its applications can be found in two recent review articles by Valleau and co-workers[3,4].

LECTURE 2 : SOLUTION OF THE EQUATIONS OF MOTION

For the sake of simplicity let us suppose initially that the system of interest is made up of identical, structureless particles of mass m. Then the equations of motion which must be solved are the set of N second-order differential equations of the type

$$m\dot{\mathbf{r}}_i(t) = \mathbf{F}_i(t) \tag{1}$$

where $\mathbf{F}_i$ is the total force acting on a particle i which at time t is located at a position $\mathbf{r}_i(t)$. In the simplest case of central pair forces, $\mathbf{F}_i$ can be written as

$$\mathbf{F}_i = - \sum_j (\mathbf{r}_{ij}/r_{ij}) \frac{d\phi(r_{ij})}{dr_{ij}} \tag{2}$$

where $\phi(r)$ is the pair potential, which we suppose is continuous, and $\mathbf{r}_{ij} = \mathbf{r}_i - \mathbf{r}_j$. Because $\mathbf{F}_i$ is a function of all $\mathbf{r}_j$ (or at least of all $\mathbf{r}_j$ for which r_{ij} is less than the radius of the truncation sphere), the equations of motion of different particles are coupled together and some difficulty might be anticipated in achieving a stable solution. Fortunately, however, this does not appear to be a serious problem.

The application of molecular dynamics to systems with continuous potentials was first described by Rahman[5] in a paper on an argon-like liquid published in 1964. The algorithm used there was a predictor-corrector scheme of the form

$$\begin{aligned} &(a)\ \mathbf{r}_i(t+h) = \mathbf{r}_i(t-h) + 2\dot{\mathbf{r}}_i(t) \\ &(b)\ \dot{\mathbf{r}}_i(t+h) = \dot{\mathbf{r}}_i(t) + \frac{1}{2} h\,[\ddot{\mathbf{r}}_i(t+h)+\ddot{\mathbf{r}}_i(t)] \\ &(c)\ \mathbf{r}_i(t+h) = \mathbf{r}_i(t) + \frac{1}{2} h\,[\dot{\mathbf{r}}_i(t+h)+\dot{\mathbf{r}}_i(t)] \end{aligned} \tag{3}$$

Knowing $\mathbf{r}_i(t-h)$ and $\dot{\mathbf{r}}_i(t)$, a rough prediction of the coordinates $\mathbf{r}_i(t+h)$ can be obtained from (a); the predicted coordinates are then used to determine $\ddot{\mathbf{r}}_i(t+h)$ from Newton's equation (1); then (b) is used to obtain a first prediction for the velocity $\dot{\mathbf{r}}_i(t+h)$ and substitution into (c) leads to a corrected value for $\mathbf{r}_i(t+h)$. Higher accuracy can be achieved by iteration; $\ddot{\mathbf{r}}_i(t+h)$ is recalculated from the corrected coordinates and a better approximation to $\mathbf{r}_i(t+h)$ is obtained from (b) and (c). In practice two or three iterations are sufficient.

Though the accuracy attained is high, methods of this type have the disadvantage that the forces must be calculated more than once at each time step. This is clearly uneconomic, since the calculation of the forces is by far the most time-consuming part of the computations. A simpler algorithm is therefore needed and the one currently used by most workers is the central-difference predictor popularized by Verlet[6]. As before, let the coordinates of particle i at time t be $\underset{\sim}{r}_i(t)$. Then its coordinates at times $t \pm h$ can be found from a Taylor expansion about $\underset{\sim}{r}_i(t)$:

$$\underset{\sim}{r}_i(t+h) = \underset{\sim}{r}_i(t) \pm h\dot{\underset{\sim}{r}}_i(t) + \frac{1}{2}h^2\ddot{\underset{\sim}{r}}_i(t) \pm (h^3/6)\frac{d^3}{dt^3}\underset{\sim}{r}_i(t) + O(h^4) \tag{4}$$

By adding the two expressions we obtain a prediction for $\underset{\sim}{r}_i(t+h)$

$$\underset{\sim}{r}_i(t+h) = -\underset{\sim}{r}_i(t-h) + 2\underset{\sim}{r}_i(t) + (h^2/m)\underset{\sim}{F}_i(t) \tag{5}$$

where the error is of order h^4. On the other hand, by subtracting the two expressions, we obtain an estimate for the velocity at time t

$$\dot{\underset{\sim}{r}}_i(t) = (1/2h)\ [\underset{\sim}{r}_i(t+h) - \underset{\sim}{r}_i(t-h)] \tag{6}$$

The error is now of order h^3, but the velocities play no part in the integration of the equations of motion, that is to say in the calculation of the trajectories ($\underset{\sim}{r}_i(t)$ as a function of t) of the particles.

Although no corrector is used, the time step h may be taken as large as in the more elaborate scheme described in eqns.(3). Moreover, the rather limited tests which have been carried out suggest [7] that the central-difference algorithm is at least as stable, if not more so, than a variety of higher order algorithms in which the prediction of $\underset{\sim}{r}_i(t+h)$ is based not only on the coordinates $\underset{\sim}{r}_i(t)$ and $\underset{\sim}{r}_i(t-h)$ but also on $\underset{\sim}{r}_i(t-2h)$, $\underset{\sim}{r}_i(t-3h)$... etc. Numerical analysts seem to find this surprising, but from the point of view of economy it is certainly a comforting result.

The discussion so far has been a little loose, since we have not indicated how the "accuracy" of a particular method is to be assessed. Luckily there are some objective tests available, since in an isolated system or a system with periodic boundary conditions the total energy (potential plus kinetic) and total linear momentum are constants of the motion. In practice some fluctuation is inevitable, but this must fall within acceptable limits and there should be no perceptible drift. The fluctuation in total energy is sensitively dependent on the choice of time step. As h is increased, the fluctuations build up in size; eventually an overall upward drift in energy may develop, though there will still be fluctuations. A monotonic upward drift, however, is a virtually infallible sign that there is an error in the program. This is frequently associated with a mismatch between energies and forces, as would happen if we failed to differentiate the potential function correctly. In such a case, of course, the system is no longer a conservative one.

If the fluctuations in total energy are of order one part in 10^4 it is usually safe to conclude that everything is working satisfactorily, though larger fluctuations are sometimes acceptable. The central-difference algorithm automatically conserves total linear momentum, but allowance must be made for accumulated round-off errors. These lead typically to fluctuations in the velocity of the centre of mass of the system which are of order one part in 10^8 of the mean speed of a particle; to achieve this accuracy it is usually necessary to work in double precision.

It is worth adding here that the apparent accuracy of the central-difference algorithm can be improved if the velocity is calculated in a different way. For example, a Taylor expansion about $\underset{\sim}{r}_i(t+h)$ yields

$$\underset{\sim}{r}_i(t) = \underset{\sim}{r}_i(t+h) - h\dot{\underset{\sim}{r}}_i(t+h) + \frac{1}{2}h^2\ddot{\underset{\sim}{r}}_i(t+h) - (h^3/6)\frac{d^3}{dt^3}\underset{\sim}{r}_i(t+h) + O(h^4) \qquad (7)$$

Then if we approximate the third derivative appearing in eqn.(7) by writing

$$\frac{d^3}{dt^3}\underset{\sim}{r}_i(t+h) = (1/h)\,[\ddot{\underset{\sim}{r}}_i(t+h) - \ddot{\underset{\sim}{r}}_i(t)] + O(h) \qquad (8)$$

and eliminate the second derivatives in favour of the interparticle forces, we obtain an estimate for the velocity in the form

$$\dot{\underset{\sim}{r}}_i(t+h) = (1/h)\,[\underset{\sim}{r}_i(t+h) - \underset{\sim}{r}_i(t)] + (h/6m)\,[2\underset{\sim}{F}_i(t+h) + \underset{\sim}{F}_i(t)] + O(h^3) \qquad (9)$$

The error is again of order h^3 but is numerically smaller than in (6). Thus an improved estimate is obtained for the kinetic energy, resulting in a smaller fluctuation in the total energy. The improvement is only apparent, however, since the trajectories of the particles are unaltered.

It should be clear from the discussion we have just given that a degree of care is needed in the choice of time step. Roughly speaking, h should be significantly smaller than the mean time between collisions, but this criterion has only a qualitative value since the concept of a "collision" is one which is well-defined only for systems such as hard spheres. A variety of different factors are involved: the mass of the particles, the strength of the interaction, the temperature and density, and the level of fluctuation in the total energy which is regarded as acceptable. As an example, consider the case of a Lennard-Jones fluid, for which the pair potential is given by

$$\phi(r) = 4\varepsilon[(\sigma/r)^{12} - (\sigma/r)^{6}] \tag{10}$$

where σ is the collision diameter, i.e. $\phi(\sigma) = 0$, and ε is the maximum depth of the attractive bowl. On introducing the reduced units $r^* = r/\sigma$ we can rewrite the central-difference formula for $\underset{\sim}{r}_i(t+h)$ as

$$\underset{\sim}{r}_i^*(t+h) = -\underset{\sim}{r}_i^*(t-h) + 2\underset{\sim}{r}_i^*(t) - h^2(48\varepsilon/m\sigma^2)(\underset{\sim}{r}_{ij}^*/r_{ij}^*) \sum_{j\neq i} [(1/r_{ij}^*)^{13} - \tfrac{1}{2}(1/r_{ij}^*)^{7}] \tag{11}$$

Thus the quantity

$$\tau = (m\sigma^2/48\varepsilon)^{1/2} \tag{12}$$

appears as a natural unit of time. This suggests a general rule, applicable to other potentials, that for a given accuracy the time step increases as the linear dimensions of the system and as the square root of the mass, and decreases as the square root of the interaction strength (or temperature). For the Lennard-Jones fluid a value $h \simeq 0.03\tau$ proves to be satisfactory, corresponding in the case of argon ($\varepsilon/k_B = 119.8$ K, $\sigma = 3.405$ Å) to $h \simeq 10^{-14}$sec.

Similar ideas can easily be applied to systems of molecules if these are regarded as being made up of independent atoms, the intramolecular forces being designed to maintain the geometry of the molecule. In general this is not the best way to proceed, since the choice of time step is in that case dictated by the time scale of the molecular vibrations rather than by slower (and for the most part more interesting) translational and rotational motion. In terms of computing costs it is more economic to work with rigid molecules, but this introduces new problems. The conventional approach to the solution of the equations of motion of a rigid body involves a separation into internal and centre-of-mass coordinates. A considerable degree of simplification is possible when the molecules are linear and a number of satisfactory schemes have been developed for use in this particular case [8,9,10]. The equations of motion for non-linear molecules are much more complicated, since one has to treat both the rotation of the molecule about its principal axes and the rotation of these axes themselves[11]. Such complications can be avoided by adopting a recently developed and rather ingenious method[12] in which the equations of motion for the individual atoms are solved subject to constraints imposed by the geometry of the molecule.

To illustrate the application of the method of constraints, consider a triatomic molecule in which each bond is of length s and each atom (labelled 1, 2 and 3) is of mass m. The total force acting on atom 1, say, is the sum of three terms: $\underset{\sim}{F}_1(t)$, the force due to surrounding molecules; a force of constraint $\underset{\sim}{f}_{12}(t)$ which ensures that the bond vector $\underset{\sim}{r}_{12}(t)$ remains of fixed length s; and a force $\underset{\sim}{f}_{13}(t)$ which preseves the bond between atoms 1 and 3. As similar arguments apply to atoms 2 and 3, the equations of motion of the molecule can be written as

$$\begin{aligned} m\ddot{\underset{\sim}{r}}_1(t) &= \underset{\sim}{F}_1(t) + \underset{\sim}{f}_{12}(t) + \underset{\sim}{f}_{13}(t) \\ m\ddot{\underset{\sim}{r}}_2(t) &= \underset{\sim}{F}_2(t) + \underset{\sim}{f}_{21}(t) + \underset{\sim}{f}_{23}(t) \\ m\ddot{\underset{\sim}{r}}_3(t) &= \underset{\sim}{F}_3(t) + \underset{\sim}{f}_{31}(t) + \underset{\sim}{f}_{32}(t) \end{aligned} \qquad (13)$$

The forces of constraint must be directed along the corresponding bond vector, so in general we may write that

$$\underset{\sim}{f}_{ij}(t) = \lambda_{ij}\underset{\sim}{r}_{ij}(t) \qquad (i \neq j) \qquad (14)$$

where λ_{ij} is an unknown constant. Moreover, from the law of action and reaction, we know that

$$\begin{aligned} \underset{\sim}{f}_{ji}(t) &= \lambda_{ji}\underset{\sim}{r}_{ji}(t) \\ &= -\underset{\sim}{f}_{ij}(t) \end{aligned} \qquad (15)$$

Thus $\lambda_{ij} = \lambda_{ji}$. If we now apply the central-difference algorithm to the solution of (13) we find that

$$\begin{aligned} \underset{\sim}{r}_1(t+h) &= \underset{\sim}{r}_1'(t+h) + (h^2/m)\,[\lambda_{12}\underset{\sim}{r}_{12}(t)+\lambda_{13}\underset{\sim}{r}_{13}(t)] \\ \underset{\sim}{r}_2(t+h) &= \underset{\sim}{r}_2'(t+h) + (h^2/m)\,[-\lambda_{12}\underset{\sim}{r}_{12}(t)+\lambda_{23}\underset{\sim}{r}_{23}(t)] \\ \underset{\sim}{r}_3(t+h) &= \underset{\sim}{r}_3'(t+h) + (h^2/m)\,[-\lambda_{13}\underset{\sim}{r}_{13}(t)-\lambda_{23}\underset{\sim}{r}_{23}(t)] \end{aligned} \qquad (16)$$

where $\underset{\sim}{r}_i'(t+h)$ are the predicted coordinates in the absence of any constraint.

The set of equations (16) must now be solved subject to the three conditions that

$$\begin{aligned} |\underset{\sim}{r}_{12}(t+h)|^2 = |\underset{\sim}{r}_{13}(t+h)|^2 &= |\underset{\sim}{r}_{23}(t+h)|^2 \\ &= s^2 \end{aligned} \qquad (17)$$

This leads to three simultaneous equations for the coefficients λ_{12}, λ_{13} and λ_{23} which may be written schematically in the form

$$\left(\frac{2h^2}{m}\right)\begin{vmatrix} 2\mathbf{r}_{12}(t)\cdot\mathbf{r}_{12}'(t+h) & \mathbf{r}_{13}(t)\cdot\mathbf{r}_{12}'(t+h) & -\mathbf{r}_{23}(t)\cdot\mathbf{r}_{12}'(t+h) \\ \mathbf{r}_{12}(t)\cdot\mathbf{r}_{13}'(t+h) & 2\mathbf{r}_{13}(t)\cdot\mathbf{r}_{13}'(t+h) & \mathbf{r}_{23}(t)\cdot\mathbf{r}_{13}'(t+h) \\ -\mathbf{r}_{12}(t)\cdot\mathbf{r}_{23}'(t+h) & \mathbf{r}_{13}(t)\cdot\mathbf{r}_{23}'(t+h) & 2\mathbf{r}_{23}(t)\cdot\mathbf{r}_{23}'(t+h) \end{vmatrix} \begin{vmatrix} \lambda_{12} \\ \lambda_{13} \\ \lambda_{23} \end{vmatrix}$$

$$= \begin{vmatrix} s^2 - |\mathbf{r}_{12}'(t+h)|^2 \\ s^2 - |\mathbf{r}_{13}'(t+h)|^2 \\ s^2 - |\mathbf{r}_{23}'(t+h)|^2 \end{vmatrix} + O(h^4) \qquad (18)$$

the terms of order h^4 being quadratic functions of the λ_{ij}.

Having derived the basic relations linking the unknown coefficients λ_{ij} we can now summarize the steps involved in carrying through the full calculation :

a) the coordinates in the absence of any constraint are computed;

b) the constant terms appearing in eqn.(18) are evaluated;

c) the terms of order h^4 in (18) are set equal to zero and the resulting system of linear equations is solved to yield a first estimate for the λ_{ij};

d) the approximate λ_{ij} obtained in step c) are used to estimate the terms of order h^4 in (18) and the system of equations is solved again to yield improved values for λ_{ij};

e) step d) is repeated until convergence is achieved. In general three to four iterations are sufficient to maintain the bondlengths constant to within one part in 10^6.

The great merit of the method is its simplicity. The equations of motion can be solved in cartesian form, no internal coordinates appear and the computer code is not very different from that used for an atomic system. The iterations required to solve equations such as (18) add a negligible amount to the total computing time, except when the molecule is very large. In the general case a system of q simultaneous equations must be solved, where q is the number of constraints, but many elements in the matrix are likely to be zero and some speeding up of the solution is usually possible. The constraints themselves can be imposed either on bondlengths or on bond angles. The method is therefore ideally suited for use with molecules in which a number of internal rotations are permitted. Application of the constraints introduces an additional error into the calculation of the trajectories, but this can be shown to be of the same order as that arising from the algorithm used to predict the coordinates in the absence of any constraints[12], at least for algorithms based on a Taylor expansion in time.

The last class of computational problems which we shall discuss are those concerned with the calculation of the various static and dynamic properties of the system. A typical simulation is initiated by placing the N particles inside the basic cell in some arbitrary way, often in a lattice-type arrangement; this defines the coordinates $\underset{\sim}{r}_i(0)$. The coordinates at time $\underset{\sim}{r}_i(-h)$ are then obtained by allocating to each particle a velocity which is chosen randomly but in such a way that the total kinetic energy corresponds to the temperature of interest and the total linear momentum is zero. The system is then allowed to evolve for an "equilibration" period of a few hundred steps, during which time it will occasionally be necessary to scale the velocities of the particles so as to maintain the total kinetic energy at its required value. Thereafter the system is left undisturbed and the accumulation of data can begin.

Quantities which are routinely calculated during the course of an "experiment" include the mean potential energy of the system, the pressure and temperature, and the radial distribution function (or functions). The latter can be obtained by constructing a histogram which records the number of pairs of particles separated by distances lying within equal intervals. The temperature, as we have already indicated, is determined by the mean kinetic energy and the pressure is calculated from the mean value of the intermolecular virial. Certain other properties of interest are determined by the mean-square fluctuation in observable quantities. For example, the specific heat can be calculated from the fluctuations in kinetic energy, but the accuracy which can be attained is rather low, of order 10%. By contrast, it is easy to obtain the mean potential energy to an accuracy of 1% or better.

In the context of a computer "experiment" the term "mean value" is equivalent to "time average". Thus if the value of a dynamical variable A at time t is A(t), its mean value over an "experiment" lasting n integration steps (excluding equilibration) is

$$\langle A \rangle = \frac{1}{n+1} \sum_{k=0}^{n} A(kh) \tag{19}$$

The coordinates and momenta of every particle at each step in the integration are usually stored on magnetic tape. The data so recorded can subsequently be analysed to yield, in particular, a variety of time correlation functions. It is worth adding that the computer time required for detailed analysis of the data often exceeds that used in the "experiment" itself.

LECTURE 3: COMPUTER "EXPERIMENTS" ON SIMPLE LIQUIDS

We must now turn away from the discussion of technical problems and look at some of the ways in which computer "experiments" have been used in the study of the microscopic properties of liquids. We shall begin by reviewing what has been learned about the dynamic properties of simple, non-conducting monatomic liquids, in particular of hard spheres, of argon-like liquids modelled by Lennard-Jones or similar potentials, and of systems representing liquid alkali metals, for which the potential represents an effective interaction between neutral pseudoatoms. The progression from hard spheres to rare-gas liquids to liquid alkali metals involves a softening of the repulsive part of the potential ($\sim r^{-12}$ in argon, $\sim r^{-6}$ in sodium or potassium) and the appearance of a long-range tail which in argon is weakly attractive but in liquid metals may be oscillatory. The dynamic properties of primary interest are the transport coefficients (self-diffusion D, shear viscosity η, bulk viscosity ξ and thermal conductivity λ) and the autocorrelation functions of certain collective variables, namely the longitudinal and transverse currents and the particle density. The definitions of these collective variables have already been given in the lectures of Hansen.

Any transport coefficient can be written as the time integral of an appropriate time correlation function. Equations such as these are known as Green-Kubo formulae, which for our present purposes may be written in the general form

$$\alpha = \int_0^\infty \langle A(t)A(o) \rangle \, dt \qquad (20)$$

where A is a dynamical variable (in practice a flux) and α is the corresponding transport coefficient. For one-component monatomic systems we can then make the identification shown in Table I, where m is the particle mass, E_i is the total energy of particle i and $\underset{\sim}{r}_i \equiv (x_i, y_i, z_i)$.

The Green-Kubo formulae provide a convenient route for the calculation of transport coefficients from data obtained in a molecular dynamics "experiment". In the calculation of D the statistical error can be greatly reduced by averaging over the trajectories of individual particles. No such possibility exists in the case of η, ξ and λ, since the corresponding variable A is a function of the coordinates and momenta of all particles in the system; the attainable precision is therefore very much lower and extremely long runs ($\sim 10^5$ steps) are needed to achieve acceptable results. It is for this reason that self-diffusion has been studied in much greater detail than other transport processes. Transport coefficients can also be determined by measuring the response of the system to a weak applied field of appropriate character, an approach

Table I: Dynamical variables and associated transport coefficients

A(t) name	A(t) definition	α
particle velocity (any component)	$\dot{x}_i(t)$	D
stress tensor (any off-diagonal component)	$m\frac{d}{dt}\sum_i x_i(t)\dot{y}_i(t)$	$Vk_B\eta T$
stress tensor* (any diagonal component)	$m\frac{d}{dt}\sum_i x_i(t)\dot{x}_i(t)$	$Vk_B(\frac{4}{3}\eta + \xi)T$
energy current (any component)	$\frac{d}{dt}\sum_i x_i(t)E_i(t)$	$Vk_B\lambda T^2$

*But excluding the invariant part, $\langle A(0)\rangle$.

which has been exploited in a variety of ways in molecular dynamics calculations [13, 14, 15, 16] and which we shall briefly mention again in the last lecture.

So far as hard spheres are concerned, the qualitative behaviour of the velocity autocorrelation function $Z(t)$ ($= \langle\dot{x}_i(t)\dot{x}_i(0)\rangle$) can be summarized as follows [17]. At low to intermediate densities it is a monotonic decreasing function of t, with a weak but well-defined tail extending for upwards of thirty mean collision times. At higher densities the behaviour is more complicated, Z(t) becoming negative after only a short time. Negative autocorrelation means that at time t a particle is on average moving in a direction opposite to that in which it was moving at the earlier time t = 0. At high densities it describes a "backscattering" effect, corresponding to the fact that collisions between nearest neighbours lead on average to a reversal of velocity into a comparatively narrow range of angles. Thereafter Z(t) passes through a minimum and levels off into an extended tail which in this case is negative. Similar behaviour is observed for the Lennard-Jones fluid [5, 18] and for a wide range of more complicated systems.

The appearance of a persisting long-time tail in Z(t) implies that highly collective effects make an appreciable contribution to the process of self-diffusion and the suggested mechanism whereby this comes about is as follows [19]. The initial motion of a tagged particle creates around it a "vortex" or "backflow", causing a retarded current of particles to develop in the direction of the initial velocity. At low densities, where the initial direction of motion is likely to persist, the effect of this current is to reduce the drag on the particle, pushing it onwards in the initial direction and enhancing the diffusion coefficient. At high densities, however, the initial direction of motion is on average soon reversed. In this case the retarded current gives rise to an added drag at later times and thus to an extended negative region in Z(t) and a consequent decrease in the diffusion coefficient. At very large times an enhancement in the forward direction can again be expected, as in the low density case, but the effect is likely to be undetectable.

That the "vortex" model provides basically the correct explanation has been confirmed in striking fashion by analysis of the velocity field which develops around a moving particle in a fluid of hard disks (a two-dimensional system). A vortex pattern quickly develops, which after only a few collisions per particle turns out to match very closely that obtained by numerical solution of the Navier-Stokes equation of hydrodynamics. This makes it natural to associate the persisting tail in Z(t) with a coupling between the motion of single particles and the hydrodynamic modes of the system, showing that even an apparently simple dynamical process such as self-diffusion is in reality a very complicated phenomenon. These results also make it reasonable to expect the detailed behaviour of Z(t) at large t to be influenced by the magnitude of the phenomenological coefficients entering the Navier-Stokes equation, notably the shear viscosity. This is true, as we shall shortly see.

A quantitative estimate of the contribution of many-body dynamical correlations to the transport coefficients of hard spheres can be gained by comparing the computer results [17] with those obtained by solution of the Enskog equation, since in Enskog's theory (itself based on the Boltzmann equation) the velocity correlations built up by successive collisions are ignored. The deviations in the case of thermal conductivity are very small, and the corresponding autocorrelation function (see Table I) turns out to have no significant long-time tail. Presumably, therefore, cooperative effects play only a minor rule. In the case of self-diffusion, however, there is a marked enhancement in D at intermediate densities, arising from the positive tail in Z(t), followed by a rapid fall as crystallization is approached, that is to say in the region in which Z(t) has a substantial negative portion. The decrease in D at high densities is matched by an equally dramatic rise in the shear viscosity. The net result is that the product $D\eta$ remains almost constant. In fact Stokes' law with "slip" boundary conditions

$$D\eta = k_B T/2\pi\sigma \tag{21}$$

where σ is the hard-sphere diameter, is satisfied within a few percent at all except the very lowest densities. The enhancement of the shear viscosity near crystallization can be associated with the development of a pronounced positive tail in the stress autocorrelation function $\eta(t)$ (see Table I). The same effect has been found in the Lennard-Jones fluid near its triple point [20]. The results obtained for the hard-sphere fluid [17] are summarized in Table II.

Levesque and Verlet [18] have analyzed the behaviour of the velocity autocorrelation function in the Lennard-Jones fluid in a somewhat different way by introducing the concept of a generalized frequency-dependent friction coefficient. In the classic theory of Brownian motion the rate of decay of Z(t) is given in terms of a friction coefficient ξ in the form

$$\dot{Z}(t) = -\xi Z(t) \qquad (22)$$

which has the solution

$$Z(t) = Z(0)\exp(-\xi t) \qquad (23)$$

with $Z(0) = k_B T/m$. From Table I it follows that ξ and D are related by the expression

$$\xi = k_B T/mD \qquad (24)$$

If, however, we associate a certain memory with the motion of the Brownian particle, we may replace eqn.(22) by a generalized Langevin equation in which the frictional force is written as a convolution in time :

Table II: Transport coefficients of the hard-sphere fluid relative to the predictions of Enskog's theory* [17].

ρ/ρ_o	D/D_E	η/η_E	λ/λ_E	$2\pi(\frac{D\eta\sigma}{k_B T})$
0.010	1.02	1.01	0.98	
0.050	1.04	1.00	0.99	
0.200	1.16	0.99	0.97	0.98
0.333	1.34	1.02	1.00	0.92
0.500	1.27	1.10	1.07	0.99
0.555	1.15	1.10	1.03	0.92
0.625	0.84	1.44	1.05	0.94
0.667	0.58	2.16	1.05	1.06

*Subscript E ; ρ_o is the number density at close packing

$$\dot{Z}(t) = -\int_0^t \xi(t-t')\, Z(t')\, dt' \qquad (25)$$

or, in terms of Fourier-Laplace transforms, as

$$\tilde{Z}(\omega) = \frac{k_B T}{m} \frac{1}{-i\omega + \tilde{\xi}(\omega)} \qquad (26)$$

Thus the function $\xi(t)$ can be recognised as the memory function discussed in the lectures of Hansen. Though difficult to calculate, it can be given a rigorous statistical mechanical definition in terms of a suitable projection operator [21].

Eqn.(25) can be inverted in such a way as to extract the memory function $\xi(t)$ from molecular dynamics results on Z(t). It turns out that $\xi(t)$ consists of two parts, the characteristic time scales differing by a factor of order ten. The first is short range in time and quasi-gaussian in shape. It seems safe to conclude that this part of the generalized friction coefficient represents the effects of binary collisions. In fact, if we represent this short-lived part by a delta-function in time and neglect the weaker, long-lived part, we recover the simple Langevin equation (22). The second contribution to $\xi(t)$ is a weak and long-lived quasi-exponential tail which is positive for states close to the triple point (representing a positive contribution to the friction) and negative elsewhere. It is this part of the memory function which is responsible for the negative plateau observed in Z(t) at high densities and the long positive tail at low densities and high temperatures.

Before leaving this discussion of the velocity autocorrelation function it is worth mentioning a curious and interesting result: it appears that at very long times Z(t) decays to zero from above as $t^{-3/2}$ or, more generally, as $t^{-d/2}$, where d is the dimensionality of the system [19, 22]. This is the famous "long-time tail" phenomenon much discussed by theoreticians. A number of elegant theories have been developed to account for the $t^{-3/2}$ behaviour, but qualitatively it can be very simply explained [19] on the basis of the "vortex" model already discussed. Suppose that at time t = 0 particle i has a velocity component $\dot{x}_i(0)$. After some short time τ, as a result of collisions between particles, particle i will have shared its initial momentum with the ρV_τ (ρ is the number density) particles occupying a volume V_τ centred on i, so that

$$\dot{x}_i(\tau) \simeq \dot{x}_i(0)/\rho V_\tau \qquad (27)$$

Then any subsequent decay in $\dot{x}_i$ can be largely ascribed to the growth in V_τ due to the spread of the velocity field around i . This growth is dominated by the transverse component of the velocity field, which diffuses at a rate determined by the kinematic shear viscosity $\nu = \eta/\rho m$ (units: $cm^2 sec^{-1}$). Thus if R_τ is the radius of V_τ, we can write that $R_\tau \sim (\nu t)^{1/2}$, so that $V_\tau \sim (\nu t)^{3/2}$ and consequently $Z(t) \sim (\nu t)^{-3/2}$. Once again, therefore, we find a link between diffusive and viscous effects. A more complete calculation, taking account of the diffusion of particle i itself, shows that

$$Z(t) \sim [(D + \nu)t]^{-3/2} \tag{28}$$

Levesque and Ashurst [22] have made a very careful computation of Z(t) in a Lennard-Jones-like fluid at a density approximately two-thirds that at the triple point. They find unmistakeably the onset of a $t^{-3/2}$ behaviour after approximately eighteen mean collision times, with a numerical coefficient which agrees with that predicted on theoretical grounds within the statistical error on the calculation. This type of "experiment" is difficult to carry out successfully, not only because the effect itself is very weak but also because of the recurrence time problem discussed in the first lecture, which makes it necessary to work with a very large system. The $t^{-3/2}$ behaviour at large t means that the Fourier transform of Z(t) should vary as $\omega^{-1/2}$ at small ω. In principle such an effect is detectable by incoherent neutron scattering techniques, but the probability of doing so seems at present rather small.

LECTURE 4: MORE ON SIMPLE LIQUIDS

In the last lecture we looked at the role played by cooperative effects in determining certain of the dynamical properties of liquids and in particular at the coupling of self-diffusion (a single-particle property) to the other (collective) dynamical modes of the system. Now we must turn to the problem of describing the collective behaviour of the system in a more direct way, in terms of currents and densities.

The collective dynamical modes of a one-component monatomic liquid are conveniently analyzed in terms of a) the transverse current autocorrelation function $C_t(k,t)$ and b) the density-density autocorrelation function $F(k,t)$ and it Fourier transform $S(k,\omega)$, the dynamic structure factor. The latter is related in a simple way to the Fourier transform of the longitudinal current autocorrelation function $C_\ell(k,t)$, so that the study of the longitudinal current fluctuations adds nothing new. As all these functions have been introduced in the lectures of Hansen it is unnecessary to discuss any of their formal properties here, except to remark that in an isotropic system all wavelength-dependent time correlation functions are functions only of wavenumber k rather than wavevector.

We shall begin by looking at the transverse current fluctuations. It is easy to show [23] that in the hydrodynamic limit the transverse current autocorrelation function behaves as

$$\dot{C}_t(k,t) + (\eta/\rho m)k^2 C_t(k,t) = 0 \tag{29}$$

or

$$\begin{aligned} C_t(k,t) &= C_t(k,0)\exp(-\eta k^2 t/\rho m) \\ &= \omega_o^2 \exp(-\nu k^2 t) \end{aligned} \tag{30}$$

where, as before, ν is the kinematic shear viscosity and $\omega_o^2 \equiv C_t(k,0) = k^2(k_BT/m)$. Hydrodynamically, therefore, the spectrum of transverse current fluctuations, i.e. $C_t(k,\omega)$, the Fourier transform of $C_t(k,t)$, is a Lorentzian curve centred at $\omega = 0$ with a half-width at half peak height equal to νk^2. This corresponds to a purely "diffusive" mode; transverse current fluctuations decay without propagating at a rate determined by the kinematic shear viscosity, with a lifetime which becomes infinite in the limit $k \to 0$.

At small k the hydrodynamic calculation gives results which are qualitatively in agreement with molecular dynamics results[20] on the Lennard-Jones fluid. But, as we can see from Figure 1, the computer "experiment" also shows that at a sufficiently large

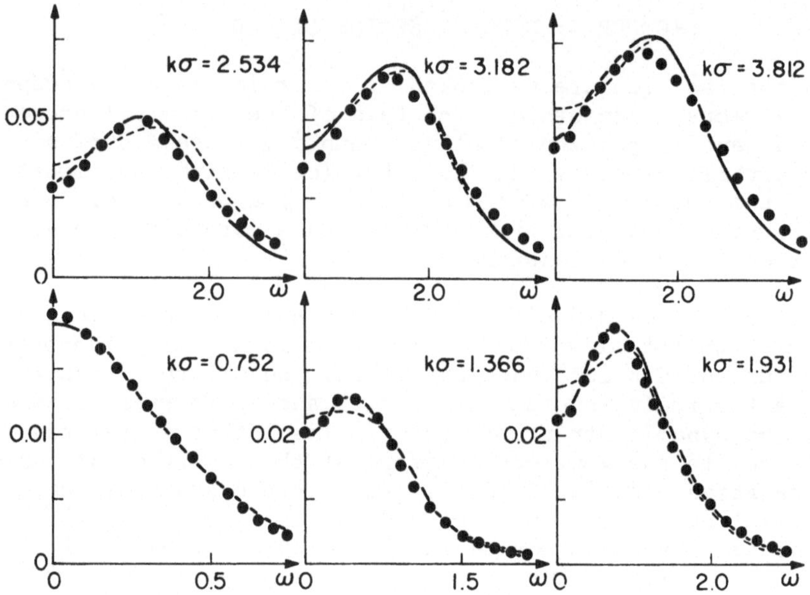

Figure 1: Spectrum of transverse current fluctuations in the Lennard-Jones fluid near the triple point [20]. The dots are molecular dynamics results, the dashed line is calculated from the approximation (36) and the full line includes the contribution from a slowly decaying tail in $\eta(k,t)$. The unit of frequency is τ^{-1}, with τ defined by eqn.(12); σ is the parameter in the Lennard-Jones potential (10).

wavenumber a well-defined peak appears at non-zero frequency. The presence of such a peak indicates that under appropriate conditions the liquid is able to support a propagating shear wave. This occurs at wavelengths comparable with the interparticle spacing, at which point we are no longer justified in ignoring the rigidity of the liquid. Converting to units appropriate to liquid argon, we estimate the value of the critical wavenumber below which hydrodynamic-type behaviour sets in to be approximately 0.3 Å^{-1}, corresponding to a wavelength of approximately 20 Å, or roughly six times the nearest-neighbour distance. In an intermediate range of k the peak frequencies are of similar magnitude to those of transverse phonons in solid argon, which strengthens the evidence for identification of the peaks as propagating shear waves. As k increases still further, molecular interactions become less important and the peak eventually disappears, the spectrum taking on the form predicted for an ideal gas.

It is not difficult to find a qualitative explanation for

onset of shear waves. Let us suppose that a shearing force is applied to a liquid. The strain at a point (x,y,z) within the liquid can be expressed in terms of the displacement $\underset{\sim}{r}$ at that point and the rate of strain in terms of the velocity $\dot{\underset{\sim}{r}}$. If the flow is purely viscous, the shearing stress, T^{xz} say, is proportional to the rate-of-strain tensor and can be written in approximate form as

$$T^{xz} = \eta \frac{\partial}{\partial t} \left(\frac{\partial r_x}{\partial z} + \frac{\partial r_z}{\partial x} \right) \tag{31}$$

which corresponds to the hydrodynamic description. On the other hand, if the force is applied suddenly, the instantaneous displacement will be determined by the stress through an elastic stress-strain relation

$$T^{xz} = G \left(\frac{\partial r_x}{\partial z} + \frac{\partial r_z}{\partial x} \right) \tag{32}$$

where G is a high-frequency modulus of rigidity.

We can now interpolate between the extremes represented by the Eqns.(31) and (32) by making a so-called viscoelastic approximation

$$\left(\frac{1}{\eta} + \frac{1}{G}\frac{\partial}{\partial t} \right) T^{xz} = \frac{\partial \dot{r}_x}{\partial z} + \frac{\partial \dot{r}_z}{\partial x} \tag{33}$$

On taking the Fourier-Laplace transform of (33) we find that the viscoelastic approximation is equivalent to replacing η in eqn.(31) by a complex, frequency-dependent shear viscosity given by

$$\tilde{\eta}(\omega) = \frac{G}{-i\omega + G/\eta} \tag{34}$$

or, in time space, by

$$\eta(t) = G \exp(-t/\tau_M) \tag{35}$$

The constant $\tau_M = \eta/G$ is called the Maxwell relaxation time; the limit $\omega\tau_M \ll 1$ corresponds to viscous flow and $\omega\tau_M \gg 1$ to the propagation of elastic waves. The form of (35) suggests an obvious generalization to arbitrary wavenumber of the form

$$\eta(k,t) = G(k) \exp[-t/\tau_M(k)] \tag{36}$$

where G(k) is related to the second moment of $C_t(k,\omega)$ and may thus be calculated exactly, and $\tau_M(k)$ can be treated as an adjustable parameter. This is a typical example of a generalized hydrodynamic prescription. Furthermore, the function $\eta(k,t)$ is nothing more than the memory function for transverse current fluctuations (see the lectures of Hansen) and in the limit $k \to 0$ can be identified [23] with the stress autocorrelation function $\eta(t)$ introduced in the previous lecture.

Empirically, $\tau_M(k)$ is found to decrease with increasing k. The best fit which can be obtained with the approximation (36) is shown in Figure 1 and is clearly far from perfect. In particular, the shear wave peaks are significantly too broad and flat. However, we know that the stress autocorrelation function has a tail which decays very slowly and we may reasonably suspect that the transverse current fluctuations at small wavenumbers can be adequately represented only if a similar, slowly decaying tail is included in the memory function $\eta(k,t)$. This turns out to be true. Taking a two-term representation for $\eta(k,t)$ similar in form to that used by Levesque and Verlet [18] to represent the generalized friction coefficient $\xi(t)$ (see eqn. (25)), an excellent fit to the computer data can be obtained, as seen in Figure 1. This behaviour is by no means unique to the Lennard-Jones fluid [24].

We now turn to the somewhat more complicated problem of describing the density fluctuations. In the hydrodynamic limit the dynamic structure factor $S(k,\omega)$ consists of three Lorentzian lines, the Rayleigh line centred at $\omega = 0$ and two Brillouin lines at $\omega = \pm c_s k$, where c_s is the adiabatic speed of sound. The central line represents a purely thermal mode and can be identified with the decay of entropy fluctuations. The two shifted components correspond to propagating sound waves and are analagous to the longitudinal phonons in a solid; their width is determined partly by the kinematic longitudinal viscosity, $(\frac{4}{3}\eta + \xi)/\rho m$, and to a lesser extent by the thermal diffusivity $\lambda/\rho c_p$ (c_p is the specific heat per particle). What is remarkable is that molecular dynamics calculations show that the three-peak structure persists down to wavelengths comparable with the nearest-neighbour separation, both in the Lennard-Jones fluid [20] and in liquid rubidium [25].

In the case of rubidium, similar effects have been observed in neutron scattering experiments [26]. Some results for the Lennard-Jones fluid are shown in Figure 2, and in Figure 3 we plot the dispersion of the Brillouin peak in liquid rubidium. The molecular dynamics results on rubidium are in extremely good agreement with the neutron scattering measurements. It is clear from these figures that a propagating sound wave persists to larger wavenumbers in rubidium than in an argon-like liquid. This fact is presumably related to the differences in the corresponding pair potentials, perhaps to the relatively greater harshness of the repulsive forces in the Lennard-Jones fluid, but the problem is not yet fully understood.

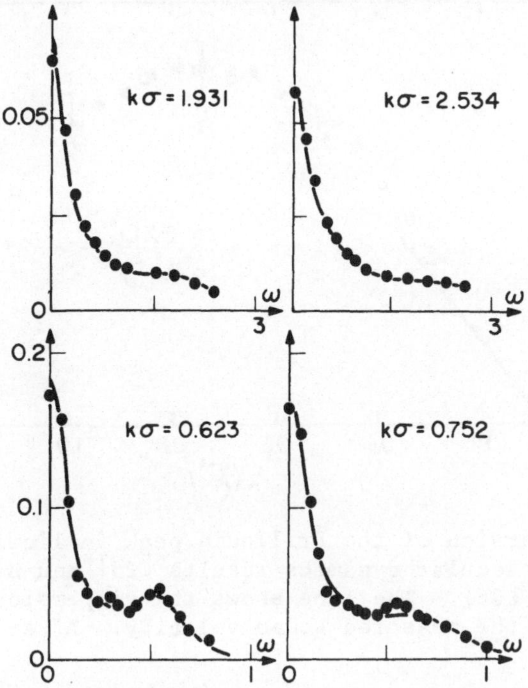

Figure 2: Dynamic structure factor of the Lennard-Jones fluid near the triple point [20]. The dots are molecular dynamics results and the curves are obtained from a generalized hydrodynamic theory (see text). The unit of frequency is τ^{-1}, with τ defined by eqn.(12); σ is the parameter in the Lennard-Jones potential (10).

Given the success of the generalized hydrodynamic description of the transverse current fluctuations, it is natural to attempt a similar treatment of the density fluctuations, introducing a larger set of frequency and wavenumber dependent transport coefficients [20]. The calculation is too complicated to enter into in any detail here, but there is one very important point to be made: the spectrum of density fluctuations at small wavenumbers can be adequately represented (as shown in Figure 2) by a generalization of hydrodynamics only if the generalized shear viscosity has the same form, including the long-time tail in $\eta(k,t)$, as that required to reproduce the detailed form of the transverse current spectrum. The generalized shear viscosity therefore emerges as a function having real physical significance and its calculation on a microscopic basis is clearly a theoretical problem of considerable importance. Similar conclusions emerge from attempts to calculate the velocity autocorrelation function by generalizing Stokes' law to non-zero frequencies [18,27]. To

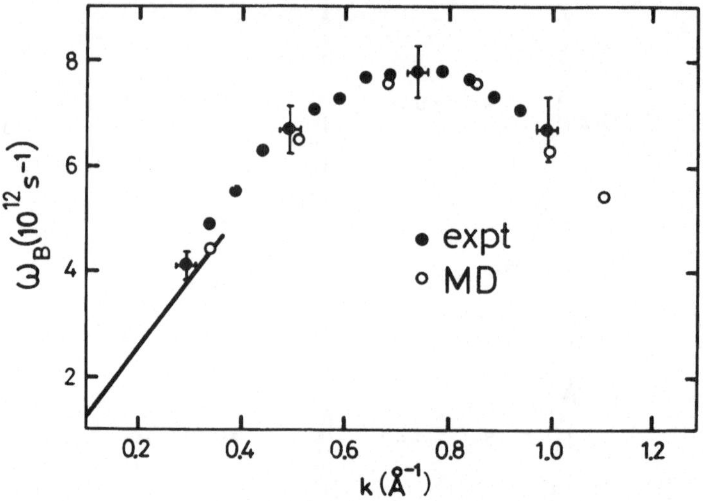

Figure 3: Dispersion of the Brillouin peak in liquid rubidium, obtained from molecular dynamics results [25] and neutron scattering experiments [26]. The line shows the dispersion calculated on the basis of the measured sound velocity. After Copley and Lovesey [28].

reproduce the long-time behaviour of the velocity autocorrelation function it is necessary to introduce a frequency-dependent shear viscosity (cf. the relation between D and η in the simple Stokes' law, eqn.(21)) having essentially the same properties as the function required to describe the fluctuations in density and transverse current.

So far we have concentrated on applications which are largely of theoretical interest. Before leaving the subject of non-conducting liquids, it seems appropriate to mention at least one investigation which uses computer simulation as a tool for studying the extent to which a relatively simple potential model can account for the properties of some real liquid. A good example of this approach is the work of Cheung and Powles [9, 29] on liquid nitrogen. The model adopted initially was the so-called atom-atom potential, in which the interaction between two nitrogen molecules is assumed to consist of identical Lennard-Jones interactions between the four pairs of interaction centres associated with the atoms in the two molecules. In later work [29] the atom-atom potential was supplemented by a superimposed quadrupole-

quadrupole interaction. The static properties of nitrogen (thermodynamics and structure) are very well reproduced by the atom-atom potential alone, as indeed is the self-diffusion coefficient. Addition of the quadrupolar interaction gives some small but not insignificant improvement, the structure factor in particular being notably closer to the experimental results. The structure factors at 68 K and 79 K are plotted in Figure 4. The improvement achieved by including the quadrupolar interaction may appear rather small, but in fact it represents a considerable achievement, since the structure factor at high densities is notoriously insensitive to details of the intermolecular potential. The reorientational time correlation functions prove more difficult to reproduce, and significant discrepancies remain even when the quadrupole is added; these discrepancies are particularly marked at high temperatures. The potential model could be further improved by allowing for anisotropy in the dispersion interactions, but it is unclear whether the extra computational effort would be worthwhile.

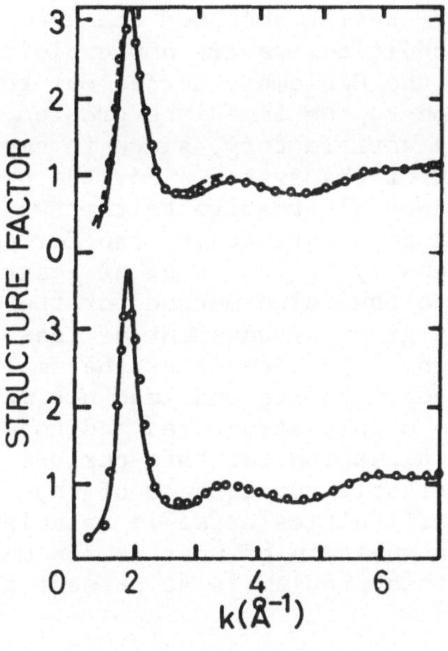

Figure 4: Structure factor of liquid nitrogen at 68 K (lower) and 79 K (upper). The dashed line shows the molecular dynamics results based on the atom-atom potential alone, and the circles show the effects of including the quadrupolar interaction. The full lines are experimental results (X-rays at 68K, neutrons at 79K). After Cheung and Powles [29].

LECTURE 5 : IONIC FLUIDS

We want to look now at the way in which computer "experiments" have contributed to our understanding of the properties of ionic systems. The behaviour of ionic fluids is in many respects very different from that of liquids such as argon, since the long-ranged nature of the Coulomb potential introduces many new and interesting features. Unfortunately, it also creates certain technical complications in the molecular dynamics calculations. In particular, it is no longer appropriate to truncate the force law in the way we described earlier. Instead we must compute explicitly the interaction of a given particle, i say, not only with every other particle j within the central cell, but also with all periodic images of j. An infinite lattice sum of this type can be evaluated by the method of Ewald, whereby the slowly convergent sum in r^{-1} is converted into two series which separately are rapidly convergent. One series is a sum over pairs of ions in real space, and the other is a sum over the reciprocal lattice of the periodic array of cells. The method is the same as that used in the evaluation of Madelung sums for ionic crystals, and the details can be found in elementary textbooks on solid-state physics. What is important to emphasize here is that we are, in effect, making a virtue of necessity; having accepted the need for employing a periodic boundary condition, we are now exploiting the periodicity in order to sum the Coulombic forces out to infinite range. This is a neat solution to the immediate problem, but it is obviously not entirely satisfactory, since it emphasizes the artificial periodicity of the system. For the present, however, there seems to be no good alternative to the Ewald method. One might hope to obtain a good estimate of the force acting on ion i by summing over all ions lying in a cube of length L centred on i, this being analogous to the Evjen method for the calculation of lattice sums. For any given arrangement of ions the result is quite good, but at high charge densities the small errors introduced are apparently accumulative and lead ultimately to configurations which are too highly structured and too strongly bound. There is no obvious explanation for this curious behaviour, but it seems to be an inevitable consequence of that type of energy summation. Similar difficulties arise in calculations on strongly polar systems, and the question of what is the most effective method of overcoming this problem is at present a matter of some debate [3, 30, 31, 32].

Perhaps the simplest imaginable ionic fluid is the one-component plasma, or OCP, made up of identical ions moving in a uniform and rigid neutralizing background, and interacting only through Coulombic forces. One would expect to see, in such a system, Coulombic effects at their most dramatic. The OCP has only a single fluid phase, since there are no attractive forces, and in that respect it is similar to other systems of particles interacting through inverse-power potentials of the form $\varepsilon(\sigma/r)^n$ (the limit $n \to \infty$ corresponds to hard spheres). The OCP provides a fair model for the conduction electrons in a metal and it is also of some importance in astrophysical applications, but the main motivation for studying such a system by means of a computer "experiment" is for the insight such calculations might yield into the general problem of fluids with long-ranged forces. Furthermore, we can expect the properties of the OCP to be reflected in those of systems of more practical importance, molten salts being an obvious example.

The excess thermodynamic properties of the OCP do not depend separately on temperature and density, but only on the dimensionless variable

$$\Gamma = (Ze)^2/ak_BT \tag{37}$$

where Z is the ionic charge in units of e and $a = (3/4\pi\rho)^{1/3}$ is the so-called ion-sphere radius; the value of the Coulombic coupling parameter Γ is a measure of the potential energy of the system relative to the kinetic energy. From Monte Carlo calculations the OCP is known to undergo a transition to a body-centred cubic crystal at a value of Γ of approximately 155.

Theoretical treatments of the dynamics of the OCP are agreed in predicting that the Coulombic interaction organises the motion of the ions in such a way that at long wavelengths the collective dynamical properties are dominated by a well-defined density oscillation at a frequency close to the plasma frequency ω_p, defined as

$$\omega_p^2 = 4\pi\rho(Ze)^2/m \tag{38}$$

This plasmon mode, as it is often called, is in some respects analagous to the second wave mode in a neutral fluid, but has a much longer lifetime. Molecular dynamics calculations have confirmed the existence of these strongly collective effects [33], but more surprisingly they have also shown that for large values of Γ the motion of single ions is strongly influenced by the plasma oscillations. In Figure 5 we show the results obtained for the velocity autocorrelation function Z(t) at several values of Γ. For $\Gamma = 1$, Z(t) decays quasiexponentially, behaviour typical of a dilute gas. As Γ increases, however, we see the appearance of a superimposed oscillation at a frequency close to ω_p, the oscillations becoming more pronounced at still larger values of Γ. This means that the motion of single particles becomes very strongly coupled to the collective oscillations in the system. As we have already seen, the coupling of single particle and collective modes is also a feature of the dynamics of argon-like liquids, but in the OCP the effects are very much more pronounced.

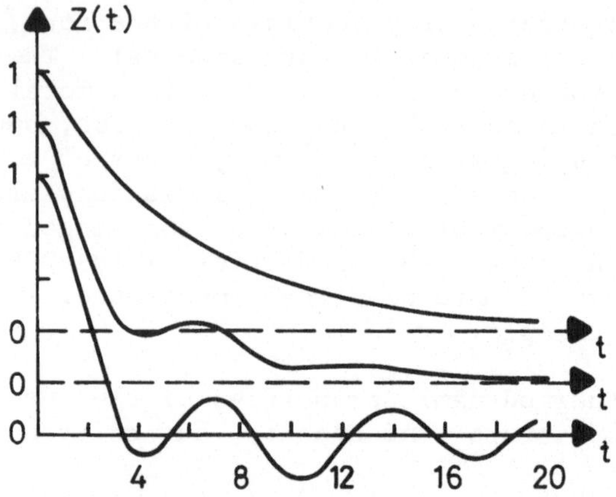

Figure 5: Normalized velocity autocorrelation function of the OCP at (reading from top to bottom) Γ = 1, 10 and 110 [33]. The unit of time is $\sqrt{3}\ \omega_p^{-1}$

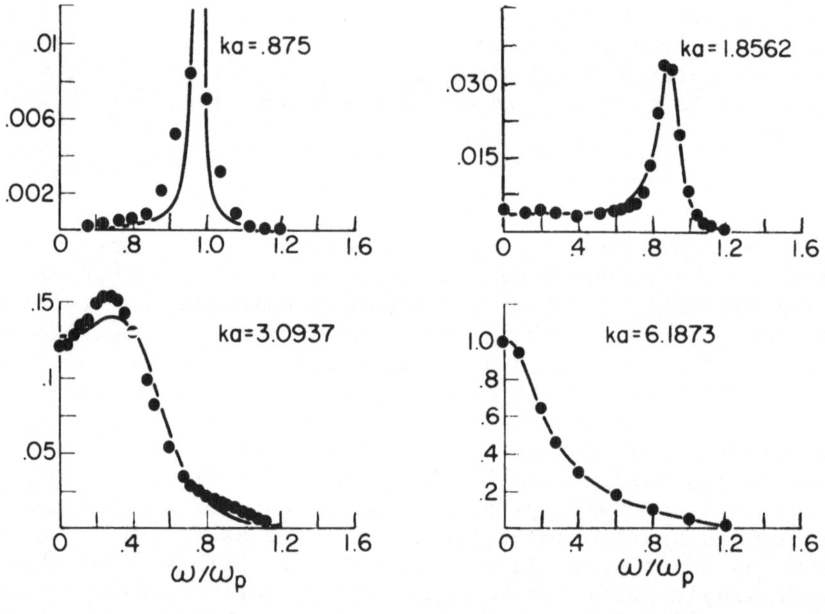

Figure 6: Dynamic structure factor of the OCP at Γ = 110 [34]. The dots are molecular dynamics results and the curves show the predictions of a viscoelastic-type theory (see Lecture 4).

In Figure 6 we show some of the molecular dynamics results [34] for the dynamic structure factor $S(k,\omega)$ at $\Gamma = 110$. At the smallest wavenumber the spectrum is dominated by a very sharp peak centred around a frequency close to ω_p, providing conclusive evidence for the existence of a propagating plasmon mode. By contrast with the case of neutral fluids, there is very little intensity near $\omega = 0$; in other words the equivalent of the Rayleigh peak, if it exists at all, is very weak. On increasing k the peak broadens and shofts to lower frequencies and simultaneously a low-frequency sidewing develops. Eventually, at sufficiently large k, the plasmon peak disappears and the spectrum takes on the gaussian-type shape characteristic of free-particle behaviour.

Qualitatively similar results are obtained at other values of Γ, down to $\Gamma \simeq 10$, though for a given value of ka the collective mode becomes more strongly damped as Γ decreases. However, between $\Gamma = 10$ and $\Gamma = 1$ there is a change in the sign of the dispersion, the peak frequency $\omega(k)$ becoming an increasing function of k.

To what extent can the unusual dynamical properties of the OCP by understood on general theoretical grounds? In considering this problem we must be careful to distinguish between a) the so-called kinetic regime or weak-coupling limit, $\Gamma \ll 1$, where the mean time between collisions is much greater than the period of a plasma oscillation, and b) the hydrodynamic regime, $\Gamma \gg 1$, where the reverse is true. The extent to which either kinetic theory or hydrodynamics is successful in the intermediate range of Γ is, of course, a matter for detailed investigation.

The Vlasov theory is the simplest form of kinetic theory which can be usefully applied to the OCP. The Vlasov approximation is equivalent to replacing the two-body distribution function in the first equation of the BBGKY hierarchy by a product of single-particle distribution functions. It represents a typical "collisionless" approximation, since the effect of interactions is taken account of only in an averaged, mean-field sense. In practice [34] it yields a dispersion curve of the form

$$\omega(k) = \omega_p[1 + \tfrac{1}{2}(ka)^2/\Gamma] + O(k^4) \tag{39}$$

in fair agreement with the molecular dynamics results at $\Gamma = 1$. At larger Γ, where the dispersion is negative, the Vlasov approximation breaks down completely, and even at $\Gamma = 1$ there is only qualitative agreement between the theoretical and measured line-shapes.

The hydrodynamic theory of the OCP [35] is based on the completely opposite assumption that each ion experiences many collisions during the period of a plasma oscillation, thereby maintaining local thermodynamic equilibrium. The calculation is similar to that for neutral fluids, the main difference being the appearance in the Navier-Stokes equation of a term representing an electric force, the local electric field being related to the local density fluctuation by Poisson's equation. However, the result for $S(k,\omega)$ itself is very much altered from that obtained for uncharged systems, though there is still a three-peak structure. In the first place, the frequency of the propagating mode now tends to a non-zero value, ω_p, in the limit $k \to 0$, with a dispersion law given by

$$\omega(k) = \omega_p[1 + \Delta k^2] + O(k^4) \tag{40}$$

The coefficient Δ is determined by static properties of the OCP, and in practice is positive for $\Gamma \lesssim 3$ and increasingly negative for larger Γ, corresponding (as required) to a change in sign of the dispersion. Second, the integrated intensity of the thermal (Rayleigh) peak relative to that of the plasmon doublet vanishes as k^2, so that the mechanical plasmon mode is completely dominant at small k. The results of the computer "experiments" at large Γ are therefore consistent with the hydrodynamic calculation, notwithstanding the fact that the propagating plasma oscillation is not a hydrodynamic mode in the accepted sense.

As we have already pointed out, much of the interest in the properties of the OCP arises from points of theory. Molten salts, on the other hand form a class of ionic liquids of some practical importance to both chemists and physicists. In the OCP the assumed rigidity of the background means that mass and charge fluctuations are equivalent. In molten salts, by contrast, the two are independent, at least in the long-wavelength limit. The possibility therefore exists of detecting both the sound-wave propagation typical of neutral fluids and the plasma oscillation characteristic of the OCP. Such modes, if present, would be observable as peaks in the spectra, respectively, of mass and charge fluctuations, and may be regarded as the liquid-state analogues of the longitudinal acoustic and optic phonons in an ionic crystal.

If we are interested in investigating only the general character of the dynamics of molten salts, rather than in making detailed calculations for a specific system, it is sufficient to work with a very simple force law. A tolerably realistic potential model can be devised quite easily, since the only essential ingredient, apart from the Coulombic interaction, is a short-range repulsive force between ions with charges of opposite sign. The detailed form of the short-range repulsion between ions of like charge is much less important, since close contacts between ions of like charge are comparatively rare, at least when all ions are roughly equal in size. Long-range dispersion forces, though undoubtedly present in real molten salts, are also of relatively little importance in comparison with the Coulombic terms. Thus the interaction between two ions of either species in a binary monovalent salt can be adequately represented by a potential of the form

$$\phi(r) = \frac{e^2}{r_o} \left[\frac{1}{n}\left(\frac{r_o}{r}\right)^n \pm \frac{r_o}{r}\right] \tag{41}$$

a representation which has the advantage of allowing thermodynamic properties to be scaled on a corresponding-states basis. The length parameter r_o is in fact the separation at which the cation-anion potential has its minimum. The choice of the exponent n is somewhat arbitrary. For the alkali chlorides the value $n = 9$ seems to be best , in which case one finds that a fair fit to experimental thermodynamics data on NaCl can be achieved by setting $r_o = 2.34$ Å [36]. If the ionic masses are taken as equal, the model is a completely symmetric one, the two species differing only in the sign of their charges. In this particularly simple case there is no distinction (apart from a constant factor) between mass and number densities.

The dynamical properties of the symmetric model have been studied by a computer "experiment" carried out near triple-point conditions [37]. In terms of the parameters introduced in discussing the properties of the OCP, the temperature and density at which the calculations were made correspond to $\Gamma = 64.6$ and $a = 0.866\ r_o$.

In Figure 7 we plot the "experimental" results for the velocity autocorrelation function $Z(t)$ of the symmetric model. There appears to be no remnant of any plasma oscillation, and the curve resembles very closely that obtained for argon-like liquid, as discussed in an earlier lecture. We see again a region of strong negative autocorrelation at intermediate times and the type of long-lived negative plateau that we know to be associated with the correlated motion of a relatively large group of particles. Presumably, therefore, there exists the same vortex-like backflow around a diffusing particle. Here, however, since the nearest neighbours of a given ion are predominantly ions of opposite charge (the charge-ordering effect), a backflow of particles is in fact a backflow of charge. Whereas such motion may contribute to the process of diffusion, it contributes much less to that of electrical conduction. Macroscopically this leads to a deviation of the observed electrical conductivity σ from that expected on the basis of the observed self-diffusion coefficients, D_+ and D_-, as expressed by the Nernst-Einstein relation

$$\sigma = \frac{1}{2}\,(Ne^2/Vk_BT)\ (D_+ + D_-) \tag{42}$$

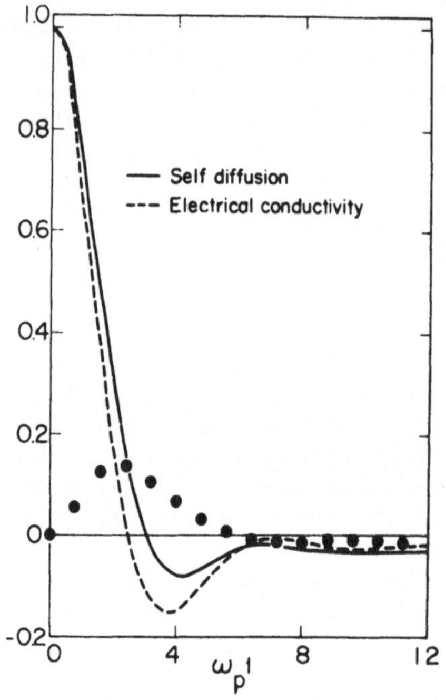

Figure 7: Normalized autocorrelation functions of velocity (full curve) and electrical current (dashed curve) for the symmetric molten salt [37]. The dots show the difference between the two curves and represent the contribution from cross-correlations of velocity (see text).

Microscopically this relation can be derived by assuming that Z(t) is proportional to J(t), the autocorrelation function of the fluctuating electrical current, which in turn would be true if correlations of the type $\langle \dot{x}_i(t)\dot{x}_j(o)\rangle$ were negligible for all $i \neq j$. This assumption is quite unjustified, as Figure 7 shows. Cross-correlations make a significant contribution to J(t), their maximum effect being felt after a time approximately equal to the interval between correlations. In the case in question the true conductivity is 19% smaller than that predicted by the rule (42), a result for the so-called Nernst-Einstein deviation which is in good agreement with experimental data on the alkali chlorides: 15% (RbCl), 18% (NaCl) and 23% (CsCl). It follows that it is unnecessary to postulate the existence of long-lived ion pairs in order to account for the observed deviations from the Nernst-Einstein relation.

Some of the results obtained [37] for $S_{QQ}(k,\omega)$, the spectrum of charge fluctuations, are displayed in Figure 8. At small k there is a well-defined optic-type peak at high frequency and another peak at $\omega \doteq 0$, so that $S_{QQ}(k,\omega)$ appears as the superposition of two symmetrical bell-shaped curves, corresponding to two distinct relaxation processes. Obviously the situation is more complex than in the OCP, where the spectrum consists of a sharp high-frequency peak and a weak low-frequency tail. As k increases, the "optic" peak moves to lower frequencies, the dispersion being much stronger than in the high-density plasma. Finally, at $ka \simeq 3$, the "optic" peak disappears and the spectrum narrows dramatically; thereafter it broadens steadily with increasing k, gradually taking on the free-particle shape. Thus charge fluctuations continue to propagate down to a wavelength of approximately twice the nearest neighbour separation, whereas sound wave propagation apparently ceases at a wavelength which is at least four times larger, since no trace of a Brillouin is seen, even at the lowest accessible wavenumber.

These results on the charge fluctuations are somewhat surprising, since the hydrodynamic (long-wavelength) behaviour of a

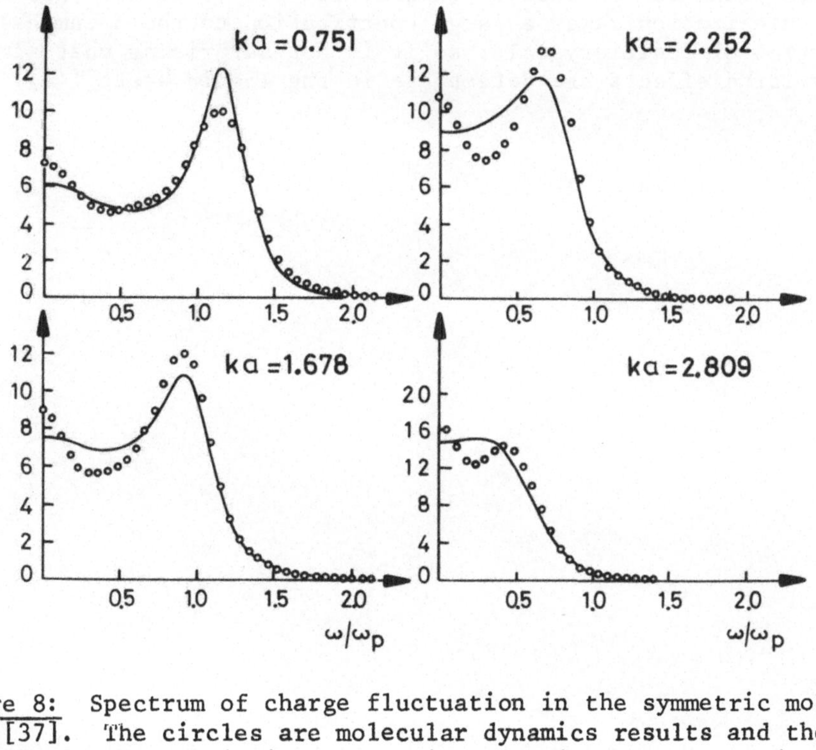

Figure 8: Spectrum of charge fluctuation in the symmetric molten salt [37]. The circles are molecular dynamics results and the curves show the predictions of a viscoelastic-type theory (see Lecture 4).

molten salt [38, 39] is no different from that of a neutral binary mixture. In particular, $S_{QQ}(k,\omega)$ (which in the neutral case is the spectrum of concentration fluctuations) is a single Lorentzian curve centred at $\omega = 0$ with a width determined by the interdiffusion coefficient, i.e. the electrical conductivity. The "optic" peak should therefore vanish in the hydrodynamic limit. This is not an unexpected conclusion, since a long-wavelength disturbance sees only neutral entities (NaCl molecules), not individual charges. The molecular dynamics results do indicate some weakening of the "optic" mode as k becomes small, and it would clearly be interesting to extend the calculations to significantly larger wavelengths. This, however, would be an expensive undertaking.

A large number of molecular dynamics calculations based on more realistic interionic potentials have also been carried out, and the results compared with experimental data on alkali halides, alkaline earth halides and alkali nitrates. Much of this work has recently been reviewed by Sangster and Dixon [40]. Systematic deviations observed in certain of the transport coefficients [16] can be ascribed to the neglect of ionic polarization. (The calculations of transport and coefficients were made by the linear response technique alluded to in Lecture 3.) It is well-known that ionic polarization makes a large contribution to the dynamical properties of ionic crystals, so it is not surprising that similarly large effects are detectable in the liquid state [41].

REFERENCES

[1] B.J. Alder and T.E. Wainwright, J. Chem. Phys. 27, 1208 (1957).
[2] M. Metropolis, A.W. Rosenbluth, M.N. Rosenbluth, A.N. Teller and E. Teller, J. Chem. Phys. 21, 1087 (1953).
[3] J.P. Valleau and S.G. Whittington in "Modern Theoretical Chemistry", vol. 5A, ed. B.J. Berne (Plenum Press, New York: 1977).
[4] J.P. Valleau and G.M. Torrie in "Modern Theoretical Chemistry", vol. 5A, ed. B.J. Berne (Plenum Press, New York: 1977).
[5] A. Rahman, Phys. Rev. 136, A405 (1964).
[6] L. Verlet, Phys. Rev. 159, 98 (1967).
[7] D. Beeman, J. Comput. Phys. 20, 130 (1976).
[8] J. Barojas, D. Levesque and B. Quentrec, Phys. Rev. A7, 1092 (1973).
[9] P.S.Y. Cheung and J.G. Powles, Molec. Phys. 30, 921 (1975).
[10] K. Singer, A. Taylor and J.V.L. Singer, Molec. Phys. 33, 1757 (1977).
[11] A. Rahman and F.H. Stillinger, J. Chem. Phys. 55, 3336 (1971).
[12] J.P. Ryckaert, G. Ciccotti and H.J.C. Berendsen, J. Comput. Phys. 23, 327 (1977).
[13] E.M. Gosling, I.R. McDonald and K. Singer, Molec. Phys. 26, 1475 (1973).
[14] W.T. Ashurst and W.G. Hoover, Phys. Rev. Letters 31, 206 (1973).
[15] G. Ciccotti and G. Jacucci, Phys. Rev. Letters 35, 789 (1975).
[16] G. Ciccotti, G. Jacucci and I.R. McDonald, Phys. Rev. 13A , 426 (1976).
[17] B.J. Alder, D.M. Gass and T.E. Wainwright, J. Chem. Phys. 53, 3813 (1970).
[18] D. Levesque and L. Verlet, Phys. Rev. A2, 2514 (1970).
[19] B.J. Alder and T.E. Wainwright, Phys. Rev. A1, 18 (1970).
[20] D. Levesque, L. Verlet and J. Kürkijarvi, Phys. Rev. A7, 1690 (1973).
[21] H. Mori, Prog. Theor. Phys. 33, 423 (1965).
[22] D. Levesque and W.T. Ashurst, Phys. Rev. Letters 33, 277 (1974).
[23] J.P. Hansen and I.R. McDonald, "Theory of Simple Liquids" (Academic Press, London : 1976).
[24] E.M. Adams, I.R. McDonald and K. Singer, Proc. Roy. Soc. A351, 37 (1977)
[25] A. Rahman, Phys. Rev. A9, 1667 (1974).
[26] J.R.D. Copley and J.M. Rowe, Phys. Rev. A9, 1656 (1974).
[27] R. Zwanzig and M. Bixon, Phys. Rev. A2, 2005 (1970).
[28] J.R.D. Coplay and S.W. Lovesey, Rep. Prog. Phys. 38, 461 (1975).
[29] P.S.Y. Cheung and J.G. Powles, Molec. Phys. 32, 1383 (1976).
[30] D.J. Adams and I.R. McDonald, Molec. Phys. 32, 931 (1976).
[31] D. Levesque, G.N. Patey and J.J. Weis, Molec. Phys. 34, 1077 (1977).

[32] A.J.C. Ladd, Molec. Phys. 33 1039 (1977).
[33] J.P. Hansen, E.L. Pollock and I.R. McDonald, Phys. Rev. Letters 32, 277
[34] J.P. Hansen, I.R. McDonald and E.L. Pollock, Phys. Rev. A11, 1025 (1975).
[35] P. Vieillefosse and J.P. Hansen, Phys. Rev. A12, 1106 (1975).
[36] D.J. Adams and I.R. McDonald, Physica 79B, 159 (1975).
[37] J.P. Hansen and I.R. McDonald, Phys. Rev. A11, 2111 (1975).
[38] P. Vieillefosse, J. Physique 38, L43 (1977).
[39] P.V. Giaquinta, M. Parrinello and M.P. Tosi, Phys. Chem. Liq. 5, 305 (1976).
[40] M.J. Sangster and M. Dixon, Adv. Phys. 25, 247 (1976).
[41] G. Jacucci, I.R. MacDonald and A. Rahman, Phys. Rev. A13, 1581 (1976).

INTERATOMIC AND INTERMOLECULAR FORCES

A.D. Buckingham

University Chemical Laboratory, Lensfield Road

Cambridge, CB2 1EW, United Kingdom

LECTURE 1

GENERAL SURVEY

1.1 Introduction

The equilibrium and non-equilibrium properties of matter reflect the interatomic and intermolecular forces. Elementary considerations of the mere existence of liquids lead us to conclude that molecules attract one another when they are far apart. Also the fact that liquids and solids have finite densities of the magnitude observed under normal conditions means that molecules repel at short range. There is therefore a balance point of zero force or minimum energy. These basic truths are represented in Figure 1.1 which shows the interaction energy u(R) of two spherical atoms as a function of their separation R. The separation R_e at the minimum is a measure of the diameter of an atom. In the case of two interacting argon atoms the well depth ϵ is 0.196×10^{-20} J and $R_e = 3.76 \times 10^{-10}$ m.

The existence of a potential energy function u(R) is dependent on the Born-Oppenheimer approximation [1]; we fix the nuclei in a particular configuration and let u(R) be the difference in energy of the system in that configuration from its value when the separation $R \rightarrow \infty$. There are some interesting effects resulting from a breakdown of the Born-Oppenheimer approximation (particularly when there are electronic degeneracies) but for the purposes of studying liquids we may safely employ it. The accuracy of the approximation can be gauged from the following results:

(i) the Rydberg constant for the H atom is reduced by 0.054 per cent (59.8 cm^{-1}) on changing the nuclear mass from infinity to that of a proton.

(ii) the clamped-nuclei non-relativistic dissociation energy of H_2 is D_e = 38292.83 cm^{-1} [2]; the relativistic correction takes it to 38292.30 cm^{-1} [3] and the experimental value is $38295.6 < D_e < 38297.6$ cm^{-1} [4].

(iii) the dipole moment of HD is 5.85×10^{-4}D = 19.5×10^{-34} Cm [5] and it arises solely from the breakdown of the approximation, since in the clamped-nuclei model HD is electrically centrosymmetric. The dipole may have the sense H^+D^- [6].

It should also be noted that the Born-Oppenheimer approximation improves as the mass of the nuclei increases - the inaccuracies are greatest for protons.

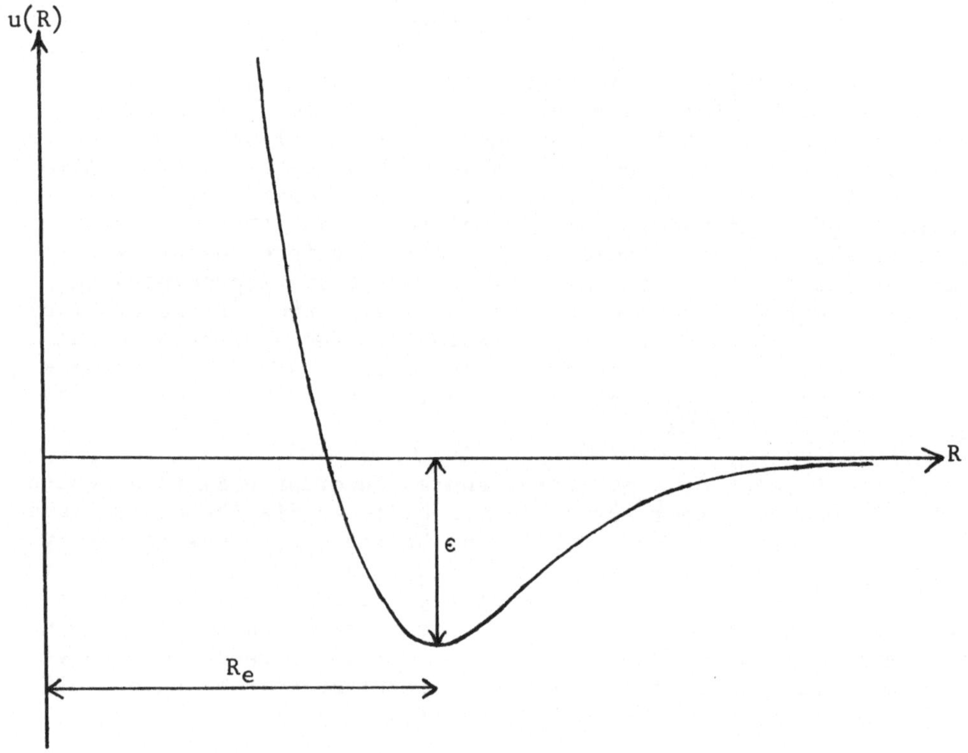

Figure 1.1. The interaction energy u(R) of two spherical molecules as a function of their separation R. The well-depth is ε and R_e the equilibrium separation.

The number of variables upon which the interaction energy depends increases rapidly as the molecular size increases. Thus, for an atom and a diatomic molecule (for example HCl + Ar) there are three, which may be taken to be the separation R of the atom from the centre of mass of the molecule, the angle θ between the molecular axis and the intermolecular vector R and the distance r between the two nuclei in the molecule (see Figure 1.2).

For two diatomic molecules there are six variables (the intermolecular separation R, the angles θ_1 and θ_2 between the molecular axes and the line of centres, the angle ϕ between the planes containing the vector R and the molecules, and the two intramolecular distances r_1 and r_2). In the general case of two interacting non-linear polyatomic molecules, the interaction energy depends on $3(N_1 + N_2)-6$ independent variables, where N_1 and N_2 are the number of atoms in molecules 1 and 2; six of these variables describe the relative position and orientation of the two molecules and the remaining $(3N_1 - 6) + (3N_2 - 6)$ are the internal vibrational co-ordinates of the two molecules. In some problems the vibrational motion may not be of interest and the effective number of degrees of freedom may be reduced. However, if one is interested in the effects of the interaction on the vibrations, then it is essential to retain the internal degrees of freedom.

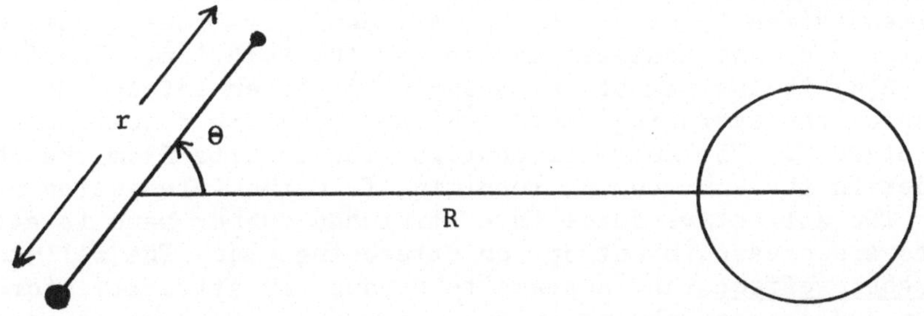

Figure 1.2. The three co-ordinates R, θ, r in the interaction of a diatomic molecule and an atom.

1.2 What is a Molecule and What is a Force?

Since we are discussing intermolecular forces, it is essential that we know what we mean by a molecule and what we mean by a force [7]. A molecule is a group of atoms (or a single atom) whose binding energy is sufficiently large to permit it to interact with its environment without losing its identity. Thus the hydrogen molecule H_2 may conveniently be classified as a molecule but Ar_2 is better considered as a bound pair of argon atoms. The potential energy functions for both H_2 and Ar_2 are of the form shown in Figure 1.1 but in Ar_2 the well-depth is only about one-half the thermal energy kT at room temperature, so Ar_2 is readily dissociated on collision. The infrared spectrum of H_4 has been studied at about 20 K by Watanabe and Welsh [8]; two of the bonds in H_4 are very similar to that in H_2 and it is convenient to consider H_4 as $(H_2)_2$, that is, a bound complex of two H_2 molecules. If a molecule is tightly bound, its thermally populated vibrational and rotational states have similar structure and properties; however, with a shallow, anharmonic potential well, as in Ar_2 and $(H_2)_2$, the structure and properties vary substantially from state to state. In a non-rigid molecule, such as ammonia (NH_3), 1,2-dichloroethane ($ClH_2C - CH_2Cl$), or a polypeptide, there may be only a small change in energy with a large change in an internal co-ordinate; considerable interest may exist in changes in the conformation of such a molecule with changes in the environment.

And what is a force? In the collision of two monatomic molecules, the force is equal to $-du/dR$ and there is no difficulty. Similarly, there is a torque $-\partial u/\partial\theta$ acting on a diatomic molecule colliding with an atom, as in Figure 1.2. But what is the force between two ions in an aqueous solution? It is convenient to consider a potential of average force A(R), which is a Helmholtz free energy, and is the mean interaction energy of the two ions at a fixed separation R, averaged over all configurations of all other molecules and ions that are present in the solution. The free energy A(R) is the sum of an energy u(R) and an entropy term -TS(R). Because of the averaging, both u(R) and S(R) are functions of the temperature T. The entropic contribution results from the change in order in the environment resulting from the interaction of the pair. The attractive force in a stretched rubber band is attributable to a decrease in entropy on stretching [9]. The well-known hydrophobic effect that appears to produce an attractive force between hydrocarbon chains in aqueous media [10] depends on the entropy S(R), for the decrease in entropy on forming a cage of water molecules is apparently less in the case of a nearby pair of solutes than when they are far apart.

1.3 Classification of Intermolecular Forces

The significant forces between molecules have an electric origin. It is true that other sources exist, such as magnetic and gravitational interactions, but these can normally be neglected. In considering the nature of intermolecular potentials, it is helpful to attempt to separate various contributions. The development of a general theory of interaction energies provides a basis for this separation.

The primary separation to be made is to divide the interactions into two classes, short-range and long-range. The former decrease exponentially with increasing R and result from overlap of the electronic wavefunctions describing the isolated molecules. Long-range interactions vary as R^{-m} at large R, where m is a positive integer. Thus the long-range energy of interaction of two ions varies as R^{-1} and of two dipoles in fixed relative orientations as R^{-3}; the corresponding forces vary as R^{-2} and R^{-4}. At long-range the indistinguishability of electrons resulting in their exchange between the molecules plays an insignificant role in determining the interaction and it is possible to consider the electrons as if they were associated with one or other of the molecules. The n-electron wavefunction, where $n = n_1 + n_2$, need not be anti-symmetrized with respect to exchange of electrons between molecules 1 and 2, for such an anti-symmetrization leads to short-range interactions. Thus long-range forces can be related to the properties of the free molecules, such as the moments of the electric charge distribution and the polarizabilities.

Short-range forces may be attractive or repulsive, but at small separations they are always repulsive. Long-range forces may likewise be attractive or repulsive; for spherical atoms, as in Ar_2, the long-range force is attractive.

The Hellmann-Feyman theorem [11] requires that the forces on the nuclei may be evaluated by classical electrostatics from a knowledge of the charge distribution. At long-range, the attractive force between two argon atoms is associated with a build-up of electron charge in the region between the nuclei. The atoms acquire equal and opposite electric dipole moments proportional to R^{-7} at large R. The attractive force varying as R^{-7} does not result from the interaction of these dipoles but rather from the force exerted on each nucleus by the distorted electron distribution in its own atom [12]. This charge distortion at long-range results from intermolecular electron correlation. In the short-range overlap region, it is no longer necessary to invoke a redistribution of charge to explain the force, although such a redistribution does occur.

range	type	attractive (-) or repulsive (+)	additive or non-additive
short	overlap (Coulomb and exchange)	$\mp$	non-additive
long	electrostatic	$\mp$	additive
	induction	-	non-additive
	dispersion	-	nearly additive
	resonance	$\mp$	non-additive
	magnetic	$\mp$	(weak)

Table 1. Classification of molecular interaction energies. The primary separation is into short-range and long-range effects. The long-range energy is further divided into five distinct types. The attractive or repulsive nature of the forces and the additivity or non-additivity of the interactions is shown.

1.4 Electrostatic Energy

A secondary classification of the long-range interactions into several distinct types is helpful (see Table 1). The simplest, and for some systems, such as electrolyte solutions, the most important, long-range interaction is the electrostatic energy, $u_{\mathrm{electrostatic}}$, which is the interaction energy of the unperturbed charge distributions of the molecules. The electrostatic forces may be attractive or repulsive and are additive in the sense that for three molecules 1, 2, 3

$$u_{123} = u_{12} + u_{23} + u_{31} \tag{1.1}$$

$u_{\mathrm{electrostatic}}$ is determined by the charge distribution of the free molecules and it may be evaluated in general by performing an integration over the space of each molecule; however, if the separation between the molecules is large compared to their dimensions, the multipole expansion may conveniently be employed [13,14].

1.5 Induction Energy

The induction energy, $u_{\mathrm{induction}}$, is the energy resulting from the distortion of the electronic clouds by the static electric field due to the other molecules. Like the electrostatic energy, it is absent in the case of inert gas atoms. The main contribution to $u_{induction}$ is due to the induced dipoles:

$$u_{induction} = -\frac{1}{2} \sum_i \underset{\sim}{\alpha}^{(i)} : \underset{\sim}{F}^{(i)2} - \ldots \qquad (1.2)$$

where $\underset{\sim}{\alpha}^{(i)}$ is the polarizability tensor of molecule i and $\underset{\sim}{F}^{(i)}$ is the electric field at the origin of this molecule resulting from the charge distribution of all the other molecules. Thus, in the interaction of an ion and a spherical atom, the induction energy at long range is $-\frac{1}{2}\alpha q^2 R^{-4}(4\pi\epsilon_o)^{-2}$ where $q R^{-2}(4\pi\epsilon_o)^{-1}$ is the field strength at the centre of the atom due to the ion of charge q. For a spherical atom $\underset{\sim}{\alpha}$ is isotropic, so the polarizability is expressible in terms of the unit tensor of rank two (the Kronecker delta, $\delta_{\alpha\beta}$):

$$\alpha_{\alpha\beta} = \alpha\, \delta_{\alpha\beta}, \qquad\qquad \alpha = \frac{1}{3}\alpha_{\gamma\gamma} = \frac{1}{3}\left(\alpha_{xx}+\alpha_{yy}+\alpha_{zz}\right) \quad (1.3)$$

Since the induction energy results from distortion of a charge distribution in response to an external field, it is always negative for molecules in their ground electronic states (the variation principle requires that such a distortion must lower the energy of the ground state). From equation (1.2) it is clear that $u_{induction}$ is not additive:

$$F^{(i)2} = \sum_{j \neq i} F^{(ij)2} + \sum_{j \neq i} \sum_{k \neq i,j} \underset{\sim}{F}^{(ij)} \cdot \underset{\sim}{F}^{(ik)} \qquad (1.4)$$

where $\underset{\sim}{F}^{(ij)}$ is the field at i due to the charge distribution of molecule j. The second contribution to (1.4) is responsible for the non-additivity. Thus the dipole induction energy of an atom at the mid-point between two ions of charge q is zero, although the induction energy of the atom with each of the ions is negative.

1.6 Dispersion Energy

The dispersion energy, $u_{dispersion}$, results from intermolecular correlation in the fluctuation in the electronic co-ordinates of the molecules. It is a consequence of the quantum-mechanical nature of the electron. If the electron were treated as a classical particle, its position could be specified and there would be an electrostatic energy for each electronic configuration; for two spherical atoms the classical electrostatic energy would average to zero in first-order but would lead to a temperature-dependent attractive energy in second-order because of the Boltzmann favouring of the configurations of lower energy. Temperature-dependent forces of this nature were discussed by Keesom in 1921 [15]. The origin of the binding energy of the liquid and solid inert gases was a major problem in the 1920s; the binding energy of argon is of the same order of magnitude as that of the dipolar species HCl. It was known at that time that Ar had no dipole moment, but Debye [16] suggested that it may have a quadrupole which would produce an induction energy. After the advent of quantum mechanics it became clear that the inert gas atoms are spherical in their ground states and therefore have no permanent electric moments of any order. In 1926, Wang [17] showed that there is an interaction energy between two H atoms that varies as R^{-6} at large R. In 1930 London [17] gave a sound description of this long-range interaction using second-order perturbation theory, and he provided a useful approximate formula for evaluating its strength. He pointed out a link between this interaction and optical dispersion and hence gave the name dispersion energy. All molecules exhibit dispersion interactions, although they are absent in the interaction of a proton and an atom.

1.7 Resonance Energy

The resonance energy, $u_{resonance}$, is the additional energy that results from a lifting of degeneracy by the interaction. The degeneracy may arise because one of the molecules is in a degenerate state, as in the interaction of an excited H atom with a principal quantum number of 2 with an ion or molecule, or the degeneracy may result from the possibility of an exchange of excitation between two like molecules, as in the case of an H atom in the ground 1s state near an excited H atom in the $2p_z$ state, when the long-range interaction energy varies as R^{-3}. The lifting of the degeneracy by the interaction produces two or more potential surfaces which may lie above or below zero. A sum over all the surfaces produces zero in the long-range limit, although in any particular collision $u_{resonance}$ produces either an attractive or repulsive interaction according to the quantum numbers describing the state of the pair.

Resonance energy may be important in determining the widths of rotational lines in the pure-rotation and rotation-vibration spectra of small gaseous molecules. For example, the width of a line in the vibration-rotation spectrum of gaseous HCl is greater in the pure gas than for HCl in DCl at the same pressure, since resonant dipolar interactions between HCl molecules differing by one in their rotational quantum numbers occur in the pure gas but not in the mixture [18]. The fact that the lifetime of a vibrationally excited C_6H_6 molecule is significantly shorter in liquid C_6H_6 than in C_6D_6 is also a consequence of the resonance interaction energy [19].

1.8 Magnetic Interactions

Finally, there are the long-range magnetic interactions. These are normally neglected since they are so weak compared to the corresponding electric interaction. The magnetostatic dipole-dipole energy is typically 10^{-4} times the electrostatic dipole-dipole energy, and the energy resulting from induced or fluctuating magnetic moments is of the order of 10^{-8} of the corresponding electrical energy (these figures follow from the fact that magnetic dipole moments are of the order of 1 Bohr magneton = 0.9274×10^{-20} e.m.u. = 0.9274×10^{-23} A m^2, while electric dipole moments are of the order of 1 debye = 10^{-18} e.s.u. = 3.336×10^{-30} C m). In optically active species, where the molecules may exist in distinguishable right- and left-handed forms, there is a coupling of the fluctuating electric and magnetic moments giving rise to a weak dispersion energy that is dependent on the handedness of the molecules; that is, there is a small component of the R^{-6} dispersion energy that changes sign on changing one of the molecules into its mirror image [20]. There are in addition small relativistic contributions to the interaction energy. These have been reviewed by Hirschfelder and Meath [21] and for most purposes relating to the study of normal liquids they are negligible. However, there may be significant relativistic effects when heavy atoms are present. For example, the mean radius of the 6s atomic orbital in the mercury atom is reduced by 15 per cent as a result of the relativistic increase in the effective mass of the electron when it is near the nucleus and the well-depth ϵ in Hg_2 has been calculated to be only 45% of the non-relativistic value [22].

1.9 Short-Range Interactions

When the overlap of the electron clouds of interacting molecules is significant, it is essential that the total wavefunction be antisymmetric with respect to exchange of all pairs of electrons. If we do not impose this condition then the wavefunction does not represent observable states of the system, for it does not reflect the principle of indistinguishability of electrons [23].

A possible route to short-range potentials is through applications of self-consistent-field theory to the interacting system at fixed nuclear configurations and to the free molecules. The interaction energy is then the difference between the calculated energies, but unlike the total energy (which, by the variational principle, is an upper bound to the true energy) it is not in general bound. This approach can give useful results for short-range repulsive energies but if only a single configuration is employed there can be no electron correlation and hence no dispersion energy at long-range. Since the dispersion energy is the sole source of attraction between inert-gas atoms, it is to be expected that the Hartree-Fock potential curve for these systems should have no minimum. Minima have sometimes been obtained but these result from the incompleteness of the basis set and can be eliminated by the addition of "ghost orbitals" in the calculations on the separate molecules to compensate for the extension of the basis set for the pair [25].

In the region of electron overlap, when intermolecular exchange is important, the identity of the interacting molecules is lost and they are merged into a "supermolecule". It is therefore unlikely to be helpful to seek a theory of short- and intermediate-range intermolecular forces which relate the interaction to the properties of the free molecules. However, in the long-range region such a general theory can be formulated and it will be developed in Lecture 2.

References

[1] M. Born and J.R. Oppenheimer, Ann. Physik, 84, 457 (1927); M. Born and H. Huang, "Dynamical Theory of Crystal Lattices", (Oxford University Press, 1954), pp. 166 and 402.

[2] W. Kołos and L. Wolniewicz, J. Chem. Phys., 49, 404 (1968).

[3] W. Kołos and L. Wolniewicz, J. Chem. Phys., 41, 3663 (1964).

[4] G. Herzberg, J. Mol Spectros., 33, 147 (1970).

[5] M. Trefler and H.P. Gush, Phys. Rev. Lett., 20, 703 (1968).

[6] P.R. Bunker, J. Mol. Spectros., 46, 119 (1973).

[7] H.C. Longuet-Higgins, Discuss. Faraday Soc., 40, 7 (1965).

[8] A. Watanabe and H.L. Welsh, Canad. J. Phys., 1965, 43, 818; 1967, 45, 2859.

[9] "A Discussion on Rubber Elasticity", Proc. Roy. Soc A, 351, 295-406 (1976).

[10] C. Tanford, "The Hydrophobic Effect", (Wiley, New York, 1973).

[11] H. Hellmann, "Quantenchemie", Demticke, Leipzig, 1937, p.285; R.P. Feynman, Phys. Rev., 56, 340 (1939).

[12] J.O. Hirschfelder and M.A. Eliason, J. Chem. Phys., 47, 1164 (1967).

[13] J.O. Hirschfelder, C.F. Curtiss and R.B. Bird, "Molecular Theory of Gases and Liquids", (Wiley, New York, 1954).

[14] A.D. Buckingham in "Intermolecular Interactions: From Diatomics to Biopolymers" (Ed. B. Pullman), (Wiley, Chichester, 1977), Chap. 1.

[15] W.H. Keesom, Physik. Z., 22, 129 (1921).

[16] P. Debye, Physik. Z., 22, 302 (1921).

[17] F. London, Z. Physik, 63, 245 (1930); Trans. Faraday Soc., 33, 8 (1937).

[18] C.G. Gray, J. Chem. Phys., 61, 418 (1974).

[19] J.E. Griffiths, M. Clerc and P.M. Rentzepis, J. Chem. Phys., 60, 3824 (1974).

[20] C. Mavroyannis and M.J. Stephen, Molec. Phys., 5, 629 (1962).

[21] J.O. Hirschfelder and W.J. Meath, Adv. Chem. Phys., 12, 3 (1967).

[22] N.C. Pyper, I.P. Grant and R.B. Gerber, Chem. Phys. Lett., 49, 479 (1977).

[23] P. Claverie, Int. J. Quantum Chem., 5, 273 (1971).

[24] H. Margenau and N.R. Kestner, "Theory of Intermolecular Forces", 2nd Edn., (Pergamon Press, Oxford, 1971).

[25] N.S. Ostlund and D.L. Merrifield, Chem. Phys. Letters, 39, 612 (1976).

LECTURE 2

LONG-RANGE FORCES

2.1 Introduction

In Lecture 1 we divided molecular interactions into short- and long-range effects and we further subdivided the long-range interactions into five distinguishable categories (the electrostatic, induction, dispersion, resonance and magnetic contributions). In this second lecture we employ quantum-mechanical perturbation theory to formulate a general theory of long-range interaction energies. We shall exploit the fact that at long-range the particles retain their identity so that we can hope to relate the interaction to the properties of the free molecules.

First we shall illustrate some of the features of long-range interactions by considering a simple model [1].

2.2 Coupled One-Dimensional Oscillators

Consider two identical one-dimensional harmonic oscillators of mass m and force constant K (see Figure 2.1). The Hamiltonian for each oscillator is of the form $\frac{1}{2m}p^2 + \frac{1}{2}Kz^2$ where p is the momentum operator and z the displacement of the oscillator from equilibrium. If we add an interaction linear in the displacement of each oscillator the Hamiltonian of the pair is

$$\mathcal{H} = \frac{1}{2m}\left[p_1^2 + p_2^2\right] + \frac{1}{2}K\left[z_1^2 + z_2^2 + 2cz_1z_2\right] \tag{2.1}$$

This can be re-arranged to the separable form

$$\mathcal{H} = \frac{1}{4m}\left[(p_1+p_2)^2 + (p_1-p_2)^2\right] + \frac{1}{4}K\left[(1+c)(z_1+z_2)^2 + (1-c)(z_1-z_2)^2\right] \tag{2.2}$$

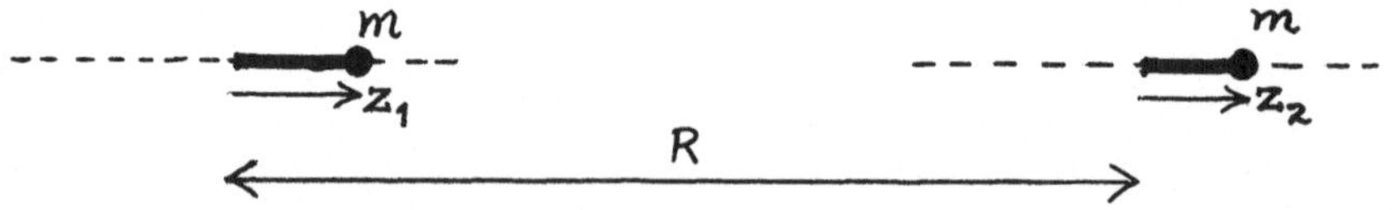

Figure 2.1. A pair of identical coupled linear oscillators. The interaction potential energy is proportional to z_1z_2 and the Hamiltonian is given by equation (2.1). If the oscillators are considered to be electric dipoles of magnitude qz_1 and qz_2, then the coupling constant c in (2.1) is negative and equal to $-2(q^2/KR^3)(4\pi\epsilon_0)^{-1}$.

The oscillations are therefore split into two normal vibrations of angular frequencies ω_+ and ω_- where

$$\omega_{\pm} = \omega_0 (1 \pm c)^{\frac{1}{2}} \tag{2.3}$$

and $\omega_0 = (K/m)^{\frac{1}{2}}$ is the angular frequency of an isolated oscillator.

The partition function (which yields the equilibrium thermodynamic properties of a macroscopic system) for a harmonic oscillator is

$$Q_{vibration} = \sum_{v=0}^{\infty} \exp[-v\hbar\omega/kT] \tag{2.4}$$

If $\hbar\omega/kT < 1$ this can be summed to give $Q_{vibration} = [1-\exp(-\hbar\omega/kT)]^{-1}$ and in the classical limit ($\hbar\omega/kT \rightarrow 0$) this is $kT/\hbar\omega$. Hence the classical partition function for our coupled oscillators is

$$Q_{classical} = \frac{kT}{\hbar\omega_+}\,\frac{kT}{\hbar\omega_-} = \left(\frac{kT}{\hbar\omega_0}\right)^2 (1-c^2)^{-\frac{1}{2}} \tag{2.5}$$

and the corresponding Helmholtz free energy is

$$A = -kT \ln Q = A_0 + \tfrac{1}{2} kT \ln(1-c^2) \tag{2.6}$$

The change $A - A_0 = \Delta A$ on coupling the oscillators is <u>negative</u>, giving an attractive force $-dA/dR$. This force is purely entropic, for the entropy change is

$$\Delta S = -\left(\frac{\partial \Delta A}{\partial T}\right)_V = -\tfrac{1}{2} k \ln(1-c^2) \tag{2.7}$$

and $\Delta A = -T\Delta S$. The change in internal energy is therefore zero in the classical limit.

However, in reality, the oscillators possess zero-point energy and this changes on coupling the pair. Hence, for the ground state

$$\begin{aligned} u - u_0 &= \tfrac{1}{2}\hbar\,(\omega_+ + \omega_- - 2\omega_0) \\ &= \tfrac{1}{2}\hbar\omega_0\left[(1+c)^{\frac{1}{2}} + (1-c)^{\frac{1}{2}} - 2\right] \end{aligned} \tag{2.8}$$

This is negative when c is either positive or negative and vanishes when c is zero. For small c

$$u - u_0 = -\frac{1}{8}\hbar\omega_0\left[c^2 + \frac{5}{16}c^4 + \ldots\right] \qquad (2.9)$$

and we see that the interaction energy $u-u_o$ varies quadratically with the strength c of the interaction. This is the dispersion energy and it comes from correlation in the motion of the coupled quantum particles.

If the two oscillators are dissimilar there is again a lowering of the total zero-point energy that is quadratic in the coupling parameter c. However, there is an important effect on excited states resulting from the identical nature of the oscillators. There is then a resonant interaction leading to a change in zero-point energy that is linear in c. This is $u_{resonance}$ and it comes from the first-order lifting of the degeneracy by the coupling. Consider the case of a single quantum of vibrational excitation. The unperturbed excited state (corresponding to $R = \infty$) is degenerate since the excitation may be considered to be on either of the oscillators. However, the coupling lifts the degeneracy and the interaction energy is either

$$u'_+ = \frac{1}{2}\hbar\omega_- + \frac{3}{2}\hbar\omega_+ - 2\hbar\omega_0 = \hbar\omega_0\left[\frac{1}{2}(1-c)^{\frac{1}{2}} + \frac{3}{2}(1+c)^{\frac{1}{2}} - 2\right] \qquad (2.10)$$

or

$$u'_- = \frac{3}{2}\hbar\omega_- + \frac{1}{2}\hbar\omega_- - 2\hbar\omega_0 = \hbar\omega_0\left[\frac{3}{2}(1-c)^{\frac{1}{2}} + \frac{1}{2}(1+c)^{\frac{1}{2}} - 2\right] \qquad (2.11)$$

The mean interaction energy $\frac{1}{2}(u'_+ + u'_-)$ is therefore precisely twice that in the ground state. There is a large splitting equal to

$$u'_+ - u'_- = \hbar\omega_0\left[(1+c)^{\frac{1}{2}} - (1-c)^{\frac{1}{2}}\right] \qquad (2.12)$$

For small c this is equal to

$$\mathcal{U}'_+ - \mathcal{U}'_- = \hbar\omega_0 \left[c + \frac{1}{8} c^3 + \dots \right] \tag{2.13}$$

which varies linearly with c at large separations R. This resonance energy would not be present if the interacting oscillators possessed different fundamental frequencies.

2.3 General Theory of Long-Range Forces [2-4]

Consider the interaction of molecule a in the electronic state m_a with molecule b in state m_b. We shall suppose that the effects of electron exchange are negligible, so this treatment will be limited to the long-range regime. The electrons may therefore be considered to be associated with a single molecule.

The Hamiltonian of the pair is

$$\mathcal{H} = \mathcal{H}^{(a)} + \mathcal{H}^{(b)} + \mathcal{H}' \tag{2.14}$$

where $\mathcal{H}^{(a)}$ and $\mathcal{H}^{(b)}$ are the clamped-nuclei Hamiltonians of the free molecules a and b and $\mathcal{H}'$ is the interaction. If we confine our considerations to Coulombic interactions then

$$\mathcal{H}' = (4\pi\epsilon_0)^{-1} \sum_{i,j} e_i^{(a)} e_j^{(b)} R_{ij}^{-1} \tag{2.15}$$

where R_{ij} is the distance between the ith charge $e_i^{(a)}$ (either a nucleus or an electron) in molecule a and the jth charge $e_j^{(b)}$ in molecule b. A general theory of long-range intermolecular forces is obtained by treating $\mathcal{H}'$ in equation (2.14) as a perturbation to $\mathcal{H}^{(a)} + \mathcal{H}^{(b)}$. The unperturbed wavefunctions are the eigenfunctions of $\mathcal{H}^{(a)} + \mathcal{H}^{(b)}$ and are simple products of the wavefunctions of the free molecules a and b. The perturbed wavefunction may be expressed in terms of these unperturbed states by standard quantum-mechanical theory and is

$$\Psi_{m_a m_b} = m_a m_b - \sum_{p_a, p_b}{}' \frac{\langle p_a p_b | \mathcal{H}' | m_a m_b \rangle}{W_{p_a} - W_{m_a} + W_{p_b} - W_{m_b}} p_a p_b + \dots \tag{2.16}$$

where the symbol $\sum'_{p_a,p_b}$ deonotes a sum over all the unperturbed states with the exception of $m_a m_b$. The first-order perturbed wavefunction therefore consists of $m_a m_b$ and small amounts of all other unperturbed states $p_a p_b$ that are mixed with $m_a m_b$ by the perturbation $\mathcal{H}'$; the extent of the admixture is equal to the off-diagonal matrix element $\langle p_a p_b|\mathcal{H}'|m_a m_b\rangle$ divided by the difference $-(W_{p_a}-W_{m_a}+W_{p_b}-W_{m_b})$ in the energies of the unperturbed states. The perturbed and unperturbed states are taken to be orthonormal. The energy of the pair of molecules in the perturbed state $\psi_{m_a m_b}$ is

$$W_{m_a m_b} = \langle \psi_{m_a m_b}|\mathcal{H}|\psi_{m_a m_b}\rangle$$

$$= W_{m_a} + W_{m_b} + \langle m_a m_b|\mathcal{H}'|m_a m_b\rangle - \sum_{p_a,p_b}{}' \frac{|\langle p_a p_b|\mathcal{H}'|m_a m_b\rangle|^2}{W_{p_a}-W_{m_a}+W_{p_b}-W_{m_b}} + \ldots \quad (2.17)$$

If $m_a m_b$ is degenerate and if this degeneracy is lifted by $\mathcal{H}'$, then the zero-order wavefunction $m_a m_b$ is chosen so that $\mathcal{H}'$ is diagonal. The first-order energy $\langle m_a m_b|\mathcal{H}'|m_a m_b\rangle$ is the unperturbed expectation value of $\mathcal{H}'$ and is the sum of the electrostatic energy $u_{\text{electrostatic}}$ and the resonance energy $u_{\text{resonance}}$:

$$u_{\text{electrostatic}} + u_{\text{resonance}} = \langle m_a m_b|\mathcal{H}'|m_a m_b\rangle \quad (2.18)$$

If $m_a m_b$ is non-degenerate or if $\mathcal{H}'$ does not lift the degeneracy to first order (and this is the normal situation) then $u_{\text{electrostatic}}$ is equal to $\langle m_a m_b|\mathcal{H}'|m_a m_b\rangle$ and it can be interpreted as the interaction of the unperturbed charge distributions of the two molecules.

The second-order energy in equation (2.17) may be separated into two distinct contributions. The first is $u_{\text{induction}}$ and consists of all those terms in which either $p_a = m_a$ with $p_b \neq m_b$, or $p_b = m_b$ with $p_a \neq m_a$:

$$u_{\text{induction}} = u^{(a)}_{\text{induction}} + u^{(b)}_{\text{induction}} \quad (2.19)$$

where

$$u^{(a)}_{\text{induction}} = -\sum_{p_a \neq m_a} \frac{|\langle p_a m_b|\mathcal{H}'|m_a m_b\rangle|^2}{W_{p_a} - W_{m_a}} \quad (2.20)$$

$$u^{(b)}_{\text{induction}} = -\sum_{p_b \neq m_b} \frac{|\langle m_a p_b | \mathcal{H}' | m_a m_b \rangle|^2}{W_{p_b} - W_{m_b}} \tag{2.21}$$

Thus $u^{(a)}_{\text{induction}}$ is due to the distortion of the electronic charge distribution of molecule a through interaction with the unperturbed molecule b. If m_a is the ground state then $W_{p_a} - W_{m_a} > 0$ for all p_a and $u^{(a)}_{\text{induction}}$ is negative; this is in accord with the variation principle which ensures that if a system in its ground state is given the freedom to distort it uses that freedom to lower its energy. However, if m_a were an excited electronic state, there could be exceptional cases in which $u^{(a)}_{\text{induction}}$ is positive.

The other contribution to the second-order energy is $u_{\text{dispersion}}$:

$$u_{\text{dispersion}} = -\sum_{\substack{p_a \neq m_a \\ p_b \neq m_b}} \frac{|\langle p_a p_b | \mathcal{H}' | m_a m_b \rangle|^2}{W_{p_a} - W_{m_a} + W_{p_b} - W_{m_b}} \tag{2.22}$$

and it is negative when both molecules a and b are in their ground states. The dispersion energy results from matrix elements of $\mathcal{H}'$ that are off-diagonal in the electronic eigenfunctions of both molecules and it is therefore attributable to the simultaneous fluctuations in the charge distributions in each molecule.

To proceed further we need to relate $u_{\text{electrostatic}}$, $u_{\text{induction}}$, and $u_{\text{dispersion}}$ to properties of the free molecules a and b in the states m_a and m_b. We could follow Longuet-Higgins [3] and employ the diagonal and off-diagonal elements of the charge density operator $\rho(\mathbf{r})$

$$\rho(\mathbf{r}) = e\left[\sum_{\nu} Z_\nu \, \delta(\mathbf{r} - \mathbf{r}_\nu) - \sum_i \delta(\mathbf{r} - \mathbf{r}_i)\right] \tag{2.23}$$

where the summation $\sum_i$ extends over all the i electrons of the system and $\sum_\nu$ over all the nuclei whose charges are Z_ν and positions $\mathbf{r}_\nu$.

$\delta(\mathbf{r} - \mathbf{r}_\nu)$ is the Dirac delta function which is zero unless $\mathbf{r} = \mathbf{r}_\nu$ and is such that

$$\int f(\underset{\sim}{r}') \, \delta(\underset{\sim}{r} - \underset{\sim}{r}') \, d\underset{\sim}{r}' = f(\underset{\sim}{r}) \tag{2.24}$$

The diagonal matrix elements of $\rho_a(\underset{\sim}{r})$, $\langle p_a | \rho_a(\underset{\sim}{r}) | p_a \rangle$, are the total charge density of molecule a in the particular electronic state p_a. This charge density is discontinuous at the nuclei because of the inclusion of the nuclear charges in $\rho(\underset{\sim}{r})$ in (2.23). The off-diagonal elements of $\rho_a(\underset{\sim}{r})$, $\langle p_a | \rho_a(\underset{\sim}{r}) | m_a \rangle$, may conveniently be called transition densities [3] and they are related to the change in charge density when the molecule undergoes the transition from state m_a to p_a.

The interaction Hamiltonian may be written in the form

$$\mathcal{H}' = (4\pi\epsilon_0)^{-1} \iint \frac{\rho_a(\underset{\sim}{r}_a) \, \rho_b(\underset{\sim}{r}_b)}{|\underset{\sim}{r}_a - \underset{\sim}{r}_b|} \, d\underset{\sim}{r}_a \, d\underset{\sim}{r}_b \tag{2.25}$$

The electrostatic energy for a system in the non-degenerate electronic state $m_a m_b$ is

$$u_{\text{electrostatic}} = (4\pi\epsilon_0)^{-1} \iint \frac{\langle m_a | \rho_a(\underset{\sim}{r}_a) | m_a \rangle \langle m_b | \rho_b(\underset{\sim}{r}_b) | m_b \rangle}{|\underset{\sim}{r}_a - \underset{\sim}{r}_b|} d\underset{\sim}{r}_a \, d\underset{\sim}{r}_b \tag{2.26}$$

and is therefore determined by the charge densities of the molecules in their unperturbed states. The second-order interaction energies involve the transition densities. For example

$$u^{(a)}_{\text{induction}} = -(4\pi\epsilon_0)^{-2} \iiiint \frac{\langle m_b | \rho_b(\underset{\sim}{r}_b) | m_b \rangle \langle m_b | \rho_b(\underset{\sim}{r}_b') | m_b \rangle \, \alpha(\underset{\sim}{r}_a, \underset{\sim}{r}_a') \, d\underset{\sim}{r}_a d\underset{\sim}{r}_a' d\underset{\sim}{r}_b d\underset{\sim}{r}_b'}{|\underset{\sim}{r}_a - \underset{\sim}{r}_b| \; |\underset{\sim}{r}_a' - \underset{\sim}{r}_b'|} \tag{2.27}$$

where

$$\alpha(\underset{\sim}{r}_a, \underset{\sim}{r}_a') = \sum_{p_a \neq m_a} \frac{\langle m_a | \rho_a(\underset{\sim}{r}_a) | p_a \rangle \langle p_a | \rho_a(\underset{\sim}{r}_a') | m_a \rangle}{W_{p_a} - W_{m_a}} \tag{2.28}$$

is a measure of the response of the charge density at $\mathbf{r}'_a$ in molecule a in state m_a to an infinitesimal change in the potential at $\mathbf{r}_a$. However, we know little about such molecular properties and shall not pursue this approach.

2.4 The Multipole Expansion of $\mathcal{H}'$

Consider the configuration of charges $e_i^{(a)}$ and $e_j^{(b)}$ in Figure 2.2. The interaction Hamiltonian is

$$\mathcal{H}' = (4\pi\epsilon_0)^{-1} \sum_{i,j} e_i^{(a)} e_j^{(b)} R_{ij}^{-1}$$

$$= \sum_i e_i^{(a)} \phi_i^{(a)} = \sum_j e_j^{(b)} \phi_j^{(b)} \tag{2.29}$$

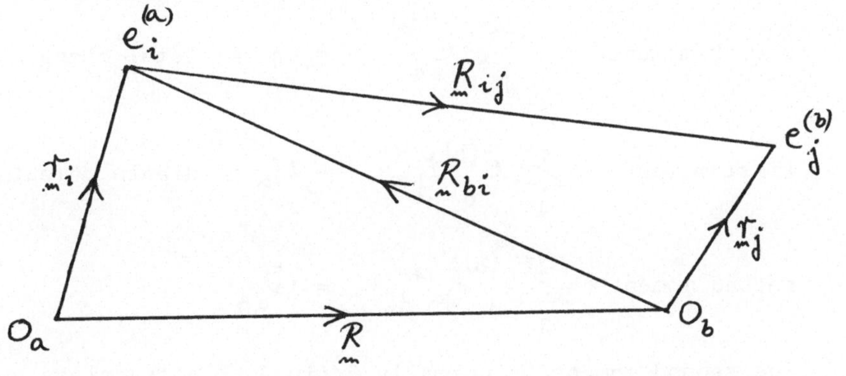

Figure 2.2. The positions of the point charges $e_i^{(a)}$ and $e_j^{(b)}$ in molecules a and b. O_a and O_b are origins fixed in molecules a and b.

where $\phi_j^{(b)} = (4\pi\epsilon_0)^{-1}\sum_i e_i^{(a)} R_{ij}^{-1}$ is the potential at the jth charge of molecule b due to all the charges in molecule a. This potential may be expanded in a Taylor series about the origin O_b fixed in molecule b (this origin would normally be the centre of mass of molecule b but it could be some other point, such as the electrical centre in HD which is midway between the two nuclei; the total interaction energy must be independent of any choice of origin):

$$\mathcal{H}' = \sum_j e_j^{(b)} \phi_j^{(b)}$$

$$= \sum_j e_j^{(b)} \left[\phi_0^{(b)} + (\nabla_\alpha \phi^{(b)})_0 r_{j\alpha} + \tfrac{1}{2} (\nabla_\alpha \nabla_\beta \phi^{(b)})_0 r_{j\alpha} r_{j\beta} + \ldots \right] \quad (2.30)$$

We have used the usual tensor summation convention in which a repeated tensor suffix implies a summation over all three components; for example

$$(\nabla_\alpha \phi)_0 r_{j\alpha} \equiv (\nabla_x \phi)_0 r_{jx} + (\nabla_y \phi)_0 r_{jy} + (\nabla_z \phi)_0 r_{jz} .$$

Equation (2.30) can be expressed in terms of the electric moment operators of molecule b:

$$\text{zeroth moment} = \sum_j e_j^{(b)} = q = \text{total change} \quad (2.31)$$

$$\text{first moment} = \sum_j e_j^{(b)} r_{j\alpha} = \mu_\alpha = \text{dipole moment} \quad (2.32)$$

$$\text{second moment} = \sum_j e_j^{(b)} r_{j\alpha} r_{j\beta} = Q_{\alpha\beta} \quad (2.33)$$

The second moment is normally defined as a traceless quantity and we introduce the quadrupole moment

$$\Theta_{\alpha\beta} = \sum_j e_j^{(b)} \left(\tfrac{3}{2} r_{j\alpha} r_{j\beta} - \tfrac{1}{2} r_j^2 \delta_{\alpha\beta} \right) = \tfrac{3}{2} Q_{\alpha\beta} - \tfrac{1}{2} Q_{\gamma\gamma} \delta_{\alpha\beta} \quad (2.34)$$

With these definitions for second moments, the expectation value of Q_{zz} is non-zero for a spherical atom whereas that of Θ_{zz} is zero. Equation (2.30) now becomes

$$\mathcal{H}' = q^{(b)}\phi_0^{(b)} - \mu_\alpha^{(b)} F_\alpha^{(b)} - \frac{1}{3}\Theta_{\alpha\beta}^{(b)} F_{\alpha\beta}^{(b)} - \dots$$

$$- \frac{1}{1.3.5\dots(2n-1)} \xi^{(n)(b)}_{\alpha\beta\dots\nu} F^{(b)}_{\alpha\beta\dots\nu} - \dots \qquad (2.35)$$

where the nth moment is

$$\xi^{(n)(b)}_{\alpha\beta\dots\nu} = (-1)^n (n!)^{-1} \sum_j e_j r_j^{2n+1} \frac{\partial}{\partial r_{j\alpha}} \frac{\partial}{\partial r_{j\beta}} \cdots \frac{\partial}{\partial r_{j\nu}} r_j^{-1} \qquad (2.36)$$

$F_\alpha^{(b)} = -(\nabla_\alpha \phi^{(b)})_0$ is the electric field strength at the origin O_b due to the instantaneous charge distribution of molecule a; $F_{\alpha\beta} = \nabla_\alpha F_\beta = -(\nabla_\alpha \nabla_\beta \phi^{(b)})_0$ is the field-gradient at O_b, and so on.

The moment operators $\xi^{(n)}$ are symmetric with respect to interchange of any pair of suffixes (for example $\Theta_{\alpha\beta} = \Theta_{\beta\alpha}$) and they are reduced to zero on contraction (for example $\xi^{(n)}_{\alpha\alpha\dots\nu} = 0$) since $\nabla_\alpha \nabla_\alpha r^{-1}$ is zero if $r \neq 0$. Only the first non-vanishing moment is independent of the choice of origin. For example, the dipole moment of an ion varies linearly with the origin (see Figure 2.3). Thus on moving the origin through r' from O to O′, $r'_j = r_j - r'$ and $\mu' = \mu$ if and only if $q = 0$:

$$\mu'_\alpha = \sum_j e_j r'_{j\alpha} = \sum_j e_j (r_{j\alpha} - r'_\alpha) = \mu_\alpha - q r'_\alpha \qquad (2.37)$$

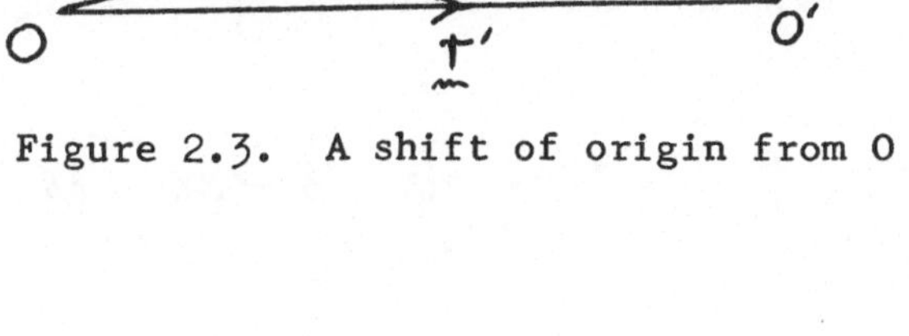

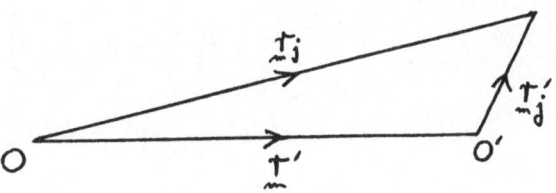

Figure 2.3. A shift of origin from O to O′.

To complete the multipolar expression for $\mathcal{H}'$ it is necessary to relate $\phi_0^{(b)}$ to the electric moments of molecule a. This may be achieved by expanding R_{bi} in Figure 2.2 as a Taylor series:

$$\phi_0^{(b)} = (4\pi\epsilon_0)^{-1} \sum_i e_i^{(a)} R_{bi}^{-1}$$

$$= (4\pi\epsilon_0)^{-1} \sum_i e_i^{(a)} \left[R^{-1} + \left(\frac{\partial}{\partial t_{i\alpha}} R_{bi}^{-1} \right)_0 t_{i\alpha} + \frac{1}{2} \left(\frac{\partial}{\partial t_{i\alpha}} \frac{\partial}{\partial t_{i\beta}} R_{bi}^{-1} \right)_0 t_{i\alpha} t_{i\beta} + \ldots \right]$$

$$= (4\pi\epsilon_0)^{-1} \sum_i e_i^{(a)} \left[R^{-1} - t_{i\alpha} \nabla_\alpha R^{-1} + \frac{1}{2} t_{i\alpha} t_{i\beta} \nabla_\alpha \nabla_\beta R^{-1} - \ldots \right]$$

$$= q^{(a)} T - \mu_\alpha^{(a)} T_\alpha + \frac{1}{3} \Theta_{\alpha\beta}^{(a)} T_{\alpha\beta} - \ldots + (-1)^n \frac{1}{1.3.5\ldots(2n-1)} \xi_{\alpha\beta\ldots\nu}^{(n)(a)} T_{\alpha\beta\ldots\nu} + \ldots \quad (2.38)$$

where $T_{\alpha\beta\ldots\nu} = (4\pi\epsilon_0)^{-1} \nabla_\alpha \nabla_\beta \ldots \nabla_\nu R^{-1}$ is a tensor which is proportional to $R^{-(n+1)}$, is symmetric with respect to interchange of any pair of suffixes, and is reduced to zero on contraction (that is, $T_{\alpha\alpha\ldots\nu} = 0$). The first few elements of the T-tensor are:

$$T = (4\pi\epsilon_0)^{-1} R^{-1} \quad (2.39)$$

$$T_\alpha = -(4\pi\epsilon_0)^{-1} R_\alpha R^{-3} \quad (2.40)$$

$$T_{\alpha\beta} = (4\pi\epsilon_0)^{-1} \left(3 R_\alpha R_\beta - R^2 \delta_{\alpha\beta} \right) R^{-5} \quad (2.41)$$

$$T_{\alpha\beta\gamma} = -3(4\pi\epsilon_0)^{-1} \left[5 R_\alpha R_\beta R_\gamma - R^2 \left(R_\alpha \delta_{\beta\gamma} + R_\beta \delta_{\gamma\alpha} + R_\gamma \delta_{\alpha\beta} \right) \right] R^{-7} \quad (2.42)$$

The electric field $\underset{\sim}{F}^{(b)}$ is minus the gradient of $\phi^{(b)}$ at O_b,

$$F_\alpha^{(b)} = -\nabla_\alpha \phi_0^{(b)} = -q^{(a)} T_\alpha + \mu_\beta^{(a)} T_{\alpha\beta} - \frac{1}{3} \Theta_{\beta\gamma}^{(a)} T_{\alpha\beta\gamma} + \ldots \quad (2.43)$$

$$F_{\alpha\beta} = -\nabla_\alpha \nabla_\beta \phi_0^{(b)} = -q^{(a)} T_{\alpha\beta} + \mu_\gamma^{(a)} T_{\alpha\beta\gamma} - \frac{1}{3} \Theta_{\gamma\delta}^{(a)} T_{\alpha\beta\gamma\delta} + \ldots \quad (2.44)$$

Combining equations (2.35) and (2.38) gives the multipolar expression for $\mathcal{H}'$:

$$\mathcal{H}' = T q^{(a)} q^{(b)} + T_\alpha \left(q^{(a)} \mu_\alpha^{(b)} - q^{(b)} \mu_\alpha^{(a)} \right) + T_{\alpha\beta} \left(\frac{1}{3} q^{(a)} \Theta_{\alpha\beta}^{(b)} + \frac{1}{3} q^{(b)} \Theta_{\alpha\beta}^{(a)} - \mu_\alpha^{(a)} \mu_\beta^{(b)} \right) + \ldots \quad (2.45)$$

$$= \sum_{n=0}^{\infty} \sum_{n'=0}^{\infty} (-1)^{n'} \frac{1}{1.3.5\ldots(2n-1)} \frac{1}{1.3.5\ldots(2n'-1)} T_{\alpha\beta\ldots\nu\alpha'\beta'\ldots\nu'} \xi_{\alpha\beta\ldots\nu}^{(n)(b)} \xi_{\alpha'\beta'\ldots\nu'}^{(n')(a)} \quad (2.46)$$

A movement of the origins O_a and O_b changes each term in $\mathcal{H}'$ but leaves $\mathcal{H}'$ invariant. The multipole expansion is accurate provided the intermolecular distance R is large compared to the dimensions of each molecule. In most applications one retains only the first few non-vanishing contributions and one relies on the correctness of the asymptotic behaviour of the series. Kreek, Pan and Meath [5] have discussed the validity of the expansion in R^{-2} of the dispersion energy of atoms; they have shown that outisde the region of validity of the expansion, inclusion of more terms leads to a poorer approximation.

References

[1] H.C. Longuet-Higgins, Discussions Faraday Soc., 40, 7 (1965).

[2] J.S. Dahler and J.O. Hirschfelder, J. Chem. Phys., 25, 986 (1956).

[3] H.C. Longuet-Higgins, Proc. Roy. Soc. A, 235, 537 (1956).

[4] A.D. Buckingham, Adv. Chem. Phys., 12, 107 (1967).

[5] H. Kreek, Y.H. Pan and W.J. Meath, Molecular Physics, 19, 513 (1970).

LECTURE 3

THE EVALUATION OF LONG-RANGE INTERACTION ENERGIES

3.1 Introduction

In Lecture 2 we formulated a general theory of the long-range interaction energy of two molecules. Perturbation theory was used to express the electrostatic, resonance, induction and dispersion energies in terms of the matrix elements of the interaction Hamiltonian $\mathcal{H}'$ for ground and excited states of the free molecules. In this lecture we shall use the multipole expansion of $\mathcal{H}'$ (equations (2.45) and (2.46)) to relate these general expressions to observable properties of the free molecules.

3.2 The Electrostatic Energy

If the electronic states m_a and m_b of molecules a and b are non-degenerate, the electrostatic interaction energy is given by equation (2.18):

$$u_{\text{electrostatic}} = \langle m_a m_b | \mathcal{H}' | m_a m_b \rangle \tag{3.1}$$

If the multipole expansion (2.46) for $\mathcal{H}'$ is inserted into (3.1), we obtain

$$u_{\text{electrostatic}} = \sum_{n=0}^{\infty} \sum_{n'=0}^{\infty} (-1)^{n'} \frac{2^{n+n'} n!\, n'!}{(2n)!\, (2n')!} T_{\alpha\beta\ldots\nu\alpha'\beta'\ldots\nu'} \xi^{(n)(m_b)}_{\alpha\beta\ldots\nu} \xi^{(n')(m_a)}_{\alpha'\beta'\ldots\nu'} \tag{3.2}$$

where

$$\xi^{(n')(m_a)}_{\alpha'\beta'\ldots\nu'} = \langle m_a | \xi^{(n')(a)}_{\alpha'\beta'\ldots\nu'} | m_a \rangle \tag{3.3}$$

is the n'th permanent multipole moment of molecule a in the state m_a. For n = 0, 1, and 2 (that is, for the charge, dipole and quadrupole) there are sound experimental techniques for determining these moments for molecules in their ground electronic states [1]. The charge $q^{(a)}$ is independent of state. The dipole of a molecule in an excited electronic state may be measured through the optical Stark effect [2]. The first few terms of (3.2) are

$$
\begin{aligned}
u_{\text{electrostatic}} = {} & q^{(a)}\left[T q^{(b)} + T_{\alpha}\mu_{\alpha}^{(m_b)} + \tfrac{1}{3}T_{\alpha\beta}\Theta_{\alpha\beta}^{(m_b)} + \tfrac{1}{15}T_{\alpha\beta\gamma}\Omega_{\alpha\beta\gamma}^{(m_b)} + \ldots\right] \\
& - \mu_{\alpha}^{(m_a)}\left[T_{\alpha} q^{(b)} + T_{\alpha\beta}\mu_{\beta}^{(m_b)} + \tfrac{1}{3}T_{\alpha\beta\gamma}\Theta_{\beta\gamma}^{(m_b)} + \tfrac{1}{15}T_{\alpha\beta\gamma\delta}\Omega_{\beta\gamma\delta}^{(m_b)} + \ldots\right] \\
& + \tfrac{1}{3}\Theta_{\alpha\beta}^{(m_a)}\left[T_{\alpha\beta} q^{(b)} + T_{\alpha\beta\gamma}\mu_{\gamma}^{(m_b)} + \tfrac{1}{3}T_{\alpha\beta\gamma\delta}\Theta_{\gamma\delta}^{(m_b)} + \tfrac{1}{15}T_{\alpha\beta\gamma\delta\epsilon}\Omega_{\gamma\delta\epsilon}^{(m_b)} + \ldots\right] \\
& - \tfrac{1}{15}\Omega_{\alpha\beta\gamma}^{(m_a)}\left[T_{\alpha\beta\gamma} q^{(b)} + T_{\alpha\beta\gamma\delta}\mu_{\delta}^{(m_b)} + \tfrac{1}{3}T_{\alpha\beta\gamma\delta\epsilon}\Theta_{\delta\epsilon}^{(m_b)} + \tfrac{1}{15}T_{\alpha\beta\gamma\delta\epsilon\phi}\Omega_{\delta\epsilon\phi}^{(m_b)} + \ldots\right] \\
& + \ldots
\end{aligned}
\tag{3.4}
$$

where $\Omega_{\alpha\beta\gamma} \equiv \xi_{\alpha\beta\gamma}^{(3)}$ is the octopole moment.

If the molecules are axially symmetric there is at most one independent moment of any order [3]. The nth multipole moment is defined as the expectation value of $\xi_{zz\ldots z}^{(n)}$ where the z axis is the molecular axis of symmetry. The electrostatic energy (3.4) can then be written in terms of the relative orientations of a and b (see Figure 3.1).

$$
\begin{aligned}
u_{\text{electrostatic}} = {} & (4\pi\epsilon_0)^{-1} q^{(a)}\Big[q^{(b)}R^{-1} + \mu^{(m_b)}\cos\theta_b R^{-2} + \Theta^{(m_b)}\left(\tfrac{3}{2}\cos^2\theta_b - \tfrac{1}{2}\right)R^{-3} \\
& \qquad + \Omega^{(m_b)}\left(\tfrac{5}{2}\cos^3\theta_b - \tfrac{3}{2}\cos\theta_b\right)R^{-4} + \ldots\Big] \\
& + (4\pi\epsilon_0)^{-1}\mu^{(m_a)}\Big[q^{(b)}\cos\theta_a R^{-2} + \mu^{(m_b)}(2\cos\theta_a\cos\theta_b + \sin\theta_a\sin\theta_b\cos\phi)R^{-3} \\
& \qquad + 3\Theta^{(m_b)}\left\{\cos\theta_a\left(\tfrac{3}{2}\cos^2\theta_b - \tfrac{1}{2}\right) + \sin\theta_a\sin\theta_b\cos\theta_b\cos\phi\right\}R^{-4} + \ldots\Big] \\
& + (4\pi\epsilon_0)^{-1}\Theta^{(m_a)}\Big[q^{(b)}\left(\tfrac{3}{2}\cos^2\theta_a - \tfrac{1}{2}\right)R^{-3} + 3\mu^{(m_b)}\Big\{\cos\theta_b\left(\tfrac{3}{2}\cos^2\theta_a - \tfrac{1}{2}\right) + \\
& \qquad \sin\theta_a\cos\theta_a\sin\theta_b\cos\phi\Big\}R^{-4} + \tfrac{3}{4}\Theta^{(m_b)}\Big\{1 - 5\cos^2\theta_a - 5\cos^2\theta_b + \\
& \qquad 17\cos^2\theta_a\cos^2\theta_b + 2\sin^2\theta_a\sin^2\theta_b\cos^2\phi + \\
& \qquad 16\sin\theta_a\cos\theta_a\sin\theta_b\cos\theta_b\cos\phi\Big\}R^{-5} + \ldots\Big] \\
& + \ldots
\end{aligned}
\tag{3.5}
$$

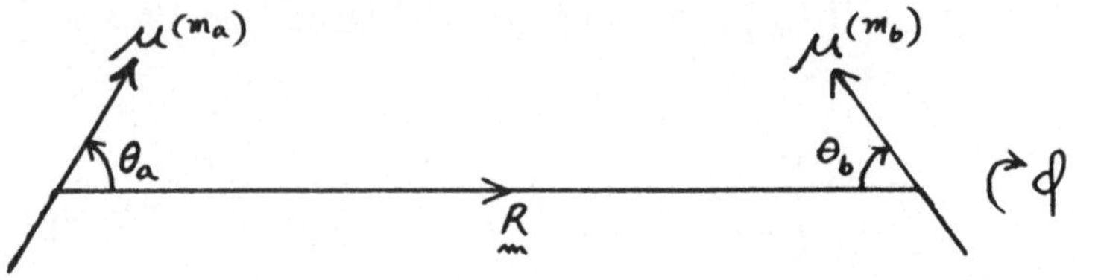

Figure 3.1. The relative orientations of molecules a and b. The angle ϕ is the angle between the planes formed by the two molecular axes with the line of centres $\underset{\sim}{R}$.

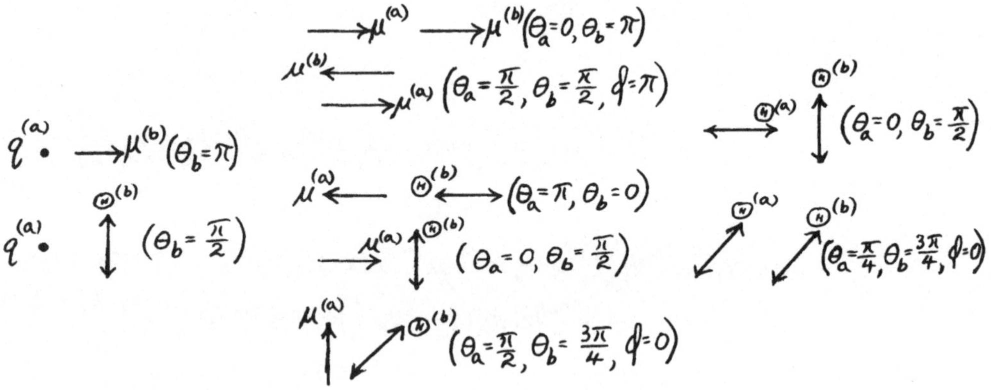

Figure 3.2. Favourable relative orientations for a pair of linear molecules with charge-dipole, charge-quadrupole, dipole-dipole, dipole-quadrupole and quadrupole-quadrupole interactions. All the electric moments q, μ and Θ are taken to be positive.

3.3 Molecules in External Fields - Polarizability

The induction and dispersion energies result from the second-order perturbed energy and therefore depend on off-diagonal matrix elements of $\mathcal{H}'$ (see equations (2.20) and (2.22)). When the multipole expansion of $\mathcal{H}'$ is used, the longest-range contributions involve the square of the transition dipole moment, since the total charges $q^{(a)}$ and $q^{(b)}$ are constant and therefore possess no off-diagonal elements (that is, there are no fluctuations in the total charge of a molecule in long-range interaction with another). Transition dipoles also determine the intensities of allowed spectral transitions as well as the polarizability $\underset{\sim}{\alpha}$ which is a measure of the dipole induced in a molecule by an external field of unit strength. Polarizabilities are readily measured through the refractive index.

For a single molecule fixed in orientation in a uniform external field $\underset{\sim}{F}$, the Hamiltonian is

$$\mathcal{H}(\underset{\sim}{F}) = \mathcal{H} - \mu_\alpha F_\alpha \tag{3.6}$$

and the energy can be written as a power series in $\underset{\sim}{F}$ [4]:

$$W_m(\underset{\sim}{F}) = W_m - \mu_\alpha^{(m)} F_\alpha - \frac{1}{2}\alpha_{\alpha\beta}^{(m)} F_\alpha F_\beta - \frac{1}{6}\beta_{\alpha\beta\gamma}^{(m)} F_\alpha F_\beta F_\gamma - \ldots \tag{3.7}$$

where $\mu_\alpha^{(m)}$ is the permanent dipole moment, $\alpha_{\alpha\beta}^{(m)}$ the polarizability and $\beta_{\alpha\beta\gamma}^{(m)}$ the first hyperpolarizability of the molecule in the state m. Perturbation theory leads to the following expressions for $\mu_\alpha^{(m)}$, $\alpha_{\alpha\beta}^{(m)}$ and $\beta_{\alpha\beta\gamma}^{(m)}$:

$$\mu_\alpha^{(m)} = \langle m|\mu_\alpha|m\rangle \tag{3.8}$$

$$\alpha_{\alpha\beta}^{(m)} = \sum_{p\neq m} \frac{\langle m|\mu_\alpha|p\rangle\langle p|\mu_\beta|m\rangle + \langle m|\mu_\beta|p\rangle\langle p|\mu_\alpha|m\rangle}{W_p - W_m} = \alpha_{\beta\alpha}^{(m)} \tag{3.9}$$

$$\begin{aligned}\beta_{\alpha\beta\gamma}^{(m)} &= S(\alpha\beta\gamma)\sum_{p\neq m}\left[\sum_{q\neq m}\frac{\langle m|\mu_\alpha|p\rangle\langle p|\mu_\beta|q\rangle\langle q|\mu_\gamma|m\rangle}{(W_p-W_m)(W_q-W_m)} - \frac{\langle m|\mu_\alpha|m\rangle\langle m|\mu_\beta|p\rangle\langle p|\mu_\gamma|m\rangle}{(W_p-W_m)^2}\right]\\ &= \beta_{\alpha\gamma\beta}^{(m)} = \beta_{\beta\gamma\alpha}^{(m)} = \beta_{\beta\alpha\gamma}^{(m)} = \beta_{\gamma\alpha\beta}^{(m)} = \beta_{\gamma\beta\alpha}^{(m)}\end{aligned} \tag{3.10}$$

$S(\alpha\beta\gamma)$ implies a sum of all permutations of the tensor components, thus $S(\alpha\beta\gamma)X_{\alpha\beta\gamma} = X_{\alpha\beta\gamma} + X_{\alpha\gamma\beta} + X_{\beta\gamma\alpha} + X_{\beta\alpha\gamma} + X_{\gamma\alpha\beta} + X_{\gamma\beta\alpha}$. The static polarizability and hyperpolarizabilities are symmetric with respect to interchange of any pair of suffixes. The dipole moment in the field is obtained by differentiating $W_m(\underset{\sim}{F})$ with respect to $\underset{\sim}{F}$:

$$\mu_\alpha^{(m)}(\underset{\sim}{F}) = -\frac{\partial W_m(\underset{\sim}{F})}{\partial F_\alpha} = \mu_\alpha^{(m)} + \alpha_{\alpha\beta}^{(m)} F_\beta + \frac{1}{2}\beta_{\alpha\beta\gamma}^{(m)} F_\beta F_\gamma + \ldots \quad (3.11)$$

Thus $\underset{\sim}{\alpha}^{(m)}$ determines the linear polarization of a molecule by an external field, and $\underset{\sim}{\beta}^{(m)}$ the initial departure from a linear polarization law. If the molecule is centrosymmetric, the energy is unaffected by reversing $\underset{\sim}{F}$ so all odd powers of $\underset{\sim}{F}$ in (3.7) then vanish.

Typical values for these parameters in a small polar molecule are

$\mu \approx 1\text{D}$	$= 10^{-18}$ e.s.u.	$= 3.33564 \times 10^{-30}$ C m	
$(4\pi\epsilon_0)^{-1}\alpha \approx 1\text{Å}^3$	$= 10^{-24}$ e.s.u.	$=$	$10^{-30}\,\text{m}^3$
$(4\pi\epsilon_0)^{-2}\beta$	$\approx \pm 10^{-30}$ e.s.u.	$= \pm 0.29979 \times 10^{-30}\,\text{C}^{-1}\,\text{m}^5$	

The quantity $(4\pi\epsilon_0)^{-1}\alpha$ may be compared to the volume of the molecule. The various induced dipoles in equation (3.11) are of the same order of magnitude when the field strength $F \approx 1$ atomic unit $= ea_0^{-2}(4\pi\epsilon_0)^{-1}$ $= 1.7153 \times 10^7$ e.s.u. $= 5.1423 \times 10^{11}$ V m^{-1}.

When the external field is non-uniform, as in the vicinity of a polar molecule, the Hamiltonian becomes $\mathcal{H} - \mu_\alpha F_\alpha - \frac{1}{3}\Theta_{\alpha\beta}F_{\alpha\beta} - \ldots$ and the energy may be expressed as a power series in F_α, $F_{\alpha\beta}$, etc.:

$$W_m(\underset{\sim}{F}) = W_m - \mu_\alpha^{(m)} F_\alpha - \frac{1}{3}\Theta_{\alpha\beta}^{(m)} F_{\alpha\beta} - \frac{1}{15}\Omega_{\alpha\beta\gamma}^{(m)} F_{\alpha\beta\gamma} - \ldots$$
$$- \frac{1}{2}\alpha_{\alpha\beta}^{(m)} F_\alpha F_\beta - \frac{1}{3}A_{\alpha\beta\gamma}^{(m)} F_\alpha F_{\beta\gamma} - \frac{1}{6}C_{\alpha\beta\gamma\delta}^{(m)} F_{\alpha\beta} F_{\gamma\delta} - \ldots \quad (3.12)$$

Only those contributions that depend upon linear distortion of the molecule by the field and field-gradient have been retained. The dipole and quadrupole moments of the molecule in the field are

$$\mu_\alpha^{(m)}(\underset{\sim}{F}) = -\frac{\partial W_m(\underset{\sim}{F})}{\partial F_\alpha} = \mu_\alpha^{(m)} + \alpha_{\alpha\beta}^{(m)} F_\beta + \frac{1}{3}A_{\alpha\beta\gamma}^{(m)} F_{\beta\gamma} + \ldots \quad (3.13)$$

$$\Theta_{\alpha\beta}^{(m)}(\underset{\sim}{F}) = -3\frac{\partial W_m(\underset{\sim}{F})}{\partial F_{\alpha\beta}} = \Theta_{\alpha\beta}^{(m)} + A_{\gamma\alpha\beta}^{(m)} F_\gamma + C_{\alpha\beta\gamma\delta}^{(m)} F_{\gamma\delta} + \ldots \quad (3.14)$$

$\underset{\sim}{C}$ is the quadrupole polarizability; atoms and molecules of all symmetries possess this property. The dipole-quadrupole polarizability $\underset{\sim}{A}$ determines both the dipole induced by a field-gradient and the quadrupole induced by a field - it is origin-dependent and vanishes for centrosymmetric molecules if the centre is chosen as the origin. The higher polarizabilities $\underset{\sim}{A}$, $\underset{\sim}{C}$, etc. may be thought of as describing the distribution of polarizable matter in the molecule, just as $\underset{\sim}{\mu}$, $\underset{\sim}{\Theta}$, etc. describe the distribution of charge. Perturbation theory provides the following equations for $\underset{\sim}{A}$ and $\underset{\sim}{C}$:

$$A^{(m)}_{\alpha\beta\gamma} = \sum_{p\neq m} \frac{\langle m|\mu_\alpha|p\rangle\langle p|\Theta_{\beta\gamma}|m\rangle + \langle m|\Theta_{\beta\gamma}|p\rangle\langle p|\mu_\alpha|m\rangle}{W_p - W_m} = A^{(m)}_{\alpha\gamma\beta} \tag{3.15}$$

$$C^{(m)}_{\alpha\beta\gamma\delta} = \frac{1}{3}\sum_{p\neq m} \frac{\langle m|\Theta_{\alpha\beta}|p\rangle\langle p|\Theta_{\gamma\delta}|m\rangle + \langle m|\Theta_{\gamma\delta}|p\rangle\langle p|\Theta_{\alpha\beta}|m\rangle}{W_p - W_m} = C^{(m)}_{\gamma\delta\alpha\beta} \tag{3.16}$$

On inserting equations (2.37) into (3.9), (3.10) and (3.15) it may be shown that on moving the origin through $\underset{\sim}{r}'$ from 0 to 0',

$$\alpha'_{\alpha\beta} = \alpha_{\alpha\beta}, \qquad \beta'_{\alpha\beta\gamma} = \beta_{\alpha\beta\gamma} \tag{3.17}$$

$$A'_{\alpha\beta\gamma} = A_{\alpha\beta\gamma} - \frac{3}{2} r'_\beta \alpha_{\gamma\alpha} - \frac{3}{2} r'_\gamma \alpha_{\alpha\beta} + r'_\delta \alpha_{\delta\alpha}\delta_{\beta\gamma} \tag{3.18}$$

The number of independent components of the various polarizabilities for all appropriate symmetries has been evaluated [5]. For a tetrahedron, such as CH_4, there is just one α (so that $\alpha_{\alpha\beta} = \alpha\,\delta_{\alpha\beta}$) and one A which is $A_{xyz} = A_{xzy} = A_{yxz} = A_{yzx} = A_{zxy} = A_{zyx}$ where the x,y,z axes are perpendicular to the faces of a cube in which the corners of the tetrahedron are at the points (1,1,1), (1,-1,-1), (-1,1,-1), (-1,-1,1) (see Figure 3.3). A_{xyz} is independent of the origin in a tetrahedron, as may be confirmed from (3.18). For a model in which the tetrahedral molecule is comprised of four anisotropically polarizable bonds

$$A = A_{xyz} = \frac{4}{\sqrt{3}}\left(\alpha_{\parallel} - \alpha_{\perp}\right) R_0 \tag{3.19}$$

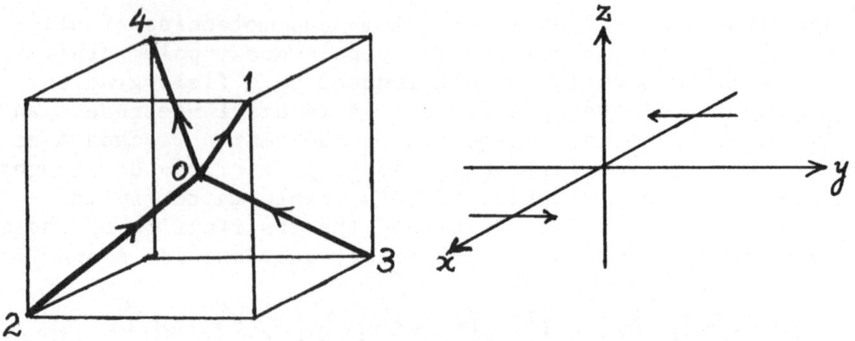

Figure 3.3. A tetrahedron enclosed in a cube. In the field-gradient F_{xy} illustrated on the right, the anisotropic polarizability along each of the four bonds of the tetrahedron contributes positively to the induced dipole μ_z. Bonds 1 and 2 have positive displacement along the x axis and therefore experience positive fields F_y, while 3 and 4 experience a negative F_y.

where $\alpha_{\parallel}-\alpha_{\perp}$ is the anisotropy of a bond and R_o is the distance of this anisotropic group from the centre. Thus for a small tetrahedral molecule $(4\pi\epsilon_0)^{-1}A \sim 1$ Å^4. For CH_4, $(4\pi\epsilon_0)^{-1}|A|$ has been estimated to be $2.35 \times 10^{-40} m^4$ [6].

3.4 Molecules in Time-Dependent Fields

If the external field is time-dependent, the molecule undergoes a time-dependent distortion which can be described by frequency-dependent polarizabilities. If the external field oscillates at an angular frequency ω

$$F_\beta = F_\beta^{(0)} \exp(-i\omega t) \tag{3.20}$$

then the complex induced dipole proportional to F_β is

$$\mu_\alpha = \tilde{\alpha}^{(m)}_{\alpha\beta}(\omega)\, F_\beta \tag{3.21}$$

From time-dependent perturbation theory [7,4]

$$\tilde{\alpha}_{\alpha\beta}^{(m)}(\omega) = \sum_{p} \frac{\omega_{pm}\left[\langle m|\mu_\alpha|p\rangle\langle p|\mu_\beta|m\rangle + \langle m|\mu_\beta|p\rangle\langle p|\mu_\alpha|m\rangle\right]}{\hbar\left(\omega_{pm}^2 - \omega^2 - i\omega\Gamma_{pm}\right)} \tag{3.22}$$

where $\hbar\omega_{pm} = W_p - W_m$ and Γ_{pm} is the width at half the maximum height of the absorption centred at ω_{pm}. The real part of $\tilde{\alpha}_{\alpha\beta}^{(m)}(\omega)$ determines the optical refractivity of the system and the imaginary part the absorption coefficient. If the molecules have an electronic angular momentum (as in a sodium atom in its ^2S ground state), or if they are in an external magnetostatic field, then there is an anti-symmetric contribution to $\tilde{\alpha}_{\alpha\beta}^{(m)}(\omega)$ depending on the imaginary part of $\langle m|\mu_\alpha|p\rangle\langle p|\mu_\beta|m\rangle$. This exists if the molecules lack symmetry under time reversal [5]:

$$\tilde{\alpha}_{\alpha\beta}^{(m)}(\omega) = \alpha_{\alpha\beta}^{(m)}(\omega) - i\alpha_{\alpha\beta}'^{(m)}(\omega) = \alpha_{\beta\alpha}^{(m)}(\omega) + i\alpha_{\beta\alpha}'^{(m)}(\omega) \tag{3.23}$$

3.5 The Induction Energy

The induction energy results from distortion of the electron clouds by the electric field and field-gradients due to the charge distribution of neighbouring molecules. It is therefore dependent upon static molecular polarizabilities. For the pair of molecules a and b in the unperturbed states m_a and m_b

$$\mathcal{U}_{induction}^{(b)} = -\frac{1}{2}\alpha_{\alpha\beta}^{(m_b)}F_\alpha^{(b)}F_\beta^{(b)} - \frac{1}{3}A_{\alpha\beta\gamma}^{(m_b)}F_\alpha^{(b)}F_{\beta\gamma}^{(b)} - \frac{1}{6}C_{\alpha\beta\gamma\delta}^{(m_b)}F_{\alpha\beta}^{(b)}F_{\gamma\delta}^{(b)} \cdots \tag{3.24}$$

where $F_\alpha^{(b)}$ and $F_{\alpha\beta}^{(b)}$ are the field and field-gradient at the origin in molecule b due to the permanent charge distribution of molecule a in state m_a. They may be expressed in terms of the permanent multipole moments of a through equations (2.43) and (2.44).

The induction energy is easy to evaluate, since molecular static polarizabilities are readily accessible. As noted in Lecture 1, it must be negative when m_b is the ground state of b. But the induction energy is rarely predominant, and in many instances it can be neglected in comparison to the electrostatic and dispersion energies.

3.6 The Dispersion Energy

Unlike the induction energy, the dispersion energy does not separate easily into single-centre contributions. If precise formulae are required it is necessary to consider the frequency-dependent polarizabilities, or rather the polarizabilities at imaginary frequencies. Their relationship to the dispersion energy rests on the identities (for $A>0$, $B>0$)

$$\frac{1}{A+B} = \frac{2}{\pi}\int_0^\infty \frac{AB}{(A^2+u^2)(B^2+u^2)}\,du = \frac{2}{\pi}\int_0^\infty \frac{u^2}{(A^2+u^2)(B^2+u^2)}\,du \tag{3.25}$$

The polarizability at the imaginary frequency $\omega = iu$ is

$$\tilde{\alpha}^{(m)}_{\alpha\beta}(iu) = \sum_p \frac{2\omega_{pm}\,\mathrm{Re}\{\langle m|\mu_\alpha|p\rangle\langle p|\mu_\beta|m\rangle\} - 2u\,\mathrm{Im}\{\langle m|\mu_\alpha|p\rangle\langle p|\mu_\beta|m\rangle\}}{\hbar(\omega_{pm}^2+u^2)}$$

$$= \alpha^{(m)}_{\alpha\beta}(iu) - i\alpha'^{(m)}_{\alpha\beta}(iu) = \alpha^{(m)}_{\beta\alpha}(iu) + i\alpha'^{(m)}_{\beta\alpha}(iu) \tag{3.26}$$

The dependence of $\tilde{\alpha}^{(m)}_{\alpha\beta}(iu)$ on the line width Γ_{pm} has been dropped since there are no poles in $\tilde{\alpha}^{(m)}_{\alpha\beta}(iu)$; actually $\alpha^{(m)}_{\alpha\beta}(iu)$ is a positive, monotonically decreasing function of u. The singular behaviour of $\tilde{\alpha}^{(m)}_{\alpha\beta}(\omega)$ at $\omega = \omega_{pm}$ is removed on going to imaginary frequencies.

From (2.22) and (2.45) the dipole dispersion energy is

$$u_{\mathrm{dispersion}}(R^{-6}) = -T_{\alpha\beta}T_{\gamma\delta}\sum_{\substack{p_a\neq m_a\\ p_b\neq m_b}} \frac{\langle m_a m_b|\mu_\alpha^{(a)}\mu_\beta^{(b)}|p_a p_b\rangle\langle p_a p_b|\mu_\gamma^{(a)}\mu_\delta^{(b)}|m_a m_b\rangle}{\hbar(\omega_{p_a m_a}+\omega_{p_b m_b})} \tag{3.27}$$

Applying to (3.27) the identities (3.25) yields

$$u_{\mathrm{dispersion}}(R^{-6}) = -\frac{\hbar}{2\pi}T_{\alpha\beta}T_{\gamma\delta}\int_0^\infty \tilde{\alpha}^{(m_a)}_{\alpha\gamma}(iu)^*\,\tilde{\alpha}^{(m_b)}_{\beta\delta}(iu)\,du \tag{3.28}$$

which has achieved the desired separation. If either molecule is symmetric under time-reversal, the α' contribution to (3.28) vanishes; it is small and negative for a pair of alkali metal atoms having the same spin and is positive when the spins are opposed. If the molecules are isotropically polarizable, so that (1.3) applies at all frequencies, (3.28) simplifies to

$$u_{dispersion}(R^{-6}) = -(4\pi\epsilon_0)^{-2}\,\frac{3\hbar}{\pi}R^{-6}\int_0^\infty \alpha^{(m_a)}(iu)\,\alpha^{(m_b)}(iu)\,du \qquad (3.29)$$

This is the expression that has been applied to the inert gases and other simple systems, using either variational calculations of the frequency-dependent polarizabilities or semi-empirical methods employing observed frequencies and intensities together with sum rules [8,9]. Very reliable values are available for C_6, the coefficient of R^{-6} in the dispersion energy [9] and bounds have been imposed on C_6 [10]. The higher coefficients C_7, C_8, etc. in the dispersion energy can similarly be expressed in terms of integrals over all imaginary frequencies of products of the appropriate dipole or higher polarizabilities of each molecule.

It is clear that a rigorous description of dispersion energy could not be based on static polarizabilities, for the interaction results from a correlation in the rapidly fluctuating electronic coordinates. However, London [11] provided a very useful approximate expression relating $u_{dispersion}$ to the product of the static polarizabilities of each molecule. His formula will be considered in the next Lecture.

References

[1] A.D. Buckingham in "Physical Chemistry. An Advanced Treatise", Vol. 4 (Ed. H. Eyring, W. Jost and D. Henderson), (Academic Press, New York, 1970), Chap. 8.

[2] A.D. Buckingham, International Review of Science, Phys. Chem. Series 1, Vol. 3, (Ed. D.A. Ramsay), (Butterworths, London, 1972), Chap. 3.

[3] L. Jansen, Physica, 23, 599 (1957).

[4] A.D. Buckingham, Adv. Chem. Phys., 12, 107 (1967).

[5] A.D. Buckingham in "Intermolecular Interactions: From Diatomics to Biopolymers" (Ed. B. Pullman), (Wiley, Chichester, 1977), Chap. 1.

[6] P. Isnard, D. Robert and L. Galatry, Mol. Phys., 31, 1789 (1976).

[7] L.I. Schiff, "Quantum Mechanics", 3rd edition, (McGraw-Hill, New York, 1968), Section 35.

[8] A. Dalgarno, Adv. Chem. Phys., 12, 143 (1967).

[9] G. Starkschall and R.G. Gordon, J. Chem. Phys., 54, 663 (1971); 56, 2801 (1972).

[10] P.W. Langhoff, R.G. Gordon and M. Karplus, J. Chem. Phys., 55, 2126 (1971).

[11] F. London, Trans. Faraday Soc., 33, 8 (1937).

LECTURE 4

LONG-RANGE AND SHORT-RANGE INTERACTIONS

4.1 Introduction

In the previous lecture, accurate equations were given for the electrostatic, induction and dispersion energies of a pair of molecules sufficiently far apart for the multipole expansion to be applicable. The equation for the dispersion energy involved an integration over all imaginary frequencies of the product of the polarizabilities at imaginary frequencies. Such polarizabilities, particularly the quadrupole and higher ones, are not available for most molecules, so there is a need for a simpler theory.

4.2 The Average-Energy Approximation

London [1] showed that the average-energy approximation of Unsöld [2] could be used to advantage in the equation for the dispersion energy obtained by substituting the multipole expansion (2.46) into (2.22). Separation into single-centre contributions can be achieved by replacing the energy denominators in (2.22) as follows:

$$\frac{1}{W_{pa}-W_{ma}+W_{pb}-W_{mb}} = \frac{U_a U_b (1+\Delta)}{(U_a+U_b)(W_{pa}-W_{ma})(W_{pb}-W_{mb})} \tag{4.1}$$

where

$$\Delta = \frac{U_a^{-1}-(W_{pa}-W_{ma})^{-1}+U_b^{-1}-(W_{pb}-W_{mb})^{-1}}{(W_{pa}-W_{ma})^{-1}+(W_{pb}-W_{mb})^{-1}} \tag{4.2}$$

The energies U_a and U_b are average excitation energies for the free molecules a and b. They are commonly equated to the first ionization potentials. Equation (4.1) is exact, but the simplification comes from neglecting Δ.

We then obtain [3]

$$u_{dispersion} = -\frac{U_a U_b}{4(U_a+U_b)}\Big[T_{\alpha\beta}T_{\gamma\delta}\,\alpha^{(m_a)}_{\alpha\gamma}\alpha^{(m_b)}_{\beta\delta} + \frac{2}{3}T_{\alpha\beta}T_{\gamma\delta\epsilon}\left(\alpha^{(m_a)}_{\alpha\gamma}A^{(m_b)}_{\beta\delta\epsilon} - A^{(m_a)}_{\beta\delta\epsilon}\alpha^{(m_b)}_{\alpha\gamma}\right)$$
$$+ \frac{1}{3}T_{\alpha\beta\gamma}T_{\delta\epsilon\phi}\left(\alpha^{(m_a)}_{\alpha\delta}C^{(m_b)}_{\beta\gamma\epsilon\phi} + C^{(m_a)}_{\beta\gamma\epsilon\phi}\alpha^{(m_b)}_{\alpha\delta} - \frac{2}{3}A^{(m_a)}_{\alpha\epsilon\phi}A^{(m_b)}_{\delta\beta\gamma}\right)$$
$$-\frac{2}{9}T_{\alpha\beta}T_{\gamma\delta\epsilon\phi}A^{(m_a)}_{\alpha\gamma\delta}A^{(m_b)}_{\beta\epsilon\phi} + \dots\Big] \tag{4.3}$$

where α, A, C, ... are the static polarizabilities of molecules a and b (see equations (3.9), (3.15) and (3.16)).

If U_a and U_b are set equal to the lowest allowed excitation energies, equation (4.3) gives a lower bound to the magnitude of C_6, where C_n is the coefficient of R^{-n} in the dispersion energy.

For a pair of identical spherical atoms, equation (4.3) gives for C_6 and C_8

$$u_{dispersion} = -(4\pi\epsilon_0)^{-2}\frac{3U\alpha^2}{4R^6}\left(1 + \frac{10C}{\alpha R^2} + \dots\right) \tag{4.4}$$

where $C = \frac{1}{5}C_{\alpha\beta\alpha\beta}$. The ratio C_8/C_6 is therefore $10C/\alpha$ in this approximation. For the hydrogen atom [3]

$$(4\pi\epsilon_0)^{-1}\alpha = \frac{9}{2}a_0^3, \quad (4\pi\epsilon_0)^{-1}C = \frac{15}{2}a_0^5 \tag{4.5}$$

where a_0 is the Bohr radius, so the ratio is $(50/3)a_0^2$; the accurate value is $19.14\ a_0^2$ [4].

The approximate equation (4.3) would be expected to be less reliable for molecules in excited electronic states, since the average excitation energy approximation may fail when some transitions are associated with positive and others with negative energy differences. However, for the lowest triplet level, the equation should be useful since $\mathcal{H}'$ does not mix states of different multiplicity.

4.3 Short-Range Interactions

As noted in Lecture 1, when the overlap of the electron clouds of molecules a and b is significant, it is essential that the total wavefunction be antisymmetric with respect to exchange of all pairs of electrons. One way to ensure that the wavefunction be antisymmetric is to choose a basis set for a variational calculation that is antisymmetric and such a basis set could be the antisymmetrized product functions $\mathcal{A} p_a p_b$ where $\mathcal{A}$ is the operator which antisymmetrizes with respect to intermolecular exchange of electrons. The set of functions $\mathcal{A} p_a p_b$ are not orthogonal at separations at which overlap is significant and cannot therefore be eigenfunctions of a Hamiltonian. Normal quantum-mechanical perturbation theory is therefore not applicable and because the basis is overcomplete there is no unique transformation to an orthogonal set. However, it is possible to perform a variational calculation with a trial wavefunction which is a sum of a finite number of terms of the set $\mathcal{A} p_a p_b$. A simpler technique is to use the usual self-consistent field theory [5] to obtain the best one-electron wavefunctions for the interacting molecules and to evaluate the interaction energy by subtracting the energy computed for the separate molecules.

Accurate variational calculations, involving extensive configuration interaction, have been performed on He_2 [6-8]. The minimum in the potential has a depth of 15×10^{-23}J (ϵ/k = 11 K) at a separation of 3.0×10^{-10}m, and for $R > 4 \times 10^{-10}$m the results agree with the energy C_6R^{-6} deduced from the known value of C_6.

A variational calculation including intermolecular but not intramolecular correlation has been performed on Ne_2 [9], yielding a well depth of 54×10^{-23}J (ϵ/k = 39.2 K) at 3.08×10^{-10}m and long-range behaviour closely following C_6R^{-6}. However, if there is no intraatomic correlation one might expect the computed long-range behaviour to correspond to a Hartree-Fock C_6 which would differ from the true C_6 by the factor $(\alpha_{\text{Hartree-Fock}}/\alpha_{\text{true}})^2 = 0.77$.

In the region of small overlap, the interaction energy is small compared to the total energy and it is tempting to seek a perturbation scheme for calculating the interaction. Unfortunately there are difficulties stemming from the need to antisymmetrize the total wavefunction with respect to exchange of electrons between the molecules; these problems have been considered by Hirschfelder [10]. A number of perturbation schemes have been suggested for the short- and intermediate ranges and their effectiveness in describing the energy of the ground and lowest triplet states of H_2 have been tested [11]. It was found that some gave satisfactory results but none offered any computational advantage over a full variational calculation. The perturbation approach to the calculation of intermolecular forces in the intermediate region has been reviewed by Certain and Bruch [12]. Gerratt has recently applied his spin-coupled formalism to this problem [13].

If the unperturbed wavefunction is taken to be the antisymmetrized product $\Psi_0 = \mathcal{A} m_a m_b$, where m_a and m_b are the unperturbed states of the separate molecules a and b, then the first-order energy $W^{(1)}_{m_a m_b} = \langle \Psi_0 | \mathcal{H} - W_{m_a} - W_{m_b} | \Psi_0 \rangle / \langle \Psi_0 | \Psi_0 \rangle$ separates into a Coulomb and an exchange contribution. The antisymmetrizer $\mathcal{A}$ may be written

$$\mathcal{A} = \left[\frac{(n_i)!\,(n_j)!}{(n_i + n_j)!} \right]^{\frac{1}{2}} (1 + P) \tag{4.6}$$

where

$$P = -\sum_{i,j} P_{ij} + \sum_{i,j} \sum_{\substack{i'>i \\ j'>j}} P_{ij} P_{i'j'} - \ldots \tag{4.7}$$

where n_i and n_j are the numbers of electrons in the separate molecules a and b and P_{ij} exchanges the ith electron of a with the jth of b. The first-order energy is

$$W^{(1)}_{m_a m_b} = \frac{\langle m_a m_b | \mathcal{H}' | m_a m_b \rangle + \langle P m_a m_b | \mathcal{H}' | m_a m_b \rangle}{1 + \langle P m_a m_b | m_a m_b \rangle} \tag{4.8}$$

The first term in the numerator is the Coulomb interaction and at large separations it gives the electrostatic energy. For spherical atoms it is non-zero in the overlap region and it is negative when the overlap is small. The second term in the numerator of (4.8) is the exchange interaction. For inert-gas atoms it is positive and larger in magnitude than the Coulomb energy.

To illustrate these results consider a pair of hydrogen atoms with parallel electron spins. Ψ_0 is taken to be the Heitler-London spatial wavefunction,

$$\Psi_0 = \frac{1}{\sqrt{2}} \left[s_a(1)\, s_b(2) - s_a(2)\, s_b(1) \right] \tag{4.9}$$

where s_a and s_b are 1s hydrogen-atom orbitals centred on a and b. The Coulomb energy is then

$$u_{\text{Coulomb}} = \frac{\langle s_a(1)\, s_b(2) | \mathcal{H}' | s_a(1)\, s_b(2) \rangle}{1 - S^2} = \frac{\langle s_a(1) s_b(2) | -R_{b1}^{-1} - R_{a2}^{-1} + R_{12}^{-1} + R^{-1} | s_a(1) s_b(2) \rangle}{1 - S^2}$$

$$= \frac{-2 \langle s_a(1) | R_{b1}^{-1} | s_a(1) \rangle + \langle s_a(1) s_b(2) | R_{12}^{-1} | s_a(1) s_b(2) \rangle + R^{-1}}{1 - S^2} \quad (4.10)$$

where the overlap integral is

$$S = \langle s_a(1) | s_b(1) \rangle \quad (4.11)$$

The exchange energy is

$$u_{\text{exchange}} = \frac{-\langle s_a(1)\, s_b(2) | \mathcal{H}' | s_a(2)\, s_b(1) \rangle}{1 - S^2}$$

$$= \frac{2S \langle s_a(1) | R_{b1}^{-1} | s_b(1) \rangle - \langle s_a(1)\, s_b(2) | R_{12}^{-1} | s_a(2) s_b(1) \rangle - S^2 R^{-1}}{1 - S^2} \quad (4.12)$$

Both u_{Coulomb} and u_{exchange} vary as S^2 when S is small, so in this approximation the energy varies as $\exp(-2R/a_o)$ at large R [14].

In going to the second-order of perturbation, there appears to be no natural way to proceed. Murrell and Shaw [15] used the perturbed wavefunction

$$\Psi = \Psi_o + \sum_{p_a, p_b}{}' c_{p_a p_b}\, p_a p_b \quad (4.13)$$

in which the perturbation to Ψ_o is a simple product function. This enabled them to introduce various second-order energies such as those due to induction, dispersion and exchange-polarization.

References

[1] F. London, Trans. Faraday Soc., 33, 8 (1937).

[2] A. Unsöld, Z. Physik, 43,563 (1927).

[3] A.D. Buckingham, Adv. Chem. Phys., 12, 107 (1967).

[4] L. Pauling and J.Y. Beach, Phys. Rev., 47, 686 (1935).

[5] C.C.J. Roothaan, Rev. Mod. Phys., 23, 69 (1951).

[6] D.R. McLaughlin and H.F. Schaefer, Chem. Phys. Lett., 12, 244 (1971).

[7] P. Bertoncini and A.C. Wahl, J. Chem. Phys., 58, 1259 (1973).

[8] B. Liu and A.D. McLean, J. Chem. Phys., 59, 4557 (1974).

[9] W.J. Stevens, A.C. Wahl, M.A. Gardner and A.M. Karo, J. Chem. Phys., 60, 2195 (1974).

[10] J.O. Hirschfelder, Chem. Phys. Lett., 1, 325 (1967).

[11] P.R. Certain, J.O. Hirschfelder, W. Kołos and L. Wolniewicz, J. Chem. Phys., 49, 24 (1968).

[12] P.R. Certain and L.W. Bruch, International Review of Science, Phys. Chem. Series 1, Vol. 1 (Ed. W. Byers Brown), (Butterworths, London, 1972), Chap. 4.

[13] J. Gerratt, Proc. Roy. Soc. A, 350, 363 (1976).

[14] H. Margenau and N.R. Kestner, "Intermolecular Forces", 2nd Edition, (Pergamon Press, Oxford, 1971), Chap. 3.

[15] J.N. Murrell and G. Shaw, J. Chem. Phys., 46, 1768 (1967).

LECTURE 5

NON-ADDITIVITY AND THE DETERMINATION OF INTERMOLECULAR FORCES

5.1 Introduction

Our earlier lectures were devoted to a theoretical description of intermolecular forces. Equations were derived for the electrostatic, resonance, induction and dispersion energies in the long-range limit. Short-range interactions were considered in the previous lecture. In this lecture we shall consider the question of non-additivity. We shall also describe how actual intermolecular potentials are determined.

5.2 Non-Additivity

In many theories of liquids and solids, the total interaction energy is assumed to be pairwise additive, that is, $u = \sum_{i<j} u_{ij}$. However, as noted in Lecture 1, this is not precise and there are many-body interactions that may sometimes be important. For example, the transient dipole moment in gaseous or liquid argon must be associated with at least three atoms, since single atoms and pairs of atoms are centrosymmetric; observations of the weak collision-induced far-infrared absorption in the pure inert gases would therefore be of interest in the study of three-body interactions.

The earliest calculation of three-body forces was that of Axilrod and Teller [1] who evaluated the long-range dipole-dipole-dipole dispersion energy of three atoms. It is a third-order perturbed energy and takes the form

$$\Delta u_{abc} = D\left(1 + 3\cos\theta_a \cos\theta_b \cos\theta_c\right) R_a^{-3} R_b^{-3} R_c^{-3} \tag{5.1}$$

where R_a, R_b, R_c are the sides and θ_a, θ_b, θ_c the interior angles of the triangle formed by the atoms. The constant D can be expressed in terms of the polarizabilities of the free atoms a, b and c at imaginary frequencies [2]:

$$D = (4\pi\epsilon_0)^{-3} \frac{3\hbar}{\pi} \int_0^\infty \alpha_a(iu)\,\alpha_b(iu)\,\alpha_c(iu)\, du \tag{5.2}$$

For atomic hydrogen, D = 21.64 atomic units, while C_6 (the coefficient of R^{-6} in the dispersion energy of a pair of atoms) is -6.4990 atomic units. For three atoms at the vertices of an equilateral triangle the ratio of the three-body to the two-body energy is

$$\frac{\Delta u}{u} = \frac{\frac{11}{8} D R^{-9}}{-3C_6 R^{-6}} = -\frac{11}{24}\frac{D R^{-3}}{C_6} \qquad (5.3)$$

For atomic hydrogen at R = 7 atomic units (3.7×10^{-10} m) $\frac{\Delta u}{u} = -0.0044$. For argon at R = 3.76×10^{-10}m, $\frac{\Delta u}{u} = -0.010$.

Equation (5.1) for the long-range triple-dipole energy has been widely used in allowing for the effect of three-body interactions on the properties of simple liquids and solids [3]. The sublimation energy at OK of argon is 7.74 kJ mol^{-1}; nearest-neighbour pair interactions contribute approximately 97% of this binding energy while more distant neighbours contribute 10% and three-body effects -7% [3]. The lattice spacing at OK and zero external pressure is nearly equal to R_e for the heavier inert gases because further-neighbour attractive forces tend to cancel the effects of zero-point vibrations and three-body repulsion.

There may be significant short-range non-additive forces. The strong short-range repulsion between overlapping molecules must be affected by the distortion due to interaction with a third molecule, whether it be in long- or short-range interaction with one of the others.

5.3 The Effect of a Medium

A medium of relative permittivity (or dielectric constant) ϵ_r reduces the electrostatic energy of interaction of two molecules immersed in it by a factor of ϵ_r. McLachlan [4] examined the influence of a medium on dispersion energy and distinguished five effects relating to the interaction of molecules in a liquid:

(i) the force depends on the free energy $A(\underset{\sim}{R}) = U(\underset{\sim}{R}) - TS(\underset{\sim}{R})$ as discussed in Lecture 1;

(ii) the fixed molecules disturb the local order of the liquid;

(iii) the electromagnetic forces which would exist in free space are modified;

(iv) Archimedes' principle is involved because a molecule can move only by displacing the liquid from its path;

(v) the electron clouds overlap those of the medium and change the properties of the molecule.

The presence of polarizable matter between interacting molecules may <u>increase</u> their mutual potential energy [5]. For example, if a spherical atom b of polarizability α_b is at the point midway between a pair of charges +q and −q at a separation R, the interaction energy is

$$u_{abc} = -(4\pi\epsilon_0)^{-2} q^2 R^{-1}\left[1 + 32(4\pi\epsilon_0)^{-1}\alpha_b R^{-3}\right] \tag{5.4}$$

If the charges were of the same sign the induced dipole of b would vanish and the repulsive force between the charges would not be affected by α_b (actually there would be a field gradient $F_{zz} = -2F_{xx} = -2F_{yy} = -(4\pi\epsilon_0)^{-1}\, 32\, q\, R^{-3}$ at b and it would induce a quadrupole $C_b F_{zz}$ in b, leading to an induction energy $-\frac{1}{4} C_b F_{zz}^2 = -(4\pi\epsilon_0)^{-2}\, 256\, C_b q^2 R^{-6}$).

There is much interest in solvent effects on the interaction and properties of molecules. Theoretical approaches vary from those employing the reaction field of Onsager [6,7] to the "supermolecule" approach of Pullman[8] in which the interacting molecules are treated as a single large quantum-mechanical system.

In addition to affecting the interaction energies of molecules, the medium may have a pronounced influence on optical, electric and magnetic properties. Figure 5.1 shows how the mean polarizability of two helium atoms varies with their separation [9].

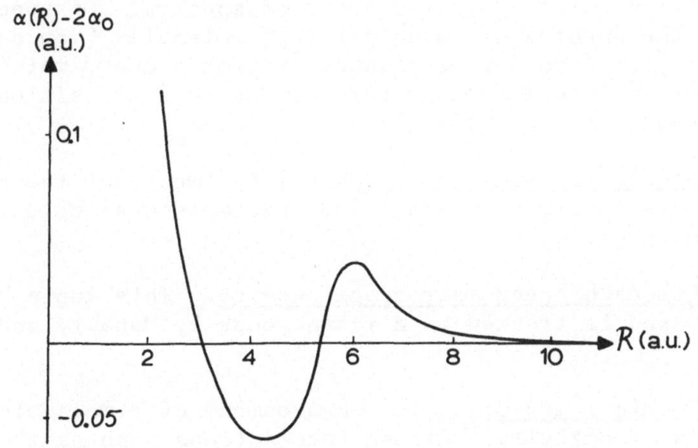

Figure 5.1. The change in the mean polarizability of He_2 as a function of the separation R. For a free He atom $\alpha_0 = 1.3832$ atomic units. The minimum in the energy is at R = 5.7 a.u.

5.4 Manifestations of Intermolecular Forces

Molecular interactions influence the properties of matter in the bulk and at a microscopic level. These influences are important sources of experimental information about intermolecular forces.

(i) Equilibrium properties of imperfect gases. Early studies, beginning with van der Waals, concentrated on the p - V - T relationships of gases. The second and third virial coefficients remain an important source of information about two- and three-body forces. Other properties such as dielectric virial coefficients [10] can expose the angle-dependence of interaction energies.

(ii) Non-equilibrium properties of fluids. While equilibrium properties are determined by the partition function, transport-properties depend on the dynamics of molecular collisions. However, in simple gases these properties can be an important source of information about intermolecular forces and torques.

(iii) Structure and properties of crystals and conformations of molecules. The structure of a crystal, and properties such as compressibility, heat of sublimation, and lattice vibrations, depend upon the forces between the molecules or ions. Similarly the conformations of molecules, and the tertiary and quaternary structure of macromolecules, is determined by the interaction of non-bonded atoms. Progress is being made in understanding and predicting these properties, usually with the aid of simple atom-atom empirical potentials [11,12].

(iv) Spectroscopic properties. Molecular interactions affect the frequency, intensity and line shape of spectra. Of special interest are the spectra of bound pairs of molecules (van der Waals molecules) [13,14], and double-resonance spectroscopy, which probes the ability of an intermolecular force to cause a transition between particular quantum states [15].

(v) Molecular-beam scattering. This important technique can provide reliable information about the intermolecular potentials of simple molecules [3,16].

(vi) Forces between macroscopic bodies. This topic has been reviewed [17] and is treated in a recent book by Mahanty and Ninham [18].

(vii) Chemical effects. The environment of a molecule may effect chemical reactivity. Strong interactions such as those associated with hydrogen bonding [19] and charge transfer [20] can be particularly effective.

(viii) Computer simulation. This is the topic of another course of Lectures at this meeting. Computer "experiments" can provide a means of refining simple intermolecular potentials [21].

References

[1] B.M. Axilrod and E. Teller, J. Chem. Phys., 11, 299 (1943).

[2] A. Dalgarno, Adv. Chem. Phys., 12, 143 (1967).

[3] J.A. Barker in "Rare Gas Solids" (Eds. M.L. Klein and J.A. Venables), (Academic Press, New York, 1976), Chap. 4.

[4] A.D. McLachlan, Discussions Faraday Soc., 40, 239 (1965).

[5] A.D. Buckingham, Phil. Trans. Roy. Soc. B, 272, 5 (1975).

[6] L. Onsager, J. Amer. Chem. Soc., 58, 1486 (1936).

[7] B. Linder, Adv. Chem. Phys., 12, 225 (1967).

[8] A. Pullman in "The World of Quantum Chemistry" (Ed. R. Daudel and B. Pullman), (Reidel, Dordrecht, 1974), p. 239.

[9] A.D. Buckingham in "The World of Quantum Chemistry" (Ed. R. Daudel and B. Pullman), (Reidel, Dordrecht, 1974), p. 253.

[10] H.G. Sutter in "Dielectric and Related Molecular Processes", Vol. 1, (Ed. M. Davies), (Spec. Period. Rep. Chem. Soc. London, 1972), p. 65.

[11] A.I. Kitaigorodsky, "Molecular Crystals and Molecules", (Academic Press, New York, 1973).

[12] A.J. Hopfinger, "Conformational Properties of Macromolecules", (Academic Press, 1973).

[13] H.L. Welsh, International Review of Science, Phys. Chem. Series 1, Vol. 3 (Ed. D.A. Ramsay), (Butterworths, London, 1972), Chap. 2.

[14] B.J. Howard, International Review of Science, Phys. Chem. Series 2, Vol. 2 (Ed. A.D. Buckingham), (Butterworths, London, 1975), Chap. 3.

[15] T. Oka, Adv. At. and Mol. Phys., 9, 127 (1973).

[16] "Molecular Beam Scattering", Faraday Discussions Chem. Soc. London, 55 (1973).

[17] J.N. Israelachvili and D. Tabor, Prog. Surface and Membrane Sci., 7, 1 (1973).

[18] J. Mahanty and B.W. Ninham, "Dispersion Forces", (Academic Press, London, 1976).

[19] "The Hydrogen Bond. Recent Developments in Theory and Experiment" (Ed. P. Schuster, G. Zundel and C. Sandorfy), (North-Holland, Amsterdam, 1976).

[20] R.S. Mulliken and W.B. Person, "Molecular Complexes", (Wiley, New York, 1969).

[21] P.S.Y. Cheung and J.G. Powles, Molec. Phys., 32, 1383 (1976).

ELECTRONIC THEORY OF THE THERMODYNAMICS
AND STRUCTURE OF LIQUID METALS

R. Evans

H.H. Wills Physics Laboratory, University of Bristol
Tyndall Avenue, Bristol U.K.

1. INTRODUCTION

Over the last six or seven years signficant progress has been made in understanding the structure and the thermodynamic properties of liquid metals. This has resulted from a conjugation of the electron theory of metals and the theory of classical fluids. These lecture notes will describe some recent work in this area. A comprehensive review is not intended. The selection of topics reflects the author's own interest.

In the terminology of liquid state physics,liquid metals are usually classified 'simple liquids'. Like the rare gases they are monatomic and at temperatures close to their melting points have structure factors which can be modelled,in the crudest approximation, by hard spheres of an appropriately chosen diameter. They exhibit neither the complications of strong directional bonding characteristic of molecular and associated liquids nor the difficulties associated with long-range Coulomb forces which beset the polar liquids. Liquid metals are, however, two component systems. If we assume that the core electrons are rigidly fixed to the nuclei then an elemental metal will consist of N ions and NZ 'conduction' electrons where Z is the number of electrons outside the core. Any theory of the interionic forces, structure, thermodynamics etc. of the liquid must depend on the electronic structure of the conduction band. Conversely, the details of the electronic structure will be dependent on the arrangement of ions in the liquid. In general a proper theoretical treatment of these problems constitutes a formidable programme and indeed most studies of liquids <u>model</u> the interatomic potentials by some convenient functional form. We shall see that only for a particular class of metals has the programme been carried through with any real success.

The notes are arranged as follows: in § 2 we briefly review some of the electronic and electrical properties of liquid metals. The concept of a weak electron-ion interaction (the pseudopotential) is introduced and this is used to discuss perturbation theories for the electrical transport properties and the electronic density of states in simple liquid metals. The difficulties of extending this model to other metals are described. In § 3 we apply the pseudopotential approach to the evaluation of the free energy of a liquid metal. The theory of effective pairwise interionic potentials is developed and various examples of these are presented. The application of thermodynamic perturbation theories to liquid metals is described in § 4 and the results of calculations of the entropy and structure factor based on these theories are discussed in detail. In § 5 the problem of calculating the isothermal compressibility for a liquid metal is examined. A theory for the long wavelength limit of the structure factor, which goes beyond the random phase approximation, is presented. The surface tension of liquid metals is discussed in § 6 and the possibility of inverting experimental structure factor data to obtain interionic potentials is considered in § 7. Finally in § 8 we briefly mention some other topics of current interest in this field.

2. ELECTRONIC PROPERTIES AND ELECTRONIC STRUCTURE

From the outset it is useful to divide metals into different classes according to their electronic structure in the crystalline state.

Simple metals are those without d or f electrons in the conduction band and whose core levels lie at energies well below the conduction band. For these metals Z is just the chemical valence. There is strong evidence that in the crystalline state the electronic structure of these metals is nearly-free-electron like [1]. The alkalis, Mg, Zn, Cd, Hg, Al, Ga, In, Sn and, perhaps, Tl and Pb fall into this class.

The Noble metals Cu, Ag and Au have narrow, filled d bands lying 1 or 2 eV below the Fermi energy while the Transition metals have unfilled d states at the Fermi energy. The Rare-earths (lanthanides) and the Actinides also have tightly bound d bands but have additional complications due to the presence of highly localised outer f states. The latter cannot easily be incorporated into the usual one-electron band theory. The electronic structure of these three classes is in no sense free-electron like.

The alkaline earths Ca, Sr and Ba have unfilled d bands above the Fermi energy. Hybridisation between these bands and the nearly-free-electron bands has significant effects on the electronic structure in the neighbourhood of the Fermi energy. Consequently these metals can be thought of as intermediate between simple and transition metals.

2.1 Some information from experiment.

Liquid metals are characterized by their large electrical and thermal conductivities. When a simple metal melts its electrical conductivity σ is reduced by roughly a factor of two and in the transition and rare earth metals the change is much smaller. The resistivities of all liquid metals, at temperatures close to their melting points T_M, lie in the range 10-300 μ.ohm.cm. The semiconductors Si and Ge and the semi-metals As, Sb and Bi increase their conductivities by several orders of magnitude and behave like simple metals in the liquid state [2].

In solid metals the Hall coefficient R_H is a sensitive probe of the geometry of the Fermi surface. When the divalent metal Cd melts R_H changes sign from positive (hole-like conduction) to negative (electron conduction) and this reflects the change from a complicated Fermi surface to a spherical one in the isotropic liquid. For all the simple liquid metals R_H is negative and temperature independent and for the great majority this coefficient takes its free electron value $-(n_0e)^{-1}$ where e is the electronic charge and n_0 is the number of conduction electrons per unit volume. The latter is obtained from the valence, thus liquid Ge has four conduction electrons and liquid Sb has five. The noble metals have values of R_H corresponding to one conduction electron. In liquid transition and rare earth metals R_H can have either sign [3] so Hall coefficient measurements do not provide direct information about the number of current carrying electrons.

The thermal conductivity λ is difficult to measure accurately for liquid metals but recently accurate measurements of the ratio λ/σ have been performed for a few, low melting-point simple metals [3,4]. These have shown that the Wiedemann-Franz law is obeyed i.e. $\lambda = L\sigma T$ where T is the absolute temperature and L is the Lorenz number as calculated for free electrons. In other words the thermal conductivity is almost completely electronic in origin.

The magnetic susceptibilities of liquid simple metals are positive (paramagnetic) or negative (diamagnetic) depending on the particular element. The noble metals are diamagnetic. It is clear, however, that in all these systems there is a substantial Pauli–Landau contribution from the nearly-free conduction electrons which acts to cancel the large diamagnetic ion-core contribution. The susceptibilities of Mn, Fe, Co, Ni and the rare-earths are typically two orders of magnitude larger than those of the paramagnetic simple metals and show little change on melting [3]. This indicates that the d and f electronic structure is rather insensitive to the extra disorder induced by melting.

These experimental results suggest a) that in the simple liquid metals the electronic structure, at least in the neighbourhood of the Fermi energy, should show little departure from free-electron behaviour and b) that in the noble, transition and rare-earth metals we might expect that the tightly bound d and f electrons are not seriously affected by the liquid disorder so the d and f parts of the electronic density of states should not be drastically changed by melting.

2.2 Pseudopotential treatment of the electron-ion interaction.

An explanation of point a) can be found in the pseudopotential theory of electrons in metals. We suppose that the total potential experienced by a single electron can be written as a sum of contributions from each ion:

$$V(\underset{\sim}{r}) = \sum_{\ell=1}^{N} v(\underset{\sim}{r} - \underset{\sim}{R}_{\ell}) \tag{2.1}$$

where $\underset{\sim}{R}_{\ell}$ denotes the position of the ℓth ion. (Of course we need to define exactly what we mean by the one-electron potential and this is discussed in Appendix 2. For the purpose of the present discussion we can think of V as a Hartree-like self consistent potential). In general $v(\underset{\sim}{r})$ will exhibit screened Coulomb behaviour outside the core due to screening by conduction electrons and within the core v is strongly attractive. The conduction electron wave functions oscillate rapidly in the core, in order to orthogonalise themselves to the core eigenfunctions. The valence eigenvalue spectrum, however, does not depend on the details of these oscillations. It is determined by matching the logarithmic derivatives at the core radius. Thus, if one is only concerned with the valence eigenvalues, the true potential v can be replaced by a pseudopotential v_{ps}^{sc} (the superscript *sc* implies suitable screening) which will have the same logarithmic derivatives at the core radius but no eigenstates of lower energy. If, furthermore, the pseudopotential is small in the core region, then it should be possible to treat it as a weak perturbation and this leads to considerable simplications. Techniques for constructing pseudopotentials are well-documented [1,5] and there is considerable expertise, gained from solid state research, regarding their usefulness for calculating electronic properties [6]. For most of simple metals the pseudopotentials turn out to be weak and many meaningful calculations of the crystalline band structure, Fermi surface, electrical transport coefficients etc. have been carried out.

For noble and transition elements this is not the case. The pseudopotential is extremely energy dependent, spatially non-local and large so perturbation theory is inappropriate. This is most easily expressed in the language of scattering theory. If we take the potential v to be spherically symmetrical inside a given radius, which should be greater than that of the core, and average the potential outside to some constant value, E_0, we can describe the scattering properties of the ion by phase shifts. In most of the simple metals the phase shifts are small modulo π which means that the core of the ion offers only a small scattering cross-section to the conduction electrons in the appropriate energy range. The Born approximation is appropriate. In the noble metals the d phase shift exhibits resonant behaviour i.e. it increases rapidly from 0 to $\sim \pi$ at an energy of a few eV above the energy zero E_0. In the transition metals similar resonant behaviour occurs but at higher energies. The scattering cross-sections are very large at the resonant energies and cannot be evaluated using perturbation theory. Any attempt to model these features in terms of a pseudo-potential rapidly runs into difficulties so it is better to work with the real potential v or equivalently its logarithmic derivatives or phase shifts [7]. The alkaline earth metals have fairly large d phase shifts at energies close to the Fermi energy [8]. Several of the simple metals have sizeable s and p phase shifts - especially the four and five valent elements - and it is difficult to construct reliable, weak pseudopotentials for these.

2.3 Electronic structure and electrical transport in liquid simple metals.

The central quantity of interest in the study of the electronic structure of liquid metals is the one-electron density of states n(E) or more accurately the configurationally averaged density of states $\langle n(E) \rangle$. In the liquid state one cannot define a 'band structure' $E(\underset{\sim}{K})$ since Bloch's theorem is inapplicable and the wave vector $\underset{\sim}{K}$ is not a good quantum number. The density of states is well-defined and Green function techniques have been developed to calculate this quantity [9,10] (see Appendix 1 for a brief discussion of the theory). For the simple metals, where pseudo-potentials are appropriate, the expansions of the Green function in powers of the (pseudo) potential can be expected to converge and several meaningful calculations have been performed [10]. Provided the relevant series are truncated at second order in V_{ps}^{sc} the configurational averages only require the pairwise distribution function. Higher order terms involve the higher order atomic distribution functions and hence must be approximated. In second order the corrections to the free electron $E^{\frac{1}{2}}$ behaviour are found to be small (a few percent at most) and $\langle n(E) \rangle$ has somewhat less structure than is obtained from the equivalent calculations in crystalline metals. The liquid, of course, shows no Brillouin zone effects.

While the results of the pseudopotential approach would seem to lend strong support to a nearly-free-electron model, we should remember that, in some sense, it was introduced to do just this! Reality may be different. Photoemission spectroscopy data on liquid In, Al, Sn and Hg cannot easily be reconciled with an almost free electron density of states [11] and in liquid Ga, Tl, Pb and Bi there is evidence for pronounced structure in $\langle n(E) \rangle$ at energies below the respective Fermi energies⁺.

The interpretation of photoemission data is, however, a difficult and delicate business ($\langle n(E) \rangle$ is not measured directly) and it would be premature to abandon the nearly-free-electron approach on this evidence alone- especially when electrical transport results seem to point to the contrary.

The electrical conductivity was the first liquid metal property to be treated in the pseudopotential approach. In Born approximation it is easy to show that

$$\sigma^{-1} = \frac{3\pi\Omega_0}{4e^2\hbar v_F^2 K_F^4} \int_0^{2K_F} dK\, K^3 \, |v_{ps}^{sc}(K)|^2 S(K) \qquad (2.2)$$

where K_F is the radius of the free electron Fermi sphere ($K_F^3 = 3\pi^2 n_0$), V_F is the Fermi velocity, Ω_0 is the atomic volume and the other symbols have their usual meaning. The static liquid structure factor S(K) enters since at liquid metal temperature inelastic scattering is not important. This well-known Ziman [12] formula has proved remarkably successful for interpreting experimental data [2]. It gives a good qualitative account of the variation of σ and $d\sigma/dT$ through the simple metals and also provides explanation of why σ is reduced by about a factor of two on melting. The results of calculations based on the formula are usually in reasonable quantitative agreement with experiment. The extension of the theory to binary alloys is straight forward [2] and σ can be expressed in terms of the concentration, the pseudopotentials of each species and the three partial structure factors. This has been very useful for explaining the concentration dependence of σ and $d\sigma/dt$ in many different alloy systems [3]. An expression for the thermoelectric power can also be derived and this has also proved useful [2].

The corresponding analysis for the Hall coefficient is much more difficult and has only recently been convincingly carried through [13]. It turns out that the corrections to the free-electron value are of third order in the pseudopotential and are thus expected to be very small. For some of the heavy polyvalent metals i.e. Pb, Tl and Bi, the spin-orbit interaction becomes significant and it is argued that these can substantially reduce R_H below the free-electron value. The experimental results for

+ J. Wotherspoon (private communication)

these metals are still uncertain [3].

2.4 Electronic structure and electrical transport in liquid noble, transition and rare-earth metals.

When the ions are strong scatterers of the electrons the lack of lattice periodicity leads to much greater theoretical difficulties for the calculation of electronic structure than was the case for simple metals. For perfect crystals band theorists have developed various non-perturbative methods of solving the one-electron Schrödinger equation. All methods treat the scattering problem at a single ion exactly and then solve the appropriate multiple scattering problem. The first step entails the calculation of the phase shifts (or the single-site transition matrix) corresponding to the potential v while the second is made feasible by the imposition of Bloch boundary conditions. In the liquid metal the multiple scattering problem is extremely complex. Several theories exist [14]. These are based on different assumptions concerning the higher order atomic distribution functions and the relevance of particular classes of diagrams in the multiple scattering expansions. There are few calculations based on realistic models so it is difficult to judge what progress, if any, has been achieved. (This state of the art should be contrasted with that existing in the theory of random, substitional, crystalline alloys. There the CPA (coherent potential approximation) has gained general acceptance as the 'best' starting point for practical calculations of the electronic structure of 'real' alloys [15]. The CPA is strictly a single site approximation which ignores all correlations between sites. Because of the topological disorder of the liquid state any sensible starting approximation must include at least pairwise correlations and this is a much harder problem).

On the experimental side there is photoemission data for the liquid noble metals and some of their alloys [11] and soft X-ray band spectra for liquid Cu, Fe, Co and Ni [16]. Comparison of the various spectra with those for the corresponding solid metals indicates that the d bands are little changed on melting. There is a substantial amount of magnetic susceptibility and electrical transport data for liquid Pd, several of the 3d transition metals and their alloys [3,17] and the rare earth metals and their alloys [3,18]. This can, in principle, yield much useful information about the electronic structure in the neighbourhood of the Fermi energy. However a discussion of the results and their interpretation lies outside the scope of these notes.

At this stage is should be clear that by comparison with the simple metals our understanding of the electronic structure and properties of liquid noble and transition metals is limited. Consequently we are in no position to make a systematic attack on our original programme in these systems. We do not yet have a formalism capable of evaluating the free energy of the electronic system as a function of the coordinates of the ions and, as we shall see, this must be a basic ingredient of any theory of the thermodynamics and structure. For simple liquid metals, on the other hand, we might hope that conventional perturbation theory could be applied to the calculation of the free energy of the interacting system of electrons and ions. The success of the pseudopotential treatment of electronic properties suggests that this is a sensible philosophy and this is the approach we shall pursue through most of this article.

3. THEORY OF EFFECTIVE INTERIONIC POTENTIAL AND THERMODYNAMIC PROPERTIES OF SIMPLE LIQUID METALS

This presentation is similar to that of Ashcroft and Stroud [19].

3.1 From a two-component to a quasi one-component fluid.

As earlier, we suppose the liquid has a volume $\mathcal{V}$ and contains N ions of valence Z so that the total number of conduction electrons is NZ. The basic Hamiltonian is

$$H = \sum_i \frac{p_i^2}{2m} + \frac{e^2}{2}\sum_{i\neq j}\frac{1}{|\underset{\sim}{r}_i-\underset{\sim}{r}_j|} + \sum_{i,\ell} v_{ps}(\underset{\sim}{r}_i-\underset{\sim}{R}_\ell) + \sum_\ell \frac{p_\ell^2}{2M} + \frac{1}{2}\sum_{\ell\neq\ell'} W(|\underset{\sim}{R}_\ell-\underset{\sim}{R}_{\ell'}|) \tag{3.1}$$

$$= \quad H_e \quad + \quad H_{ee} \quad + \quad H_{ei} \quad + \quad H_K \quad + \quad H_{ii}$$

where $\{\underset{\sim}{r}_i\}$ and $\{\underset{\sim}{R}_\ell\}$ denote electronic and ionic coordinates, $\{\underset{\sim}{p}_i\}$ and $\{\underset{\sim}{P}_\ell\}$ the corresponding momenta and m and M the masses. v_{ps} is the 'bare' unscreened pseudopotential of a single ion. For simplicity this will be taken to be a local function of r. Inside the core it is assumed small and outside it will behave like $-Ze^2/r$. (The relationship between v_{ps} and the screened pseudopotential v_{ps}^{sc} which was used to describe the one-electron states in §2, is given in Appendix 2). W is the bare ion-ion interaction which is taken to be pairwise. For large separations W(R) will be Coulombic but at short range it should include contributions from Van der Waals attraction and Born-Mayer repulsion between the ion cores. Provided the ion cores are small and the ions are not highly polarizable both of these contributions will be insignificant for the separations of interest in liquid metals so we simply set $W = Z^2e^2/R$.

The Helmholtz free energy F corresponding to this Hamiltonian is

$$F = -\frac{1}{\beta}\ln Q_N \tag{3.2}$$

where $\beta = 1/K_BT$ and the canonical partition function is

$$Q_N = \mathrm{Tr}\exp(-\beta H) \tag{3.3}$$

The trace runs over both electronic and ionic states of the system but in the adiabatic approximation these can be separated

$$Q_N = \frac{1}{N!h^{3N}}\int d\underset{\sim}{R}_1 \ldots d\underset{\sim}{R}_N \int d\underset{\sim}{P}_1 \ldots d\underset{\sim}{P}_N \; \mathrm{Tr}_e \exp(-\beta H) \tag{3.4}$$

where Tr_e refers to a complete set of electronic states corresponding to a particular ionic configuration. If the ions obey classical statistics (an excellent approximation for liquid metals) the integration over momenta is trivial and the partition function simplifies:

$$Q_N = \left(\frac{2\pi M}{h^2\beta}\right)^{3N/2} Z_p$$

with the configurational partition function given by

$$Z_p = \frac{1}{N!}\int d\underset{\sim}{R}_1 \ldots d\underset{\sim}{R}_N \exp(-\beta H_{ii}) \left\{\mathrm{Tr}_e \exp[-\beta(H_e + H_{ee} + H_{ei})]\right\} \tag{3.5}$$

- since H_{ii} is independent of electronic coordinates. The term in brackets is just $\exp[-\beta F'(\underset{\sim}{R}_1, \ldots \underset{\sim}{R}_N)]$ where $F'(\underset{\sim}{R}_1, \ldots \underset{\sim}{R}_N)$ is the Helmholtz free energy of an electron system interacting in the presence of an external potential described by H_{ei}. The latter corresponds to a fixed ionic configuration $(\underset{\sim}{R}_1, \ldots \underset{\sim}{R}_N)$. F' is precisely the quantity we argued we should be able to calculate for simple metals. Once F' has been evaluated (in some approximation scheme) the electronic degrees of freedom no longer explicitly appear and the problem is reduced to that of a classical fluid in which the ions move in an effective interaction potential given by

$$\Phi = H_{ii} + F'. \tag{3.6}$$

3.2 Evaluation of the free energy of the electron system.

In order to calculate F' we begin by considering a homogeneous interacting electron gas of density $n_0 = NZ/V$, in the presence of a uniform compensating charge background. The potential of the latter is

$$V_+(\underset{\sim}{r}) = -e^2\int \frac{d\underset{\sim}{r}' \; n_0}{|\underset{\sim}{r} - \underset{\sim}{r}'|} \tag{3.7}$$

where the integral is over the total volume. Let the free energy of this system be F_{eg} (this energy includes the contribution from the self-energy of the positive background). We now perturb this system by replacing the potential $V_+(\underset{\sim}{r})$ by the total electron-ion pseudopotential. The change in potential is then

$$\delta V(\underset{\sim}{r}) = \sum_{\ell} v_{ps}(\underset{\sim}{r} - \underset{\sim}{R}_{\ell}) - V_+(r) \tag{3.8}$$

The free energy of the final system F' can then be obtained via the Hellman-Feynman theorem:

$$F' = F_{eg} - \frac{e^2}{2}\iint \frac{d\underset{\sim}{r}\, d\underset{\sim}{r}'\, n_0^2}{|\underset{\sim}{r} - \underset{\sim}{r}'|} + \int_0^1 d\lambda \int d\underset{\sim}{r}\; \delta V(\underset{\sim}{r}) n(\underset{\sim}{r};\lambda) \tag{3.9}$$

where $n(\underset{\sim}{r};\lambda)$ is the electron density corresponding to the system in which the external potential is $V_+(\underset{\sim}{r}) + \lambda\delta V(\underset{\sim}{r})$. Until now we have not required v_{ps} to be weak so the theory could equally well be applied to strong-scattering metals. In order to procede, however, we need to evaluate $n(\underset{\sim}{r};\lambda)$. If $\delta V(\underset{\sim}{r})$ is small we can use linear response theory to calculate this quantity [20]:

$$n(\underset{\sim}{r};\lambda) = n_0 + \int d\underset{\sim}{r}'\chi^e(|\underset{\sim}{r} - \underset{\sim}{r}'|)\;\; \lambda\delta V(\underset{\sim}{r}') \tag{3.10}$$

where χ^e is the interacting density reponse function of the homogeneous electron gas - for density n_0. (Strictly speaking $n(\underset{\sim}{r};\lambda)$ is the pseudo-charge density if δV refers to pseudopotentials). Fourier transforming we have:

$$\begin{aligned} n(\underset{\sim}{q};\lambda) &= \lambda\chi^e(q)\;\delta V(\underset{\sim}{q}) && q \neq o \\ &= NZ && q = o \end{aligned} \tag{3.11}$$

and substituting into equ. (3.9) gives

$$\begin{aligned} F' &= F_{eg} - \frac{e^2}{2}\iint \frac{d\underset{\sim}{r}\, d\underset{\sim}{r}'\, n_0^2}{|\underset{\sim}{r} - \underset{\sim}{r}'|} + \frac{1}{V}\int_0^1 d\lambda \Big(NZ\delta V(q = o) + \sum_{\underset{\sim}{q}\neq o} \lambda\chi^e(q)|\delta V(\underset{\sim}{q})|^2\Big) \\ &= F_{eg} - \frac{e^2}{2}\iint \frac{d\underset{\sim}{r}\, d\underset{\sim}{r}'\, n_0^2}{|\underset{\sim}{r} - \underset{\sim}{r}'|} + n_o\;\delta V(q = o) + \frac{1}{2V}\sum_{\underset{\sim}{q}\neq o}\chi^e(q)|\delta\; V(\underset{\sim}{q})|^2 \end{aligned}$$

$\delta V(\underset{\sim}{q})$ can be expressed in terms of the pseudopotential form factor $v_{ps}(q)$ and the ion density $\rho(\underset{\sim}{q})$:

$$\begin{aligned} \delta V(\underset{\sim}{q}) &= v_{ps}(q)\rho(\underset{\sim}{q}) && q \neq o \\ &= N w_c(q = o) && q = o \end{aligned} \tag{3.12}$$

where $w_c(q) \equiv v_{ps}(q) + \frac{4\pi Ze^2}{q^2}$

and $$\rho(\underset{\sim}{q}) = \sum_{\ell} \exp(i\, \underset{\sim}{q}.\underset{\sim}{R}_{\ell}) \tag{3.13}$$

w_c is the non-Coulombic part of the pseudopotential. The free energy F' can then be written as

$$F' = F_{eg} - \frac{e^2}{2} \iint \frac{d\underset{\sim}{r}\, d\underset{\sim}{r}'\, n_o^2}{|\underset{\sim}{r} - \underset{\sim}{r}'|} + U_1 + U_2 \tag{3.14}$$

where $U_1 = n_o N w_c(q = o)$

and

$$U_2 = \frac{1}{2V} \sum_{\underset{\sim}{q} \neq o} \chi^e(q)\, |v_{ps}(q)|^2\, \rho(\underset{\sim}{q})\, \rho(-\underset{\sim}{q}) \tag{3.15}$$

The second term in equ (3.14) is the negative of the self-energy of the uniform positive background. This is precisely cancelled by the uniform (q = o) contribution to the bare ion-ion interaction, H_{ii} leaving a Madelung contribution of the form:

$$U_M = \frac{1}{2V} \sum_{\underset{\sim}{q} \neq o} \frac{4\pi Z^2 e^2}{q^2} \left(\rho(\underset{\sim}{q})\, \rho(-\underset{\sim}{q}) - N\right) \tag{3.16}$$

Thus, in the approximation of linear response, the effective interaction potential for the ions is

$$\Phi(\underset{\sim}{R}_1, \ldots \underset{\sim}{R}_N) = F_{eg} + U_1 + U_2 + U_M . \tag{3.17}$$

This result is valid to second order in δV and constitutes the starting point for most recent studies of the structure and thermodynamic properties of liquid simple metals. Since the terms depending on the ionic coordinates U_2 and U_M involve only the product $\rho(\underset{\sim}{q})\, \rho(-\underset{\sim}{q})$ it follows that ensemble averaging introduces the pairwise distribution function but none of the higher order functions. More explicity, if we define the ensemble average of a quantity f by

$$\langle f \rangle = \frac{\int d\underset{\sim}{R}_1 \ldots d\underset{\sim}{R}_N\, f(\underset{\sim}{R}_1, \ldots \underset{\sim}{R}_N) \exp(-\beta\Phi)}{\int d\underset{\sim}{R}_1 \ldots d\underset{\sim}{R}_N \exp(-\beta\Phi)} \tag{3.18}$$

then

$$<\rho(\underset{\sim}{q})\rho(-\underset{\sim}{q})> = <\sum_{\ell,\ell'} \exp(i\underset{\sim}{q}.(\underset{\sim}{R}_\ell - \underset{\sim}{R}_{\ell'}))>$$

$$= NS(q) + N^2\delta_{q,o} \qquad (3.19)$$

where S(q) is the liquid structure factor.(We have now assumed that the liquid is homogeneous and the existence of a thermodynamic limit i.e. we take $N\to\infty$, $V\to\infty$, $N/V = \rho$, the constant ion density, at the end of the calculation). This means we can derive rather simple expressions for some of the important thermodynamic properties. For example the internal energy U:

$$U = -\frac{\partial}{\partial\beta}\left(\ell n\, Q_N\right)_V$$

$$= -\frac{\partial}{\partial\beta}\ell n\left(\left(\frac{2\pi M}{h^2\beta}\right)^{\frac{3N}{2}}\frac{1}{N!}\int d\underset{\sim}{R}_1 \ldots d\underset{\sim}{R}_N \exp(-\beta\Phi)\right)$$

$$= \frac{3N}{2} K_BT + F_{eg} + U_1 + <U_2> + <U_M>$$

$$= \frac{3N}{2} K_BT + F_{eg} + U_1 + \frac{N}{2V}\sum_{\underset{\sim}{q}\neq o} \chi^e(q)|v_{ps}(q)|^2 S(q)$$

$$+ \frac{N}{2V}\sum_{\underset{\sim}{q}\neq o}\frac{4\pi Z^2e^2}{q^2}(S(q)-1) \qquad (3.20)$$

Converting the summations over $\underset{\sim}{q}$ to integrals the internal energy per ion becomes:

$$\frac{U}{N} = \frac{3}{2}K_BT + \frac{F_{eg}}{N} + n_ow_c(q=o) + \frac{1}{4\pi^2}\int_0^\infty dq\, q^2\chi^e(q)\, |v_{ps}(q)|^2 S(q)$$

$$+ \frac{1}{4\pi^2}\int_0^\infty dq\, 4\pi Z^2e^2(S(q)-1) \qquad (3.21)$$

where the first term is the classical Kinetic energy of the ions, the fourth is often called the 'band structure energy' and the last the 'Madelung' energy. Precisely equivalent terms arise in the theory of the total energy of crystalline metals [1].

The pressure p and isothermal bulk modulus B_T can also be evaluated by the standard methods used for simple liquids. The resultant expressions involve derivatives of F_{eg}, U_1 and $\chi^e(q)$ with respect to the average electron density n_0 as well as derivatives with respect to q. The former are clearly metallic in origin and they arise because when V is changed n_0 must alter to maintain charge neutrality. We will discuss this in some detail later.

At temperatures appropriate to simple liquid metals close to their melting points the free energy of the electron gas can be approximated by the ground state energy. (The correction terms are $\sim -\frac{5\pi^2}{12}\left(\frac{K_BT}{E_F}\right)^2$ where E_F is the free electron Fermi Energy.)

The ground state energy per electron is (in Rydbergs):

$$u_{eg}(r_s) = \frac{2\cdot 21}{r_s} - \frac{0\cdot 916}{r_s} + E_{corr}(r_s) \tag{3.22}$$

where the electron radius is defined by $4\pi r_s^3/3 = n_0^{-1}$. The first term is the Kinetic energy, the second is the exchange energy and the third the correlation contribution. Several theories exist for the latter. These all yield similar results for metallic densities (i.e. for $1 < r_s < 6$ a.u.). One in common use is due to Nozières and Pines [20].

$$E_{corr}(r_s) = -0\cdot 115 + 0\cdot 031 \ln r_s \tag{3.23a}$$

and a more recent formula is that of Vashista and Singwi [21] :

$$E_{corr}(r_s) = -0\cdot 112 + 0\cdot 0335 \ln r_s - \frac{0\cdot 02}{0\cdot 1 + r_s} \tag{3.23b}$$

The response function $\chi^e(q)$ can also be approximated by its zero temperature limit and this is conveniently written in terms of the non-interacting response function $\chi^{ni}(q)$ and the effective electron-electron interaction $U_{ee}(q)$ [20] :

$$\chi^e(q) = \chi^{ni}(q)\left(1 - \chi^{ni}(q)\, U_{ee}(q)\right)^{-1} \tag{3.24}$$

In the random phase approximation U_{ee} is set equal to the direct Coulomb interaction $4\pi e^2/q^2$ (this is also called the Hartree approximation). At metallic densities this is a rather poor approximation and many schemes have been developed which go beyond this and explicity include exchange and correlation effects so that U_{ee} can be written as:

$$U_{ee}(q) = \frac{4\pi e^2}{q^2} + U_{xc}(q) \tag{3.25}$$

Some explicit formulae for the exchange and correlation correction $U_{xc}(q)$ are given in Appendix 2. χ^{ni} is simply the Lindhard function:

$$\chi^{ni}(q) = -\frac{K_F}{4\pi^2}\left(1 + \frac{(4K_F^2 - q^2)}{4K_F q}\;\ln\left|\frac{q + 2K_F}{q - 2K_F}\right|\right) \tag{3.26}$$

(in Rydberg units).

Thus, given a suitable pseudopotential and some means of evaluating the structure factor, the internal energy is readily calculated from equ. (3.21). Some of the first calculations used experimental data for S(q) and an ansatz for its density dependence, to evaluate U, p and B_T for several liquid metals [22]. This was rather unsatisfactory since a 'proper' theory should also give a prescription for S(q) so that self-consistency is maintained. It is well known that such considerations are extremely important in calculations of thermodynamic properties. Before discussing how S(q) or its real space equivalent, the radial distribution function, can be obtained from the theory it is instructive to investigate the consequences of going beyond second order in δV in the expansion of the free energy of the electron system.

For finite q the electron density can be expanded as:

$$n(\underset{\sim}{q};\lambda) = \lambda\chi^{e}(q)\;\delta V(\underset{\sim}{q}) + \lambda^2\sum_{\underset{\sim}{K}}\chi^{(2)}(\underset{\sim}{q},\underset{\sim}{K})\;\delta V(\underset{\sim}{K})\delta V(\underset{\sim}{q} - \underset{\sim}{K}) + \cdots \tag{3.27}$$

where $\chi^{(2)}$ is the second-order linear response function. It is clear that the inclusion of the term in λ^2 will introduce three body correlations $\langle\rho(\underset{\sim}{q})\,\rho(\underset{\sim}{K})\,\rho(-\underset{\sim}{q} - \underset{\sim}{K})\rangle$ and the high order terms will involve higher order correlations. The second and higher order response functions are not well understood. Consequently the usefulness of the present approach is rapidly diminished if linear response is inadequate.

3.3 Real space representation and effective pairwise potentials.

Equ. (3.17) for the effective interaction potential can be transformed into a convenient real space representation. The structure dependent terms can be written as:

$$U_2 + U_M = \frac{1}{2V}\sum_{q}\phi(q;n_o)\sum_{\ell\neq\ell'}\exp\left(i\underset{\sim}{q}.(\underset{\sim}{R}_\ell - \underset{\sim}{R}_{\ell'})\right)$$

$$+\frac{N}{2V}\sum_{\underset{\sim}{q}\neq o}\chi^{e}(q)|\,v_{ps}(q)|^2 - \frac{N(N-1)}{2V}\lim_{q\to o}\phi(q;n_o) \tag{3.28}$$

where

$$\phi(q;n_o) = \frac{4\pi\,Z^2e^2}{q^2} + \chi^{e}(q)|\,v_{ps}(q)|^2 \tag{3.29}$$

(The n_o in $\phi(q;n_o)$ signifies that χ^e depends on the average electron density). The limit in the last term of equ. (3.28) can be evaluated:

$$\phi(q = o; n_o) = 2Z\, w_c\,(q = o) - Z^2\left(\chi^{ni}(q = o)^{-1} - U_{xc}\,(q = o)\right) \tag{3.30}$$

and substituting from equ. (3.28) into equ. (3.17) we have

$$\Phi(\underset{\sim}{R}_1, \ldots \underset{\sim}{R}_N) = Nu(n_o) + \frac{1}{2}\sum_{\ell \neq \ell'} \phi(|\underset{\sim}{R}_\ell - \underset{\sim}{R}_{\ell'}|;\, n_o) \tag{3.31}$$

with

$$u(n_o) = Zu_{eg}(n_o) + \frac{Zn_o}{2}\left(\chi^{ni}(q{=}o)^{-1} - U_{xc}(q = o)\right) + \frac{1}{2V}\sum_{\underset{\sim}{q} \neq o} \chi^e(q)\,|v_{ps}(q)|^2 \tag{3.32}$$

and
$$\phi(R;n_o) = \frac{1}{(2\pi)^3}\int d\underset{\sim}{q}\, \exp(-\, i\underset{\sim}{q}.\underset{\sim}{R})\, \phi(q;n_o) \tag{3.33}$$

We have assumed both F_{eg} and χ^e take their zero temperature values. $u(n_o)$ is not structure dependent and is sometimes referred to as the self-energy of a pseudo-ion. The effective pairwise potential $\phi(R;n_o)$ combines the direct Coulomb repulsion between the ions and the effects of polarization of the ion by the electron screening cloud.

If the bare pseudopotential is non-local the physical content of ϕ remains the same but $\chi^e(q)|\,v_{ps}(q)|^2$ must be replaced by a more complicated energy-wave number characteristic [23]. Taking the q = o limit then becomes difficult and, in general, there is no simple, explicit expression for $u(n_o)$.

If the term in λ^2 is retained in equ. (3.27) the real space representation for Φ becomes considerably more complicated:

$$\Phi(\underset{\sim}{R}_1, \ldots \underset{\sim}{R}_N) = Nu'(n_o) + \frac{1}{2}\sum_{\ell \neq \ell'} \phi'\,(|\underset{\sim}{R}_\ell - \underset{\sim}{R}_{\ell'}|;\, n_o) \tag{3.34}$$
$$+ \frac{1}{3}\sum_{\ell \neq \ell' \neq \ell''} \phi^{(3)}\,(\underset{\sim}{R}_\ell;\, \underset{\sim}{R}_{\ell'};\, \underset{\sim}{R}_{\ell''};n_o)$$

Both the self energy and the pair potential are modified and an explicit three-body potential $\phi^{(3)}$ appears [24].

Equ. (3.31) is a convenient starting point for calculations of liquid state properties. If we fix the ion density (and hence the average electron density) the structure of the liquid is completely determined by the effective pairwise potential $\phi(R;n_0)$ The atomic distribution functions can be calculated using the various techniques (Monte-Carlo, molecular dynamics, perturbation theories) which were developed for the simple insulating liquids. The thermodynamic functions can be derived after the usual fashion. The internal energy is:

$$\frac{U}{N} = \frac{3}{2} K_B T + u(n_0) + \frac{\rho}{2}\int d\underset{\sim}{R} \; g(R)\,\phi(R;n_0) \qquad (3.35)$$

where g(R) is the radial distribution function evaluated for the particular density under consideration. This result differs from that for simple fluids due to the presence of the self-energy term. Other constant volume quantities such as the entropy and the specific heat will have the same form as in simple liquids - apart from very small contributions from the entropy of the free electron gas. Quantities which depend on volume derivatives are more subtle. The pressure can be calculated using the standard trick of scaling the variables:

$$p = \rho K_B T + \rho n_0 \frac{du(n_0)}{dn_0} - \frac{\rho^2}{2}\int d\underset{\sim}{R}\; g(R)\left(\frac{R}{3}\frac{\partial}{\partial R} - n_0\frac{\partial}{\partial n_0}\right)\phi(R;n_0)$$

(3.36)

and differs significantly from the familiar virial equation of state. The liquid metal is not in equilibrium under the influence of pairwise forces along; the additional terms reflect theinherent two component nature of the system. The bulk modulus is:

$$B_T = \rho K_B T + \rho\left(2n_0 \frac{du(n_0)}{dn_0} + n_0^2\frac{d^2u(n_0)}{dn_0^2}\right) + \frac{Vd}{dV}\left(\frac{\rho^2}{2}\int d\underset{\sim}{R}\; g(R)\times\right.$$

$$\left.\left(\frac{R\partial}{3\partial R} - \frac{n_0\partial}{\partial n_0}\right)\phi(R;n_0)\right) \qquad (3.37)$$

Price [25] carried out a careful study of U,p and B_T in liquid Na using a reliable pseudopotential to calculate$\phi(R;n_0)$ and a radial distribution function obtained from a computer simulation based on the same pair potential. To our knowledge this is the only detailed calculation of this kind (g(R) has been generated by several authors using various calculated pair potentials [26,27]) although Mountain [28] has recently evaluated g(R) and several portions of the equation of state using two different pair potentials for Na and K. Most authors have employed thermodynamic perturbation theories rather than full scale computer simulations.

3.4 The form of the calculated pairwise potentials

$\phi(R;n_0)$ has been calculated by many authors for many simple metals. However the pair potentials, especially in the region of the first minimum, are quite sensitive to the choice of pseudopotential and to the form of the exchange and correlation correction U_{XC} in the response function. Some consensus of what are 'good' pseudopotentials and what constitutes a reasonable treatment of exchange and correlation has emerged [23].

In figure 1 we plot pairwise potentials for liquid Na calculated for different choices of U_{XC}. Both the depth of the first minimum and the short-range repulsive part vary somewhat. In the polyvalent metals ϕ is more sensitive to U_{XC} (see figure 2 where results for Al are presented).

This sensitivity results from a strong cancellation between the direct Coulomb repulsion and the polarisation term. In Al, for $R \sim 5$ a.u. the Coulomb part is ~ 3 Ryd but the resultant $\phi(R;n_0)$ is only $\sim 10^{-3}$ Ryd.

The depth of the first minimum does not appear to vary in any systematic way from element to element. It is usually in the range $1 - 4 \times 10^{-3}$ Ryd and these are typical thermal energies $\sim K_B T_M$. It does not correlate with the observed cohesive energies (heat of vaporization). Since the latter are $15 - 40\ K_B T_M$ this implies that most of the binding energy of the metal resides in the self-energy term and explicit calculations for a wide range of simple metals [29] show the pairwise contribution in equ. (3.35) is only a small fraction of the total. This is completely different from the liquid rare gases.

In figure 3 the density dependence of the pair potential is illustrated. The depth of the first minimum varies strongly with n_0 but the short-range repulsive force is relatively insensitive. The repulsive part of $\phi(R;n_0)$ exhibits a systematic variation from element to element. This is illustrated in figure 4 where pairwise potentials for the alkali series are plotted. On proceeding down this column in the periodic series the first minimum of ϕ shifts to progressively larger separations reflecting the increasing size of the metallic atoms. We should note that the first zero of the pair potential is much larger than the diameter of the ion core in metals. For example in Na the diameter of the core is < 4 a.u. while the zero of the potential occurs at ~ 6 a.u. The repulsive force in the metal arises from the incomplete screening of the direct Coulomb interaction between the ions at small R. This is to be contrasted with the rare gases where the short range forces are due repulsion between closed electron shells. From figure 4 we also see that the repulsive part

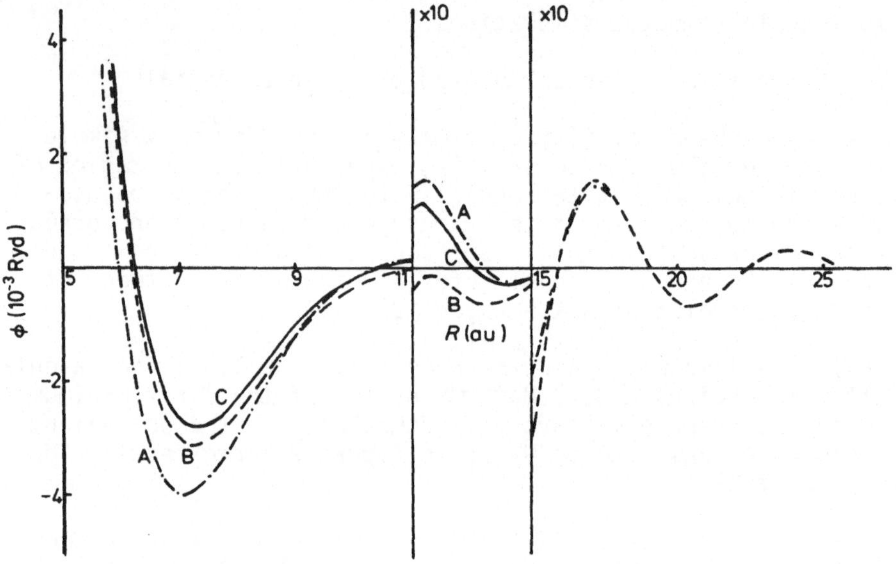

Figure 1. Pair potentials for liquid Na calculated with different forms of exchange and correlation correction U_{xc}: curve A(chain) from Vashista and Singwi; curve B(broken) from Kleinman; curve C (full) from Singwi et al. (From ref. [23]).

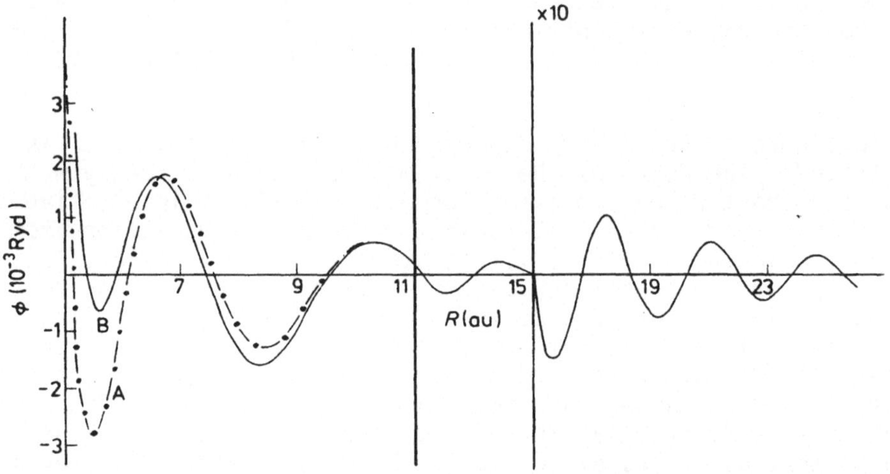

Figure 2. Pair potentials for liquid Al calculated with different forms of exchange and correlation correction U_{xc}: curve A(chain) Vashista and Singwi; curve B (full) Kleinman. (From ref. [23])

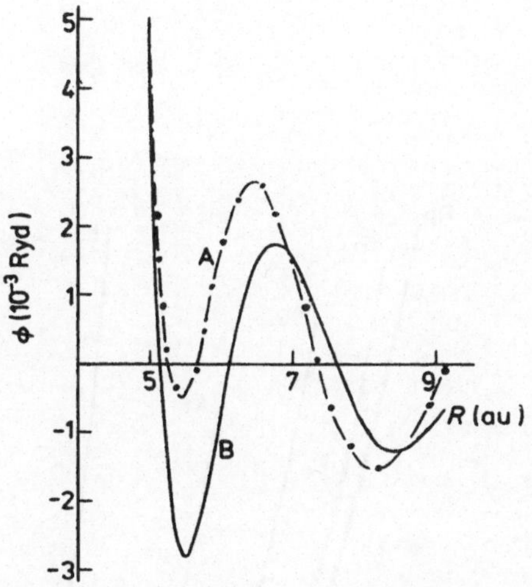

Figure 3. Pair potentials for Al calculated at densities corresponding to the low-temperature solid (curve A, chain) and the liquid (curve B, full). (From ref. [23].)

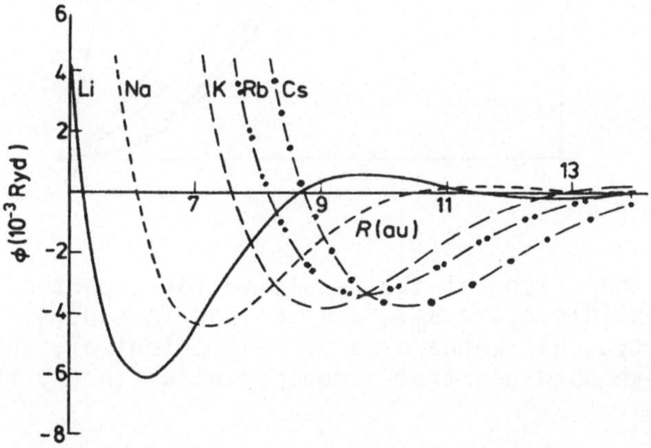

Figure 4. Pair potentials calculated for the liquid alkali metals. (From ref. [23].)

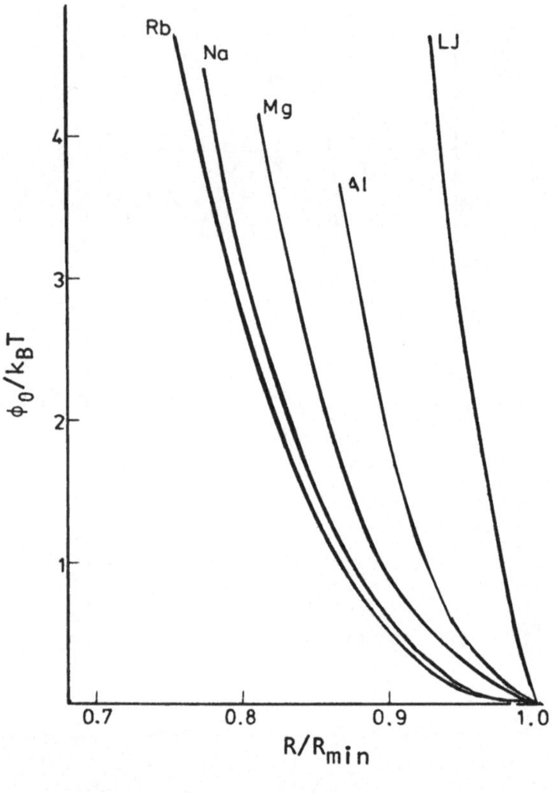

Figure 5. A comparison of the repulsive force part of various pairwise potentials. ϕ_0 and R_{min} are defined in equ.(4.11). Cure L J refers to a 'typical' Lennard-Jones 6-12 potential; the other potentials were obtained from pseudopotential theory. (From ref.[23]).

of ϕ 'softens' on going from Li towards Cs i.e. as the average electron density is reduced. This softening is further illustrated in figure 5 where the repulsive part of ϕ is plotted for several metals and for a typical Lennard-Jones 6-12 potential. The latter is much harder than the metallic potentials. It is significant that the pair potentials for Mg and Al are considerably harder than those for the two alkali metals and we will return to this point later.

In the next chapter we shall see that the systematic variation of the size of the metallic atoms has important consequences for both the entropy and the structure factor in liquid metals.

4. THE APPLICATION OF THERMODYNAMIC PERTURBATION THEORIES

Although the 'metallic' pairwise potentials described above differ in detail from those used to describe say rare-gas liquids they still posess a strongly repulsive part and a fairly weak attractive tail and so it is not surprising that perturbation theories, which were originally developed for simple liquids, have been applied to the present model of a liquid metal. Since the thermodynamic perturbation theories have been comprehensively reviewed [30,31,32] here we concentrate on their application to metals but, for the sake of clarity, we include some background material.

All theories exploit the similarity between the structure factor of a simple liquid (or liquid metal) and that of a hard sphere with an appropriately chosen hard sphere diameter. The difference between the true pairwise potential and the hard sphere potential is then treated as a perturbation. All the relevant thermodynamic properties can then be evaluated provided the corresponding quantities for the hard sphere system are known. The method of choosing the hard sphere diameter and the detailed structure of the perturbation expansion differ from theory to theory. For pure metals the variational approach introduced by Mansoori and Canfield for simple liquids [33], has been widely used [34-41, 23] and the extension to alloys has also been employed [39, 40, 42-44] (An excellent review of this approach and its applications is given by Ashcroft and Stroud [19]). The scheme of Weeks,Chandler and Anderson (WCA) [45,46] has been used to investigate the structure factors [47,48, 23, 24] and some thermodynamic properties [23] of pure liquid metals.

4.1 The variational approach.

We divide the total interaction potential Φ of equ. (3.31) into a part Φ_0 corresponding to a reference system and a part Φ_1 corresponding to a perturbation:

$$\Phi = \Phi_0 + \Phi_1 \tag{4.1}$$

It is well-known that the free energy of the true system satisfies the following inequality:

$$F \leqslant F_0 + \langle \Phi_1 \rangle_0 \tag{4.2}$$

where F_0 is the free energy of the reference system at the same density ρ and $\langle \; \rangle_0$ denotes a configuration average over the reference system. This result is often called the Gibbs-Bogoliubov inequality. Since $\phi(R;n_0)$ has a strongly repulsive part it is convenient to introduce a reference system of hard-spheres of diameter d:

$$\Phi_0 = \tfrac{1}{2} \sum_{\ell \neq \ell'} \phi_{hs}(|\underset{\sim}{R}_\ell - \underset{\sim}{R}_{\ell'}|) \tag{4.3}$$

where $\phi_{hs}(R) = \infty \quad R \leqslant d$

$\qquad\qquad = 0 \quad R > d$

It follows that

$$F \leqslant F_{hs} + Nu(n_0) + \frac{N}{2}\rho \int d\underset{\sim}{R} \; g_{hs}(R) \; \phi(R;n_0) \tag{4.4}$$

since for the hard sphere fluid $g_{hs} = 0$ for $R \leqslant d$. F_{hs} is the free energy of a hard sphere fluid of density ρ. The diameter d is then used as a variational parameter to minimize the R.H.S. of equ.(4.4). The minimum value is then taken as the free energy of the true system. Since $Nu(n_0)$ is structure independent this procedure is entirely equivalent to that used for simple, insulating liquids. The resultant estimate of the free energy satisfies

$$\left(\frac{\partial F}{\partial d}\right)_{V,T} = 0.$$

or alternatively

$$\left(\frac{\partial F}{\partial \eta}\right)_{V,T} = 0. \tag{4.5}$$

where $\eta = \frac{\pi d^3}{6}\rho$ is the packing fraction of the hard spheres. The entropy is given by

$$S = -\left(\frac{\partial F}{\partial T}\right)_V$$

$$= \left(\frac{\partial F}{\partial T}\right)_{V,\eta} - \left(\frac{\partial F}{\partial \eta}\right)_{V,T}\left(\frac{\partial \eta}{\partial T}\right)_V \qquad (4.6)$$

and the second term vanishes by virtue of equ.(4.5). In the present approximation the only explicit temperature dependence in F is in F_{hs} so S takes a particularly simple form:

$$S_{VAR} = -\left(\frac{\partial F_{hs}}{\partial T}\right)_{V,\eta} \qquad (4.7)$$

which is just the entropy of the hard sphere fluid of density ρ in which d is fixed via equs. (4.4) and (4.5). (There will be a contribution to S from the free energy of the electron gas but this almost two orders of magnitude smaller than the ionic contribution in simple metals).

Using this result we can write the internal energy as:

$$\frac{U}{N} = \frac{3}{2} K_B T + u(n_o) + \frac{\rho}{2}\int d\underset{\sim}{R}\; g_{hs}(R)\, \phi(R;n_o) \qquad (4.8)$$

The pressure can also be evaluated [49]:

$$p = \rho K_B T + \rho n_o \frac{du(n_o)}{dn_o} - \frac{\rho^2}{2}\int d\underset{\sim}{R}\; g_{hs}(R) \left(\frac{R}{3}\frac{\partial}{\partial R} - n_o \frac{\partial}{\partial n_o}\right) \phi(R;n_o) \qquad (4.9)$$

Thus the variational approach is equivalent to replacing the true g(R) in the exact equs. (3.35) and (3.36) by its hard sphere equivalent - with the diameter chosen so equ. (4.5) is satisfied.

In table 1 we present values of the effective hard sphere diameter, packing fraction and excess entropy S_E calculated by Kumaravadivel and Evans [23] using the variational approach.

These calculations employed the Watts and Henderson [50] $g_{hs}(r)$ and the Carnahan and Starling [51] analytical form for F_{hs}. The latter is

$$F_{hs} = F_i + NK_B T\, f_{cs}(n) \qquad (4.10)$$

with

$$f_{cs}(\eta) = (4\eta - 3\eta^2) \,/\, (1 - \eta)^2$$

Table 1. Packing fractions η, effective hard-sphere diameters d, and excess entropies S_E calculated by the variational approach (VAR) and the WCA approach for temperatures near the melting point (From ref. [23]).

Metal	d^{VAR} (a.u.)	d^{WCA} (a.u.)	η^{VAR}	η^{WCA}	VAR $-S_E/K_B$	WCA $-S_E/K_B$	$-S_E/K_B$ experiment
Cs	9·14	9·15	0·48	0·48	4·54	3·78	3·56
Rb	8·35	8·46	0·47	0·49	4·33	3·73	3·63
K	7·88	7·91	0·48	0·49	4·54	3·84	3·45
Na	6·35	6·36	0·48	0·48	4·54	3·83	3·45
Cd	5·00	4·90	0·42	0·39	3·42	2·82	4·00
Mg	5·41	5·41	0·47	0·47	4·33	3·68	3·29
Zn	4·64	4·72	0·47	0·49	4·33	3·80	3·78
Ga	4·83	4·94	0·46	0·49	4·13	4·05	4·62
Al	4·77	4·80	0·45	0·46	3·94	3·42	3·49
Sn	5·31	5·37	0·41	0·42	3·26	3·02	4·08

F_i is the free energy of the ideal gas at the same density and temperature.

The overall level of agreement between the theoretical and experimental values of S_E is reasonable. However it is significant that in the alkali metals the calculated values of $-S_E$ are larger than the experimental values. For all of the metals listed in table 1 the calculated packing fractions lie between 0·41 and 0·48. Since it is well known that the hard sphere structure factor evaluated for $\eta \sim 0{\cdot}45$, at the appropriate density, gives a reasonably accurate description of the first peak of the experimental structure factor in most liquid metals, it is clear that these calculations predict sensible model structure factors as well as entropies. For some other metals [23] the calculated values of η were $>0{\cdot}5$ leading to unphysical entropies. This implies that the calculated pair potentials were non-realistic for these latter metals (see also [37]).

4.2 The WCA approach.

This perturbation theory stresses the role played by the

repulsive force part of the pairpotential. WCA separate the potential into a part which gives rise to a repulsive force and a remainder which contains all the attractive force. For a Lennard-Jones fluid this is straight-forward but for liquid metals such a division is complicated by the long range Friedel oscillations in $\phi(R;n_o)$. We ignore such complications and write

$$\phi(R;n_o) = \phi_0(R;n_o) + \phi_1(R;n_o) \tag{4.11}$$

where

$$\phi_0(R;n_o) = \begin{cases} \phi(R;n_o) - \phi_{min} & R \leqslant R_{min} \\ 0 & R > R_{min} \end{cases}$$

$$\phi_1(R;n_o) = \begin{cases} \phi_{min} & R \leqslant R_{min} \\ \phi(R;n_o) & R > R_{min} \end{cases}$$

R_{min} is the position of the first minimum in the pair potential and ϕ_{min} is the value of the potential at that point. Consider a reference system of the same density and temperature as the true system but in which the potential of interaction is

$$\frac{1}{2} \sum_{\ell \neq \ell'} \phi_0(|\underset{\sim}{R}_\ell - \underset{\sim}{R}_{\ell'}|;\, n_o)$$

Let the free energy of this system be F_o and the radial distribution function be $g_o(R)$. The free energy of the true system can then be obtained by introducing the usual coupling constant algorithm:

$$F = F_o + Nu(n_o) + \frac{N}{2}\rho \int_0^1 d\lambda \int d\underset{\sim}{R} \quad g(r;\lambda) \quad \phi_1(R;n_o) \tag{4.12}$$

where $g(R;\lambda)$ is the radial distribution function in the system whose potential of interaction is Φ^T where

$$\Phi^T(\underset{\sim}{R}_1, \ldots \underset{\sim}{R}_N) = \frac{1}{2} \sum_{\ell \neq \ell'} [\Phi_0(|\underset{\sim}{R}_\ell - \underset{\sim}{R}_{\ell'}|;n_o) + \lambda\phi_1(|\underset{\sim}{R}_\ell - \underset{\sim}{R}_{\ell'}|;n_o)] + \lambda Nu(n_o) \tag{4.13}$$

with $o \leqslant \lambda \leqslant 1$. F can be re-expressed as:

$$F = F_o + Nu(n_o) + \frac{N}{2}\rho \; \phi_1(q = o;n_o) + \frac{N}{2(2\pi)^3}\int d\underset{\sim}{q}\; h_o(q)\phi_1(q;n_o)$$

$$+ \frac{N}{2(2\pi)^3}\int_0^1 d\lambda \int d\underset{\sim}{q} \left(h(q;\lambda) - h_0(q)\right)\phi_1(q;n_0) \qquad (4.14)$$

where $h(R) = g(R) - 1$, h_0 refers to the reference system and

$$h(q) = \rho\int d\underset{\sim}{R}\ \exp\ (i\underset{\sim}{q}.\underset{\sim}{R})\ h(R).$$

The liquid structure factor is given by

$$S(q) = h(q) - 1.$$

The essential hypothesis of the WCA treatment is that for a dense fluid the behaviour of $S(q)$, even for fairly small wavevectors, should be determined by the repulsive force part of the pairwise potential. This is an assertion that the density correlations in the fluid are almost entirely due to excluded volume effects - provided the average density is large enough. WCA suggest $\rho\sigma^3$ should be $\gtrsim 0{\cdot}65$, where σ is the position of the first zero of pair potential. For liquid metals at temperatures near T_M, $\rho\sigma^3$ is typically $\sim 0{\cdot}95$. If we accept this hypothesis it follows that $h(q;\lambda) \approx h_0(q)$ - except, perhaps, at low values of q, so the last term in equ. (4.14) should be negligible. The free energy is reduced to

$$F = F_0 + Nu(n_0) + \frac{N}{2}\rho\int d\underset{\sim}{R}\ g_0(R)\ \phi_1(R;n_0) \qquad (4.15)$$

The form of this result is identical to equ. (4.4) but the reference system is not hard spheres. Further analysis is required to express F_0 and $g_0(R)$ in terms of hard-sphere quantities. F_0 is expanded in a functional Taylor series about F_{hs}:

$$F_0 = F_{hs} + \int d\underset{\sim}{R}\ \frac{\delta F[f]}{\delta f(R)}\bigg|_{hs} \Delta f(R) + \frac{1}{2}\iint d\underset{\sim}{R}\, d\underset{\sim}{R}'\ \frac{\delta^2 F[f]}{\delta f(R)\delta f(R')}\bigg|_{hs} \Delta f(R)\Delta f(R')$$

$$+ \ \cdots\cdots\cdots \qquad (4.16)$$

where $f(R) = \exp\ (-\beta\phi(R)) - 1$ is the Mayer function and

$$\Delta f(R) = \exp\ (-\beta\phi_0(R;n_0)) - \exp\ (-\beta\phi_{hs}(R)). \qquad (4.17)$$

Provided the hard sphere diameter d is chosen sensibly the 'blip' function $\Delta f(R)$ is only finite over a short range $\mathfrak{J}d$ in the neighbourhood of $R = d$. $\mathfrak{J}$ clearly depends on the softness of the repulsive part of the pair potential ϕ_0 - the softer the potential the larger is $\mathfrak{J}$. Higher order terms involve higher powers of $\mathfrak{J}$. The functional derivatives can be evaluated [46]. The first one is simple:

$$\frac{\delta F[f]}{\delta f(R)} = -\frac{NK_BT\rho}{2} y(R) \tag{4.18}$$

where $y(R) = \exp(\beta\phi(R))\, g(R)$

while the second derivative involves three and four body distribution functions. Thus the free energy expansion is of the form:

$$F_0 = F_{hs} - \frac{N}{2} K_BT\rho \int d\underset{\sim}{R}\; y_{hs}(R)\, \Delta f(R) + \text{higher order terms} \tag{4.19}$$

An obvious choice for the hard sphere diameter d is to require that the first order term in equ. (4.19) vanishes i.e.

$$\int d\underset{\sim}{R}\; y_{hs}(R)\; \Delta f(R) = 0 \tag{4.20}$$

For any realistic repulsive potential d is a decreasing function of temperature and decreases slowly with increasing density at constant temperature. With this choice of d equ. (4.19) becomes:

$$F_0 = F_{hs}[1 + 0(\xi^4)] \tag{4.21}$$

i.e. no terms of $0(\xi^2)$ appear [46] . The higher order terms should be negligible provided the repulsive potential is steep enough. The radial distribution function $g_0(R)$ can be generated from equ. (4.19) using equ. (4.18) and

$$y_0(R) = y_{hs}(R) + \text{higher order terms}$$

thus

$$g_0(R) = \exp(-\beta\phi_0(R;n_0))y_{hs}(R)[1 + 0(\xi^2)] \tag{4.22}$$

On Fourier transforming we obtain the structure factor of the reference system:

$$S_0(q) = S_{hs}(q) + \rho\int d\underset{\sim}{R}\; \exp(i\underset{\sim}{q}.\underset{\sim}{R})\; y_{hs}(R)\, \Delta f(R) + 0(\xi^2) \tag{4.23}$$

where $S_{hs}(q)$ is the hard sphere structure factor.

From equ. (4.15) we find the free energy of the metal is

$$F = F_{hs} + Nu(n_0) + F_A \tag{4.24}$$

where the 'attractive' contribution F_A is given by

$$F_A = \frac{N\rho}{2}\int d\underset{\sim}{R}\ \exp\left(-\beta\phi_0(R;n_0)\right)y_{hs}(R)\phi_1\ (R;n_0)$$

$$= 2\pi N\rho\int_0^{R_{min}} dR\ R^2\left(\exp(-\beta\phi_0(R;n_0))-1\right)y_{hs}(R)\ \phi_1(R;n_0)$$

$$+ 2\pi N\rho\int_0^{\infty} dR\ R^2 y_{hs}(R)\phi_1(R;n_0)$$

$$= 2\pi N\rho\int_d^{\infty} dR\ R^2\ g_{hs}\ (R)\phi_1\ (R;n_0) \qquad (4.25)$$

and we have used equs. (4.11) and (4.20) and the fact that $y_{hs}(R) = g_{hs}(R)$ for $R>d$.

This version of the WCA theory neglects any effects of the attractive tail ϕ_1 on the structure of the liquid metal. Such an approximation should be valid at high temperatures and so it is usually referred to as the HTA (high temperature approximation). For a Lennard-Jones fluid the HTA gives an excellent quantitative description of the thermodynamic properties not only at high temperatures but also at high densities for temperatures as low as the triple point. It also gives a good qualitative account of the structure factor over this range. Provided the long range Friedel oscillations are not important, and the repulsive potentials not too soft, we might expect a similar degree of success for liquid metals near their melting points since we earlier argued the attractive part of the potential was rather weak.

[More sophisticated theories, which remedy some of the defects of the HTA at lower densities, have been developed for simple fluids [30-32] but to our knowledge these have not been used for calculations on liquid metals.]

It is instructive to compare the entropy calculated from the HTA approximation with that from the variational method:

$$S_{WCA} = -\left(\frac{\partial F_{hs}}{\partial T}\right)_{V,\eta} - \left(\frac{\partial F_{hs}}{\partial \eta}\right)_{V,T}\left(\frac{\partial \eta}{\partial T}\right)_V - \left(\frac{\partial F_A}{\partial T}\right)_V \qquad (4.26)$$

By comparing with equ. (4.7) we see that there are two additional terms arising from a) the variation of d with temperature and b) the attractive part of the pairwise potential.

In table 1 we have compared the packing fractions and excess entropies calculated by the WCA approach with those obtained from the variational procedure (based on the same pair potentials). The values of η from each scheme are very close. The WCA values of the excess entropy are, with a few exceptions, in better agreement with experiment than the results from the variational approach. The hard-sphere contribution to $-S_E$ is reduced by the sum of the terms in $(\partial\eta/\partial T)_V$ and $(\partial F_A/\partial T)_V$ [23] and for the alkali metals, in particular, the WCA values of S_E are very close to experiment. Constant volume specific heats were also calculated by Kumaravadivel and Evans [23] in the WCA scheme. The results are in fair agreement with experiment.

In figure 6 we compare the results obtained from the WCA procedure, equ. (4.23), with the hard sphere structure factor for liquid Na. The experimental data are also plotted. The large q oscillations in $S_0(q)$ are more damped than those of $S_{hs}(q)$ and this simply reflects the softness of $\phi_0(R;n_0)$. The experimental results are damped in a similar fashion. The WCA procedure leads to spurious behaviour at values of $q \leqslant 1{\cdot}4$ Å^{-1} which implies the Fourier transform of $y_{hs}(R)\ \Delta f(R)$ in equ. (4.23) over estimates the correction to the hard sphere result in this region. This in turn implies the repulsive potential is soft. A better test of the WCA theory is to compare $S_0\ (q)$ with the S(q) generated by a computer simulation using the same potential. Such a comparison is shown in figure 7 for liquid Rb. The results are in reasonable agreement for large q and both exhibit weaker oscillations than the corresponding hard sphere result. For small values of q the WCA treatment again shows spurious behaviour. In figures 8 and 9 we compare the calculated $S_0(q)$ for liquid Mg and Al with the experimental data for these metals. The overall level of agreement between theory and experiment is reasonable and the low q discrepancies are less pronounced than in Na and Rb.

In figure 10 we plot the radial distribution function $g_0(R)$ obtained from the WCA method for Rb and compare it with the result of the computer simulation.

It is clear from these results that the HTA gives a much less accurate desciption of the structure of liquid metals than is the case for Lennard-Jones fluids [45,46] . This does not mean that ignoring the attractive part of the potential is a bad approximation. It is more probable that the 'blip' function expansion of $g_0(R)$ is much less convergent for the soft, metallic potentials and hence the approximation of equ. (4.22) is poorer in metals than in Lennard-Jones fluids. For Mg and Al the repulsive potentials are harder than in the alkalis (figure 5) and the HTA is more accurate. We should also note that the HTA for F_0 may still be very accurate (even if it is poor for g_0) since the correction terms in the former are $O\ (\xi^4)$.

In the long wavelength limit it follows from equs. (4.20) and (4.23) that

$$S_0\ (0) = S_{hs}(0) \tag{4.27}$$

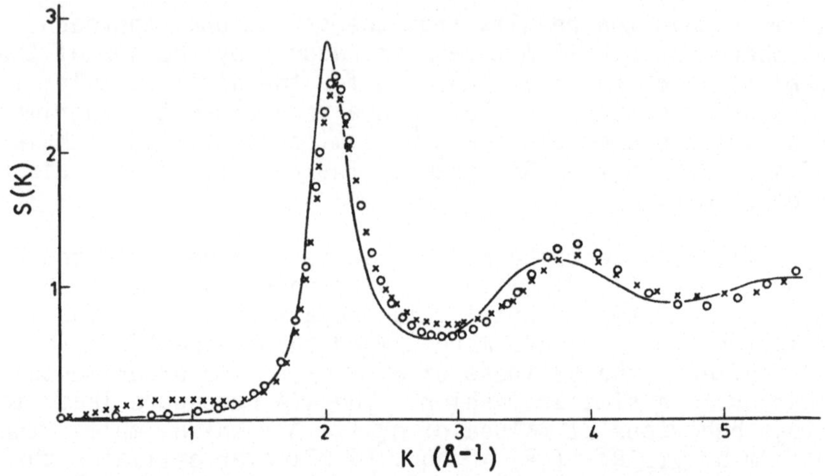

Figure 6. The liquid structure factor of Na calculated from the WCA approach using a pair potential obtained from pseudopotential theory. The crosses denote the WCA calculation; the open circles denote the Percus-Yevick hard sphere structure factor and the full curve represents the experimental results.

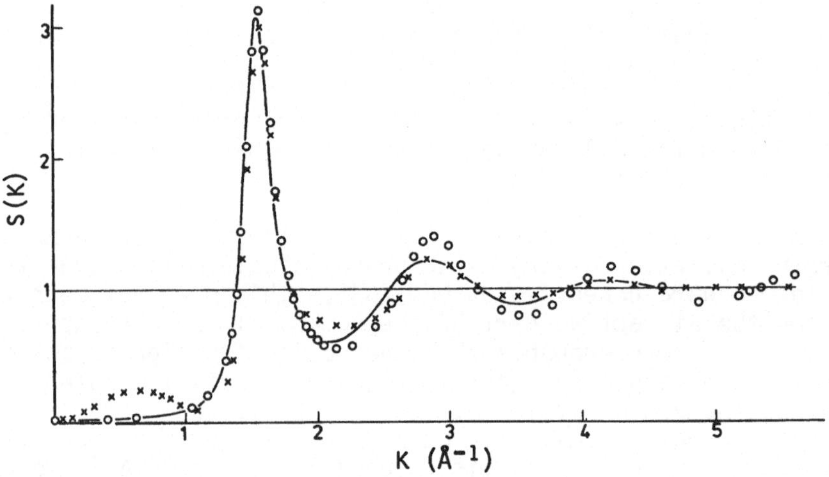

Figure 7. The liquid structure factor of Rb calculated from the WCA approach using a pair potential obtained from pseudopotential theory. The crosses denote the WCA calculation; the open circles denote the Percus-Yevick hard sphere structure factor and the full curve is the result of a molecular dynamics calculation using the same pair potential. (From ref [23]).

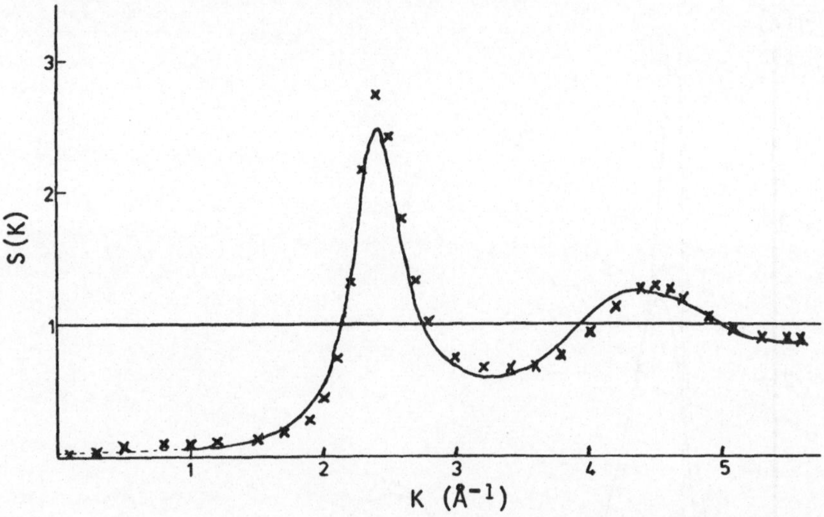

Figure 8. The liquid structure factor of Mg calculated from the WCA approach using a pair potential obtained from pseudopotential theory. The crosses denote the WCA calculation and the full curve represents the experimental results. (From ref. [23])

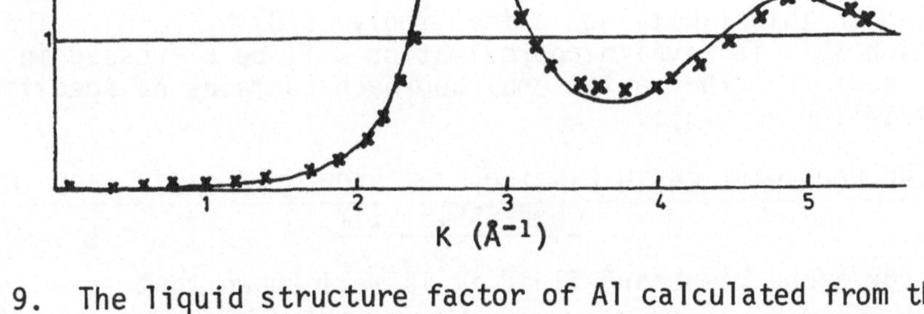

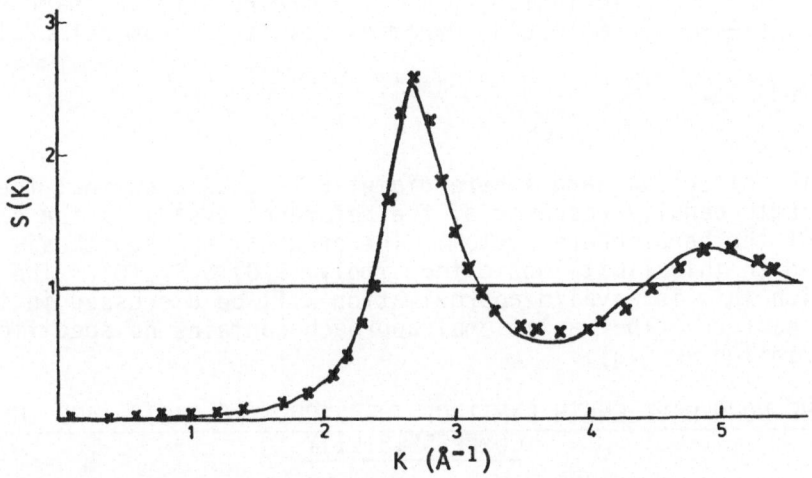

Figure 9. The liquid structure factor of Al calculated from the WCA approach using a pair potential obtained from pseudopotential theory. The crosses denote the WCA calculation and the full curve represents the experimental results (From ref. [23])

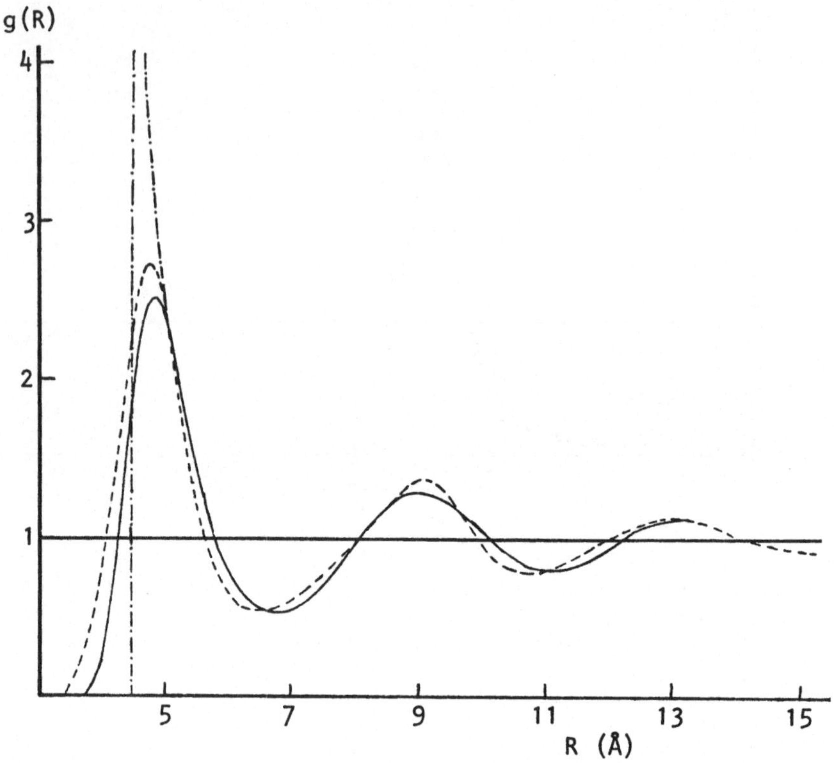

Figure 10. The radial distribution function of liquid Rb: (- - - -) WCA calculation, (–·–·–·) Percus-Yevick hard sphere result; (————) molecular dynamics result. (From ref. [23])

i.e. the effective hard sphere diameter is <u>chosen</u> so that the long wavelength density response of the reference system is the same as that of the hard sphere system. The original WCA hypothesis (if it applied in this limit) would then imply $S(0) \approx S_{hs}(0)$. The extent to which this is a valid approximation will be discussed in the next chapter. (The variational approach contains no specific prescription for $S(q)$).

5. THE LONG WAVELENGTH LIMIT OF THE STRUCTURE FACTOR AND THE COMPRESSIBILITY

For a one component fluid it is well known that

$$S(o) = \frac{N}{V} K_B T \chi_T \qquad (5.1)$$

where χ_T is the isothermal compressibility. This result can be generalized to multicomponent systems to obtain the q = o limit of the various partial structure factors in terms of thermodynamic properties. In charged fluids the requirement of charge neutrality simplifies these relations and for a liquid metal with ions of charge Z it is easy to show [52 - 54]

$$S(o) = \frac{N}{V} K_B T \chi_T \tag{5.2a}$$

$$S_{ee}(o) = \frac{NZ}{V} K_B T \chi_T \tag{5.2b}$$

$$S_{ei}(o) = \frac{NZ^{\frac{1}{2}}}{V} K_B T \chi_T \tag{5.2c}$$

where χ_T is the compressibility of the complete electron-ion system. S refers to the ion-ion, S_{ee} the electron-electron and S_{ei} the electron-ion structure factor. Thus an exact theory of the structure and thermodynamics of a liquid metal should satisfy the compressibility sum rule i.e. χ_T can be evaluated either from the long wavelength limit of S(q) or by taking the volume derivative of the pressure. In an approximate treatment of the electron-ion system the compressibilities obtained from the two different prescriptions will, in general, differ. We will return to this important problem after we have considered the long wavelength limit in some detail.

5.1 Failings of the hard sphere model.

We saw in §4 that the HTA of the WCA theory indicates that S(o) should be roughly equal to the hard sphere equivalent. The latter can be calculated using equ. (5.1). If the Carnahan and Starling result, equ. (4.10), is used for the free energy of the hard sphere system we find

$$S_{hs}(o) = (1 - \eta)^4 \Big/ \left((1 + 2\eta)^2 + \eta^4 - 4\eta^3 \right) \tag{5.3}$$

which differs slightly from the well-known Percus-Yevick result

$$S^{PY}_{hs}(o) = (1 - \eta)^4 \Big/ (1 + 2\eta)^2 \tag{5.4}$$

Since the packing fractions η calculated for liquid metals, and, incidentally, for a Lennard-Jones 6-12 model of liquid Ar, all lie close to 0·45, the HTA would imply $S(o) \approx 0{\cdot}025$ for these liquids. However, the experimental results, derived from measured compressibilities, range between about 0·053 for the liquid rare gases and 0·006 for liquid Ga and Hg. Only for the alkalis and liquid Mg is S(o) reasonably close to 0·025 and the polyvalent metals are considerably less compressible than their equivalent hard sphere system.

This failure of the HTA in the long wavelength limit is not unexpected since it is precisely this limit which should be strongly influenced by the attractive part of the pair potential. In order to study the effects of attraction we could employ the more sophisticated perturbation theories referred to earlier[30, 32] but these were primarily developed for use at finite wavevectors. Here we present an alternative theory which is aimed specificly at the small q density response.

5.2 Linear response theory and the direct correlation function.

For systems obeying classical statistics the structure factor is simply related to the static density response function $\chi(q)$:

$$S(q) = -\frac{1}{\beta\rho}\chi(q) \tag{5.5}$$

As usual χ is written in terms of the non-interacting (ideal gas) response function $\chi^i(q) = -\beta\rho$ and an effective interaction $U(q)$:

$$\chi(q) = \chi^i(q)\left(1 - \chi^i(q)\,U(q)\right)^{-1} \tag{5.6}$$

(This is precisely analogous to the electron problem - see equ. (3.24).)

If we consider a liquid in which the total potential energy can be expressed as a sum of pairwise potentials, the crudest (random phase) approximation sets $U(q) \equiv \phi(q)$ where $\phi(R)$ is the pairwise potential. In this approximation the structure factor reduces to

$$S(q) = (1 + \rho\beta\phi(q))^{-1} \tag{5.7}$$

which is hopeless for quantitative purposes since it takes no account of correlations.

We now suppose that $\phi(R)$ can be divided into a reference part $\phi_0(R)$ and a perturbation part $\phi_1(R)$:

$$\phi(R) = \phi_0(R) + \phi_1(R) \tag{5.8}$$

Let the structure factor corresponding to the reference system of the same density ρ and temperature T as the true system be $S_0(q)$ then

$$S_0(q) = (1 + \rho\beta\,U_0(q))^{-1} \tag{5.9}$$

where U_0 is the effective interaction corresponding to ϕ_0. The structure factor of the true system can then be expressed in terms of that of the reference system:

$$S(q) = S_0(q) \left(1 + \rho\beta S_0(q)\,(U(q) - U_0(q))\right)^{-1} \tag{5.10}$$

Clearly this manipulation is only useful if a) we have an independent theory for $S_0(q)$ (and hence $U_0(q)$) and b) we have some independent way of calculating the difference $U(q) - U_0(q)$. If this difference is, in some sense, small it may be more amenable to approximation schemes than the full $U(q)$. It is this suggestion which we shall investigate.

The simplest approximation replaces the difference by the difference in pair potentials:

$$U(q) - U_0(q) \approx \phi(q) - \phi_0(q) = \phi_1(q) \tag{5.11a}$$

and the corresponding structure factor is

$$S_{RPA}(q) = S_0(q)\left(1 + \rho\beta S_0(q)\,\phi_1(q)\right)^{-1} \tag{5.11b}$$

For convenience we shall refer to this approximation as the RPA. If ϕ_0 is chosen to be the short range repulsive part of ϕ then the attractive perturbation $\phi_1(q)$ will be large for q close to the first peak in $S_0(q)$ (This will be true for any sensible division of ϕ into repulsive and attractive parts). Consequently $S_{RPA}(q)$ will differ strongly from $S_0(q)$ and the corresponding radial distribution function $g_{RPA}(R)$ can be unphysical at small R. Thus the RPA is rather poor at finite q and the various perturbation theories [32] modify it in rather arbitrary ways. Such theories constitute serious attempts to put back important correlations which are missing in the RPA.

It is convenient to introduce the Ornstein-Zernike direct correlation function c(R). This is defined as usual via:

$$g(R) - 1 = c(R) + \rho\int d\underset{\sim}{R}' \left(g(R') - 1\right) c\,(|\underset{\sim}{R} - \underset{\sim}{R}'|) \tag{5.12}$$

or alternatively,

$$\rho c(q) = 1 - 1/S(q) \tag{5.13}$$

Using equs. (5.5) and (5.6) we can make the identification:

$$c(q) \equiv -\beta U(q) \tag{5.14}$$

In this language, the central quantity to be calculated is the difference between the direct correlation functions of the true and reference systems, $c(q) - c_0(q)$. We seek to evaluate this difference for small q by considering the free energy of the corresponding inhomogeneous fluids.

5.3 The direct correlation function in terms of the free energy of an inhomogeneous fluid.

In order to proceed we use the formal relation[55] :

$$c(\underset{\sim}{R}_1, \underset{\sim}{R}_2) = -\beta \frac{\delta^2 \mathcal{F}[\rho]}{\delta\rho(\underset{\sim}{R}_1)\,\delta\rho(\underset{\sim}{R}_2)} + \frac{1}{\rho(\underset{\sim}{R}_1)}\,\delta(\underset{\sim}{R}_1 - \underset{\sim}{R}_2) \qquad (5.15)$$

where $\mathcal{F}[\rho]$ is a unique functional of the density ρ of the inhomogeneous fluid. For the equilibrium density $\mathcal{F}$ is the free energy of the system minus the direct contribution from any external potential which may be present. The direct correlation function $c(\underset{\sim}{R}_1, \underset{\sim}{R}_2)$ satisfies the analogue of the Ornstein-Zernike equation for an inhomogeneous system:

$$g(\underset{\sim}{R}_1, \underset{\sim}{R}_2) - 1 = c(\underset{\sim}{R}_1, \underset{\sim}{R}_2) + \int d\underset{\sim}{R}_3\, c(\underset{\sim}{R}_1, \underset{\sim}{R}_3)\,\rho(\underset{\sim}{R}_3)(g(\underset{\sim}{R}_3, \underset{\sim}{R}_2) - 1) \qquad (5.16)$$

where we introduced the analogue of the radial distribution function, $g(\underset{\sim}{R}_1, \underset{\sim}{R}_2)$. This is defined by:

$$\rho^{(2)}(\underset{\sim}{R}_1, \underset{\sim}{R}_2) \equiv \rho(\underset{\sim}{R}_1)\,\rho(\underset{\sim}{R}_2)\, g(\underset{\sim}{R}_1, \underset{\sim}{R}_2) \qquad (5.17)$$

where $\rho^{(2)}(\underset{\sim}{R}_1, \underset{\sim}{R}_2)$ is the pairwise distribution function. For a homogeneous system $\rho(\underset{\sim}{R}_1) = \rho(\underset{\sim}{R}_2) = \rho$ and $g(\underset{\sim}{R}_1, \underset{\sim}{R}_2) = g(|\underset{\sim}{R}_1 - \underset{\sim}{R}_2|)$ and equ. (5.16) reduces to equ. (5.12).

The free energy of the true system (with pair potential ϕ) can be expressed in terms of the free energy $\mathcal{F}_0$ of the reference system (pair potential ϕ_0) using the coupling constant algorithm:

$$\mathcal{F} = \mathcal{F}_0 + \tfrac{1}{2}\int_0^1 d\lambda \iint d\underset{\sim}{R}_1\, d\underset{\sim}{R}_2\, \rho^{(2)}(\underset{\sim}{R}_1, \underset{\sim}{R}_2; \lambda)\,\phi_1(R_{12}) \qquad (5.18)$$

where $R_{12} = |\underset{\sim}{R}_1 - \underset{\sim}{R}_2|$ and $\rho^{(2)}(\underset{\sim}{R}_1, \underset{\sim}{R}_2; \lambda)$ is the pairwise distribution function for a system in which the atoms interact via a pair potential $\phi_0(R) + \lambda\phi_1(R)$. Both $\mathcal{F}_0$ and $\rho^{(2)}$ are to be evaluated assuming the density in the respective systems is the same as that in the true system i.e. $\rho(\underset{\sim}{R})$. (Equ (5.18) reduces to the familiar result when the fluid is homogeneous). From equ. (5.15) it follows:

$$c(R_{12}) - c_0(R_{12}) = -\left.\frac{\beta\delta^2(\mathcal{F}[\rho] - \mathcal{F}_0[\rho])}{\delta\rho(\underset{\sim}{R}_1)\,\delta\rho(\underset{\sim}{R}_2)}\right|_{\rho} \qquad (5.19)$$

where $|_{\rho}$ indicates the densities $\rho(\underset{\sim}{R}_1)$ and $\rho(\underset{\sim}{R}_2)$ should be set equal to ρ, the density of the homogeneous fluid, after taking the functional derivative. Thus we have the formal result:

$$c(R_{12}) - c_o(R_{12}) = -\frac{\beta}{2} \left. \frac{\delta^2}{\delta\rho(\underset{\sim}{R}_1)\delta\rho(\underset{\sim}{R}_2)} \int_0^1 d\lambda \iint d\underset{\sim}{R}_1 d\underset{\sim}{R}_2 \rho^{(2)}(\underset{\sim}{R}_1, \underset{\sim}{R}_2; \lambda) \phi_1(R_{12}) \right|_\rho \tag{5.20}$$

which is valid for any system interacting via pairwise potentials. To make progress some approximation scheme for $\rho^{(2)}(\underset{\sim}{R}_1, \underset{\sim}{R}_2; \lambda)$ must be introduced. The crudest model ignores all correlations and sets

$$\rho^{(2)}(\underset{\sim}{R}_1, \underset{\sim}{R}_2; \lambda) = \rho(\underset{\sim}{R}_1)\, \rho(\underset{\sim}{R}_2) \tag{5.21}$$

The resultant differentiation is trivial and gives

$$c(R_{12}) - c_o(R_{12}) = -\beta\phi_1(R_{12}) \tag{5.22}$$

which is just the RPA (see equ. 5.11a). This derivation identifies the precise nature of this approximation.

To go beyond the RPA we expand in powers of $\phi_1(R_{12})$. In lowest order:

$$c(R_{12}) - c_o(R_{12}) = -\frac{\beta}{2} \left. \frac{\delta^2}{\delta\rho(\underset{\sim}{R}_1)\delta\rho(\underset{\sim}{R}_2)} \iint d\underset{\sim}{R}_1 d\underset{\sim}{R}_2 \rho_o^{(2)}(\underset{\sim}{R}_1, \underset{\sim}{R}_2) \phi_1(R_{12}) \right|_\rho \tag{5.23}$$

where $\rho_o^{(2)}(\underset{\sim}{R}_1, \underset{\sim}{R}_2) \equiv \rho^{(2)}(\underset{\sim}{R}_1, \underset{\sim}{R}_2; \lambda=0)$ is the pairwise distribution function of the reference system. Since we are particularly interested in the difference between the direct correlation functions for small wave vectors we need only consider small departures of $\rho_o^{(2)}(\underset{\sim}{R}_1, \underset{\sim}{R}_2)$ from its homogeneous limit $\rho_o^{(2)}(R_{12})$:

$$\rho_o^{(2)}(\underset{\sim}{R}_1, \underset{\sim}{R}_2) = \rho_o^{(2)}(R_{12}) + \tfrac{1}{2}\left(\rho(\underset{\sim}{R}_1) - \rho + \rho(\underset{\sim}{R}_2) - \rho\right)\frac{d\rho_o^{(2)}}{d\rho}(R_{12})$$

$$+ \tfrac{1}{2}\left(\rho(\underset{\sim}{R}_1) - \rho\right)\left(\rho(\underset{\sim}{R}_2) - \rho\right)\frac{d^2\rho_o^{(2)}}{d\rho^2}(R_{12}) + \ldots \tag{5.24}$$

where the derivatives are to be evaluated at density ρ. Substituting into equ. (5.23) and performing the functional differentiation gives

$$c(R_{12}) - c_o(R_{12}) = -\frac{\beta}{2}\phi_1(R_{12}) \frac{d^2\rho_o^{(2)}}{d\rho^2}(R_{12}) \tag{5.25}$$

(This result was recently derived by Henderson and Ashcroft [56] from a similar line of argument.) Fourier transforming yields

$$c(q) - c_o(q) = \frac{-\beta}{2(2\pi)^3} \int d\underset{\sim}{K}\, \phi_1(K) \frac{d^2 \rho_o^{(2)}}{d\rho^2} (|\underset{\sim}{q} - \underset{\sim}{K}|) \quad (5.26)$$

For $q = o$ this result is exact to first order in ϕ_1. It may also be reasonably accurate for small, finite q but for larger q the expansion of equ. (5.24) becomes meaningless and a more sophisticated treatment of the inhomogeneous fluid is required. Since $\rho_o^{(2)}(R_{12}) \equiv \rho^2 g_o(R_{12})$ the next obvious approximation is to ignore the density dependence of $g_o(R_{12})$ and then equ. (5.25) reduces to

$$c(R_{12}) - c_o(R_{12}) = -\beta\phi_1(R_{12})\, g_o(R_{12}) \quad (5.27)$$

which in reciprocal space gives

$$c(q) - c_o(q) = -\beta\phi_1(q) - \frac{\beta}{\rho(2\pi)^3} \int d\underset{\sim}{K}\phi_1(K)(S_o(|\underset{\sim}{q}-\underset{\sim}{K}|)-1) \quad (5.28)$$

The second term involving S_o, the structure factor of the reference system, gives an explicit correction to the RPA result which should yield a rough estimate of the importance of correlation effects at low q. Of course, if one has a theory of the density dependence of g_o equ. (5.26) can be used directly. A detailed investigation of these schemes is currently in progress (Evans and Schirmacher to be published) to learn whether the RPA is reasonably accurate near $q = o$.

5.4 Application of the RPA for various liquids.

We assume for the present that the RPA is fairly good for long wavelengths and we suppose that $S_o(q)$ refers to the reference system of the HTA i.e. to the repulsive force part of the pair potential. From equs. (4.27) and (5.11b) it follows that

$$S(o) \approx S_{hs}(o) \left(1 + \rho\beta S_{hs}(o)\phi_1(q = o)\right)^{-1} \quad (5.29)$$

with the hard sphere diameter determined as usual by equ. (4.20).

For a Lennard-Jones fluid $\phi_1(R)$ is negative for all R so $\phi_1(q = o)$ is negative and quite large. This implies S(o) should be considerably enhanced over the hard sphere value of $\sim 0{\cdot}025$–as is indeed observed experimentally for rare gas fluids. (This argument is essentially the same as that of Gaskell [57]). In liquid metals $\phi_1(R;n_o)$ exhibits Friedel oscillations for large R and $-\beta\phi_{min}$ is roughly unity and we can expect a more varied behaviour of $\phi_1(q = o;n_o)$. Preliminary calculations indicate that this quantity is very small in magnitude for the alkalis but in the polyvalent metals, where the Friedel oscillations are more pronounced, $\phi_1(q = o;n_o)$ can be positive and fairly large. Further work is required to see whether this explanation of the variation of S(o) through the liquid metals is correct †. It may turn out that even in the polyvalent metals S(o) obtained from an exact treatment of the pair potential takes a value close to the hard sphere result. Most previous authors[22, 25] have assumed this to be the case and looked for another explanation of the magnitude of the observed compressibilities.

5.5 Long waves versus differentiation of the pressure.

As we remarked earlier in this chapter, in an approximate theory of liquid metals the compressibility sum rule will not be satisfied. In the present model the expansion of the total energy in powers of the pseudopotential is truncated at second order. Calculating the bulk modulus from equ. (3.37) then includes all terms of first and second order in v_{ps}. It can be shown (see Appendix 3) that the method of long waves based on the effective pairwise potential $\phi(R;n_o)$ will omit some of the terms which appear from the differentiation of the pressure. These missing terms correspond to density derivatives of the electron response function χ^e. They can be formally included in the method of long waves but only by explicitly incorporating terms which are of third and fourth order in v_{ps} before taking the q = o limit.

Jones [58] has argued that the difference between the compressibilities calculated by the two different methods should not be too important for the alkali metals but might be large for a polyvalent metal such as Al. χ_T from long waves should be larger than that obtained by differentiating the pressure. The analagous problem for crystalline metals has been discussed by several authors [59 - 63] and they have reached similar conclusions. However, while it is clear in principle that differentiating the pressure is the better thing to do, in practice this method also suffers some disadvantages.

For equilibrium, the pressure as calculated from equ. (3.36) (or its eqivalent in a thermodynamic perturbation theory) should be essentially zero. Any ab-initio calculation is unlikely to give this result at the observed density since p involves the sum of large terms with differing signs. Consequently most authors have followed Ashcroft and Langreth [64] and modified the q = o limit of the pseudopotential (equ (3.12)) so that an additional density dependent term $\Delta u(n_o)$ appears in the self-energy. Since the adjustment is strictly at q = o the pairwise potential can be left unmodified. $\Delta u(n_o)$ is chosen so that p = o at the observed equilibrium density. This rather arbitrary procedure does lead to values of χ_T which are in rather good agreement with experiment for several metals [22] but the contribution to B_T from $\Delta u(n_o)$ is very important. Furthermore the density derivatives of the self energy contain terms which are extremely sensitive to both the choice

† See Evans and Schirmacher(to be published)

of pseudopotential and the exchange and correlation correction U_{XC} [29]. These terms are precisely the ones which do not appear in the method of long waves. Such considerations, coupled with the difficulty of evaluating the radial distribution function at different densities make it hard to assess the validity of the existing calculations of χ_T in liquid metals and it is for this reason that we have advocated a more detailed examination of the long wavelength limit.

6. THE SURFACE TENSION OF LIQUID METALS

The liquid-vapour surface tension γ varies enormously from liquid metal to liquid metal. For Cs near its melting point $\gamma \sim 70$ dyn cm^{-1}, which is similar to water, for Al $\gamma \sim 900$ dyn cm^{-1} while the surface tension of the transition metal Re is about 2700 dyn cm^{-1}[65]. In spite of the practical importance of surface pheonomena in liquid metals relatively little theoretical work has been aimed at understanding the magnitude of γ and its variation through the periodic table [66].

6.1 Predictions of the present model.

In the last chapter we outlined some of the difficulties associated with the calculation of thermodynamic quantities which depend on volume derivatives of the free energy. Since the surface tension of a liquid is given by

$$\gamma = \left(\frac{\partial F}{\partial A}\right)_{N,V,T} \tag{6.1}$$

where A is the area of the (planar) interface, the problems which arise in the calculation of say p and χ_T should not arise in the calculation of γ. Indeed it is easy to show that for the present model of a liquid mdetal (i.e. that derived in § 3) the surface tension is identical in form to that which is derived for simple, insulating liquids. The argument is outlined below.

Equ. (3.31) for the effective interaction potential Φ is valid for any configuration of N ions, including an inhomogeneous distribution which occurs at a planar surface, provided the volume of the sytem is fixed at V and the total number of conduction electrons is fixed at NZ. The differentiation of the partition function implied by equ. (6.1) can be carried out by scaling the variables [66] and the surface tension written as:

$$\gamma = \frac{1}{2}\int_{-\infty}^{\infty} dz_1 \int d\underset{\sim}{R}_{12} \, \frac{(x_{12}^2 - z_{12}^2)}{R_{12}} \, \frac{\partial \phi}{\partial R_{12}}(R_{12};n_0) \, \rho^{(2)}(\underset{\sim}{R}_{12};z_1) \tag{6.2}$$

where $\rho^{(2)}(\underset{\sim}{R}_{12};z_1) \equiv \rho^{(2)}(\underset{\sim}{R}_1,\underset{\sim}{R}_2)$ is the pairwise distribution function in the inhomogeneous liquid. The surface is chosen to be perpendicular to the z direction. The Kirkwood-Buff [67] result for the surface tension of a simple liquid is identical to equ. (6.2) with $\phi\ (R;n_0)$ replaced by the appropriate pairwise potential. The surface energy U^s (defined as the surface excess internal energy per unit area) can also be evaluated:

$$U^s = \tfrac{1}{2}\int_{-\infty}^{0} dz_1 \int d\underset{\sim}{R}_{12}\,\phi(R_{12};n_0)\left(\rho^{(2)}(\underset{\sim}{R}_{12};z_1) - \rho_\alpha^2 g_\alpha(R_{12})\right) \tag{6.3}$$
$$+ \tfrac{1}{2}\int_{0}^{\infty} dz_1 \int d\underset{\sim}{R}_{12}\,\phi(R_{12};n_0)\left(\rho^{(2)}(\underset{\sim}{R}_{12};z_1) - \rho_\beta^2\, g_\beta(R_{12})\right)$$

where ρ_α and ρ_β are the densities of the co-existing bulk liquid and bulk vapour phases and g_α and g_β are the corresponding radial distribution functions. It is assumed that the Gibbs dividing surface is located at z = o. The surface energy is then related to the surface tension via the standard result:

$$U^s = \gamma - T\,\frac{\partial\gamma}{\partial T} \tag{6.4}$$

The density $\rho(z)$ satisfies the equation:

$$K_B T\,\frac{\partial\rho(z_1)}{\partial z_1} = \int d\underset{\sim}{R}_{12}\,\frac{z_{12}}{R_{12}}\,\frac{\partial\phi(R_{12};n_0)}{\partial R_{12}}\,\rho^{(2)}(\underset{\sim}{R}_{12};z_1) \tag{6.5}$$

with the boundary conditions

$$\rho(-\infty) = \rho_\alpha$$
$$\rho(\infty) = \rho_\beta \tag{6.6}$$
$$\int d\underset{\sim}{R}\,\rho(z) = N$$

where the integral is over the whole volume.

It is important to note that the average electron density n_0 is fixed at the outset so $u(n_0)$ and $\phi(R;n_0)$ are given once and for all at the beginning of the analysis. The co-existing densities should be determined by equating the pressures and Gibbs free energies of the bulk phases calculated with the assumption that the self-energy and pair potential are density independent. Similarly g_α and g_β should be evaluated at densities ρ_α and ρ_β using the same fixed pair potential $\phi(R;n_0)$. The obvious choice for n_0 is the measured average electron density of the bulk liquid metal.

Certainly for a crystalline surface in vacuo this would be the natural selection but other possibilities exist for the liquid vapour problem.

Given n_0 the calculation of the surface properties is exactly equivalent to that for simple liquids. The only difference lies in the form of the pair potential.

Various approximate schemes exist for simple liquids [68]. These schemes fall into two categories: a) $\rho^{(2)}(R_{12};z_1)$ is approximated by $\rho(z_1)\,\rho(z_2)$x a suitable average of the bulk radial distribution functions and equ.(6.5) is used to find an approximate solution for $\rho(z)$. This is then used in equ. (6.2) to calculate γ. b) the free energy F of the inhomogeneous fluid is written as an approximate functional of $\rho(z)$ using thermodynamic perturbation theories and an approximation for $\rho^{(2)}$ similar to that in a). F is minimized with respect to a particular functional form of $\rho(z)$ and γ is given directly. Both procedures have been used to calculate the surface tension for liquid Ar and the theories give a very good account of both the magnitude and temperature dependence. A few computer simulations have been reported [69] and the results for $\gamma(T)$ are in quite close agreement with those obtained from the approximation schemes.

In § 3 we stressed that the pair potentials $\phi(R;n_0)$ calculated for liquid metals are not strongly binding and the depth of the first minimum is typically $\sim K_BT_M$. For liquid Ar the well-depth is somewhat bigger than K_BT_M. Since the melting points of the simple metals are about 3 - 12 x that of Ar, we might expect, the present theory to predict surface tensions between 40 - 150 dyn cm^{-1} for liquid metals (γ for liquid Ar near its triple point is $\sim$13 dyn cm^{-1}). The measured surface tensions are much larger. Detailed calculations [70] confirm the failure of the theory. For example a calculation of the surface energy based on equ. (6.3), using realistic approximations for $\rho^{(2)}$ and ρ, gives results for the alkalis which are roughly half those found experimentally while for other simple metals U^s (calculated) is only 10 - 30% of U^s (experimental). Essentially equivalent results and conclusions have been obtained for the surface energy of crystalline metals [71]. Thus the model which worked well for many of the <u>bulk</u> thermodynamic properties is completely inadequate for the surface problem.

6.2 Beyond linear screening theory.

At the surface of a liquid metal we expect the ions to adopt a density profile $\rho(z)$ similar to that in simple liquids. Near the melting point $\rho(z)$ should fall rapidly, over a few atomic diameters, from its bulk liquid value to its bulk vapour value. The averaged conduction electron density n(z) should vary in a rather similar fashion. The difference between $\rho(z)$ and n(z) will produce the electrostatic dipole barrier. In the model described in § 6.1 n(z) is simply the appropriate average of

$$n(\underset{\sim}{q}) = \chi^{e}(q) \quad v_{ps}(q)\, \rho(\underset{\sim}{q}) \tag{6.7}$$

where χ^e is evaluated at density n_o and $\rho(\underset{\sim}{q})$ is the Fourier component of the ion density. It is unlikely that this linear screening approximation is adequate when $\rho(z)$ varies rapidly - as it does at the surface. The perturbation caused by replacing the uniform positive background by a strongly inhomogeneous density of ions, i.e. δV (r) in equ. (3.8), is too large to be treated as a weak perturbation.

For crystalline metal surfaces at T = 0°K a serious attempt has been made to improve upon the linear screening approximation [72]. The theory closely follows that presented in § 3.2 but rather than starting from uniform 'jellium' the lattice is first replaced by a positive distribution of charge $\rho_o(z)$:

$$\begin{aligned} \rho_o(z) &= \bar{n} \qquad & z \leqslant o \\ &= o & z > o \end{aligned} \tag{6.8}$$

where $\bar{n}$ is the bulk average conduction electron density. The electron density $n_o(z)$ and the surface energy of this jellium model is then calculated using the Hohenberg-Kohn-Sham (See Appendix 2) theory. This is a fully self-consistent, non-linear theory of the interacting inhomogeneous electron gas. The positive charge distribution is then replaced by a half-lattice of ions and the accompanying change in surface energy is calculated to first order in the difference potential

$$\delta V'(\underset{\sim}{r}) = \sum_{\ell'} v_{ps}(\underset{\sim}{r} - \underset{\sim}{R}_{\ell'}) + \int \frac{d\underset{\sim}{r}'\, \rho_o(z')}{|\underset{\sim}{r}-\underset{\sim}{r}'|} \tag{6.9}$$

where ℓ' refers to half-lattice sites. (This corresponds to retaining the first three terms in equ. (3.11)). The contribution from the direct ion-ion interaction is also included.

This procedure yields rather good estimates of the surface energy for most of the simple metals for which it has been applied [72] (It should be noted that surface energies of a crystal are difficult to measure so 'experimental' estimates are obtained by extrapolating data on $\gamma(T)$ for liquid metals down to T = 0°K!) The theory is, however, rather incomplete. Although $\delta V'(\underset{\sim}{r})$ has smaller Fourier components than $\delta V(\underset{\sim}{r})$, these are still quite large, especially in the polyvalent metals at small q, and subsequent work [71] indicated that second order contributions to the surface energy (terms $O(\delta V'^2)$) could destroy the good agreement between theory and experiment for Al. Since the second order term requires the density response function of the inhomogeneous electron gas, which is not well understood, it is difficult to test the convergence of the expansion.

In the calculations of the surface energy the crystal structure is usually taken as that given by experiment and no attempt is made to calculate the equilibrium positions of the ions i.e. surface relaxation is not usually investigated. In principle this could be done since the procedure is quite general and leads to an effective interaction potential Φ' which is similar to but more complicated than $\Phi(\underset{\sim}{R}_1, \ldots \underset{\sim}{R}_N)$ obtained in §3.2. To our knowledge this approach has not been applied to liquid metals. The complexity of Φ' makes ensemble averaging rather difficult and there is no simple real space representation to aid the interpretation.

6.3 A pseudo-atom model.

An alternative, but more empirical, approach to the surface tension problem was developed by the present author [66]. This model assumes that the effective interaction potential for the ions can be written as:

$$\Phi^s(\underset{\sim}{R}_1, \ldots \underset{\sim}{R}_N) = \sum_{\ell} u(n_0(z_\ell)) + \frac{1}{4}\sum_{\ell \neq \ell'}\left[\phi(R_{\ell\ell'};n_0(z_\ell)) + \phi(R_{\ell\ell'};n_0(z_{\ell'}))\right] \quad (6.10)$$

where $R_{\ell\ell'} = |\underset{\sim}{R}_\ell - \underset{\sim}{R}_{\ell'}|$ and $n_0(z)$ is the electron density obtained from a jellium calculation as described above. The physical content is clear. An ion in the surface is assumed to have a different self-energy, due to the reduced electron density, from an ion in the bulk liquid metal. Similarly the pairwise potential between two ions is allowed to vary with the local density of conduction electrons. (The second term of equ. (6.10) is constructed so as to retain symmetry between labels ℓ and ℓ'). For a bulk metal Φ^s reduces to Φ as obtained via the linear screening approximation since $n_0(z) \equiv n_0$ in this case. $u(n_0)$ and $\phi(R;n_0)$ are to be evaluated as previously. This model, like the other models suggested above, is incorrect in the low-density, vapour limit. For the vapour a nearly-free-electron theory is clearly inappropriate and the potential energy would be better represented by a sum of density independent potentials as in rare-gases or molecular fluids. However, since the surface tension of liquid metals is much larger than that of insulating liquids the dominant forces should be metallic in character and we ignore the complications at low density.

Explicit expressions for both γ and U^s can be derived for this model [66]:

$$\gamma = -\int_{-\infty}^{\infty} dz_1\, z_1 \frac{\partial u(n_o(z_1))}{\partial z_1}\rho(z_1) - \frac{1}{2}\int_{-\infty}^{\infty} dz_1 \int d\underset{\sim}{R}_{12}\, z_1 \frac{\partial \phi(R_{12}; n_o(z_1))}{\partial z_1}\rho^{(2)}(\underset{\sim}{R}_{12}; z_1)$$

$$+\frac{1}{2}\int_{-\infty}^{\infty} dz_1 \int d\underset{\sim}{R}_{12} \frac{(x_{12}^2 - z_{12}^2)}{R_{12}} \frac{\partial \phi(R_{12}; n_o(z_1))}{\partial R_{12}} \rho^{(2)}(\underset{\sim}{R}_{12}; z_1) \tag{6.11}$$

and

$$U^s = \int_{-\infty}^{0} dz_1 \left(u(n_o(z_1))\rho(z_1) - u(n_\alpha)\rho_\alpha\right) + \int_0^\infty dz_1 \left(u(n_o(z_1))\rho(z_1) - u(n_\beta)\rho_\beta\right)$$

$$+\frac{1}{2}\int_{-\infty}^{0} dz_1 \int d\underset{\sim}{R}_{12} \left(\phi(R_{12}; n_o(z_1))\rho^{(2)}(\underset{\sim}{R}_{12}; z_1) - \phi(R_{12}; n_\alpha)\rho_\alpha^2 g_\alpha(R_{12})\right)$$

$$+\frac{1}{2}\int_{0}^{\infty} dz_1 \int d\underset{\sim}{R}_{12} \left(\phi(R_{12}; n_o(z_1))\rho^{(2)}(\underset{\sim}{R}_{12}; z_1) - \phi(R_{12}; n_\beta)\rho_\beta^2 g_\beta(R_{12})\right) \tag{6.12}$$

The electron density is constrained so that

$$\int d\underset{\sim}{R}\;\; n_o(z) = NZ$$

and for simplicity the bulk values are taken to be

$$n_o(-\infty) = n_\alpha$$

$$n_o(\infty) = n_\beta = 0$$

where n_α is the measured value in the bulk liquid metal. The bulk ion densities ρ_α and ρ_β should then be found from the conditions for co-existence but since the effective potential energy Φ^s is inappropriate for the vapour it is assumed that $\rho_\alpha = n_\alpha/Z$ and $\rho_\beta = 0$ i.e. the vapour is ignored. The analogue of equ. (6.5) is

$$K_B T \frac{\partial \rho(z_1)}{\partial z_1} = -\rho(z_1)\frac{\partial u(n_o(z_1))}{\partial z_1} - \frac{1}{2}\int d\underset{\sim}{R}_{12} \frac{\partial \phi(R_{12}; n_o(z_1))}{\partial z_1}\rho^{(2)}(\underset{\sim}{R}_{12}; z_1)$$

$$+\frac{1}{2}\int d\underset{\sim}{R}_{12} \frac{z_{12}}{R_{12}}\left(\frac{\partial \phi(R_{12}; n_o(z_1))}{\partial R_{12}} + \frac{\partial \phi(R_{12}; n_o(z_1 + z_{12}))}{\partial R_{12}}\right) \tag{6.13}$$

which illustrates the various forces acting to produce the ion density profile.

Kumaravadivel and Evans [70] used equ. (6.12) to calculate the surface energy of several liquid metals. They approximated $\rho^{(2)}(\underset{\sim}{R}_{12}; z_1)$ by $\rho(z_1)\rho(z_2)\, g_\alpha(R_{12})$ and assumed that the ion density $\rho(z_1)$ closely follows the jellium electron density $n_o(z_1)$.

In fact the magnitude of U^s is not over sensitive to the details of $\rho(z_1)$. For the alkali metals the calculated results were in good agreement with experiment while for some of the polyvalent metals the calculations over estimated U^s. The contribution to the surface energy from the self-energy terms is large and completely dominates over the pairwise contributions in the polyvalent metals. The WCA perturbation theory was later generalised to inhomogeneous liquids and applied to this model of the liquid metal surface [73]. Calculations of the equilibrium ion density profile and surface tension of liquid Na, K and Al were performed [73]. The theory predicts very 'steep' profiles i.e. the transition zone should be complete over one or two atomic diameters. The results for γ are in reasonable agreement with experiment. Again the self-energy terms contribute a large fraction of the total surface tension. These terms are also responsible for the very sharp ion density profile since the free energy is lowered if the ions are situated where the electron density $n_0(z)$ is largest ($u(n_0)$ is lower for bigger n_0).

The temperature dependence of the surface tension is of particular interest in liquid metals. Although the experimental situation is somewhat confused [65] there is evidence [74] for inversion of the surface tension characteristic, i.e. positive values of $\partial\gamma/\partial T$, in liquid Zn. Croxton [75] has speculated at length on possible explanations for such behaviour and argued that a liquid metal might have an oscillatory ion density profile under equilibrium conditions. For such a situation the surface excess entropy might be negative leading to a positive $\partial\gamma/\partial T$. While the existence of a stable oscillatory density profile in simple liquids seems to have been ruled out by recent computer studies [69, 76] the possibility is still open for liquid metals [66,73]. The surface tension characteristic clearly requires further investigation, both experimental and theoretical.

Throughout this section we have used the Kirkwood-Buff [67] formulation of the surface tension. This is a natural description when the potential energy can be written as a sum of pairwise potentials . A more general and, perhaps, more powerful treatment of surface problems can be formulated in terms of the direct correlation function (see equ. (5.15) and (5.16)). An exact formula for γ in a simple liquid has been derived [77] which does not make any assumption about the form of the potential energy and the theory makes interesting predictions concerning the nature of the pairwise correlations in the surface region [78] . It would be valuable to apply this approach to liquid metals where the two-component aspects of the system could lead to even more interesting behaviour.

7. INVERSION OF EXPERIMENTAL STRUCTURE FACTOR DATA

Throughout these notes we have been investigating the consequences of the pseudopotential theory of the energetics and structure of liquid metals and we have examined in detail the effective pairwise potentials obtained from this electronic theory. In this section we examine a different and, in fact, older approach. We assume the total effective interaction potential for the ions is of the form:

$$\Phi(\underset{\sim}{R}_1, \ldots \underset{\sim}{R}_N) = C + \tfrac{1}{2} \sum_{\ell \neq \ell'} \phi(|\underset{\sim}{R}_\ell - \underset{\sim}{R}_{\ell'}|) \qquad (7.1)$$

where C is independent of the positions of the ions but may be volume dependent. $\phi(R)$ is a pairwise potential. The aim of this approach is to extract information about ϕ directly from the structure factor S(K) as measured by X-ray or neutron diffraction experiments. (Strictly speaking only neutron data should be used since we require the ion-ion structure factor. X-rays are scattered by both conduction and core electrons so the measured X-ray intensity involves a linear combination of S(K), $S_{ei}(K)$ and $S_{ee}(K)$. For metals in which the number of core electrons is relatively small the difference between the neutron and X-ray 'structure factors' may be significant [79]). The approach was pioneered by Johnston et al. [80] who calculated pair potentials for several metals using the Percus-Yevick (PY) hypernetted-chain (HNC) and Born-Green (BG) approximate theories. Subsequently it was recognized that $\phi(R)$ as obtained from all three prescriptions was highly sensitive to small uncertainties in S(K). The attractive part of the potential is particularly sensitive to the details of S(K) for small K. Since early measurements were highly inaccurate in this region this made the inversion schemes completely unreliable [81]. Interest was revived when accurate X-ray data for liquid Na and K [82] was produced and Howells and Enderby [83] calculated what appeared to be quite realistic and numerically reliable pair potentials using the PY and HNC approximations and this data. The BG approximation yields non-realistic, strongly temperature dependent potentials [83,84] and can be dismissed as a practical scheme of inversion [85]. Further investigation [29] of the HNC and PY results showed that these were also unacceptable pair potentials i.e. if they were used to generate the structure factor or g(R) these were completely at variance with the original experimental data. (The repulsive part of both ϕ_{HNC} and ϕ_{PY} falls at large separations, see figure 11, and this implies unphysically large effective hard sphere diameters). Exactly the same inadequacies of the HNC and PY theories had been pointed out earlier [86] in a careful study of high density Lennard-Jones fluids. Thus all three 'standard' inversion procedures are useless for quantitative purposes and must therefore be abandoned.

During the last few years several new inversion schemes have been proposed. Brennan et al [87] have developed a powerful technique which although similar in spirit to the PY and HNC theories contains extra parameters which are fitted by imposing certain constraints on thermodynamical quantities calculated from the pair potential. Experimental data for the pressure, internal energy and compressibility, as well as S(K), are required as input. Their procedure seems to work remarkably well for liquid Ne; g(R) and thermodynamic quantities are well-produced by a molecular dynamics simulation based on the calculated $\phi(R)$. For liquid metals, however, the procedure has to be modified, since, as we saw earlier, only certain constant volume thermodynamic quantities depend solely on $\phi(R)$. Other quantities involve the density dependence of ϕ and the term C in equ. (7.1). Mitra et al [88] have used S(K) measured at constant density and two different temperatures along with the measured value of the specific heat C_V, to calculate a pair potential for liquid Rb. This pair potential generates a g(R) which is quite close to the measured one. Unfortunately very few measurements of the temperature dependence of S(K) at fixed density have been carried out so the application of this procedure is rather limited.

In later work Mitra [89] employed what is essentially a trial and error procedure. An (educated) guess is made for $\phi(R)$ and this is used in a computer simulation of g(R). $\phi(R)$ is then suitably modified and g(R) recalculated until reasonable agreement with experiment is obtained. Pair potentials for liquid Cu and Ni [89] have been calculated in this way. Schommers [90] has calculated a pair potential for Rb using a similar approach. Such brute force methods are necessarily heavy on computer time!

Attempts to utilize the WCA theory in an inversion procedure have recently been reported [91,92] and the results are encouraging. As we might have expected, the repulsive force part of the pair potential can be calculated quite reliably but the attractive part is quite sensitive to approximations and inaccuracies in data. For liquid Na and Rb the calculated potentials are fairly close to those obtained from pseudopotential theory while for Pb the calculated potential shows a large positive 'hump' in the long range part [92].

A simple but qualitatively successful inversion scheme has been employed by Ailawadi and Naghizadeh [93]. Their approach is based on the work of Singwi et al [94] on the interacting electron gas. The latter authors introduced a crude decoupling ansatz for the two particle probability density

$$f^{(2)}(\underset{\sim}{R}_1, \underset{\sim}{p}_1; \underset{\sim}{R}_2, \underset{\sim}{p}_2, t) \approx f^{(1)}(\underset{\sim}{R}_1, \underset{\sim}{p}_1, t)\, f^{(1)}(\underset{\sim}{R}_2, \underset{\sim}{p}_2, t)\, g(|\underset{\sim}{R}_1 - \underset{\sim}{R}_2|) \tag{7.2}$$

where g(R) is the (static) radial distribution function which is introduced in an attempt to account for short range correlations. This approximation leads to a simple expression for the effective interaction U in equ. (5.6):

$$\frac{dU(R)}{dR} = g(R) \frac{d\phi(R)}{dR} \qquad (7.3)$$

where $\phi(R)$ is the (assumed) pair potential. (For the electron gas ϕ (R) is just the Coulomb potential) . In terms of the direct correlation function we have

$$- K_B T \frac{dc(R)}{dR} = g(R) \frac{d\phi}{dR} (R) \qquad (7.4)$$

so given c(R) and g(R) from experiment this equation gives the interionic force $d\phi/dR$ directly. $\phi(R)$ is then obtained by integration. Clearly this procedure is as straight forward as the PY or HNC methods. It has been tested [93] using input data for S(K) from Rahman's computer simulation on Rb (see §4.2). In figure 11 we compare $\phi(R)$ as obtained from the inversion procedure with the original input potential of the computer simulation. Although the potentials look rather different the interionic force is reproduced rather accurately in the important region near the first minimum of $\phi(R)$. We have also plotted the 'mean-field or RPA' result for the pair potential i.e. $\phi(R) = - K_B Tc(R)$. As noted earlier [95] the mean-field result agrees remarkably well with the original potential for values of R beyond the first minimum but is poor at smaller R. It is clear that the correlations introduced by the presence of g(R) in equ. (7.4) 'soften' the harsh repulsion inherent in the mean field result (except for very small values of R where $g(R) < 1$). However the main features of the potential (obtained by inversion) are strongly governed by the shape of c(R) and the latter is clearly the crucial quantity in any theory. We note that the HNC inversion yields a potential which is considerably worse than the mean-field result. ϕ_{PY} is even less realistic. For liquid Na inversion of the experimental data using the Singwi et al theory produces a pair potential fairly close to those obtained from pseudopotential theory [93] and for liquid Ne [96] the calculated potential is very close to those obtained by Brennan et al [87] and Mitra [91].

It is, perhaps, too early to attempt to summarize this recent work on inversion since many calculations are in progress but it certainly appears that provided the experimental data is good enough to obtain an accurate c(R) it is now possible, by one procedure or another, to make appropriate adjustments to the mean-field potential in the core region and obtain realistic pair potentials. (By realistic we mean that the calculated $\phi(R)$ will generate the original g(R) or S(K) to some reasonable degree of accuracy). We stress that the experimental S(K) must be given accurately for all values of K in order to obtain c(R) with any reasonable precision. At present, it seems that only the data for Na and K strictly meet this requirement. For example Ailawadi and co-workers

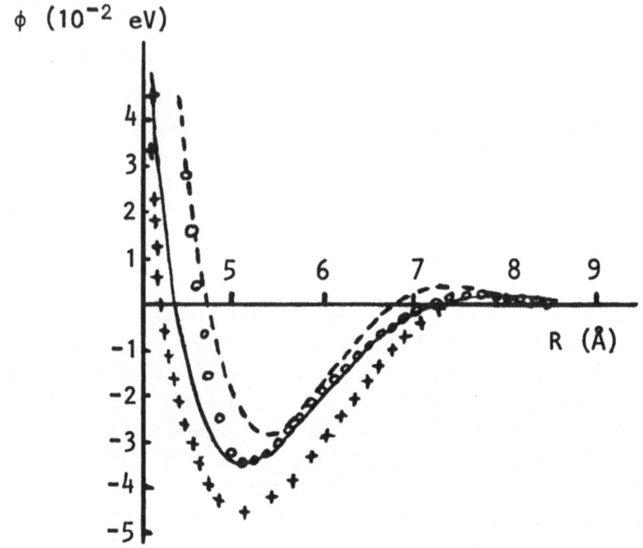

Figure 11. Pair potential for liquid Rb. The solid curve represents the original input pair potential for the computer simulation. The open circles represent $-K_BTc(R)$, the dashed curve is the HNC potential and the crosses are the result of the inversion procedure based on the Singwi et al theory. (See refs [93,95]).

[private communication] have examined several recent experimental S(K)'s for liquid Al and found these differ enormously from each other. Not surprisingly the pair potentials obtained by inversion vary enormously. Consequently we are restricted, for nearly all metals, to qualitative statements regarding the pair potentials obtained by inversion.

For Na and K all authors seem to agree that, in each metal, the repulsive force part of the pair potential is close to that obtained via pseudopotential theory - as it should be since we have previously argued the pseudopotential approach gives a good description of the structure factor and constant volume properties in these metals. The depth of the minimum of $\phi(R)$ can differ by as much as a factor of two from the corresponding pseudopotential results but is still, typically, $\sim K_BT_M$. This confirms the ideas of § 3.4 where we argued that the pairwise contributions to the binding energies were small. For Rb it is clear that the repulsive part of the pair potential is very soft - varying as slowly as R^{-4} [92] - as predicted by the electronic theory. For the other simple metals an examination of c(R) [29] indicates that the polyvalent metals should be have much harder pair potentials - as discussed in § 4.2. Mitra's inversion

[92] on Pb produces a repulsive potential which behaves like R^{-12} and the depth of the minimum is $\sim K_B T_M$.

Structure factors have been measured for the noble metals and several transition metals [97] . In general the results are fairly well represented by the hard sphere model but there are some interesting trends in g(R) and S(K) across the periodic table. It appears [97] that the pair potentials in the early transition metals e.g. Ti,V and Cr should be 'softer' than in say Fe and Ni. In other words the 'repulsivity' of $\phi(R)$ seems to increase as the number of d electrons increases across the transition metal series. The inversion of the data on Cu and Ni [89] produces repulsive potentials which vary like R^{-11} and R^{-13} respectively and in each case the depth of the minimum is $\sim K_B T_M$. S(K) has also been measured for some of the early trivalent rare earth metals and for Eu and Yb [97, 98] . The oscillations at large K are very strongly damped indicating rather soft repulsive pair potentials.

Here we should introduce a note of caution. For noble, transition and rare-earth metals the concept of an effective pairwise potential is not easily justified. When the electron-ion pseudopotential is strong (see also § 2.4) the expansion of the free energy of the electron system (equ. (3.9)) in powers of the pseudopotential will not converge in any sensible fashion. Consequently the separation of $\Phi(\underset{\sim}{R}_1, \ldots \underset{\sim}{R}_N)$ into two-body, three-body etc. potentials may not be meaningful and one should probably seek an alternative formulation. On the other hand, the short-range repulsive forces which certainly determine the structure of these liquid metals are probably well-represented in terms of pairwise potentials. If one could obtain detailed information about the repulsive forces from studies of the liquid metals this could be usefully applied to point defect calculations in the corresponding crystals.

8. OTHER TOPICS

Lack of space (and energy!) prohibits discussion of other aspects of the structure and cohesive properties of liquid metals so here we merely list some relevant topics.

There has been important work on melting in pure metals [34,34,99]. This uses the pseudopotential theory and the variational approach of § 4.1 to calcualte the free energy in the liquid phase and a similar model for the solid phase. The theory and results of calculations are reviewed in ref [19].

Young and co-workers, see e.g. [39], have analysed the entropy and specific heat data for many different metals on the basis of the variational approach.

One problem which still receives considerable attention [19, 39,100,101] concerns the occurrence of assymmetries and 'shoulders' at the first peak of the liquid structure factor. Many metals [97] show rather strong departures from the ideal 'hard sphere like' behaviour for K in this region. Such behaviour cannot be accounted for in terms of the repulsive pairwise forces described throughout this article. If the short-range repulsive potential is not smooth but consists of a hard core part plus an adjacent 'ledge' then it can obviously give rise to a shoulder on the high angle side of the first peak in S(K) [39,100]. It has been suggested [19] that such a $\phi(R)$ can occur if the ion cores are highly polarizable so that the bare ion-ion interaction W in equ. (3.1) is modified from the simple Coulombic form. On the other hand it should be signficant that liquid Si, Ge, Sn, Sb and Bi, which exhibit pronounced shoulders in S(K), show covalent bonding in the solid state. It is possible that in the liquid state the interatomic forces may retain some covalent character. This would then make a description solely in terms of pairwise potentials rather dubious.

The thermodynamic properties of binary liquid alloys are of considerable current interest and several theoretical papers have recently appeared e.g. [39,40,42 - 44] . The actual three-component system of conduction electrons plus A and B ions is reduced to an effective two component system using the argument of §3.1. If the electron-ion interactions can be modelled by weak pseudopotentials v_{ps}^{A} and v_{ps}^{B} the free energy of the electron system F' can be evaluated in linear response theory as in § 3.2. The starting system is now a homogeneous electron gas of density $\rho(c_A Z_A + c_B Z_B)$ where c_A and c_B are the concentrations of the A and B ions, Z_A and Z_B are the valences and ρ is the number density at the appropriate concentration . The total effective interaction potential for the ions can be written as in equ. (3.31) but now there are three different pairwise potentials ϕ_{AA}, ϕ_{BB} and ϕ_{AB} and each is a function of the average electron density and hence the alloy concentration. In all the published work the free energy of the alloy is estimated using the variational approach, with the reference system being a mixture of hard spheres of different diameters (for a review see ref. [19]). Quantities such as the entropy of mixing , the volume, enthalpy and entropy of formation [40] have been calculated for a few alloy systems. Since such calculations necessarily involve the computation of very small energy differences it is not surprising that the results are sensitive to small changes in the input information e.g. the pseudopotentials and densities. Nevertheless some useful results have emerged. The difficult problem of phase separation has also been attacked for alloys of the alkali metals [43, 56,19] and the results are encouraging.

We must remember, however, that in alloys the applicability and efficiency of low order perturbation theory are more severly limited than in pure metals. In most alloys the difference between the pseudopotentials v_{ps}^{A} and v_{ps}^{B} will not be 'small'. This is clearly the case in heterovalent alloys. Consequently, the use of a linear response approximation to estimate the variation of the total energy of the alloy as a function of concentration may be quite inaccurate. For this reason applications of the simple theory outlined above are usually restricted to homovalent alloys. When $|v_{ps}^{A} - v_{ps}^{B}|$ is 'large' we might expect significant charge transfer to take place between the ions and this will lead to strongly concentration dependent interionic forces. Such behaviour will be reflected in the electronic structure and electrical transport properties of the alloy. In the extreme case we can find liquid-semiconducting alloys (see e.g. [2]) which exhibit singular behaviour in their electrical properties at certain stoichiometric compositions. A proper description of the energetics in such systems lies outisde the regime of low order perturbation theory.

Acknowledgements

The author is grateful to Drs. D.A. Greenwood, R.Kumaravadivel and W. Schirmacher and Professor J.E. Enderby for stimulating discussions.

References

1. See,for example, W.A. Harrison 'Pseudopotentials in the Theory of Metals' (New York: Benjamin) 1966.

2. T.E. Faber, 'Introduction to the Theory of Liquid Metals' (London: Cambridge) 1972.

3. A recent review of experimental work is given by G. Busch and H-J. Güntherodt in Sol.St.Phys., 29, 235, 1974.

4. W.Haller, H-J. Güntherodt and G. Busch in 'Liquid Metals 1976' ed. R. Evans and D.A. Greenwood (Inst. of Phys. Conf. Series. No. 30) page 207, 1977.

5. V. Heine in Sol. St. Phys, 24, 1 1970.

6. M.L. Cohen and V. Heine in Sol. St. Phys, 24, 38, 1970, and V. Heine and D. Weaire in Sol. St. Phys. 24, 250, 1970.

7. See, for example, J.M. Ziman in Sol. St. Phys, 26, 1, 1970.

8. V.K. Ratti and R. Evans, J. Phys.F: Metal Phys., 3, L238, 1973.

9. S.F. Edwards, Proc. Roy. Soc. A267, 518, 1962.

10. An excellent review is given by L.E. Ballentine in Adv. Chem. Phys., 31. 263, 1975.

11. Some experimental results are reviewed by C. Norris in 'Liquid Metals 1976' ed. R. Evans and D.A. Greenwood (Inst. of Phys. Conf. Series, No.30) pg. 171, 1977.

12. J.M. Ziman, Phil.Mag,, 6, 1013, 1961.

13. L.E. Ballentine in 'Liquid Metals 1976' ed. R.Evans and D.A. Greenwood (Inst. of Phys. Conf. Series No 30) page 188, 1977.

14. For a short review see M. Watabe in 'Liquid Metals 1976' ed. R. Evans and D.A. Greenwood (Inst. of Phys. Conf. Series No 30) page 288, 1977. This volume also contains several other papers on this topic.

15. See, for example, B.L. Gyorffy and G.M. Stocks in 'Electrons in Finite and Infinite Structures' ed. P. Phariseau and L. Scheire (NATO Series Vol. 24: Plenum) page 144, 1977.

16. C.F. Hague in 'Liquid Metals 1976' ed. R. Evans and D.A. Greenwood.(Inst. of Phys. Conf. Series No 30) page 360, 1977.

17. H-J. Güntherodt, H.U. Künzi, M. Liard, R. Müller, R. Oberle and H. Rudin in 'Liquid Metals 1976' ed. R. Evans and D.A. Greenwood (Inst. of Phys. Conf. Series No. 30) page 342, 1977. This paper contains references to previous work. B.C. Dupree, J.E. Enderby, R.J. Newport and J.B. Van Zytveld (same volume) page 337.

18. H-J. Güntherodt, E. Hauser and H.U. Künzi in 'Liquid Metals 1976' ed. R. Evans and D.A. Greenwood (Inst. of Phys. Conf. Series No 30) page 324, 1977, and references therein.

19. N.W. Ashcroft and D. Stroud in Sol.St. Phys (to appear)

20. See for example, D. Pines and P. Nozières 'Quantum Liquids' (New York: Benjamin) 1966.

21. P. Vashista and K.S. Singwi, Phys. Rev. B, 6, 875, 1972 (Errata B6 4883).

22. M. Hasegawa and M. Watabe, J. Phys. Soc. Japan, 32, 14,1972.

23. R. Kumaravadivel and R. Evans, J. Phys. C: Sol. St. Phys. 9 3877, 1976.

24. M. Hasegawa, J. Phys. F: Metal Phys, 6, 649, 1976.

25. D.L. Price, Phys. Rev. A, 4, 358, 1971.

26. A. Rahman, Phys. Rev. A, 9, 1667, 1974.

27. For example, R.D. Murphy and M.L. Klein, Phys. Rev. A, 8, 2640, 1973.

28. R.D.Mountain in 'Liquid Metals 1976' ed. R. Evans and D.A. Greenwood (Inst. of Phys. Conf. Series. No 30) page 62, 1977.

29. R. Kumaravadivel. Thesis: University of Bristol, 1976 (unpublished).

30. J-P. Hansen and I.R. MacDonald in 'Theory of Simple Liquids' (New York: Academic Press), 1976.

31. J.A. Barker and D. Henderson in Rev. Mod. Phys., 48, 587, 1976.

32. H.C. Andersen, D. Chandler and J.D. Weeks in Adv. Chem. Phys., 34, 105, 1976.

33. G.A. Mansoori and F.B. Canfield, J. Chem. Phys, 51, 4958, 1969.

34. H.D. Jones, J. Chem. Phys. 55, 2640, 1971, and Phys. Rev. A, 8. 3215, 1973.

35. D. Stroud and N.W. Ashcroft, Phys. Rev. B. 5, 371, 1972.

36. D.J. Edwards and J. Jarzynski, J. Phys. C: Sol. St. Phys., 5, 1745, 1972.

37. I.H. Umar and W.H. Young, J. Phys. F: Metal Phys. 4, 525,1974.

38. M. Silbert, I.H. Umar, M. Watabe and W.H. Young, J. Phys.F: Metal Phys., 5, 1262, 1975.

39. W.H. Young in 'Liquid Metals 1976' ed. R. Evans and D.A. Greenwood (Inst. of Phys. Conf. Series No 30) page 1, 1977.

40. J. Hafner in 'Liquid Metals 1976' ed. R. Evans and D.A. Greenwood. (Inst. of Phys. Conf. Series No 30) page 102, 1977.

41. M. Silbert in 'Liquid Metals 1976' ed. R. Evans and D.A. Greenwood (Inst. of Phys. Conf.Series No 30) page 72, 1977.

42. D. Stroud, Phys. Rev.B,7, 4405, 1973.

43. I.H. Umar, A. Meyer, M. Watabe and W.H. Young, J. Phys.F: Metal Phys, 4, 1691, 1974.

44. I. Yokoyama, M.J. Stott, I.H. Umar and W.H. Young in 'Liquid Metals 1976' ed. R. Evans and D.A. Greenwood (Inst. of Phys. Conf. Series No. 30) page 95, 1977.

45. J.D. Weeks, D.Chandler and H.C. Andersen, J. Chem. Phys, 54, 5237, 1971.

46. H.C. Andersen, J.D. Weeks and D. Chandler, Phys. Rev. A,4, 1597, 1971.

47. J.H. Wehling, Wei-Mei Shyu and G.D. Gaspari, Phys. Lett., 39A, 59, 1972.

48. M. Hasegawa and M. Watabe, J. Phys. Soc. Japan, 36, 1510,1974.

49. M. Watabe and W.H. Young, J. Phys. F: Metal Phys., 4, L29, 1974.

50. R.O. Watts and D. Henderson, Molec. Phys., 16, 217, 1969.

51. N.F. Carnahan and K.E. Starling, J. Chem. Phys., 51, 635, 1969.

52. M. Watabe and M. Hasegawa in 'The Properties of Liquid Metals' ed. S. Takeuchi (London-Taylor and Francis) page 133, 1973.

53. J. Chihara in 'The Properties of Liquid Metals' ed. S. Takeuchi (London: Taylor and Francis) page 137, 1973.

54. P. Gray, J. Phys. F: Metal Phys., 3, L43, 1973.

55. See, for example, J.K. Percus in 'The Equilibrium Theory of Classical Fluids' ed. H.L. Frisch and J.L. Lebowitz (New York: Academic Press) page II - 33, 1964.

56. R.L. Henderson and N.W. Ashcroft, Phys. Rev. A, 13, 859, 1976.

57. T. Gaskell, J. Phys. C: Sol. St. Phys., 3 240, 1970.

58. W.Jones, J. Phys. C: Sol. St. Phys., 6, 2833, 1973.

59. P. Lloyd and C.A. Sholl, J. Phys. C: Sol.St.Phys.,1, 1620,1968.

60. D.C. Wallace, Phys. Rev., 182,778, 1969.

61. D.L. Price, K.S. Singwi and M.P. Tosi, Phys. Rev. B,2,2983, 1970.

62. R. Pynn, Phys. Rev. B, 5, 4826, 1972.

63. M.W. Finnis, J. Phys. F: Metal Phys., 4, 1645, 1974.

64. N.W. Ashcroft and D.C. Langreth, Phys. Rev.,155, 682, 1967.

65. B.C. Allen in 'Liquid Metals Chemistry and Physics' ed. S.Z. Beer (New York; Dekker) page 161, 1972.

66. R. Evans, J. Phys. C: Sol. St. Phys., 7, 2808, 1974.

67. J.G. Kirkwood and F.P. Buff, J. Chem. Phys., 17, 338, 1949.

68. For a review see S. Toxvaerd in 'Statistical Mechanics' ed. K. Singer (Specialist Periodical Reports, Chem. Soc.) Vol. 2, Chap. 4 page 256, 1975.

69. See for example, F. Abraham, D.E. Schreiber and J.A. Barker, J. Chem. Phys. 62, 1958, 1975 and J.K. Lee, J.A. Barker and G.M Pound, J. Chem. Phys. 60, 1976, 1974.

70. R. Kumaravadivel and R. Evans, J. Phys. C: Sol. St. Phys.,8, 793, 1975.

71. M.W. Finnis, J. Phys. F: Metal Phys, 5, 2227, 1975.

72. For a review see N.D. Lang in Sol.St.Phys.,28, 224, 1973.

73. R. Evans and R. Kumaravadivel, J. Phys. C: Sol. St. Phys. 9, 1891, 1976. A simplified version of the WCA approach for surfaces has recently been applied to simple liquids C.E.Upstill and R. Evans. J. Phys. C: Sol. St. Phys. 10, 2791, 1977.

74. D.W.G. White Trans. Metall.Soc.AIME, 236, 796, 1966.

75. C.A. Croxton in Adv. Phys., 22, 385, 1973.

76. S. Toxvaerd, J. Chem. Phys., 62, 1589, 1975 and G.A. Chapela, G. Saville, S.M. Thompson and J.S. Rowlinson, J. Chem. Soc. Faraday, Trans. II, 73, 1133, 1977.

77. D.G. Triezenberg and R. Zwanzig, Phys. Rev. Lett., 28, 1183, 1972. and R. Lovett, P.W. Dehaven, J.J. Vieceli and F.P.Buff, J. Chem. Phys., 58, 1880, 1973.

78. M.S. Wertheim, J. Chem. Phys., 65, 2377, 1976. This paper also lists useful references.

79. P.A. Egelstaff, N.H. March and N.C. McGill, Can.J.Phys.,52, 1651, 1974 and D.M. Straus and N.W. Ashcroft, Phys. Rev. B, 14, 448, 1976.

80. M.D. Johnston, P. Hutchinson and N.H. March, Proc. Roy. Soc.A, 282,283, 1964.

81. For a careful discussion of error analysis see L.E. Ballentine and J.C. Jones, Can. J. Phys., 51, 1831, 1973.

82. A.J. Greenfield, J. Wellendorf and N. Wiser, Phys. Rev. A, 4, 1607, 1971.

83. W.S. Howells and J.E. Enderby, J. Phys. C: Sol. St. Phys., 5, 1277, 1972.

84. R. Kumaravadivel, R. Evans and D.A. Greenwood, J. Phys. F: Metal Phys., 4, 1839, 1974,

85. See also M. Tanaka in 'Liquid Metals 1976' ed. R. Evans and D.A. Greenwood (Inst. of Phys. Conf. Series No 30) page 164, 1977.

86. D. Levesque and L. Verlet, Phys. Rev. Lett., 20, 905, 1968.

87. M. Brennan, P. Hutchinson, M.J.L. Sangster and P. Schofield, J. Phys. C: Sol. St. Phys., 7,L411, 1974.

88. S.K. Mitra, P. Hutchinson and P. Schofield, Phil. Mag., 34, 1087, 1976.

89. S.K. Mitra in 'Liquid Metals 1976' ed. R. Evans and D.A. Greenwood (Inst. of Phys. Conf. Series No 30) page 146, 1977.

90. W. Schommers, Phys. Lett., 43A, 157, 1973.

91. F. Mandel, J. Chem. Phys., 62, 2261, 1975 considered several rare gas liquids. S.K. Mitra and M.J. Gillan, J. Phys. C: Sol. St. Phys. 9, L515, 1976 tested their inversion procedure using computer simulation results for a Lennard-Jones liquid. S.K. Mitra, J. Phys. C: Sol. St. Phys., 10, 2033, 1977 calculated a potential for liquid Ne. R.E. Jacobs and H.C. Andersen Chem. Phys., 10, 73, 1975 calculated the repulsive-force part of the potential for several liquid metals.

92. S.K. Mitra (to be published)

93. N.K. Ailawadi and J. Naghizadeh, Sol. St. Comm., 20, 45, 1976.

94. K.S. Singwi, M.P. Tosi, P.H. Land and A. Sjölander, Phys. Rev., A176, 589, 1968.

95. M.I. Barker and T. Gaskell, Phys. Lett. 53A, 285, 1975.

96. R. Evans (unpublished)

97. See for example Y. Waseda in 'Liquid Metals 1976' ed. R. Evans and D.A. Greenwood (Inst. of Phys. Conf. Series No 30) page 230, 1977

98. H. Rudin, A.H. Millhouse, P. Fischer and G. Meier in 'Liquid Metals' ed. R. Evans and D.A. Greenwood (Inst. of Phys. Conf. Series No 30) page 241, 1977.

99. W. Hartmann, Phys. Rev. Lett., 26, 1640, 1971.

100. B.R. Orton in 'Liquid Metals 1976' ed. R. Evans and D.A. Greenwood (Inst. of Phys. Conf. Series No 30) page 145, 1977.

101. J.P. Badiali and C. Regnaut, Phys. Stat. Sol (b), 63, 555, 1974.

102. P.C. Hohenberg and W. Kohn, Phys. Rev., 136, B864, 1964 and W. Kohn and L.J. Sham, Phys. Rev., 140, A1133, 1965.

103. J. Hubbard, Proc. Roy. Soc. A240, 539, 1957 and A243, 336, 1958.

104. L. Kleinman,Phys. Rev. 160, 585, 1967.

105. K.S. Singwi, A. Sjölander, M.P. Tosi and R.H. Land, Phys. Rev., B1, 1044, 1970.

106. D.J.W. Geldart and R. Taylor, Can. J. Phys., 48, 167, 1970

Appendix 1 Perturbation theory for the averaged density of states.

These notes are abstracted from the review by Ballentine [10]. The one-electron Hamiltonian operator (in atomic units) is

$$H = -\nabla^2 + \sum_{\ell} V_{ps}^{sc}(\underset{\sim}{r} - \underset{\sim}{R}_{\ell}) \qquad (A.1)$$

For convenience we assume V_{ps}^{sc} is a local, energy independent function of r (in practice this may be a gross over simplification) and write

$$H = H_0 + V \qquad (A.2)$$

where $H_0 = -\nabla^2$ and V is the total pseudopotential. The corresponding Schrödinger equation is

$$H|\psi_n> = E_n |\psi_n> \qquad (A.3)$$

where the eigenfunctions and eigenvalues refer to a particular, static configuration of the ions. Rather than focussing attention on the wavefunctions it is better to work with the Green operator G(E) defined by

$$G(E) = (E-H)^{-1} \qquad (A.4)$$

$$= \sum_n \frac{|\psi n><\psi n|}{E - E_n}$$

The electronic density of states is given by

$$n(E) = \sum_n \delta(E-E_n) \qquad (A.5)$$

which can then be related to G:

$$n(E) = -\frac{1}{\pi} \operatorname{Tr} \operatorname{Im} G(E) \qquad (A.6)$$

where the trace is to be taken in some suitable representation. n(E) refers to a particular ionic configuration. Physical observables cannot depend on the details of particular configurations so the averaged density of states $<n(E)>$ and the corresponding averaged Green function $\mathcal{G}(E) \equiv <G(E)>$ are introduced.

The basis of the perturbation theory is to expand G in powers of the potential V and then perform the average term by term:

$$G = (E - H_0 - V)^{-1} \tag{A.7}$$
$$= G_0 + G_0 VG_0 + G_0 VG_0 VG_0 + \ldots\ldots$$

where $G_0 = (E - H_0)^{-1}$ is the free particle propagator and is independent of the ionic cordinates. Provided the (pseudo) potential is sufficiently weak we might expect such a series expansion to converge. For metals other than the simple ones the series will diverge and an alternative approach is required (see below).

The ensemble average of a typical term is

$$< G_0 VG_0 VG_0 > \quad = \quad G_0 < VG_0 V > G_0$$

where the average will involve only the pairwise atomic distribution function. The higher order terms require higher order distribution functions. The expansion is amenable to standard diagrammatic analysis if suitable combinations of the usual distribution functions are introduced. An important quantity is Σ the irreducible self-energy operator. This operator sums all irreducible diagrams:

Σ = 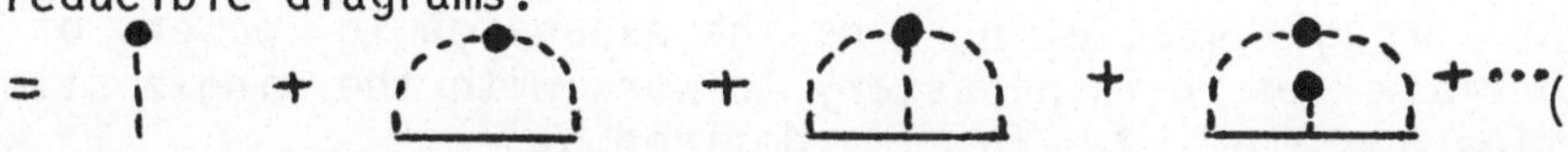 (A.8)

where a solid line represents G_0, a dashed line a V_{ps}^{sc} and the dot joining the dashed lines a 'cluster' function. The latter can be defined in terms of distribution functions [10]. The averaged Green function is then given by the infinite series

$< G >$ = 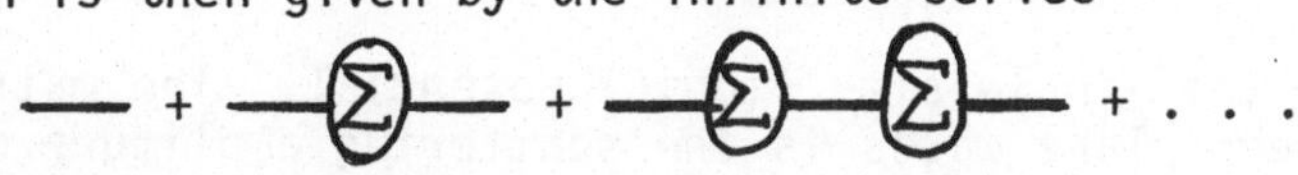

which can be formally summed to give

$$< G > = G_0 (1 - \Sigma G_0)^{-1}$$
$$= (G_0^{-1} - \Sigma)^{-1} \tag{A.9}$$

If Σ is calculated in some approximation this procedure ensures that the infinite sum of reducible terms associated with the particular Σ is included in the estimate of $< G >$. (A reducible diagram is one which is simple a product of irreducible diagrams).

The crudest calculation takes only the first two terms of equ. (A.8) as an estimate of Σ. In momentum representation this is

$$\Sigma(K,E) = V_{ps}^{sc}(q = o) + \frac{\rho}{(2\pi)^3}\int d\underset{\sim}{q} \; \frac{|V_{ps}^{sc}(q)|^2 S(q)}{E - (\underset{\sim}{K} + \underset{\sim}{q})^2} \tag{A.10}$$

where S(q) is the liquid structure factor and ρ is the number of ions per unit volume. The first term of (A.10) is a constant, the average potential, which can be eliminated. Since the averaged Green function is

$$\mathcal{G}(K,E) = (E - K^2 - \Sigma(K,E))^{-1}$$

it is easy to see that $\langle n(E) \rangle$ will depart from free-electron behaviour by terms depending on $|v_{ps}^{sc}(q)|^2 S(q)$, a quantity which is small for all q.

A more realistic approximation aims to produce a more accurate Σ by replacing the denominator in equ (A.10) by

$E - (\underset{\sim}{K} + \underset{\sim}{q})^2 - \Sigma(|\underset{\sim}{K} + \underset{\sim}{q}|,E)$ and solving the resultant self-consistent equation. This corresponds to replacing the free particle propagator G_0 by the full propagator $\langle G \rangle$ and can be described as a properly self-consistent second order perturbation theory. (The corrections to free-electron behaviour are, of course, of the same order as those mentioned above). To go beyond this approximation requires the higher order distribution functions.

For strongly scattering ions the expansion in powers of V is inappropriate and it is necessary to work with the single site transition operator t. This is defined by

$$t = v + vG_0t \tag{A.11}$$

$$= v + vG_0v + vG_0vG_0v + \ldots$$

where v is the single site (ionic) potential. The matrix element of t between plane waves is the scattering amplitude of the ion. As it can be shown [10] that Σ has an expansion in powers of t (this does not require an expansion in v to exist) one proceeds in principle as above, by seeking to truncate the expansion at low order and then improving the estimates by a self consistency argument. In practice there are severe technical difficulties to be surmounted.

Appendix 2 Some results from the theory of the interacting electron gas

These results apply at temperature T = 0^oK.

a) For the one-electron potential V in equ. (2.1) we have in mind something like the potential which enters the Hohenberg-Kohn-Sham [102] theory of the inhomogeneous electron gas. In this theory the effective single particle potential $V_{eff}[n;\underset{\sim}{r}]$ is defined by

$$V_{eff}[n;\underset{\sim}{r}] \equiv V_{Hartree}(\underset{\sim}{r}) + \frac{\delta E_{xc}[n]}{\delta n(\underset{\sim}{r})} \tag{A.12}$$

where $V_{Hartree}$ is the total electrostatic potential of the system of electrons and nuclei and $E_{xc}[n]$ is a universal functional of the electron density $n(\underset{\sim}{r})$ which accounts for all exchange and correlation contributions. The electron density is given as the solution of the self-consistent equations:

$$(-\nabla^2 + V_{eff}[n;\underset{\sim}{r}])\ \psi_i(\underset{\sim}{r}) = E_i\ \psi_i(\underset{\sim}{r}) \tag{A.13a}$$

$$n(\underset{\sim}{r}) = \sum_i |\psi_i(\underset{\sim}{r})|^2 \tag{A.13b}$$

where the summation is over the N_e lowest lying orthonormal solutions of the differential equation. N_e is the total number of electrons (core plus conduction). All the complexity of the many body problem lies in the (unknown) functional $E_{xc}[n]$. Whilst this theory gives, strictly, only a prescription for the ground state density and total energy, in the study of crystalline metals the eigenvalues E_i are often identified with the actual band structure.

b) In the pseudopotential model where the metal is described in terms of (pseudo) ions and conduction electrons, the effective single particle potential is usually taken to be

$$V_S(\underset{\sim}{r}) = \sum_\ell v_{ps}(\underset{\sim}{r} - \underset{\sim}{R}_\ell) + \int d\underset{\sim}{r}'\, U_{ee}(|\underset{\sim}{r} - \underset{\sim}{r}'|)\, n_1(\underset{\sim}{r}') \tag{A.14}$$

where v_{ps} is the bare pseudopotential of a single ion and $n_1(\underset{\sim}{r})$ is the (pseudo) density evaluated to <u>first</u> order in v_{ps}. (This is consistent with a linear response theory based on the homogeneous electron gas). Comparison with equ. (A.12) shows that the exchange and correlation part of U_{ee} is

$$U_{xc}(|\underset{\sim}{r} - \underset{\sim}{r}'|) = \left.\frac{\delta^2 E_{xc}[n]}{\delta n(\underset{\sim}{r})\, \delta n(\underset{\sim}{r}')}\right|_{n_0} \tag{A.15}$$

We write

$$V_S(\underset{\sim}{q}) = v_{ps}(q)\rho(\underset{\sim}{q}) + \left(\frac{4\pi e^2}{q^2} + U_{xc}(q)\right) n_1(\underset{\sim}{q}) \tag{A.16}$$

But

$$n_1(\underset{\sim}{q}) = \chi^e(q)\ v_{ps}(q)\rho(\underset{\sim}{q})$$

where $\rho(\underset{\sim}{q})$ is the ion density so it follows

$$V_S(\underset{\sim}{q}) = \frac{v_{ps}(q)}{\epsilon_{el}(q)} \quad \rho(\underset{\sim}{q}) \qquad (A.17)$$

where $$\frac{1}{\epsilon_{el}(q)} = 1 + \left(\frac{4\pi e^2}{q^2} + U_{xc}(q)\right) \chi^e(q)$$

Thus we can associate a 'screened' pseudopotential $v_{ps}^{sc} \equiv$ F.T. $(v_{ps}(q)/\epsilon_{el}(q))$ with each ion. ϵ_{el} is called the electron dielectric constant. It is not to be confused with the true dielectric constant $\epsilon(q)$. The latter is defined by

$$\delta V_{Hartree}(\underset{\sim}{q}) \equiv \frac{\delta V_{ext}(\underset{\sim}{q})}{\epsilon(q)} \qquad (A.18)$$

i.e. it relates the infinitesmal change in the total electrostatic potential to the corresponding infinitesmal change in whatever external potential is present in the electron system. Since

$$V_{Hartree}(\underset{\sim}{r}) = V_{ext}(\underset{\sim}{r}) + e^2 \int d\underset{\sim}{r}' \frac{n(\underset{\sim}{r}')}{|\underset{\sim}{r} - \underset{\sim}{r}'|} \qquad (A.19)$$

we have

$$\delta V_{Hartree}(\underset{\sim}{q}) = \delta V_{ext}(\underset{\sim}{q}) + \frac{4\pi e^2}{q^2} \delta n(\underset{\sim}{q}) \qquad (A.20)$$

and $\delta n(\underset{\sim}{q}) = \chi^e(q)\, \delta V_{ext}(\underset{\sim}{q})$

so

$$\frac{1}{\epsilon(q)} = 1 + \frac{4\pi e^2}{q^2} \chi^e(q) \qquad (A.21)$$

Only in the random phase approximation where $U_{xc} \equiv o$ are the two dielectric constants equal.

c) Several approximations exist for U_{xc}. These are based on different treatments of the homogeneous electron gas. The Hubbard [103] approximation only includes exchange:

$$U_{xc}(q) = -\frac{2\pi e^2}{q^2 + K_F^2} \qquad (A.22)$$

A similar form, due to Kleinman [104], attempts to include correlation:

$$U_{xc}(q) = -\pi e^2 \left(\frac{1}{q^2 + \xi K_F^2} + \frac{1}{\xi K_F^2} \right) \quad (A.23)$$

where ξ is a parameter which can be adjusted to fit the compressibility sum rule (see below). More recent theories by Singwi et al [105] and Vashista and Singwi [21] have expressed U_{xc} in the form

$$U_{xc}(q) = -\frac{4\pi e^2}{q^2} A \left(1 - \exp\left[-B(q/K_F)^2\right]\right) \quad (A.24)$$

where A and B have a rather weak dependence on the average electron density n_o. The Vashista and Singwi version has the merits of satisfying the compressibility sum rule almost exactly and yielding a physically realistic radial distribution function for the electron liquid over the metallic range of densities. Another realistic approximation is that due to Geldart and Taylor [106].

d) The compressibility χ_T^e of the uniform electron gas is related [20] to the long wavelength limit of the dielectric constant:

$$\frac{\chi_T^e}{\chi_T^{ni}} = \frac{\pi}{4K_F} \lim_{q \to o} \left(q^2 (\epsilon(q) - 1)\right) \quad (A.25)$$

where χ_T^{ni} is the compressibility of the non-interacting gas of the same density. Using equs. (A.21) and (3.24) it is easy to show

$$\frac{1}{\chi_T^e} = - n_o^2 \left(\chi^{ni}(q = o)^{-1} - U_{xc}(q = o)\right) \quad (A.26)$$

χ_T^e calculated in this way should agree with the result obtained by differentiating the pressure of the uniform electron gas i.e.

$$\frac{1}{\chi_T^e} = n_o^2 \left(2 \frac{du_{eg}}{dn_o}(n_o) + \frac{d^2 u_{eg}}{dn_o^2}(n_o) \right) \quad (A.27)$$

where u_{eg} is the ground state energy per electron (equ. (3.22)). This sum rule is often used in testing the consistency of theories for the correlation energy E_{corr} and U_{xc}.

Appendix 3 The calculation of the bulk modulus: a simplified scheme

Here we evaluate the bulk modulus B_T of a liquid metal by the method of long waves and by differentiating the pressure with respect to volume. The approximation scheme is due to Jones [58].

The pressure p given by equ. (3.36) can be rewritten:

$$p = \rho k_B T + \rho n_0 \frac{du(n_0)}{dn_0} - \frac{\rho^2}{2}\int d\underset{\sim}{R}\, \frac{R}{3}\frac{\partial\phi(R;n_0)}{\partial R} + \frac{\rho^2}{2}\int d\underset{\sim}{R}\, n_0 \frac{\partial\phi(R;n_0)}{\partial n_0}$$
$$- \frac{\rho^2}{2}\int d\underset{\sim}{R}\, (g(R)-1)\frac{R}{3}\frac{\partial\phi(R;n_0)}{\partial R} + \frac{\rho^2}{2}\int d\underset{\sim}{R}\, (g(R)-1) n_0 \frac{\partial\phi(R;n_0)}{\partial n_0} \tag{A.28}$$

Since B_T is made complicated by the appearance of the volume derivative of g(R) we simplify the analysis by setting g(R) equal to its high temperature, low density limit, i.e. we set

$$g(R) \approx 1 - \beta\phi(R;n_0) \tag{A.29}$$

Whilst this is certainly a non-realistic approximation for quantitative purposes its serves to illustrate the key points of the argument. With this assumption, and on integrating by parts, p reduces to

$$p = \rho k_B T + \rho n_0 \frac{du(n_0)}{dn_0} + \frac{\rho^2}{2}\int d\underset{\sim}{R}\, \phi(R;n_0) + \frac{\rho^2}{2}\int d\underset{\sim}{R}\, n_0 \frac{\partial\phi(R;n_0)}{\partial n_0}$$
$$- \frac{\rho^2\beta}{4}\int d\underset{\sim}{R}\, \phi^2(R;n_0) - \frac{\rho^2\beta}{4}\int d\underset{\sim}{R}\, n_0 \frac{\partial}{\partial n_0}\phi^2(R;n_0)$$

A further simplication is made by neglecting the last two terms, which are second order in ϕ, so that p is given by

$$p = \rho k_B T + \rho n_0 \frac{du(n_0)}{dn_0} + \frac{\rho^2}{2}\phi(q=0;n_0) + \frac{\rho^2}{2} n_0 \frac{\partial\phi(q=0;n_0)}{\partial n_0} \tag{A.30}$$

Using equs. (3.30) and (3.32) the self-energy can be written as:

$$u(n_0) = Z u_{eg}(n_0) - \frac{\rho}{2}\phi(q=0;n_0) + n_0 W_c(q=0) + \frac{1}{2(2\pi)^3}\int d\underset{\sim}{q}\, \chi^e(q)|v_{ps}(q)|^2 \tag{A.31}$$

and equ. (A.30) reduces to

$$p = \rho k_B T + n_0^2 \frac{du_{eg}(n_0)}{dn_0} + \rho n_0 W_c(q=0) + \frac{\rho}{2(2\pi)^3}\int d\underset{\sim}{q}\, n_0 \frac{\partial}{\partial n_0}\chi^e(q)|v_{ps}(q)|^2 \tag{A.32}$$

The bulk modulus $B_T = \rho(\partial p/\partial\rho)_T$ is then

$$B_T = \rho k_B T + \frac{1}{\chi_T^e} + 2\rho n_0 W_c(q=0) + \frac{\rho}{(2\pi)^3}\int d\underset{\sim}{q}\left(n_0\frac{\partial}{\partial n_0} + \frac{n_0^2}{2}\frac{\partial^2}{\partial n_0^2}\right)\chi^e(q)|v_{ps}(q)|^2 \tag{A.33}$$

where we have used equ. (A.27) for the compressibility of the electron gas.

In the method of long waves we require

$$S(0) = 1 + \rho \int d\underset{\sim}{R} \; (g(R) - 1)$$

and in the approximation of equ. (A.29) this is

$$S(0) = 1 - \beta\phi\,\phi(q = o; n_o) \qquad (A.34)$$

Using equ. (5.2a) the long waves bulk modulus is

$$B_T^L = \frac{1}{\chi_T} = \rho K_B T \left(1 - \rho\,\beta\phi\,(q = o; n_o)\right)^{-1}$$

which, to first order in ϕ, is

$$B_T^L = \rho K_B T + \rho^2 \phi\,(q = o; n_o) \qquad (A.35)$$

If $\phi(q;n_o)$ is evaluated to second order in the pseudopotential (as in § 3.3) $\phi(q = o;n_o)$ is given by equ. (3.30) and the bulk modulus is

$$B_T^L = \rho K_B T - n_o^2 \left(\chi^{ni}(q = o)^{-1} - U_{xc}(q{=}o)\right) + 2\rho n_o w_c(q{=}o) \qquad (A.36)$$

The second term in this equation is just the long waves result for the bulk modulus of the electron gas (see equ. A.26). It is clear that B_T^L agrees with B_T in equ. (A.33) apart from the term depending on the density derivatives of $\chi^e(q)$.

If terms of order λ^2 and λ^3 are retained in equ.(3.27) then the two-body potential $\phi(R;n_o)$ depends on the response functions $\chi^{(2)}$ and $\chi^{(3)}$. Jones [58] has shown that by taking the q = o limit on this modified potential and substituting into equ (A.35), B_T^L is identical to B_T obtained in equ. (A.33) by differentiation of the pressure. (The relevant long wavelength limits of $\chi^{(2)}$ and $\chi^{(3)}$ can be related to the density derivates of $\chi^e(q)$).

SPECTROSCOPIC METHODS FOR THE STUDY OF LOCAL DYNAMICS IN POLYATOMIC FLUIDS

F. Volino

Institut Laue-Langevin, 156X Centre de Tri
38042 Grenoble Cédex, France, and

Groupe de Dynamique des Phases Condensées
Laboratoire de Cristallographie (associé au CNRS)
Université des Sciences et Techniques du Languedoc
Place Eugène Bataillon, 34060 Montpellier Cédex, France

CONTENTS

CHAPTER 1

THE PROBLEM OF MOLECULAR MOTIONS

1.1 General

The problem of molecular motions in liquids can be approached in two complementary ways: by molecular dynamics calculations, and by spectroscopic experiments. In molecular dynamics, one considers an assembly of N (typically N = 500) rigid molecules i, assumes intermolecular pair potentials, chooses boundary and initial conditions (i.e. the volume and the energy) and solves numerically the 6N coupled equations of motion. Using these results one can in principle calculate any physical quantity associated with the system (equilibrium as well as time dependent quantities). In practice, however, the method is limited by computer memory and time; in other words, by the fact that (i) the tested volume of sample is always very small (N is always small compared to the number of particles in a real sample) and (ii) the time scale is also relatively small (the number of integration steps is necessarily finite). As a consequence, this method cannot, a priori, be very good to test long range and long time phenomena as described for example, by critical phenomena theory and hydrodynamic theory. The other limitation of the method is the restriction to pair potentials and to classical mechanics. Consequently, important phenomena such as vibrations cannot be included since a quantum description is then required. The main interest of the molecular dynamics method is to give typical results which can be compared to experimental results obtained on real liquids. The molecular dynamics problem is treated in detail by Dr. McDonald elsewhere in this book.

The other way to approach the problem of molecular motions is to use spectroscopic techniques. The idea here is to use a probe such as electromagnetic (e.m.) waves, neutrons, spins ..., prepared in a well known state, and make it interact with the degrees of freedom of the system under study. Due to the interaction, the state of the probe changes (e.g. the e.m. wave is scattered) and this change reflects the dynamical properties of the system. The result generally appears in the form of an energy spectrum which should be interpreted in terms of molecular motions. As will be seen in the next section, any spectrum is proportional to the time Fourier transform of a well defined correlation function (c.f.). A c.f. is the equilibrium ensemble average of the product of two molecular dynamical variables taken at time 0 and time t. A now widely accepted means of extracting the physical information from a spectrum is to use the concept of a dynamical model. Such a model is characterized by one or several rate equations which govern the evolution of molecular dynamical variables. These equations are used to calculate the correlation functions and, after Fourier transforming, the theoretical spectrum. This spectrum is then compared to the experimental one and the parameters of the model (jump time, gyration radius ...) eventually deduced.

1.2 Models for Molecular Motions

In the past few years, a large number of models have been developed to describe molecular translation, rotation and vibration. In this section, we list the main ones:

- for translation, self motion of the molecular centre of mass is often described in terms of the so-called Langevin model where the particle is assumed to be submitted to two forces: a viscous force and a random force. In the limits of very weak and very strong viscosity, one finds the free translation and the uniform translational diffusion, respectively. In the latter case, the motion is characterized by a single diffusion coefficient D_t (isotropic medium). On the other hand, collective translational motions are usually described in terms of longitudinal acoustic waves and thermal diffusivity in a hydrodynamic theory.

- for rotation, many models have been imagined. They can be classified roughly into inertial and stochastic models. In the inertial models, the molecules are essentially rotating, but suffer random collisions which modify their dynamical state. Such models are plausible in low density systems composed of spheroidal molecules. Among them let us mention the free rotation model (no collisions) and the extended diffusion models (so called models J and M of Gordon) where during the collision, the orientation of the molecules do not change but their angular momentum $\underset{\sim}{J}$ is randomized. In the stochastic models on the contrary, the molecules are essentially non rotating, the motions occurring by rapid rotational jumps over barriers. Such models are plausible in highly condensed systems. Among these, let us mention the Debye model (isotropic diffusion characterized by a single diffusion coefficient D_r), the Ivanov model (isotropic jumps of finite angle), the anisotropic diffusion model .... The distinction between inertial and stochastic models is in fact not clear cut, especially when the time between jumps or collisions is of the order of $(k_BT/I)^{1/2}$, the mean jump time for rotation due to thermal motion. In this case, the model should combine both aspects as is the case for example, in the Langevin model for rotation. Other models where translation and rotation are coupled have been imagined. For example, we mention the strong collision model where the molecules move freely except during instantaneous collisions which randomize both the position and the angular momentum. Concerning the collective rotational motions, some recent light scattering experiments seem to require a description in terms of coupling between shear waves and molecular rotations (cf. Section 4.4).

- finally for molecular vibrations, little is known so far. It appears however, that a great number of mechanisms can be responsible for vibrational relaxation, in particular coupling with all the other degrees of freedom: rotations, translations, and other vibrations.

This list of models is clearly not exhaustive and a considerable literature exists on the subject. In refs. [1] to [30], the reader will find a good sampling of this literature, noting that refs. [27] to [30] are reviews, rather than specalized papers.

1.3 Conclusion

At the present time where all techniques have reached a rather sophisticated state of development (lasers, high resolution and high intensity neutron spectrometers, microwave and submillimeter generators, fast electronic devices, high magnetic fields), a great number of precise experiments can be and have been performed on a variety of liquids, using all methods. A rather complete description of the state of the art can be found in the Proceedings of the 24th Annual Meeting of the Société de Chimie Physique held at Paris-Orsay, July 2-6, 1973 [1]. It appears that, in general, the experimental results are more or less satisfactorily interpreted in terms of models, as mentioned above. The fit is generally found to be good when adjusted to results obtained by one method only. When data obtained by various methods are combined (i.e. when the number of parameters is reduced), the situation becomes more complicated, and often no existing model can fit consistently all the data. It thus appears that to obtain a reliable picture of what is happening at the molecular level, it is important to collect data obtained from different methods and interpret them in a consistent way. A necessary condition for this is to know precisely the principles and the limitations of these methods, in particular the kind of motion that can be "seen" in each case, and the space and time scale which can be probed with them.

The aim of these lectures is mainly devoted to this purpose. After a general presentation of the principles of any spectroscopic measurement, the main spectroscopic methods will be described: neutron scattering, absorption and scattering of e.m. waves, nuclear magnetic resonance, spin relaxation. A few words will also be said on other less common methods. As illustrations, theoretical spectra will be calculated for the simplest models, and their characteristic features deduced in each particular case. Then, typical real results obtained for molecular liquids will be presented and their characteristic features discussed.

Clearly, the material presented in these lectures is not really new and papers dealing with the problem of comparison between the various spectroscopic methods for studying molecular motions can be found in the literature [27-30]. However, an attempt has been made to give a relatively new presentation which has the advantage of describing all cases by means of the same basic formulae. It is hoped that this will enable the specialist in only one technique to have a clear feeling for the others.

For simplicity, all the formulae have been written in the Gauss c.g.s. system of units, i.e. the system such that $\varepsilon_o = \mu_o = 1$ where ε_o and μ_o are the permittivity and permeability of free space, respectively.

CHAPTER 2

THE PRINCIPLE OF A SPECTROSCOPIC MEASUREMENT: COUPLING BETWEEN A PROBE AND A RESERVOIR

2.1 Ingredients of the Problem

Let us consider a system (e.g. a molecular liquid) composed of N particles i, at thermal equilibrium T, that we call the reservoir R. This reservoir is characterized by its Hamiltonian H_R whose eigenvalues and eigenstates are labelled $E_{m'}$ and $|m'\rangle$. The problem is to know how the molecular properties of this system vary with time. For this purpose, we consider another system (e.g. the e.m. field, the neutron field, a spin system) which we call the probe P. This probe is characterized by its Hamiltonian H_P whose eigenvalues and eigenstates are labelled E_m and $|m\rangle$. The probe is able to couple with the dynamical variables of the reservoir and this coupling is characterized by an Hamiltonian H_C. This Hamiltonian is necessarily a non trivial function of operators acting both on R and P. This is sketched in Fig. 1.

2.2 Principle of the Experiment

At initial time, we assume that the probe is in a defined dynamical state $|m\rangle$ (e.g. e.m. waves or neutrons are collimated and monochromatized). The reservoir R, being at thermal equilibrium, can be in any state $|m'\rangle$ with the probability $p_{m'}$ given by the Boltzmann law:

$$p_{m'} = \frac{1}{Z_R} \exp(-\beta E_{m'}) \tag{1}$$

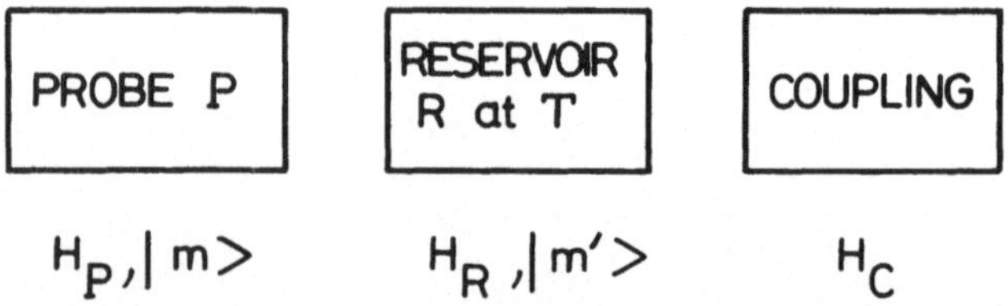

Figure 1: Sketch of a spectroscopic experiment.

with

$$Z_R = \sum_{m'} \exp(-\beta E_{m'}) \tag{2}$$

and

$$\beta = \frac{1}{k_B T} \tag{3}$$

where k_B is the Boltzmann constant.

Then, the interaction H_c is switched on. Due to this interaction, the state of the probe P can change with time from the state $|m\rangle$ to a final state $|n\rangle$. In the linear approximation (i.e. if H_c is "small" compared to H_P and H_R), and long time, quantum mechanics tells us that this change can be characterized by a probability per unit time W_{nm}. The principle of any spectroscopic experiment is to measure a quantity which is proportional to W_{nm}, as a function of $|n\rangle$ or $|m\rangle$. Since W_{nm} is a function of operators of R, its measurement yields information about what is happening in R (e.g. the molecular motions). The problem thus reduces in calculating W_{nm} and relating it to a measurable quantity.

2.3 Calculation of W_{nm}

For this calculation, the total system to be considered is the probe plus the reservoir. The corresponding eigenstates are symbolized by $|m\rangle|m'\rangle$. Let $W_{n'n\,m'm}$ be the probability per unit time that this total system changes from the state $|m\rangle|m'\rangle$ to the state $|n\rangle|n'\rangle$ due to H_c. In the linear approximation, we have:

$$W_{nm} = \sum_{m'n'} W_{nn'mm'} \; p_{m'} \tag{4}$$

where $W_{nn'mm'}$ is given by the Fermi golden rule:

$$W_{nn'mm'} = \frac{2\pi}{\hbar} \left| \langle n'|\langle m|H_c|m\rangle|m'\rangle \right|^2 \delta(E_m + E_{m'} - E_n - E_{n'}) \tag{5}$$

Let us now define the quantity $\bar{H}_c$ as the average of H_c between the initial and final states of the probe:

$$\bar{H}_c = \langle n|H_c|m\rangle \tag{6}$$

Defined in this way, $\bar{H}_c$ is an operator which acts on the states of the reservoir only. With the following definitions:

$$\hbar\omega = E_m - E_n \qquad (7)$$

$$\hbar\omega_{n'm'} = E_{n'} - E_{m'} \qquad (8)$$

and combining the above equations, the probability W_{nm} is written:

$$W_{nm} = \frac{2\pi}{\hbar^2} \sum_{m',n'} \frac{\exp(-\beta E_{m'})}{Z_R} \left|<n'|\bar{H}_c|m'>\right|^2 \delta(\omega_{n'm'}-\omega) \qquad (9)$$

Because of the presence of the δ function, such an expression is adequate to describe discrete peaks in a spectrum, as is the case for a purely quantum system. For describing more complicated systems, it is worthwhile using another equivalent expression for W_{nm}. This is obtained by using the fact that $\bar{H}_c$ is an hermitian operator and by using the integral expression for the δ function. We have:

$$\left|<n'|\bar{H}_c|m'>\right|^2 = <n'|\bar{H}_c|m'><m'|\bar{H}_c^+|n'> \qquad (10a)$$

and

$$\delta(\omega_{n'm'} - \omega) = \frac{1}{2\pi} \int_{-\infty}^{+\infty} dt \, \exp i(\omega_{n'm'} - \omega)t \qquad (10b)$$

Inserting Eqs. (10a) and (10b) in Eq. (9) and using Eq. (8), we obtain

$$W_{nm} = \frac{1}{\hbar^2} \int_{-\infty}^{+\infty} dt \sum_{n'm'} \frac{\exp(-\beta E_{m'})}{Z_R} <m'|\bar{H}_c^+|n'><n'|$$

$$\exp\left(\frac{-iE_{n'}t}{\hbar}\right)\bar{H}_c \exp\left(\frac{iE_{m'}t}{\hbar}\right)|m'> \exp(-i\omega t) \qquad (11)$$

The double sum is simply the expression of a trace in the R Hilbert space. We thus have:

$$W_{nm} = \frac{1}{\hbar^2} \int_{-\infty}^{+\infty} \mathrm{Tr}\,\{\rho_R\, \bar{H}_c^+(o)\, \bar{H}_c(t)\} \exp(-i\omega t)\, dt \qquad (12)$$

where ρ_R is the density matrix of R at thermal equilibrium:

$$\rho_R = \frac{\exp(-\beta H_R)}{\mathrm{Tr}[\exp(-\beta H_R)]} \quad (13)$$

and $\bar{H}_c(t)$ is the Heisenberg representation of operator $\bar{H}_c$:

$$\bar{H}_c(t) = \exp\left(-\frac{iH_R t}{\hbar}\right) \bar{H}_c \exp\left(\frac{iH_R t}{\hbar}\right) \quad (14)$$

The quantity

$$C_{\bar{H}_c\bar{H}_c}(t) = \mathrm{Tr}\,\{\rho_R \bar{H}_c^+(o)\,\bar{H}_c(t)\} = \langle \bar{H}_c^+(o)\,\bar{H}_c(t)\rangle \quad (15)$$

is the quantum expression for the autocorrelation function of $\bar{H}_c$. If the reservoir is classical, then $\bar{H}_c$ is a classical function of the variables of R and $C_{\bar{H}_c\bar{H}_c}(t)$ should be replaced by its classical equivalent. Defining the spectral density $C_{\bar{H}_c\bar{H}_c}(\omega)$ as the time Fourier transform of $C_{\bar{H}_c\bar{H}_c}(t)$:

$$C_{\bar{H}_c\bar{H}_c}(\omega) = \frac{1}{2\pi}\int_{-\infty}^{+\infty} C_{\bar{H}_c\bar{H}_c}(t)\exp(-i\omega t)\,dt \quad (16)$$

and using Eq. (12), the probability of transition W_{nm} is finally written:

$$W_{nm} = \frac{2\pi}{\hbar^2} C_{\bar{H}_c\bar{H}_c}(\omega) \quad (17)$$

An important point is to relate the probabilities for the direct transition W_{nm} and the inverse transition W_{mn}. Changing ω into $-\omega$ in the above equations, after a little algebra, we obtain:

$$W_{mn} = \exp(-\beta\hbar\omega)\, W_{nm} \quad (18)$$

This is the Kubo-Ayant theorem which means that, in the linear approximation, the probability per unit time for the transition from one level to another is proportional to the population of this level.

Equations (15), (16), (17) and (18) are the central ones for our purpose and will be used extensively in the following sections.

2.4 Summary

In summary, to describe any spectroscopic experiment, we are led to define the reservoir, the probe, the interaction H_c and to calculate the average value $\bar{H}_c$, its correlation function $C_{\bar{H}_c\bar{H}_c}(t)$ and the probability W_{nm}. To be complete, we must relate W_{nm} to a measurable quantity. However, this relationship is clearly dependent on the type and details of the experiment (e.g. an absorption experiment, a scattering experiment, a relaxation experiment) and no general formula can be given. We shall thus establish it for each particular case treated below. For details concerning the various theoretical concepts introduced in this section, the reader is referred to the lectures of Prof. J.P. Hansen, elsewhere in this book.

CHAPTER 3

NEUTRON SCATTERING [31-37]

In this section, we describe how neutrons can be used to study molecular motions. We first recall the properties of the neutron and associated concepts, then describe the principles of a neutron scattering experiment, deduce the relevant correlation function, relate it to the scattered intensity and give a few illustrative examples.

3.1 Properties of the Neutron

The free neutron is an elementary particle with zero charge and spin 1/2, liberated for example during the process of fission of a heavy nucleus. In a nuclear reactor, the neutrons are thermalized by the atoms of the moderator, yielding a Maxwellian distribution of velocities v peaked at some $\bar{v}$ such that the average (kinetic) energy $\bar{E}$ is

$$E = \frac{1}{2} m_n \bar{v}^2 = \frac{3}{2} k_B T \tag{19}$$

where m_n is the mass of the neutron.
Neutrons can also be considered as plane waves of wave number $\underset{\sim}{k}$ or wavelength $\lambda = 2\pi/k$. The relationships between particle and wave aspects are:

$$E = \frac{\hbar^2 k^2}{2m_n} \tag{20}$$

and

$$v = \frac{\hbar k}{m_n} \tag{21}$$

For thermal neutrons (T = 300 K), we have $\bar{E}$ = 26 meV and $\bar{\lambda}$ = 1.8 Å. It is important to note that these values are of the same order of magnitude as the intermolecular energies and molecular dimensions, respectively.

Finally, considered as quantum objects, neutrons are characterized by wave functions $|k\rangle$ such that

$$|k\rangle = (1/\sqrt{V}) \exp(i\underset{\sim}{k}\underset{\sim}{r}) \tag{22}$$

where V is the "volume of quantization" to be identified with the volume of the irradiated sample. In this volume, the density of states of momentum $\underset{\sim}{k}$ is given by [33]

$$\rho(\underset{\sim}{k}) = \frac{V}{(2\pi)^3} \tag{23}$$

Using the following expression of the volume element in spherical coordinates:

$$d\underset{\sim}{k} = k^2 dk d\Omega \tag{24}$$

where $d\Omega$ is the solid angle corresponding to $d\underset{\sim}{k}$ around $\underset{\sim}{k}$, we deduce that the number of independent neutron states between $\underset{\sim}{k}$ and $\underset{\sim}{k} + d\underset{\sim}{k}$ is

$$\rho(\underset{\sim}{k}) d\underset{\sim}{k} = \frac{V}{(2\pi)^3} k^2 dk d\Omega \tag{25}$$

3.2 Neutron-Nucleus Interaction

A neutron interacts with a nucleus via nuclear and magnetic forces. For the nuclear part, since nuclear interactions are very short range compared to the (thermal) neutron wavelength, it can be shown that the interaction potential between a neutron located at $\underset{\sim}{r}$ and a nucleus i located at $\underset{\sim}{r}_i$ can be written as

$$V(\underset{\sim}{r}) = \frac{2\pi\hbar^2}{m_n} b_i \, \delta(\underset{\sim}{r} - \underset{\sim}{r}_i) \tag{26}$$

In this expression (the so-called Fermi pseudo-potential), the scattering length b_i characterizes the interaction and is independent of neutron energy. b_i can be positive or negative according to the attractive or repulsive nature of the interaction. The theoretical calculation of b_i is very difficult and in practice it is determined experimentally. Concerning the magnetic interaction, the neutron interacts with the spins via the dipole-dipole coupling. For diamagnetic systems, it is always negligible compared to the nuclear interaction and shall not be considered in the following.

3.3 Coherent and Incoherent Scattering Lengths and Cross-sections

Consider an assembly of a given atomic species i, with many isotopes possessing a nuclear spin. The scattering length b_i will change from one atom to another, since the interaction depends on the nature of the nucleus and on the total spin state of the nucleus-neutron system.

The average $<b_i>$ of b_i over all the isotopes and spin states is called the coherent scattering length. The mean square deviation of b_i from $<b_i>$ is called the incoherent scattering length. We thus have

$$b_i^{coh} = <b_i> \qquad (27)$$

$$b_i^{incoh} = [<b_i^2> - <b_i>^2]^{\frac{1}{2}} \qquad (28)$$

From these definitions, it is clear that b^{coh} and b^{incoh} can be changed merely by modifying the relative concentration of the various isotopes. This fact is of great practical importance in neutron experiments (isotopic substitution). The coherent and incoherent scattering cross-sections are defined by

$$\sigma_i^{coh} = 4\pi \, b_i^{2\,coh} \qquad (29)$$

$$\sigma_i^{incoh} = 4\pi \, b_i^{2\,incoh} \qquad (30)$$

We list below the values of these quantities in barns (1 barn $= 10^{-24}$ cm^2) for a few common atoms:

atoms	spin	σ^{coh}	σ^{incoh}
^{16}O	0	4.2	-
^{12}C	0	5.5	-
^{14}N	1	11.6	≈ 0.3
D	1	5.6	2.0
H	$\frac{1}{2}$	1.76	79.7

3.4 Principles of a Neutron Scattering Experiment

According to section 2, in order to describe the principles of a neutron experiment we should first define the probe P, the reservoir R, and the interaction H_c. The probe P is constituted by the neutron plane waves which are characterized by their wave functions $|k\rangle$. The reservoir R is made up of N particles i located at $\underset{\sim}{r}_i$. The interaction H_c between P and R is given by Eq. (26) summed over all the particles i, namely

$$H_c = \frac{2\pi\hbar^2}{m_n} \sum_i b_i \, \delta(\underset{\sim}{r} - \underset{\sim}{r}_i) \tag{31}$$

The neutrons are generally prepared in a well defined state $|m\rangle = |k_o\rangle$ (i.e. they are collimated and monochromatized) and one looks at the probability of these neutrons to be scattered, i.e. changed into another state $|n\rangle = |k_1\rangle$.

Using Eq. (22) and (31), one easily obtains the expression for $\bar{H}_c$:

$$\bar{H}_c = \langle k_1|H_c|k_o\rangle = \frac{1}{V}\left(\frac{2\pi\hbar^2}{m_n}\right)^2 \sum_i b_i \exp(i\underset{\sim}{Q}\underset{\sim}{r}_i) \tag{32}$$

where

$$\underset{\sim}{Q} = \underset{\sim}{k}_o - \underset{\sim}{k}_1 \tag{33}$$

is the neutron momentum transfer.
Using Eq. (16), the correlation function of $\bar{H}_c$ is written:

$$C_{\bar{H}_c\bar{H}_c}(t) = \frac{1}{V^2}\left(\frac{2\pi\hbar^2}{m_n}\right)^2 \sum_{i,j} \langle b_i b_j \exp i\underset{\sim}{Q}[\underset{\sim}{r}_i(t) - \underset{\sim}{r}_j(o)]\rangle \tag{34}$$

Taking the time Fourier transform of Eq. (34), one obtains the corresponding probability transition $W_{k_1k_o}$ from Eq. (17).

We should now relate $W_{k_1k_o}$ to a measurable quantity. In practice one sends a flux of neutrons of monochromatic neutrons $\underset{\sim}{k}_o$ onto the sample and collects neutrons scattered with momenta comprised between $\underset{\sim}{k}_1$ and $\underset{\sim}{k}_1 + d\underset{\sim}{k}_1$. Let I_o be the number of incident neutrons per cm^2 and per sec. The sample is a cylinder of section S normal to $\underset{\sim}{k}_o$ and of length ℓ ($S\ell = V$). The number of incident neutrons in the volume V is $I_o(V/v_o)$ where v_o is the incident velocity. Calling I the number of neutrons scattered per second between $\underset{\sim}{k}_1$ and $\underset{\sim}{k}_1 + d\underset{\sim}{k}_1$ we have:

$$I = I_o \frac{V}{v_o} W_{k_1 k_o} \rho(\underset{\sim}{k}_1)\, d\underset{\sim}{k}_1 \tag{35}$$

Using Eqs. (21), (17), (34) and (25) and using the definition of the energy E_1 of the scattered neutron:

$$E_1 = \hbar\omega_1 = \frac{\hbar^2 k_1{}^2}{2m_n} \tag{36}$$

to calculate the dk_1, we finally obtain,

$$\frac{I}{I_o} = \frac{d^2\sigma}{d\Omega d\omega}\, d\Omega d\omega_1 = N \frac{k_1}{k_o} S(\underset{\sim}{Q},\omega)\, d\Omega d\omega_1 \tag{37}$$

with

$$S(\underset{\sim}{Q},\omega) = \frac{1}{2\pi} \int_{-\infty}^{+\infty} I(\underset{\sim}{Q},t) \exp(-i\omega t)\, dt \tag{38}$$

and

$$I(\underset{\sim}{Q},t) = \frac{1}{N} \sum_{i,j} \langle b_i b_j \exp i\underset{\sim}{Q}[\underset{\sim}{r}_i(t) - \underset{\sim}{r}_j(o)]\rangle \tag{39}$$

$d^2\sigma/d\Omega d\omega$ represents the normalized scattered intensity per unit energy and per unit solid angle and is called the double differential cross-section. $I(\underset{\sim}{Q},t)$ and $S(\underset{\sim}{Q},\omega)$ are called the intermediate scattering function and the scattering law, respectively. It is seen from Eq. (39) that neutron scattering reflects molecular motions through the variation of the position of the scattering particles. More can be said about the nature of these motions by making explicit the average in Eq. (39). This average can indeed be performed independently on $b_i b_j$ and on the exponential. The reason for this is that the actual scattering length of each nucleus is clearly independent of its position. Using Eqs. (27) and (28) in Eq. (39) we can write:

$$I(\underset{\sim}{Q},t) = I_p(\underset{\sim}{Q},t) + I_s(\underset{\sim}{Q},t) \tag{40}$$

with

$$I_p(\underset{\sim}{Q},t) = \frac{1}{N} \sum_{i,j} b_i^{coh}\, b_j^{coh} \langle \exp i\underset{\sim}{Q}[\underset{\sim}{r}_i(t) - \underset{\sim}{r}_j(o)]\rangle \tag{41}$$

$$I_s(Q,t) = \frac{1}{N}\sum_i b_i^{2\,incoh} \langle \exp iQ[\underset{\sim}{r}_i(t) - \underset{\sim}{r}_i(o)] \rangle \tag{42}$$

and by analogy

$$S(\underset{\sim}{Q},\omega) = S_p(\underset{\sim}{Q},\omega) + S_s(\underset{\sim}{Q},\omega) \tag{43}$$

These equations show that the scattering can be split into two parts: a coherent part and an incoherent part. If the system contains no hydrogen atom (e.g. a simple liquid) then $b_i^{coh} \gg b_i^{incoh}$ and the scattering is mainly coherent. $I_p(\underset{\sim}{Q},t)$ is a pair correlation function and thus the spectra mainly reflect collective atomic motions. This case is treated in detail by Dr. J.R.D. Copley elsewhere in this book and will be further commented on in section 4.4 in connection with light scattering. On the other hand, if the system is composed of molecules containing hydrogen atoms (e.g. most molecular liquids), then $b_i^{incoh} \gg b_i^{coh}$ and the scattering is mainly incoherent. $I_s(\underset{\sim}{Q},t)$ is a self correlation function and thus the spectra mainly reflect the individual atomic motions. In what follows, we consider only this latter case.

3.5 Scattering from Hydrogenous Systems: Incoherent Scattering [34-37]

Consider a system composed of identical molecules containing, for simplicity, only one hydrogen atom. Since the scattering is almost entirely incoherent, we have

$$I(\underset{\sim}{Q},t) \simeq I_s(\underset{\sim}{Q},t) = \mathrm{Const}\langle \exp iQ[\underset{\sim}{r}(t) - \underset{\sim}{r}(o)] \rangle \tag{44}$$

For simplicity, in the following we use normalized functions by making the constant equal to 1. Let us write

$$\underset{\sim}{r} = \underset{\sim}{d} + \underset{\sim}{\rho} + \underset{\sim}{u} \tag{45}$$

where $\underset{\sim}{d}$ defines the molecular centre of mass (c.o.m.), $\underset{\sim}{\rho}$ defines the position of the proton with respect to the c.o.m. and $\underset{\sim}{u}$ stands for the vibrational displacement around the average position. We have, in the classical approximation:

$$\langle \exp i\underset{\sim}{Q}(\underset{\sim}{r}-\underset{\sim}{r}_o) \rangle = \langle \exp i\underset{\sim}{Q}(\underset{\sim}{d}-\underset{\sim}{d}_o) \; \exp i\underset{\sim}{Q}(\underset{\sim}{\rho}-\underset{\sim}{\rho}_o) \; \exp i\underset{\sim}{Q}(\underset{\sim}{u}-\underset{\sim}{u}_o) \rangle \tag{46}$$

If the translations (i.e. motion of $\underset{\sim}{d}$), rotation (motion of $\underset{\sim}{\rho}$) and vibration (motion of u) are not coupled, then the averages on the three exponentials in Eq. (46) can be performed separately and we have, with evident notations:

$$I_s = I_s^{trans} \cdot I_s^{rot} \cdot I_s^{vib} \tag{47}$$

Taking the Fourier transform, we obtain

$$S_s(\underset{\sim}{Q},\omega) = S_s^{trans}(\underset{\sim}{Q},\omega) \otimes S_s^{rot}(\underset{\sim}{Q},\omega) \otimes S_s^{vib}(\underset{\sim}{Q},\omega) \tag{48}$$

This result means that, if the motions are independent, the total incoherent scattering law is the convolution product (symbol $\otimes$) of the scattering laws for the three elementary motions.

If now the molecules contain inequivalent protons, the above equations should be averaged over these protons. In this case, the spectra reflect the superposition of the motions of the different protons. A trick to separate them is to use several partially deuterated specimens in order to render invisible to neutrons successively each kind of proton. This method has been used successfully in the field of liquid crystals [42-45].

3.6 Structure of Typical Incoherent "Quasi-elastic" Scattering Spectra

In this section, we present calculations in a few simple cases in order to show the main features of incoherent spectra reflecting pure rotation, pure translation and a combination of both. Since these spectra are centered around $\omega = 0$, they are usually qualified as "quasi-elastic".

3.6.1 Rotation

Let us consider the simple case of isotropic rotational diffusion on a sphere of radius ρ. The position of the particle is characterized by a vector $\underset{\sim}{\rho}(t)$. The probability distribution G_s of its orientation Ω is governed by the following rate equation [38]

$$\frac{\partial G_s}{\partial t} = D_r \nabla_\Omega^2 G_s \tag{49}$$

where $G_s(\Omega,\Omega_o,t)$ is the probability of finding the orientation at Ω at time t if it was at Ω_o at zero time. D_r is the rotational diffusion coefficient.

The solution of this equation is

$$G_s(\Omega,\Omega_o,t) = 4\pi \sum_{\ell=0}^{\infty} \exp[-D_r\ell(\ell+1)t] \sum_{m=-\ell}^{+\ell} Y_m^\ell(\Omega) Y_m^{\ell*}(\Omega_o) \qquad (50)$$

From Eq. (44) we get:

$$I_s(\underset{\sim}{Q},t) = \frac{1}{4\pi}\int \exp i\underset{\sim}{Q}(\underset{\sim}{\rho}-\underset{\sim}{\rho}_o)\, G_s d\Omega d\Omega_o = j_o^2(Q\rho) + \sum_{\ell=1}^{\infty} (2\ell+1)\, j_\ell^2(Q\rho)\, \exp[-D_r\ell(\ell+1)t] \qquad (51)$$

and by Fourier transforming

$$S_s(\underset{\sim}{Q},\omega) = j_o^2(Q\rho)\delta(\omega) + \frac{1}{\pi}\sum_{\ell=1}^{\infty} (2\ell+1) j_\ell^2(Q\rho) \frac{D_r\ell(\ell+1)}{[D_r\ell(\ell+1)]^2+\omega^2} \qquad (52)$$

where the j_ℓ are the spherical Bessel functions.
It is seen that the scattering law (52) is composed of a sharp $\delta(\omega)$ peak superimposed on a broadened component (composed of various Lorentzians) whose width is of the order of a few D_r (Fig. 2), and whose intensity depends on Q. For more complicated models, it can be shown that these main features are conserved: a $\delta(\omega)$ peak and broadened components whose widths are Q-independent [37,39,40]. However other features (e.g. side peaks or humps) appear if the rotation becomes relatively free [38,23].

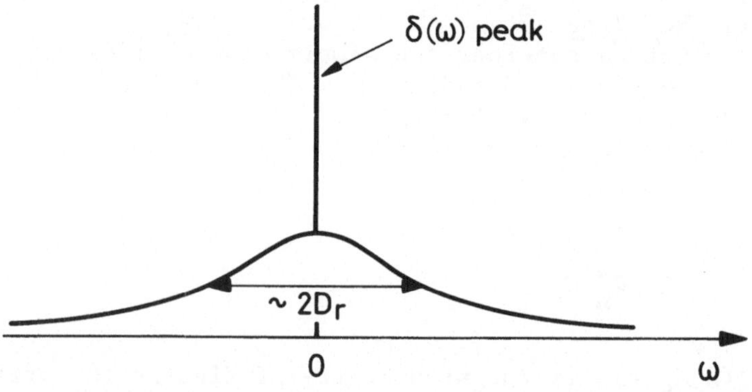

Figure 2: Theoretical incoherent neutron quasi-elastic scattering spectrum for a purely rotational diffusive motion.

3.6.2 Translation

Let us consider the simple case of isotropic translational diffusion. The probability distribution G_s of the position $\underset{\sim}{d}(t)$ of the particle is governed by the rate equation [31-36]:

$$\frac{\partial G_s}{\partial t} = D_t \nabla_d^2 G_s \tag{53}$$

where $G_s(\underset{\sim}{d},\underset{\sim}{d}_o,t)$ is the probability of finding the particle at $\underset{\sim}{d}$ at time t if it was at $\underset{\sim}{d}_o$ at zero time. D_t is the translational diffusion coefficient.

The solution of this equation is

$$G_s(\underset{\sim}{d},\underset{\sim}{d}_o,t) = (4\pi D_t t)^{-\frac{3}{2}} \exp[-(\underset{\sim}{d}-\underset{\sim}{d}_o)^2/4D_t t] \tag{54}$$

From Eq. (44) we obtain:

$$I_s(\underset{\sim}{Q},t) = \int G_s \exp i\underset{\sim}{Q}[\underset{\sim}{d}(t)-\underset{\sim}{d}(o)] \; d\underset{\sim}{d}.d\underset{\sim}{d}_o = \exp(-D_t Q^2 t) \tag{55}$$

and by Fourier transforming,

$$S_s(\underset{\sim}{Q},\omega) = \frac{1}{\pi} \frac{D_t Q^2}{(D_t Q^2)^2 + \omega^2} \tag{56}$$

The scattering law (46) is a single Lorentzian line whose width varies like Q^2 (Fig. 3). For more complicated models, one obtains

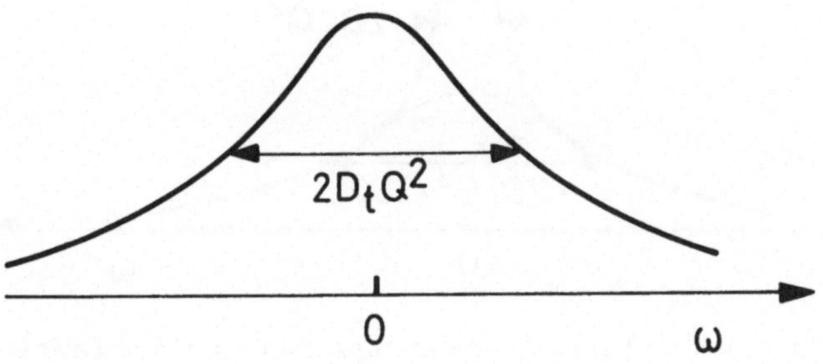

Figure 3: Theoretical incoherent neutron quasi-elastic scattering spectrum for isotropic translational diffusion.

a superposition of Lorentzian (or more complicated shapes) lines whose widths and relative intensities are Q-dependent [41; see also the paper by Prof. Fulde this book]. This is to be compared with the rotational case where only the amplitudes are Q-dependent.

3.6.3 Superposition of rotation and translation

The molecules are now assumed to undergo self-diffusion and reorientation. Assuming that these motions are independent, the total scattering law is the convolution of the rotational and translational scattering laws. It is seen that even when the two motions are described by the above mentioned simple models, the scattering law is the sum of a number of elementary curves (Lorentzians) whose widths depend both on translational and rotational parameters and whose relative amplitudes are Q-dependent (Fig. 4). It is thus a priori difficult to separate the two contributions, especially when the finite instrumental energy resolution is considered. However, as will be seen below using the concept of the Elastic Incoherent Structure Factor (EISF) and combining measurements from various instruments, this is now possible.

3.6.4 Vibrations

Finally, when vibrations are introduced, the total scattering law should be convoluted with that of vibrations, which is generally composed of sharp peaks of small intensity in the inelastic region In the quasi-elastic region, it can be shown [31-34] that this affects the scattering law through a Debye-Waller factor $\exp(-Q^2\langle u^2\rangle)$, where $\langle u^2\rangle$ is a mean square vibration amplitude. In this case we can write:

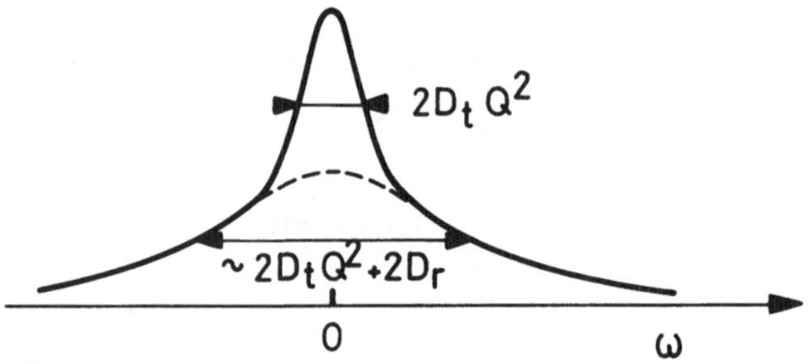

Figure 4: Theoretical incoherent neutron quasi-elastic scattering spectrum for uncoupled translational and rotational motion. The dashed lines show the separation between the pure translation contribution and the mixed ones.

$$S_s^{quasi}(\underset{\sim}{Q},\omega) \approx (S_s^{trans} \otimes S_s^{rot}) \exp(-Q^2\langle u^2\rangle) \qquad (57)$$

The Debye-Waller factor thus pictures the decrease of the total "quasi-elastic" intensity with increasing Q (i.e. the scattering angle) and gives information on the amplitude of the (fast) vibrational motions.

3.7 The Fundamental Concept of the Elastic Incoherent Structure Factor (EISF): Application to the Analysis of Real Spectra

Consider the rotational scattering law (42). The first term is the product of a function of Q, $F(Q)$, multiplied by a $\delta(\omega)$ function. This property is in fact general for any rotational model and comes from the fact that $I_s(Q,t)$ does not decay to zero where $t \to \infty$. We have:

$$F(\underset{\sim}{Q}) = I_s(\underset{\sim}{Q},t=\infty) = \int G_s(\underset{\sim}{r},\underset{\sim}{r}_o,\infty) \exp i\underset{\sim}{Q}\underset{\sim}{r}\, d\underset{\sim}{r}\, d\underset{\sim}{r}_o \qquad (58)$$

In other words, the coefficient of the $\delta(\omega)$ function is the spatial Fourier transform of final distribution of the rotating proton, averaged over all possible initial positions. It has the dimension of a structure factor and is called the Elastic Incoherent Structure Factor (EISF).

Since $F(Q)$ pictures the "trajectory" of the moving proton, if $F(\underset{\sim}{Q})$ could be extracted from spectra obtained at various Q values, one would obtain valuable information on the nature of the rotational motion performed by the proton. The way to relate the EISF to a measurable quantity is the following: we have seen that <u>any</u> rotational scattering law can be written

$$S_s^{rot}(\underset{\sim}{Q},\omega) = F(\underset{\sim}{Q})\,\delta(\omega) + \sum_n \text{other (broadened) terms} \qquad (59)$$

On the other hand, Fourier transform of Eq. (44) and integration over ω yields:

$$\int S_s(\underset{\sim}{Q},\omega)\, d\omega = I_s(\underset{\sim}{Q},0) = 1 \qquad (60)$$

Integrating Eq. (59), it is then easily seen that $F(Q)$ is <u>the fraction of the total quasi-elastic intensity contained in the purely elastic ($\delta(\omega)$) peak</u>. If the instrumental resolution $\Delta\omega$ is (much) smaller than the reorientational rate, then the real spectra have the shape sketched in Fig. 5, and the separation between the

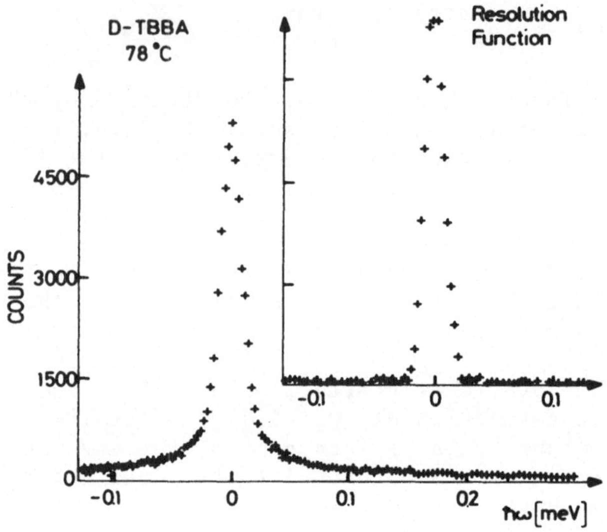

Figure 5: Neutron quasi-elastic spectrum of liquid crystalline terephtal bis butyl aniline in the smectic VI phase for Q = 1.03 $Å^{-1}$ together with the instrumental energy resolution [from ref. 42].

sharp (purely elastic) and broad components can be performed by natural extrapolation. Let $I_e(\underset{\sim}{Q})$ and $I_q(\underset{\sim}{Q})$ be the corresponding intensities (e.g. measured by graphical integration after subtraction of a flat background). We have:

$$\left(F(\underset{\sim}{Q})\right)_{exp} = \frac{I_e(\underset{\sim}{Q})}{I_e(\underset{\sim}{Q}) + I_q(\underset{\sim}{Q})} \tag{61}$$

If the two components are not well separated, then the analysis is a priori much more difficult, and in any case, less accurate. Studies of purely rotational motions in molecular crystals using these ideas are now increasing in number. Amongst them, let us mention the problems of reorientation in plastic crystals [37], in liquid crystals [42-45], and that of methyl group rotation in solid phases [44].

So far, we have considered only pure rotation. We see now that, when self-diffusion is superimposed on rotation (as it is in the case in molecular liquids), one can generalize the concept of EISF as follows. Assuming that the motions are independent, combining Eqs. (57), (59) and (56) and neglecting vibrations, we have:

$$S_s^{total}(\underset{\sim}{Q},\omega) = S_s^{rot}(\underset{\sim}{Q},\omega) \otimes S_s^{trans}(\underset{\sim}{Q},\omega) =$$

$$= F(Q) \cdot \frac{1}{\pi} \frac{D_t Q^2}{(D_t Q^2)^2 + \omega^2} + \left(\sum_n \text{other terms}\right) \otimes \frac{1}{\pi} \frac{D_t Q^2}{(D_t Q^2)^2 + \omega^2} \quad (62)$$

It is seen that the first term is the purely translation term weighted by the EISF of the rotational model. Since Eq. (60) always holds, it is seen that when diffusion is present, the EISF of the rotational model is the fraction of the total quasi-elastic intensity contained in the purely translationally broadened term (when $D_t \to 0$, Eq. (62) $\to$ Eq. (59).

In practice, it turns out that it is necessary to know D_t rather accurately in order to extract a reliable value for the EISF and consequently, a reliable picture for the rotation. The diffusion coefficient D_t can often be measured independently by the neutron method by working at sufficiently low Q (such that $D_t Q^2 \ll \tau_{rot}^{-1}$) and sufficiently high resolution (such that $\Delta\omega \ll 2D_t Q^2$). In these conditions only the translational part is seen in practice, the rotational contribution acting as a flat background. These experimental conditions can nowadays be achieved for usual liquids ($D_t \gtrsim 10^{-7}$ cm^2/sec) using the backscattering technique [36]. Once D_t is determined, one can perform experiments on a medium resolution instrument (such that $\Delta\omega \lesssim \tau_{rot}^{-1}$): typically $10 < \Delta\omega < 100$ μeV and extract the EISF from these spectra to get information about the rotational model.

A recent, and still continuing study of a molecular liquid, using these ideas has been performed on liquid cyclopropane (C_3H_6), where (i) D_t has been determined by the neutron method - and found to be in perfect agreement with spin-echo measurements - [46] and (ii) the generalized EISF was deduced, from which it could be concluded that, in this system, non-coupling between translation and rotation seems to be a good approximation [47].

3.8 Conclusion

In conclusion, incoherent neutron (quasi-elastic) scattering appears to be a powerful tool for studying molecular motions in liquids, provided that adequate experimental conditions (good resolution, high flux, low Q, partially deuterated samples) can be achieved, in order to be able to separate the various kinds of motions. In the past, this was difficult, but with the existence of novel instruments installed at the cold source of the high flux reactor of the Institut Laue-Langevin (Grenoble), this is now possible. To give a feeling of the performance of these instruments, with the backscattering technique, one can work with energy resolution between 0.2 and 1 μeV. This corresponds to a time scale of

10^{-8} sec [1 μeV ≡ 1.52 x 10^9 rd/s]. A self-diffusion coefficient between 10^{-7} and 10^{-5} cm^2/sec can easily be measured in this way. With the so-called multichopper IN5, energy resolution between 6 and 100 μeV can be obtained, depending on the value of incident wavelength between 15 and 5 Å. With these instruments, the Q range available lies between 0.05 to 1.5 $Å^{-1}$, with a worsening of the resolution for higher Q values when this is possible. It should be noted that the condition $Qa \sim \pi$, where a is a molecular dimension, can easily be reached in this range. This explains why neutrons are unique in their ability to yield information about the geometrical aspect of the molecular motions. This is generally not the case for long wavelength electromagnetic waves such as light, as we shall see now.

CHAPTER 4

ABSORPTION AND SCATTERING OF LONG WAVELENGTH ELECTROMAGNETIC WAVES: DIELECTRIC AND INFRARED ABSORPTION, RAMAN AND RAYLEIGH SCATTERING [48,49]

In this section, we describe how electromagnetic (e.m.) waves or photons of long wavelength (much greater than molecular dimensions) can be used to study molecular motions. We first recall the properties which characterize the photon, then we detail the principles of absorption and scattering experiments and give a few illustrative examples.

4.1 Properties of Electromagnetic Waves; Photons

Classically, the electromagnetic field is characterized by its electric field vector $\underset{\sim}{E}(\underset{\sim}{r})$ and magnetic field vector $\underset{\sim}{H}(\underset{\sim}{r})$, and has a wave character. A plane wave is characterized by its wave vector $\underset{\sim}{k}$, its angular frequency ω_k, and moves at the velocity of light c. These quantities are related by

$$c = \frac{\omega_k}{k} \tag{63}$$

Electromagnetic waves can also be considered as ultra relativistic particles, photons, moving at the velocity of light and of energy E_k given by

$$E_k = \hbar\omega_k \tag{64}$$

For visible light, we have $E_k \sim 2$ eV and $\lambda_k \sim 6000$ Å. At this point, it is interesting to compare these values with the corresponding ones for neutrons (section 3.1), in particular the fact that the wavelength is much greater than the usual molecular dimensions. Considered as a quantum object, the electromagnetic field must be treated within the formalism of second quantization, i.e. that the electric and magnetic fields are in fact operators. We have [48]

$$\underset{\sim}{E} = \sum_{\underset{\sim}{k},\mu} \left(c_{k\mu}\,\underset{\sim}{E}_{k\mu} + c^{+}_{k\mu}\,\underset{\sim}{E}^{*}_{k\mu}\right) \tag{65}$$

$$\underset{\sim}{H} = \sum_{\underset{\sim}{k},\mu} \left(c_{k\mu}\,\underset{\sim}{H}_{k\mu} + c^{+}_{k\mu}\,\underset{\sim}{H}^{*}_{k\mu}\right) \tag{66}$$

with

$$\underset{\sim}{E}_{k\mu} = i\sqrt{\frac{2\pi\hbar\omega_k}{V}}\ \underset{\sim}{\varepsilon}_\mu \exp(i\underset{\sim}{k}\underset{\sim}{r}) \tag{67}$$

and

$$\underset{\sim}{H}_{k\mu} = \frac{\underset{\sim}{k}}{|k|} \times \underset{\sim}{E}_{k\mu} \tag{68}$$

In these expressions, $c_{k\mu}$ and $c^+_{k\mu}$ are the so-called annihilation and creation operators and $\underset{\approx}{\varepsilon}_\mu$ is the unit vector along $\underset{\approx}{E}_{k\mu}$. ε_μ is normal to $\underset{\sim}{k}$ and is called the polarization of the wave $\underset{\sim}{k}$, and V is the volume of quantization (in fact the volume of the sample). The eigenstates $|N_{k\mu}\rangle$ of the Hamiltonian of the field are characterized by the number $N_{k\mu}$ of photons existing in each state $(\underset{\sim}{k},\mu)$

$$|N_{k\mu}\rangle = |N_{k_{1,1}}, N_{k_{2,2}} \ldots N_{k_{p,p}} \ldots . \rangle \tag{69}$$

The corresponding eigenvalues are

$$E = \sum_{\underset{\sim}{k},\mu} (N_{k\mu} + \frac{1}{2})\,\hbar\omega_k \tag{70}$$

Finally, the properties of the operators $c_{k\mu}$ and $c^+_{k\mu}$ are:

$$c_{k\mu}\,|N_{k'\mu'}\rangle = \sqrt{N_{k\mu}}\,|N_{k\mu}-1\rangle\,\delta_{kk'}\delta_{\mu\mu'} \quad \text{(annihilation of one photon } \underset{\sim}{k}\mu) \tag{71}$$

$$c^+_{k\mu}\,|N_{k'\mu'}\rangle = \sqrt{N_{k\mu}+1}\,|N_{k\mu}+1\rangle\,\delta_{kk'}\delta_{\mu\mu'} \quad \text{(creation of one photon } \underset{\sim}{k}\mu) \tag{72}$$

4.2 Long Wavelength e.m. Waves - Matter Interaction

Electromagnetic waves interact with electric charges. If the medium is neutral, at each point $\underset{\sim}{r}$, one can define a permanent dipole moment $\underset{\sim}{\mu}(\underset{\sim}{r})$, a polarizability tensor $\underset{\approx}{\alpha}(r)$ ... per unit volume. If the wavelength is sufficiently large, it can be shown that the interaction Hamiltonian H_c can be written in the form:

$$H_c = -\int_V d\underset{\sim}{r}\ [\underset{\sim}{E}(\underset{\sim}{r})\underset{\sim}{\mu}(\underset{\sim}{r}) + \frac{1}{2}\underset{\sim}{E}(\underset{\sim}{r})\ \underset{\approx}{\alpha}(\underset{\sim}{r})\ \underset{\sim}{E}(\underset{\sim}{r}) + \text{higher order terms}] \tag{73}$$

(cf. lectures of Prof. A.D. Buckingham, this book). The higher order terms include the hyperpolarizability interaction (trilinear term in the electric field), magnetic interactions, etc. For the present purpose, we shall limit ourselves to these two first terms which will give rise to dielectric and infrared absorption (linear term) and to Rayleigh and Raman scattering (bilinear term). The higher order terms could probably be treated in a similar manner.

4.3 Interaction with Permanent Dipoles: Dielectric and Infrared Absorption

4.3.1 General formalism

To study this case, we follow the general prescription of section 2. The probe is the e.m. field whose eigenstates are $|N_{k\mu}\rangle$. The reservoir is made up of N particles i (e.g. the molecules) that we assume to be point-like, located at $\underset{\sim}{r}_i$ (long wavelength approximation) and characterized by their dipole moment $\underset{\sim}{\mu}_i$. With this assumption, we can write

$$\underset{\sim}{\mu}(\underset{\sim}{r}) = \sum_i \underset{\sim}{\mu}_i \, \delta(\underset{\sim}{r} - \underset{\sim}{r}_i) \tag{74}$$

and the present relevant interaction H_c (first term of Eq. (73)) is written:

$$H_c = -\sum \int_V d\underset{\sim}{r} \, \underset{\sim}{E}(\underset{\sim}{r}) \, \underset{\sim}{\mu}_i \, \delta(\underset{\sim}{r} - \underset{\sim}{r}_i) \tag{75}$$

Using Eq. (67) in Eq. (75) and performing the integration over $\underset{\sim}{r}$ yields

$$H_c = -\sum_i \sum_{\underset{\sim}{k},\mu} (\underset{\sim}{\varepsilon}_\mu \underset{\sim}{\mu}_i) \, i\sqrt{\frac{2\pi\hbar\omega_k}{V}} \, [c_{k\mu} \exp(i\underset{\sim}{k}\underset{\sim}{r}_i) + c^+_{k\mu} \exp(-i\underset{\sim}{k}\underset{\sim}{r}_i)] \tag{76}$$

We should now calculate $\bar{H}_c$, i.e. the average value of H_c between the initial and final state of the probe. The initial state is $|N_{k_oI}\rangle$, i.e. we send photons of momentum $\underset{\sim}{k}_o$ and polarization $\underset{\sim}{\varepsilon}_I$ (a monochromatic and polarized plane wave) onto the sample. What are the possible interesting final states, i.e. those for which $\bar{H}_c$ is non zero? Clearly, they correspond to states such that the matrix elements of $c_{k\mu}$ or $c^+_{k\mu}$ are non zero. According to Eqs. (71) and (72) these are: (i) $|N_{k_oI}-1\rangle$: absorption of one photon $(\underset{\sim}{k}_o,\underset{\sim}{\varepsilon}_I)$, (ii) $|N_{k_oI}+1\rangle$: emission of one photon $(\underset{\sim}{k}_o,\underset{\sim}{\varepsilon}_I)$ and (iii) $|N_{k_oI}, 1_{k_1 s}\rangle$: emission of one photon $(\underset{\sim}{k}_1,\underset{\sim}{\varepsilon}_s)$.

The values of the matrix elements are $\sqrt{N_{k_oI}}$, $\sqrt{N_{k_oI}+1}$ and 1, respectively, and the corresponding probability of transition is thus proportional to N_{k_oI}, $N_{k_oI}+1$ and 1. If the number of incident photons is large, as is always the case in practice, then $N_{k_oI} >> 1$ and the last case (in fact, the spontaneous emission) is very weak compared to the two former ones, namely the induced absorption and emission. The correlation functions corresponding to these two latter cases are:

$$C^{\pm}_{\bar{H}_c \bar{H}_c}(t) = \frac{2\pi\hbar\omega_o}{V} N_{k_oI} < \sum_{i,j} (\underset{\sim}{\varepsilon}_I \underset{\sim}{\mu}_i(t))(\underset{\sim}{\varepsilon}_I \underset{\sim}{\mu}_j(o))$$

$$\exp\{\pm i\underset{\sim}{k}_o[\underset{\sim}{r}_i(t) - \underset{\sim}{r}_j(o)]\}> \tag{77}$$

with + for absorption and − for emission.
It is seen that, a priori, the study of induced absorption and emission of e.m. waves by such a medium can give information about the motions of $\underset{\sim}{\mu}_i$ (i.e. rotation and vibration) and on the motion of $\underset{\approx}{r}_i$ (i.e. translation). In practice, however, since k_o is very small (long wavelength approximation), the exponential decays much slower than the other terms and we can neglect its variation. In these conditions, the correlation functions for emission and absorption are the same, and we have:

$$C_{\bar{H}_c \bar{H}_c}(t) = \frac{2\pi\hbar\omega_o}{V} N_{k_oI} < \sum_{i,j} (\underset{\sim}{\varepsilon}_I \underset{\sim}{\mu}_i(t))(\underset{\sim}{\varepsilon}_I \underset{\sim}{\mu}_j(o))> \tag{78}$$

The corresponding probabilities of transition are given by Eq. (17), one with $\omega = \omega_o$ (absorption of one photon), the other with $\omega = -\omega_o$ (emission of one photon). In the linear approximation, these two probabilities are related by Eq. (18), and we have $W^{abs} > W^{emi}$. The net induced phenomenon is the absorption, as expected from general considerations, and the corresponding probability per unit time is, according to Eqs. (16), (17), (18) and (78):

$$W^t = W^{obs} - W^{emi} = \frac{2\pi}{\hbar^2}[1 - \exp(-\beta\hbar\omega_o)]\, C_{\bar{H}_c \bar{H}_c}(\omega_o) \tag{79}$$

The problem is now to relate W^t to a measurable quantity, namely the power P_a absorbed during the experiment. Let P_o be the incident power and n_{k_oI} the corresponding incident number of photons per cm^2 and per second. The sample is a cylinder of section S and length ℓ ($S\ell = V$). We have:

$$P_o = n_{k_oI} S\hbar\omega_o \tag{80}$$

The number of incident photons in volume V is

$$N_{k_o I} = \frac{n_{k_o I}}{c} V \tag{81}$$

Since the power absorbed in volume V is

$$P_a = \hbar\omega_o W^t \tag{82}$$

combining Eqs. (78) to (82) we finally obtain

$$\frac{P_a}{P_o} = N \frac{(2\pi)^2}{\hbar c} \frac{\omega_o}{S} [1 - \exp(-\beta\hbar\omega_o)] \cdot C_{\mu\mu}(\omega_o) \tag{83}$$

with

$$C_{\mu\mu}(\omega) = \frac{1}{2\pi} \int_{-\infty}^{+\infty} C_{\mu\mu}(t) \exp(-i\omega t)\, dt \tag{84}$$

and

$$C_{\mu\mu}(t) = \frac{1}{N} < \sum_{i,j} (\underset{\sim}{\varepsilon}_I \underset{\sim}{\mu}_i(o))(\underset{\sim}{\varepsilon}_I \underset{\sim}{\mu}_j(t))> \tag{85}$$

It is seen that, in the long wavelength approximation, the absorbed power is proportional to the spectral density of the fluctuation of the dipoles. More precisely, it reflects the correlation between the fluctuations of the components of the dipoles along the polarization $\underset{\sim}{\varepsilon}_I$ of the field.

An important simplification occurs if the medium can be considered as isotropic (e.g. as in a normal liquid). Then, one can average Eq. (85) over all possible orientations of $\underset{\sim}{\varepsilon}_I$ and easily obtain:

$$C_{\mu\mu}(t) = \frac{1}{N} < \sum_{i,j} \underset{\sim}{\mu}_i(o)\, \underset{\sim}{\mu}_j(t)> \tag{86}$$

Now, the electric dipole moment of a molecule is a quantity which depends on the charge distribution in the molecules. This distribution changes when the molecule vibrates. Let ν be the vibrational states and q^ν the corresponding normal coordinates. We can write:

$$\underset{\sim}{\mu} = \underset{\sim}{\mu}^o + \sum_\nu \underset{\sim}{\mu}^\nu q^\nu \tag{87}$$

with

$$\underset{\sim}{\mu}^{\nu} = \left(\frac{\partial \underset{\sim}{\mu}}{\partial q^{\nu}}\right)_{q^{\nu}=0} \tag{88}$$

$\underset{\sim}{\mu}^{o}$ is the permanent dipole and μ^{ν} the derivative dipole corresponding to vibration ν.

With these definitions, it appears that the second member of Eq. (85) or (86) can be split into various components (two for Eq. (86)):

$$C_{oo}(t) = \frac{1}{N} < \sum_{i,j} \underset{\sim}{\mu}_i^{o}(o)\, \underset{\sim}{\mu}_j^{o}(t)> \tag{89}$$

and

$$C_{\nu\nu}(t) = \frac{1}{N} < \sum_{i,j,\nu,\nu'} (\underset{\sim}{\mu}_i^{\nu}(o)\, \underset{\sim}{\mu}_j^{\nu'}(t))(q_i^{\nu}(o)\, q_j^{\nu'}(t))> \tag{90}$$

which are the relevant correlation functions for the so-called dielectric and infrared absorptions in an isotropic medium, respectively. In what follows, we detail these two cases.

4.3.2 Permanent dipoles: dielectric absorption

a) General. The correlation function is given by Eq. (29). Let $\underset{\sim}{u}_i$ and $\underset{\sim}{u}_j$ be the unit vectors along $\underset{\sim}{\mu}_i^{o}$ and $\underset{\sim}{\mu}_j^{o}$ and θ_{ij} the angle between them. If we assume that the molecules are identical, then Eq. (89) reduces to

$$C_{oo}(t) = \frac{1}{N} (\mu^{o})^{2} < \sum_{i,j} \cos\theta_{ij}(t)> \tag{91}$$

This correlation function pictures the fluctuations of the relative orientation of the permanent dipoles and thus reflects collective reorientational motions. A difficult problem is to relate this macroscopic function to monomolecular properties, namely to the single molecule correlation function $F_1(t)$ given by

$$F_1(t) = < \cos\theta_{ii}(t) > = < P_1(\cos\theta_{ii}(t)) > \tag{92}$$

where P_1 is the first order ordinary Legendre polynomial. A discussion of this problem is given in ref. [50]. A trick in approaching this problem experimentally is to dilute polar molecules in an inert solvent and make a study as a function of the dilution. If the dilution is sufficiently large, the active molecules can be considered sufficiently far from one another so that their correlations can be neglected i.e. that we have:

$$< \cos\theta_{ij}(t) > \approx 0 \quad \text{for } i \neq j \tag{93}$$

In this case, Eq. (89) can be written dropping the indices ii:

$$C_{oo}(t) = (\mu^o)^2 F_1(t) \tag{94}$$

and by Fourier transforming

$$C_{oo}(\omega) = (\mu^o)^2 F_1(\omega) \tag{95}$$

Using the fact that (i) in dielectric absorption $\hbar\omega_o << k_BT$ and (ii) $N = nS\ell$ where n is the number of dipoles per unit volume, we can write Eq. (83) as:

$$P_a = \alpha \ell P_o \tag{96}$$

where

$$\alpha = \alpha(\omega_o) = \frac{(2\pi)^2}{c} \frac{n(\mu^o)^2}{k_BT} \omega_o^2 F_1(\omega_o) \tag{97}$$

The quantity α is the absorption coefficient per unit length. Its variation with ω_o yields the dielectric absorption spectrum.

b) Structure of typical dielectric absorption spectra. As an illustration, we now present calculations of dielectric absorption spectra. We choose the case of isotropic free rotation and isotropic rotational diffusion for dilute dipoles and compare the results to a real spectrum.

(i) Isotropic rotational diffusion. In this case, the probability distribution G_s of the dipole orientation is given by Eq. (50) and we have:

$$F_1(t) = \int \cos\theta \; G_s \; d\Omega = \exp(-2D_r t) \tag{98}$$

The absorption coefficient α is thus

$$\alpha(\omega_o) \propto \omega_o^2 F(\omega_o) \propto \frac{\omega_o^2}{(2D_r)^2 + \omega_o^2} \tag{99}$$

The behaviour of $\alpha(\omega_o)$ is shown in Fig. 6. It exhibits an increase at small ω_o and saturation at large ω_o, the turning point being around $\omega_o \approx 2D_r$.

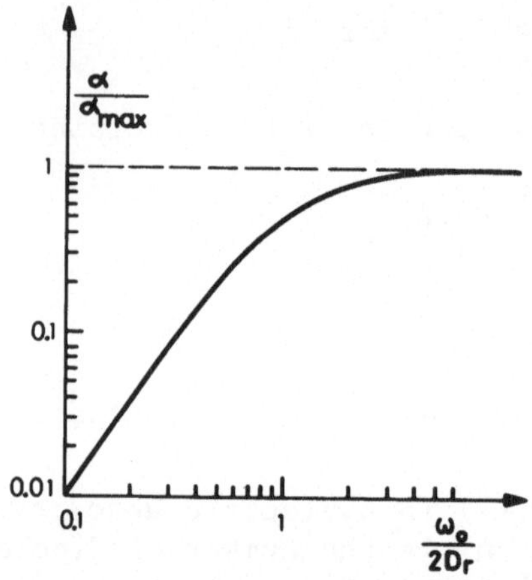

Figure 6: Theoretical dielectric absorption spectrum for dilute polar molecules undergoing isotropic rotational diffusion.

(ii) Classical free rotation. Free rotation means that the rotational energy is conserved, and thus the angular velocity Ω is constant. In this case, the angle θ varies linearly with t:

$$\theta = \Omega t \tag{100}$$

On the other hand, since the system is at thermal equilibrium, the distribution of angular velocities $p(\Omega)$ is Maxwellian

$$p(\Omega) = \frac{I}{k_B T} |\Omega| \exp(-\frac{1}{2} I\Omega^2/k_B T) \tag{101}$$

where I is the inertia momentum of the molecule. The correlation function $F_1(t)$ can then be calculated:

$$F_1(t) = \int \cos(\Omega t) \; p(\Omega) \; d\Omega \tag{102}$$

as well as its Fourier transform. The final result is

$$\alpha(\omega_o) \propto \omega_o^2 F_1(\omega_o) \propto |\omega_o|^3 \exp(-\frac{1}{2} I\omega_o^2/k_B T) \tag{103}$$

The behaviour of $\alpha(\omega_o)$ is shown in Fig. 7. It mainly exhibits a pronounced peak around $(3k_BT/I)^{1/2}$

(iii) An actual example. Figure 8 represents the dielectric absorption spectrum of a polar molecule CH_3I, dissolved in an inert

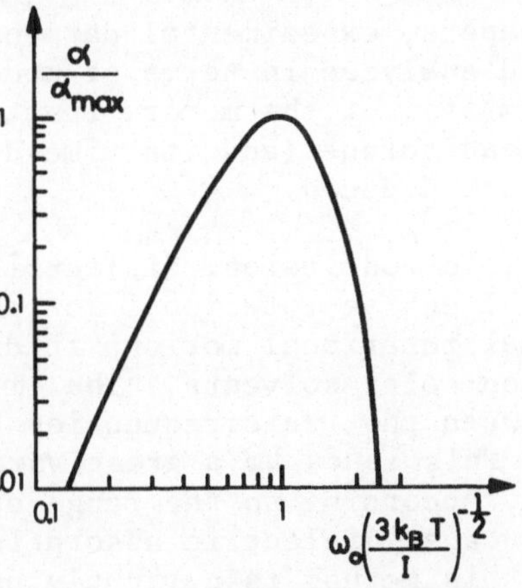

Figure 7: Theoretical dielectric absorption spectrum for dilute polar molecules undergoing classical free rotation.

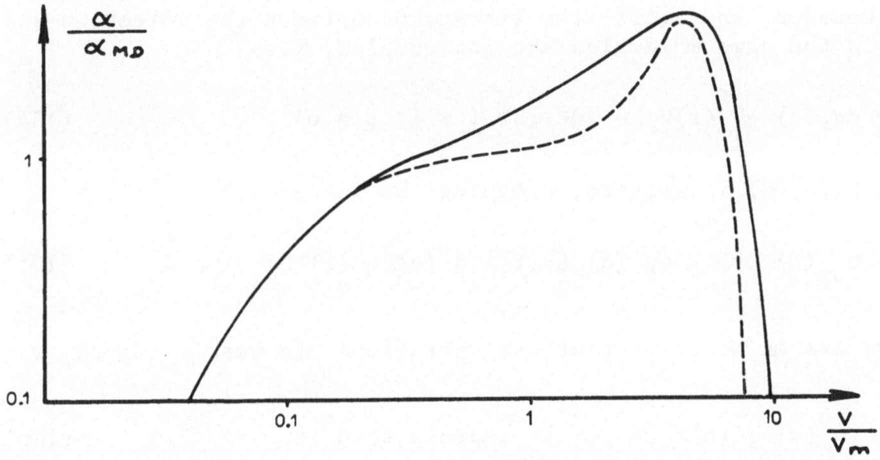

Figure 8: Dielectric absorption spectrum of 0.2M CH_3I in CCl_4 at 20°C, in reduced units. The dashed line is a possible theoretical model [from ref. 52].

solvent CCl_4 [52]. It clearly exhibits a slowly increasing part which is reminiscent of the diffusion aspect, and a peak followed by a sharp decrease which is reminiscent of the free rotation aspect. In fact, many spectra resemble this one, with one or other feature more or less pronounced. Numerous examples of dielectric studies of liquids exist in the literature, for example in ref. [1]

[50-56]. In these papers, experimental data on pure and dilute systems are shown and analyzed in terms of models. For the case of Fig. 8 [52], the formalism of the memory function is used and quantities such as the mean torque (and its time derivative) acting on the active molecule are deduced.

c) Conclusion. In conclusion, dielectric absorption can be used to study both collective rotational motions in pure polar liquids and individual rotational motions in dilute solutions of polar molecules in non-polar solvents. The ω range available in practice lies between the radiofrequencies and the far infrared via the microwaves. This leads to a great variety of different experimental set ups, according to the range of frequencies to be studied. Finally, because dielectric absorption reflects basically collective motions, this method is currently used in the field of critical phenomena, in particular in liquids (cf. the Round Table Discussion on critical phenomena, this book).

4.3.3 Derivative dipoles: infrared (IR) absorption [57]

a) General. In this case, the relevant correlation function is given by Eq. [90]. If we make the following assumptions (i) the molecules are identical, (ii) rotations and vibrations are not coupled, and (iii) the vibrations between the molecules and within the same molecules are not coupled, i.e.,

$$<q_i^\nu(o)\ q_j^{\nu'}(t)> \ = \ 0 \quad \text{for } i \neq j,\ \nu \neq \nu' \tag{104}$$

then Eq. (90) is written, dropping the index i

$$C_{\nu\nu}(t) \ = \ \sum_\nu <\underset{\sim}{\mu}^\nu(o)\ \underset{\sim}{\mu}^\nu(t)><q^\nu(o)\ q^\nu(t)> \tag{105}$$

If we assume moreover that the vibrations are weakly damped, we can write

$$<q^\nu(o)\ q^\nu(t)> \ = \ (q^\nu)^2\ \exp(-i\Omega_\nu t)\ G_\nu(t) \tag{106}$$

where q^ν is the vibration amplitude, Ω_ν the frequency of vibration ν and $G_\nu(t)$ a slowly decreasing function picturing the damping of the vibration ($G_\nu(t) \equiv 1$ for no damping). Let $\theta^\nu(t)$ be the angle between $\underset{\sim}{\mu}^\nu(o)$ and $\underset{\sim}{\mu}^\nu(t)$. We have

$$<\underset{\sim}{\mu}^\nu(o)\ \underset{\sim}{\mu}^\nu(t)> \ = \ (\mu^\nu)^2\ <\cos\theta^\nu(t)> \ = \ (\mu^\nu)^2\ F_{1\nu}(t) \tag{107}$$

and Eq. (105) is finally written:

$$C_{\nu\nu}(t) = \sum_{\nu} (\mu^{\nu} q^{\nu})^2 \exp(-i\Omega_{\nu} t) \, G_{\nu}(t) \, F_{1\nu}(t) \tag{108}$$

and after Fourier transforming

$$C_{\nu\nu}(\omega) = \sum_{\nu} (\mu^{\nu} q^{\nu})^2 \, G_{\nu}(\omega+\Omega_{\nu}) \otimes F_{1\nu}(\omega+\Omega_{\nu}) \tag{109}$$

This shows that the infrared absorption spectrum is composed of a series of absorption lines centered around the vibration frequencies and that, provided a number of more or less questionable assumptions are made, each line is proportional to the convolution of the rotational and a vibrational spectrum. Since the Ω_{ν} are generally in the infrared region of the e.m. spectrum, this phenomenon is called infrared absorption. If the vibrations are not damped $(G_{\nu}(\Omega_{\nu}+\omega) = \delta(\Omega_{\nu}+\omega))$, comparing Eqs. (95) and (109) one can roughly say that for molecules in dilute solutions the infrared spectrum reproduces the dielectric spectrum around the frequencies Ω_{ν}. Roughly because, although the absorbed power is proportional to $\omega_o[1-\exp(-\beta h\omega_o)]$ as in the dielectric case [Eq. (83)], for infrared $\omega_o \approx \Omega_{\nu}+\omega \approx \Omega_{\nu}$ and the absorption spectrum reproduces $C_{\nu\nu}(\omega)$ around Ω_{ν}.

b) Structure of typical infrared spectra. To illustrate this section and as for the case of dielectric absorption, we present theoretical I R spectra in the same simple cases as before, and assuming no damping of the vibration. For rotational diffusion, we have

$$F_1(\omega) \propto \frac{2D_r}{(2D_r)^2 + \omega^2} \tag{110}$$

and the corresponding infrared spectrum is sketched in Fig. 9: it is a single Lorentzian centered at Ω_v. For free rotation, we have

$$F_1(\omega) \propto |\omega| \exp(-\frac{1}{2} I\omega^2/k_B T) \tag{111}$$

and the corresponding spectrum is sketched in Fig. 10. It presents two maxima at $\Omega_{\nu} \pm \omega_m$ with $\omega_m = (k_B T/I)^{1/2}$.

Fig. 11 shows the infrared spectrum of the polar molecule CO diluted in liquid C_2Cl_4 [59]. It clearly exhibits features which recall the two above limiting cases. In particular, the humps on each side of the spectrum suggest that the rotation is relatively free, and its general shape suggests that the numerous assumptions made are justified for this system. In particular, the additional observation that the shapes corresponding to the transition 1-0 and 2-0 are practically the same suggests that the hypothesis of weak damping of the vibration is a good one [59].

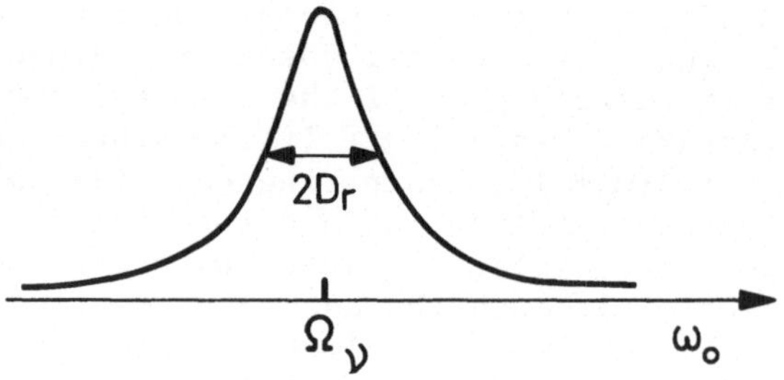

Figure 9: Theoretical infrared absorption spectrum for polar molecules undergoing isotropic rotational diffusion.

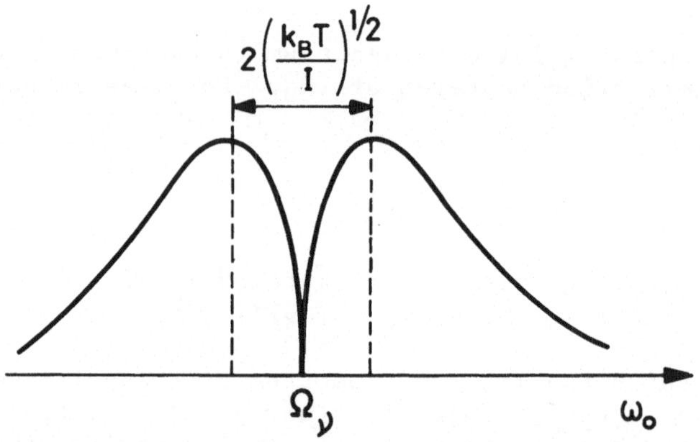

Figure 10: Theoretical infrared absorption spectrum for polar molecules undergoing classical free rotation.

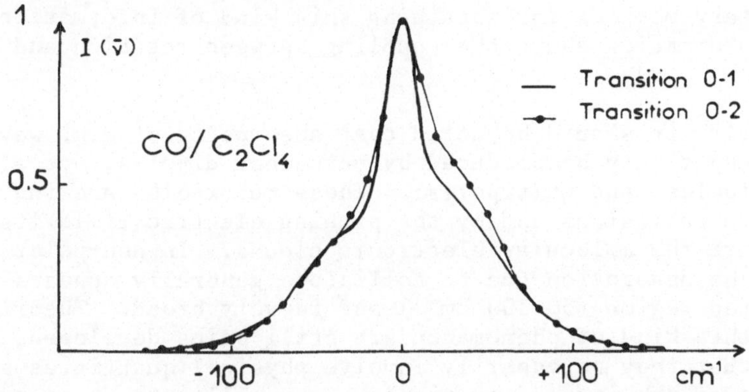

Figure 11: Infrared spectra of 0.1M CO in C_2Cl_4 [from ref. 59].

c) Discussion and conclusion. The real case described above is almost ideal: very small molecules in a dilute solution. When the molecules are bigger, the vibrations may be strongly damped and contribute significantly to the width of the spectra. In this case, an often-used trick to separate the rotational and vibrational contributions is to measure the total width of the spectra as a function of temperature, assume that the two elementary widths are additive, and assume that the vibrational width is temperature independent [58,64]. As a general rule, it is accepted that the relative contribution of vibration increases with the dimension of the molecule and may become the dominant one. An additional important complication with large molecules is that the various vibrational levels are not sufficiently far from one another so that two or several spectra corresponding to different vibrational states overlap (inhomogeneous broadening). Finally, the coupling between rotation and vibration is another important complication [68]. The literature on infrared studies of liquids is large. We mention the papers published in ref. [1] [58-65], where various experimental as well as theoretical (see also ref. [66]) results are presented.

In conclusion, it appears that infrared absorption can be used to study both rotational and vibrational relaxation in liquids, but that the separation between them (when meaningful) is difficult. Clearly, additional results from other methods are needed. Comparison between dielectric and infrared absorption spectra can, in certain cases, yield some information about the degree of cooperativity of the rotational motions. However, as we shall see below, Rayleigh and Raman scattering of light are the most suitable complementary methods for obtaining this kind of information as well as information about the coupling between rotation and vibration.

Finally, it should be noted that absorption of e.m. waves in liquids cannot only be produced by permanent dipoles, but also by induced dipoles (and multipoles). These multipoles are induced both by the collisions and by the probing electric field itself, which deform the molecular electronic clouds. In non-polar liquids, the absorption due to collisions generally appears in the far infrared region (50-100 cm^{-1}) and is very broad. Theories to describe this kind of phenomenon are still being developed, but it is clear that they necessarily involve physical quantities such as the values of the induced multipoles, the collision frequencies and consequently contain information on the specific interaction between non-dipolar molecules. A review on this subject is given in ref. [67]. Concerning the absorption due to the probing electric field, it appears as a phenomenon which superimposes on the phenomenon of scattering. This will be discussed in the next section.

4.4 Interaction with Induced Dipoles: Absorption and Scattering of Photons

4.4.1 General formalism.

As in section 3.3, here also we follow the prescription of section 2. Again, the probe is the e.m. field whose eigenstates are $|N_{k\mu}\rangle$. The reservoir is made up of N particles i (the molecules) which we assume to be point-like, located at $\underset{\sim}{r}_i$ (long wavelength approximation) but which are now characterized by their polarizability tensor $\underset{\approx}{\alpha}_i$. With this assumption we can write:

$$\underset{\approx}{\alpha}(\underset{\sim}{r}) = \sum_i \underset{\approx}{\alpha}_i \, \delta(\underset{\sim}{r} - \underset{\sim}{r}_i) \tag{112}$$

and the present relevant interaction H_c (second term of Eq. (73)) is written:

$$H_c = -\frac{1}{2} \sum_i \int_V d\underset{\sim}{r} \, \underset{\sim}{E}(\underset{\sim}{r}) \, \underset{\approx}{\alpha}_i \, \underset{\sim}{E}(\underset{\sim}{r}) \, \delta(\underset{\sim}{r} - \underset{\sim}{r}_i) \tag{113}$$

Replacing $\underset{\sim}{E}(r)$ by its expression given by Eq. (67) and performing the integration over $\underset{\sim}{r}$ yields

$$H_c = -\frac{1}{2} \sum_i \sum_{\mu\mu'} (\underset{\sim}{\varepsilon}_\mu \, \underset{\approx}{\alpha}_i \, \underset{\sim}{\varepsilon}_{\mu'}) \, F_{\mu\mu'}(\underset{\sim}{r}_i) \tag{114}$$

with

$$F_{\mu\mu'}(\underset{\sim}{r}_i) = \sum_{\underset{\sim}{k},\underset{\sim}{k}'} \frac{2\pi\hbar}{V} \sqrt{\omega_k\omega_{k'}} \{ c_{k\mu} c^+_{k'\mu'} \exp i(\underset{\sim}{k}-\underset{\sim}{k}')\underset{\sim}{r}_i$$

$$+ c^+_{k\mu} c_{k'\mu'} \exp[-i(\underset{\sim}{k}-\underset{\sim}{k}')\underset{\sim}{r}_i] - c_{k\mu} c_{k'\mu'} \exp i(\underset{\sim}{k}+\underset{\sim}{k}')\underset{\sim}{r}_i$$

$$- c^+_{k\mu} c^+_{k'\mu'} \exp[-i(\underset{\sim}{k}+\underset{\sim}{k}')]\underset{\sim}{r}_i \} \qquad (115)$$

We should now calculate $\bar{H}_c$, i.e. the average value of H_c between the initial and final states of the probe. As for absorption experiments, the initial state is $|m\rangle = |N_{k_oI}\rangle$. We should look for non-trivial final states $|n\rangle$, i.e. final states such that H_c is non-zero and such that $|n\rangle \neq |N_{k_oI}\rangle$. The problem reduces to inspecting the matrix elements of the operator $F_{\mu\mu'}(\underset{\sim}{r}_i)$ with the help of Eqs. (71) and (72). From the structure of Eq. (115), it immediately appears that there are various possible final states among which we select the three following ones:

(i) states for which two photons $(\underset{\sim}{k}_o, \underset{\sim}{\varepsilon}_I)$ are emitted or absorbed, namely $|N_{k_oI} \pm 2\rangle$. This case corresponds to induced emission and absorption of two incident photons via the induced dipoles $(|\hbar\omega| = |2\hbar\omega_o|)$, and is similar to the case described in section 4.3. The corresponding correlation functions of $\bar{H}_c$ are $(N_{k_oI} \gg 1)$

$$C^{\pm}_{\bar{H}_c\bar{H}_c}(t) = \left(\frac{\pi\hbar\omega_o}{V} N_{k_oI}\right)^2 \langle \sum_{i,j} (\underset{\sim}{\varepsilon}_I \underset{\approx}{\alpha}_i(o) \underset{\sim}{\varepsilon}_I)(\underset{\sim}{\varepsilon}_I \underset{\approx}{\alpha}_j(t) \underset{\sim}{\varepsilon}_I)$$

$$\exp\left[\pm 2i\underset{\sim}{k}_o[\underset{\sim}{r}_i(t) - \underset{\sim}{r}_j(o)]\right]\rangle \qquad (116)$$

with + for absorption and - for emission.

This equation should be compared to Eq. (77) and an analysis similar to that of section 4.3 can be made to find the probability for the net absorption and the corresponding absorption coefficient. This will not be done here, but it appears that, provided the absorption due to this phenomenon can be separated from that due to other phenomena, one can in principle obtain information on the fluctuations of the polarizability tensor (i.e. the rotations and vibrations) and on the translational motions. In practice the separation is difficult.

(ii) states for which one photon $(\underset{\sim}{k}_o, \underset{\sim}{\varepsilon}_I)$ is absorbed and a different photon which we label $(\underset{\sim}{k}_1, \underset{\sim}{\varepsilon}_s)$ is emitted, namely the states $|N_{k_oI} - 1, 1_{k_1s}\rangle$. This case corresponds to the induced emission of one photon different from the initial ones, and corresponds to the scattering phenomenon. The corresponding correlation function is $(N_{k_oI} \gg 1)$:

$$C_{\bar{H}_c\bar{H}_c}(t) = \left(\frac{2\pi\hbar}{V}\right)^2 \omega_o\omega_1 N_{k_oI} < \sum_{i,j} (\underset{\sim}{\varepsilon}_s \alpha_i(o)\underset{\sim}{\varepsilon}_I)(\underset{\sim}{\varepsilon}_s \underset{\sim}{\alpha}_j(t)\underset{\sim}{\varepsilon}_I)$$

$$\exp iQ[\underset{\sim}{r}_i(t) - \underset{\sim}{r}_j(o)]> \tag{117}$$

where

$$\underset{\sim}{Q} = \underset{\sim}{k}_o - \underset{\sim}{k}_1 \tag{118}$$

is the photon momentum transfer.
(iii) finally, states for which two photons $(\underset{\sim}{k}_1,\varepsilon_1)$ and $(\underset{\sim}{k}_2,\varepsilon_2)$ are emitted, namely $|N_{k_oI}, 1_{k_{1s}}, 1_{\underset{\sim}{k}_{2s}}>$. This case corresponds to spontaneous emission of two photons and as for the case of one photon (cf. section 4.3), its probability is very weak (1 compared to N_{k_oI} for scattering and $N^2_{k_oI}$ for absorption). We shall thus neglect this phenomenon and focus on scattering.

4.4.2 <u>The induced scattering of photons: Rayleigh and Raman scattering of light.</u>

According to section 2, the probability per unit time $W_{k_1k_o}$ that a photon is scattered into a state $(\underset{\sim}{k}_1,\underset{\sim}{\varepsilon}_s)$ when N_{k_oI} photons $(\underset{\sim}{k}_o,\underset{\sim}{\varepsilon}_I)$ are present in the sample is given by Eq. (17) where $C_{\bar{H}_c\bar{H}_c}(\omega)$ is the time Fourier transform of $C_{\bar{H}_c\bar{H}_c}(t)$ given by Eq. (117). We should relate this probability to a measurable quantity, namely the power P_1 of photons $(\underset{\sim}{k}_1,\underset{\sim}{\varepsilon}_s)$ scattered per sec. between $\underset{\sim}{k}_1$ and $\underset{\sim}{k}_1 + d\underset{\sim}{k}_1$. We have

$$P_1 = \hbar\omega_1 W_{k_1k_o} \rho(\underset{\sim}{k}_1) d\underset{\sim}{k}_1 \tag{119}$$

where the number of independent states of photons of well defined polarization $\rho(\underset{\sim}{k}_1)d\underset{\sim}{k}_1$ is given by the same equation (25) as for neutrons. Let P_o be the incident power of photons $(\underset{\sim}{k}_o,\underset{\sim}{\varepsilon}_I)$. If the sample is a cylinder of section S and length ℓ $(V = S\ell)$, the number N_{k_oI} is $(P_o/\hbar\omega_o)(V/c)$. Using the fact that $dk_1 = d\omega_1/c$, a simple calculation finally yields:

$$\frac{P_1}{P_o} = \frac{d^2\sigma}{d\Omega d\omega} d\Omega d\omega_1 = Nk_1^{\ 4}\, T(\underset{\sim}{Q},\omega)\, d\Omega d\omega_1 \tag{120}$$

where we have put, by analogy with neutrons (cf. Eqs. (38) and (39))

$$T(\underset{\sim}{Q},\omega) = \frac{1}{2\pi}\int J(\underset{\sim}{Q},t)\exp(-i\omega t)\, dt \tag{121}$$

and

$$J(\tilde{Q},t) = \frac{1}{N} < \sum_{i,j} (\tilde{\varepsilon}_s \tilde{\tilde{\alpha}}_i(o)\tilde{\varepsilon}_I)(\tilde{\varepsilon}_s \tilde{\tilde{\alpha}}_j(t)\tilde{\varepsilon}_I) \exp i\tilde{Q}[\tilde{r}_i(t)-\tilde{r}_j(o)] > \qquad (122)$$

As for neutrons, $d^2\sigma/d\Omega d\omega$ is the double differential cross-section and represents the normalized scattered intensity per unit solid angle and per unit energy, $J(\tilde{Q},t)$ is the intermediate scattering function and $T(Q,\omega)$ is the scattering law. It is seen from Eq. (122) that light scattering reflects molecular motions through the fluctuations of the position (translation) and the fluctuations of the polarizability tensor (rotation and vibrations). The general situation is thus basically the same as for neutrons where all kinds of motions are also seen. There are however significant differences:

(i) for neutrons, the correlation functions corresponding to all kinds of motions are Q dependent while for e.m. waves (in practice light), only those corresponding to translation are Q dependent (this is in fact related to the long wavelength approximation).

(ii) for neutrons, Q is typically $1\ \text{Å}^{-1}$ while for light it is typically $10^{-3}\ \text{Å}^{-1}$. Consequently, the correlation function for translation decays much more slowly ($\sim 10^{-6}$ times) for light than for neutrons. Therefore for light, the hypothesis of non-coupling between the various motions seems better justified than for neutrons (at least for individual motions), and we can reasonably separate the two statistical averages in Eq. (122). Let us focus now on the polarizability term.

As for the dipole moment, the polarizability tensor of a molecule is a physical quantity which depends on the charge distribution in the molecules, and this distribution changes when the molecules vibrate. Let ν be all the vibrational states and q^ν the corresponding normal coordinates. By analogy with Eq. (87), we can write:

$$\tilde{\tilde{\alpha}} = \tilde{\tilde{\alpha}}^o + \sum_\nu \tilde{\tilde{\alpha}}^\nu q^\nu \qquad (123)$$

with

$$\tilde{\tilde{\alpha}}^\nu = \left(\frac{\partial \tilde{\tilde{\alpha}}}{\partial q^\nu} \right)_{q^\nu=0} \qquad (124)$$

$\tilde{\tilde{\alpha}}^o$ is called the permanent polarizability tensor and $\tilde{\tilde{\alpha}}^\nu$ the derivative polarizability tensor corresponding to vibration ν. With these definitions, it appears that the second member of Eq. (122) can be split into various components. Among them, we select the two following:

$$J_{oo}(\underset{\sim}{Q},t) = \frac{1}{N} < \sum_{i,j} (\underset{\sim}{\varepsilon}_s \underset{\approx}{\alpha}_i^o(o) \underset{\sim}{\varepsilon}_I)(\underset{\sim}{\varepsilon}_s \underset{\approx}{\alpha}_j^o(t) \underset{\sim}{\varepsilon}_I)$$

$$\exp i\underset{\sim}{Q}[\underset{\sim}{r}_i(t) - \underset{\sim}{r}_j(o)] > \qquad (125)$$

and

$$J_{\nu\nu}(\underset{\sim}{Q},t) = \frac{1}{N} < \sum_{i,j,\nu,\nu'} (\underset{\sim}{\varepsilon}_s \underset{\approx}{\alpha}_i^\nu(o) \underset{\sim}{\varepsilon}_I)(\underset{\sim}{\varepsilon}_s \underset{\approx}{\alpha}_j^{\nu'}(t) \underset{\sim}{\varepsilon}_I)\ q_i^\nu(o) q_j^{\nu'}(t)$$

$$\exp i\underset{\sim}{Q}[\underset{\sim}{r}_i(t) - \underset{\sim}{r}_j(o)] > \qquad (126)$$

which are the relevant correlation functions for the so-called Rayleigh and Raman scattering respectively. In what follows, we discuss in detail these two cases, beginning for convenience with Raman scattering.

4.4.3 Derivative polarizability tensor: Raman scattering of Light.

a) General. The corresponding correlation function is $J_{\nu\nu}(\underset{\sim}{Q},t)$. The rotations appear through $\underset{\approx}{\alpha}$, the vibrations through q and the translations through $\underset{\sim}{r}$. If we can assume that these three kinds of motions are independent, and this may be a good approximation, especially for translation compared to the other motions, then the average in Eq. (126) can be split into three averages. If, moreover, we assume that the vibrations between different molecules and within the same molecules are not coupled, then Eq. (104) holds and Eq. (126) can be written:

$$J_{\nu\nu}(\underset{\sim}{Q},t) = \frac{1}{N} \sum_{i,\nu} <(\underset{\sim}{\varepsilon}_s \alpha_i^\nu(o) \underset{\sim}{\varepsilon}_I)(\underset{\sim}{\varepsilon}_s \alpha_i^\nu(t) \underset{\sim}{\varepsilon}_I)> <q_i^\nu(o)\ q_i^\nu(t)>$$

$$<\exp i\underset{\sim}{Q}[\underset{\sim}{r}_i(t) - r_i(o)] > \qquad (127)$$

Using the terminology of neutrons, we can say that in this case, the scattering is mainly incoherent since only individual motions are seen, and that this incoherence is introduced by the vibrations. Another important and always justified approximation is to assume that the exponential is ≈ 1. This approximation expresses the fact that the contribution to the broadening of the spectra due to translation is always much smaller than the instrumental energy resolution (typically 10^{-3} cm^{-1} compared to ~ 1 cm^{-1}) and that only the broadening due to rotations and vibrations is seen (typically a few cm^{-1} to a few hundred cm^{-1}). If in addition, we assume that all the molecules are identical, then the correlation function can finally be written:

$$C_{\nu\nu}(t) = \sum_{\nu} \langle(\underset{\sim}{\varepsilon}_s \underset{\approx}{\alpha}^{\nu}(o)\underset{\sim}{\varepsilon}_I)(\underset{\sim}{\varepsilon}_s \underset{\approx}{\alpha}^{\nu}(t)\underset{\sim}{\varepsilon}_I)\rangle \langle q^{\nu}(o)\, q^{\nu}(t)\rangle \quad (128)$$

To go further, we can assume, as for infrared absorption, that the vibrations are well separated from each other and weakly damped, in which case Eq. (106) holds. If we define the function $D_{SI}(t)$ as:

$$D_{SI}(t) = \langle(\underset{\sim}{\varepsilon}_S \underset{\approx}{\alpha}^{\nu}(o)\underset{\sim}{\varepsilon}_I)(\underset{\sim}{\varepsilon}_S \underset{\approx}{\alpha}^{\nu}(t)\underset{\sim}{\varepsilon}_I)\rangle \quad (129)$$

we finally have:

$$C_{\nu\nu}(t) = \sum_{\nu} (q^{\nu})^2 \exp(-i\Omega_{\nu} t)\, G_{\nu}(t)\, D_{SI}(t) \quad (130)$$

$$C_{\nu\nu}(\omega) = \sum_{\nu} (q^{\nu})^2\, G_{\nu}(\omega + \Omega_{\nu}) \otimes D_{SI}(\omega + \Omega_{\nu}) \quad (131)$$

This result shows that a Raman spectrum is composed of a series of lines centered at the vibration frequencies Ω_{ν}. It is seen that, provided some assumptions are made, each line of the Raman spectrum is the convolution of a vibrational and a rotational spectrum, as for the infrared case. There are however, two important differences: (i) the rotational correlation function is not the same, and (ii) we have here an external parameter which we do not have in infrared, namely the relative orientation of the polarization vectors $\underset{\sim}{\varepsilon}_I$ and $\underset{\sim}{\varepsilon}_S$, which allows the selection of certain components of the polarizability tensor. This polarizability tensor is usually split into its isotropic part $\bar{\alpha}\underset{\approx}{I}$ and its anisotropic part $\underset{\approx}{\beta}$:

$$\underset{\approx}{\alpha}^{\nu} = \bar{\alpha}^{\nu}\, \underset{\approx}{I} + \underset{\approx}{\beta}^{\nu} \quad (132)$$

with

$$\bar{\alpha} = \frac{1}{3}\,\mathrm{Tr}(\underset{\approx}{\alpha}) \quad (133)$$

and

$$\underset{\approx}{\beta} = \underset{\approx}{\alpha} - \frac{1}{3}\,\mathrm{Tr}(\underset{\approx}{\alpha}) \cdot \underset{\approx}{I} \quad (134)$$

$\underset{\approx}{I}$ being the unit tensor. According to the geometry of the experiment, one can pick out different combinations of the isotropic and anisotropic parts, and the corresponding spectra will reflect different aspects of the (same) motion. In this sense, Raman scattering is comparable to neutron scattering, the external parameter being the momentum transfer Q is this latter case. In the following, we discuss the calculations in a simple case, for two typical geometrical configurations: the so-called V-H and V-V configurations.

b) Application to a linear molecule. Let us consider a linear molecule, whose orientation is characterized by its unit vector $\underset{\sim}{u}$. Let us consider only the vibration q^{ν} along $\underset{\sim}{u}$. Due to the cylindrical symmetry, the polarizability tensor is diagonal in the molecular frame and contains two independent components α_A and α_B:

$$\underset{\approx}{\alpha}^{\nu} = \begin{pmatrix} \alpha_A & 0 & 0 \\ 0 & \alpha_B & 0 \\ 0 & 0 & \alpha_B \end{pmatrix} \tag{135}$$

Let u_r $(r = x,y,z)$ be the components of $\underset{\sim}{u}$ with laboratory frame. Simple algebra yields:

$$\alpha^{\nu}_{rs} = \bar{\alpha}^{\nu}\, \delta_{rs} + (\alpha_A - \alpha_B)(u_r u_s - \tfrac{1}{3}\delta_{rs}) \tag{136}$$

(i) V-H configuration: depolarized spectrum. In this configuration, we choose a scattering angle of 90°. The beam comes along Oy with $\underset{\sim}{\varepsilon}_I$ along Oz and is scattered along Ox with $\underset{\sim}{\varepsilon}_S$ along Oy. In this case we have

$$\underset{\sim}{\varepsilon}_S\, \underset{\approx}{\alpha}\, \underset{\sim}{\varepsilon}_I = \alpha_{yz} = (\alpha_A - \alpha_B)\, u_y\, u_z \tag{137}$$

and

$$D_{yz}(t) = (\alpha_A - \alpha_B)^2 < u_y(o)\, u_z(o)\, u_y(t)\, u_z(t) > \tag{138}$$

Let us assume that the medium is isotropic. Then we can average D_{yz} over all possible orientations of $\underset{\sim}{u}(o)$ or equivalently, take $\underset{\sim}{O}z$ along $\underset{\sim}{u}(o)$. Calling $\theta(t)$ and $\phi(t)$ the polar coordinates of $\underset{\sim}{u}(t)$, we have:

$$D_{yz}(t) = (\alpha_A - \alpha_B)^2 < \sin\theta(t)\, \cos\theta(t)\, \sin\phi(t) > \tag{139}$$

The function in brackets is an associated Legendre polynomial of order 2, P_2^m. Since in an isotropic medium its average is independent of m, putting $F_2(t) = \langle P_2(\cos\theta(t))\rangle$ where P_2 is the ordinary Legendre polynomial of order 2, we finally have:

$$C^{VH}_{\nu\nu}(\omega) \propto (q^{\nu})^2 (\alpha_A - \alpha_B)^2 \; G_{\nu}(\omega + \Omega_{\nu}) \otimes F_2(\omega + \Omega_{\nu}) \qquad (140)$$

This equation should be compared with Eq. (109). It appears here that the depolarized Raman spectrum reflects the same motion as the infrared absorption spectrum: vibration and rotations, except that the first order Legendre polynomial is replaced by the second order one. This shows that comparison between infrared and Raman spectra can yield valuable information on the details of the rotational model.

(ii) V-V configuration: polarized spectrum. In this configuration, the experimental geometry is the same as for V-H except that $\underset{\sim}{\varepsilon}_S$ is also along $\underset{\sim}{O}z$. In this case, we have

$$\underset{\sim}{\varepsilon}_S \, \underset{\approx}{\alpha} \, \underset{\sim}{\varepsilon}_I = \alpha_{zz} = \bar{\alpha}^{\nu} + (\alpha_A - \alpha_B)(u_z^2 - \tfrac{1}{3}) \qquad (141)$$

With the same assumptions as before, after some algebra, we obtain:

$$C^{VV}_{\nu\nu}(\omega) \propto (q^{\nu})^2 \; G_{\nu}(\omega + \Omega_{\nu}) \otimes \left[(\bar{\alpha}^{\nu})^2 + \frac{4}{45} (\alpha_A - \alpha_B)^2 \; F_2(\omega + \Omega_{\nu}) \right] \qquad (142)$$

This result shows that the polarized Raman spectrum is the sum of two terms: a purely vibration term which is proportional to $(\bar{\alpha}^{\nu})^2$ and a term identical to that obtained in the depolarized case. This is a very important result since, comparing both kinds of spectra one can immediately say something about the time scale of the various motions. If the V-H spectrum or the infrared spectrum is broad and the V-V spectrum presents a central narrow component, then we can assign with certainty this narrow component to the purely vibrational term and the broader one to rotations, and reasonably conclude that vibrations and rotations are weakly coupled. If, on the contrary, both spectra have widths of the same order, this last result is certainly not true. As an example of the former situation, Fig. 12 shows the V-V spectrum of CO in C_2Cl_2 [59]. Its width is ~ 1.5 cm^{-1}, to be compared with ~ 100 cm^{-1} for the infrared spectrum (Fig. 11). An example of the latter situation is found for HCl in CO_2 [59] where the width of the infrared spectrum is ~ 61 cm^{-1}, compared with ~ 48 cm^{-1} for the V-V + V-H Raman spectrum.

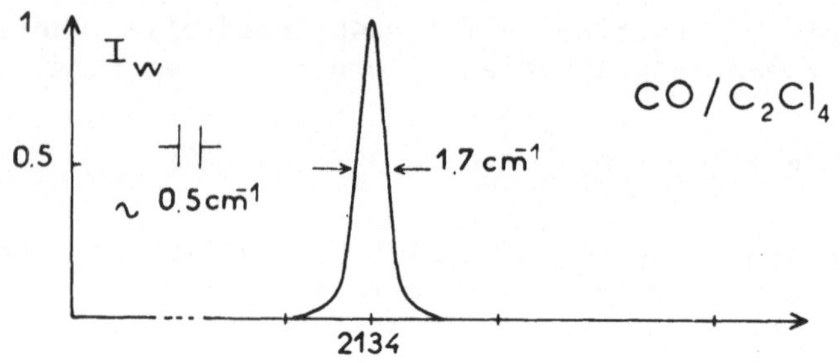

Figure 12: V-V Raman spectrum of 0.1M CO in C_2Cl_4 [from ref. 59].

c) Conclusion. In conclusion, it appears that Raman scattering of light may be a powerful method of studying vibrational and rotational motions in liquids. Due to the incoherence introduced by the vibrations, only individual motions are seen in practice, even in pure liquids. Comparison between spectra obtained in different configurations and by infrared absorption can yield valuable information about the nature, the time scale and the degree of coupling between the various motions. Although the amount of Raman data in liquids is increasing [see e.g. refs. 59, 61, 65, 69-75] the precise theoretical description of the Raman line shapes is difficult, especially due to the great number of possible vibrational relaxation processes [76]. Moverover as in the infrared case, on the experimental side the problem also becomes very difficult with large molecules, due to possible overlap of various vibrational bands (inhomogeneous broadening). Concerning the instrumental energy resolution, with the best monochromators one can reach $\Delta\omega \sim 0.2\ cm^{-1} \sim 20\ \mu eV$, i.e. the same order of magnitude as with I.L.L. neutrons. This corresponds to a time scale of 10^{-11} sec., which means that in any case only fast motions can be seen by this method. So far, we have considered experiments using continuous waves. It should be noticed that an alternative manner of conducting Raman experiments is to use picosecond pulsed techniques which can yield additional information on the nature of the relaxation processes. For details, the reader is referred to the paper by Drs. A. Lauberau and W. Kaiser, in this book.

4.4.4 Permanent polarizability tensor: Rayleigh scattering of light

a) General. In this case the relevant correlation function is given by Eq. (125), where as opposed to the Raman case, the vibrations do not appear. Consequently there is no mechanism to decouple the relative motions and the scattering is coherent in the sense that it mainly reflects the collective motions. If the translations and rotations appear on a different time scale, here also it is reasonable to separate the two averages in Eq. (125), and we have:

$$J_{oo}(\underset{\sim}{Q},t) = \frac{1}{N}\sum_{i,j} \langle(\varepsilon_{\sim S}\underset{\approx}{\alpha}{}^{o}_{i}\varepsilon_{\sim I})(\varepsilon_{\sim S}\underset{\approx}{\alpha}{}^{o}_{j}(t)\varepsilon_{\sim I})\rangle$$

$$\langle \exp iQ[\underset{\sim}{r}_i(t) - \underset{\sim}{r}_j(o)]\rangle \qquad (143)$$

The Fourier transform shows that the Rayleigh spectra are centered around the incident photon energy. As for the case of the derivative tensor, the permanent tensor can be split into an isotropic part $\bar{\alpha}^o\underset{\approx}{I}$ and an anisotropic part $\underset{\approx}{\beta}^o$. Consequently in this case also, by choosing the relative orientation of $\underset{\sim}{\varepsilon}_S$ and $\underset{\sim}{\varepsilon}_I$, one can pick out different aspects of the motion. Let us consider the same two configurations as before.

b) V-H spectrum - depolarized Rayleigh scattering. By analogy with the Raman case, the relevant tensor component, in the laboratory frame, is β_{yz} and the corresponding correlation function is

$$J^{VH}_{oo}(\underset{\sim}{Q},t) = \frac{1}{N}\sum_{i,j} \langle\beta^o_{i,yz}(o)\ \ \beta^o_{j,yz}(t)\rangle \langle\exp iQ[\underset{\sim}{r}_i(t) - \underset{\sim}{r}_j(o)]\rangle \qquad (144)$$

If the rotations are much faster than the translations, the spectrum corresponding to the correlation function of β is much broader than that corresponding to the exponential (i.e. we can take the exponential ~ 1) and the depolarized Rayleigh spectrum only reflects the (collective) rotational motions. Since, on the other hand, the depolarized Raman spectrum reflects the individual rotational (and vibrational) motions, in this case comparison between both results can give valuable information about the degree of cooperativity of the rotational motions. In practice, the Rayleigh spectrum of a pure liquid is generally found to be narrower than the corresponding Raman spectrum, showing that the collective motions are slower than the individual ones, as expected from general considerations.

c) V-V spectrum - polarized Rayleigh scattering. As for the Raman case, the corresponding correlation function contains two kinds of terms: terms related to the isotropic part of the permanent polarizability tensor and terms related to the anisotropic part. Let us consider these terms separately.
(i) The term related to $\bar{\alpha}^o$ is

$$J_{oo}^{VV}(\underset{\sim}{Q},t) = \frac{1}{N} \sum \bar{\alpha}_i^o \bar{\alpha}_j^o \langle \exp i\underset{\sim}{Q}[\underset{\sim}{r}_i(t) - \underset{\sim}{r}_j(o)] \rangle \tag{145}$$

It is seen that this function is formally identical to the neutron intermediate pair correlation correlation function $I_p(\underset{\sim}{Q},t)$ for a monoatomic liquid (Eq. (41)). This means that V-V Rayleigh light scattering and coherent neutron scattering contain a priori the same kind of information about the collective translational motions, the only difference being the very different momentum transfer range which can be reached ($\sim 10^{-3}$ Å^{-1} compared to 1 Å^{-1}). Since the space scale, and also the time scale due to a better energy resolution, are much larger with light than with neutrons, light scattering is more suitable than neutrons to test slow, large-scale phenomena such as those predicted by hydrodynamic and critical phenomena theories. In particular, hydrodynamics predicts that the spectrum contains three components, a central component called the Rayleigh line, whose width is proportional to the thermal diffusivity times Q^2, and two side lines, centered at $\omega = \pm v_s Q$ where v_s is the velocity of sound, whose width is also proportional to Q^2, and which is called the Brillouin doublet. In some special cases, a fourth central component may appear, called the Mountain line, the characteristics of which are linked to the shear viscosity properties. With neutrons, since Q is much larger, all the lines are much broader. However, neutrons are unique for testing the validity of these theories in the limits of short distances and times. For details, the reader is referred to the paper by Dr. J.R.D. Copley in this book.

(ii) The term related to $\underset{\approx}{\beta}^o$ is

$$J_{oo}^{VV}(\underset{\sim}{Q},t) = \frac{1}{N} \sum_{i,j} \langle \beta_{i,zz}^o(o)\, \beta_{j,zz}^o(t) \rangle \langle \exp i\underset{\sim}{Q}[\underset{\sim}{r}_i(t) - \underset{\sim}{r}_j(o)] \rangle \tag{146}$$

This term is comparable to the corresponding one in the V-H geometry. It generally corresponds to a broad band which mainly reflects the collective rotational motions.

The total V-V scattering being the sum of these two contributions, a polarized Rayleigh spectrum should appear as the sum of three or four sharp components superimposed on a broad band (Fig. 13). In practice, however, the feature will depend on the actual

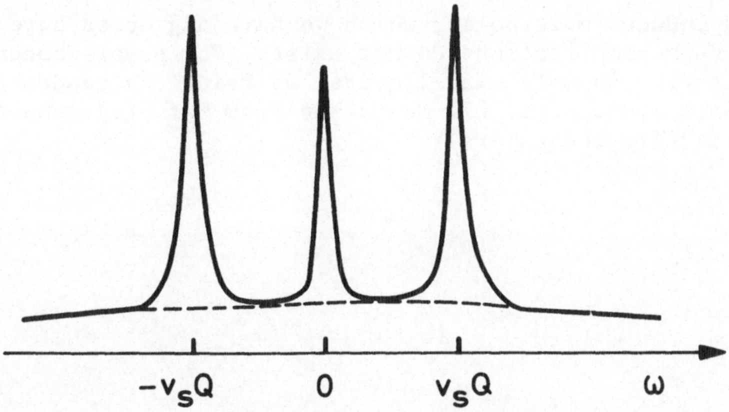

Figure 13: Theoretical V-V Rayleigh spectrum of a liquid. One recognizes the central Rayleigh line, the Brillouin doublet superimposed on a broad band.

energy resolution. Suppose one uses a monochromator, as in Raman scattering, whose resolution is of the order of 20-100 μeV. Since the frequency of the sound waves in a liquid is at most 10 GHz (∿ 40 μeV) the Brillouin doublet may not even be resolved and the spectrum would appear as a sharp peak superimposed on the broader component which reflects the collective rotational motions. Suppose now one uses a much better resolution (≈ a few MHz) such as achieved by standard Fabry-Perrot interferometers, then the triplet will be well resolved, but the broad line hardly seen. If finally it turns out that the central line is very sharp, the precise measurement of its width will require a much better resolution (≈ a few kHz) which can now be obtained by techniques of photon beating. But in this case, the energy width of the spectrum may not reach the Brillouin lines.

d) Conclusion. In conclusion, it appears that Rayleigh scattering of light is a powerful method for studying collective translational and rotational motions in liquids. It is suitable for studying long space and time scale phenomena and testing hydrodynamic theories. Progress in the technology of laser techniques has made it possible to obtain very high energy resolution, many orders of magnitude better than that which can be achieved with neutrons. Moreover, since photons are much less expensive than neutrons, we believe that for studying long range phenomena, light scattering will remain for a long time the most popular method. It must be noted, however, that the complications are greater with light, in particular due to internal field effects (the refraction index) and induced multipoles, which we have neglected here. With neutrons, such complications do not exist. For papers concerning light scattering in molecular liquids, we refer the reader to ref. [49 (textbook)], to refs. [77-82] taken from ref. [1], and to refs. [83] and [84] (review papers).

CHAPTER 5

NUCLEAR MAGNETIC RESONANCE AND SPIN-LATTICE RELAXATION [85,86]

In this section, we describe how the fact that nuclei have a spin can be used to study molecular motions. We first recall the properties of a spin in a magnetic field, and how it couples to dynamical variables. Then we describe the phenomenon of magnetic resonance absorption and deduce qualitatively the existence of three regimes of molecular motions. Finally, we define the problem of spin lattice relaxation and show theoretically and with a few examples which kinds of information on molecular motions we can obtain from this phenomenon.

5.1 Properties of a Nuclear Spin in a Constant Magnetic Field: Zeeman Interaction

It is known that a nuclear state can be characterized by its total spin I whose value can only be changed via a (high energy) nuclear transition. In condensed matter where no nuclear reaction takes place, I is well defined for each nucleus. Values of I for common nuclei are given in section 3.3. On the other hand, the spin is a vector operator noted $\underset{\sim}{I}$ acting in a Hilbert space of $2I + 1$ dimensions. The eigenstates $|I,m\rangle$ of I_z form a basis of this space. The integer m is a quantum number which can take $2I + 1$ values between $-I$ and I. The three operators I_z, I_x, I_y or I_z, $I_+ = I_x + iI_y$ and $I_- = I_x - iI_y$ have the following properties:

$$I_z \, |I,m\rangle = m \, |I,m\rangle \tag{147}$$

$$I_\pm \, |I,m\rangle = \sqrt{I(I+1) - m(m\pm 1)} \, |I,m\pm 1\rangle \quad m \neq \pm I \tag{148}$$

$$I_\pm \, |I,\pm I\rangle = 0 \tag{149}$$

Now the Wigner-Eckart theorem tells us that the matrix elements within the manifold I of all the irreducible tensor operators of the same order are proportional. In particular, this is true for the vector operators (tensors of order 1) and we have:

$$\langle I,m|\underset{\sim}{\mu}|I,m'\rangle = \hbar\gamma_I \, \langle I,m|\underset{\sim}{I}|I,m'\rangle \tag{150a}$$

where $\underset{\sim}{\mu}$ is the magnetic moment and the proportionality coefficient γ_I the gyromagnetic ratio, of the nucleus.

Eq. (150a) is usually written more simply as

$$\underset{\sim}{\mu} = \hbar\gamma_I \underset{\sim}{I} \tag{150b}$$

A nucleus interacts with a magnetic field $\underset{\sim}{H}_o$ via its magnetic moment $\underset{\sim}{\mu}$. The corresponding Hamiltonian H_Z, called the Zeeman Hamiltonian, is written, assuming H_o is along Oz:

$$H_Z = -\underset{\sim}{H}_o \underset{\sim}{\mu} = -\hbar\gamma_I H_o I_Z \tag{151}$$

From Eq. (147) it is easily seen that H_Z has $2I + 1$ equidistant energy levels E_m given by

$$E_m = -\hbar\gamma_I H_o m \tag{152}$$

the distance $|\Delta|$ between two adjacent levels being:

$$|\Delta| = |\hbar\gamma_I H_o| = \hbar\omega_L \tag{153}$$

The quantity $\omega_L = |\gamma_I H_o|$ is called the Larmor angular frequency. For usual nuclei, ω_L is typically in the megahertz range for $H_o \approx 1000$ Gauss ($\omega_L = 4.257$ MHz in $H_o = 1000$ G for the proton). It should be noted that for a given nucleus, Eq. (153) provides a new energy unit: the gauss.

5.2 Spin-Matter Interaction

A spin interacts with electric and magnetic fields. In matter, these fields are created at any point by both the distribution of the nuclear spins and by the distribution of electronic charges. One can thus imagine a number of possible interaction terms and this number is in fact rather large. For details, we refer to standard textbooks, in particular the book by Abragam [85]. For the present purpose, we describe the three most usual ones which are useful for liquids, namely the dipole-dipole, the quadrupole and the spin-rotation interactions.

5.2.1 The dipole-dipole interaction.

This interaction corresponds to the direct magnetic coupling between two magnetic moments. The corresponding Hamiltonian H_d for two spins $\underset{\sim}{I}_i$ and $\underset{\sim}{I}_j$ can be written in the following closed form

$$H_d^{ij} = \hbar^2\gamma_i\gamma_j \; \underset{\sim}{I}_i \, \underset{\approx}{D}_{ij} \, \underset{\sim}{I}_j \tag{154a}$$

The elements of the traceless, second rank tensor D_{ij} are of the form $r_{ij}^{-3} Y_2^m(\theta_{ij},\phi_{ij})$ where r_{ij}, of polar coordinates θ_{ij} and ϕ_{ij}, is the vector joining the two spins and Y_2^m are the five spherical harmonics of order 2. All the terms of H_d are thus of the form (r,s = +,-,z)

$$H^{ij}_{d,rs} \propto r_{ij}^{-3} Y_2^m I_{ir} I_{js} \tag{154b}$$

5.2.2 <u>The electric quadrupolar interaction</u>

The physics underlying this interaction is the following. The nucleus is made up of protons and neutrons to which correspond a certain electric charge density $\rho(\underset{\sim}{r})$. On the other hand, the electrons surrounding the nucleus create at the level of the nucleus an electric potential $V(\underset{\sim}{r})$. The corresponding interaction energy is

$$H_Q = \int \rho(\underset{\sim}{r})\, V(\underset{\sim}{r})\, d\underset{\sim}{r} \quad .$$

Since the charges creating $V(\underset{\sim}{r})$ are very far from the nucleus, the potential $V(\underset{\sim}{r})$ can be expanded around $\underset{\sim}{r} = 0$ and it can be shown that the relevant part of H_Q reduces to (r,s = x,y,z)

$$H_Q = \frac{1}{6} \sum_{r,s} V_{rs}\, Q_{rs} \tag{156}$$

where

$$V_{rs} = \left(\frac{\partial^2 V}{\partial x_r\, \partial x_s} \right)_{\underset{\sim}{r}=0} \tag{157}$$

is the electric field gradient tensor, and

$$Q_{rs} = \int (3x_r x_s - \delta_{rs} r^2)\, \rho(\underset{\sim}{r})\, d\underset{\sim}{r} \tag{158}$$

is the nuclear electric quadrupolar tensor. Both tensors are traceless. Now, according to the Wigner-Eckart theorem, Q_{rs} is proportional to any second order irreducible tensor, in particular the one constructed with the operators $I_r I_s$. The result is usually written as:

$$Q_{rs} = \frac{eQ}{I(2I-1)} \left[\frac{3}{2}(I_r I_s + I_s I_r) - \delta_{rs}\, I^2 \right] \tag{159}$$

where the proportionality coefficient eQ is the so-called quadrupole moment of the nucleus I. Note that $Q \neq 0$ only if

$I > 1/2$ i.e. for nuclei such as $^2D(I=1)$, $^{14}N(I=1)$.... In summary, all the terms of H_Q for a nucleus possessing a spin $\tilde{I}_i$ greater than 1/2 are of the form:

$$H^i_{Q,rs} \propto V^i_{rs} \; I_{ir} \; I_{is} \tag{160}$$

5.2.3 The spin-rotation interaction. The physics underlying this interaction is as follows. Consider a spin $\tilde{I}_i$ located on molecule m. Suppose first that the molecule is fixed. The orbital motion of the electrons in the molecule creates a magnetic field at the place of the spin $\tilde{I}_i$ producing a magnetic interaction. However this magnetic field is zero on average in the absence of an external magnetic field (quenching of the orbital momentum). In a finite magnetic field, the orbital momentum is slightly "unquenched" and this phenomenon corresponds to the so-called chemical shift interaction. We shall not consider it here.

Suppose now that the molecule is rotating. The electrons suffer an additional rotational motion and consequently create at the nucleus an extra magnetic field, which is proportional to the angular momentum $\tilde{J}_m$ of the molecule. Since this interaction is expected to be anisotropic, the corresponding Hamiltonian H_{SR} is necessarily of the form:

$$H^i_{SR} = -\hbar \, \tilde{I}_i \, \approx{C} \, \tilde{J} \tag{161}$$

where $\approx{C}$ is the so-called spin-rotation tensor. In summary, the terms of H_{SR} are of the form

$$H^i_{SR,rs} \propto C_{rs} \; I_{ir} \; J_{ms} \tag{162}$$

5.2.4 Conclusion. It is seen that the three above-mentioned interaction Hamiltonians contain spin variables and "lattice" variables (vector joining two nuclei, tensor attached to the molecular frame, molecular angular momentum). The latter necessarily fluctuate when molecular motions are present. These interactions may thus be used for studying these molecular motions. Two methods are used for this purpose: (i) absorption of electromagnetic waves by the spin system placed in a strong magnetic field and (ii) relaxation towards the "lattice" of the Zeeman energy. We describe these two methods below.

5.3 Nuclear Magnetic Resonance Absorption

5.3.1 General formalism. The problem is as follows: we

consider a system of N particles i, located at $\underset{\sim}{r}_i$ and possessing a spin $\underset{\sim}{I}_i$, placed in a strong, constant, uniform magnetic field $\underset{\sim}{H}_o$, taken as the Oz axis. We send electromagnetic waves onto this system at a frequency ω_o near ω_L and look at the possible absorption of these waves. To formulate the problem, we should follow the prescription of Chapter 2, namely find what is the probe, the reservoir and the interaction H_c. The probe is clearly the electromagnetic field characterized by the eigenstates $|N_{k\mu}>$ (cf. section 4.1). In these conditions, the reservoir R is necessarily the system made up of the spins <u>and</u> the atoms on which they sit - the lattice. The Hilbert space of the reservoir is thus the tensorial product (symbol ⊕) of the Hilbert spaces of the spins and of the lattice. In particular, the density matrix of the reservoir ρ_R is written with evident notation:

$$\rho_R = \rho_S \oplus \rho_L \tag{163}$$

and the Hamiltonian H_R is the sum of the Zeeman Hamiltonians, of the various interaction Hamiltonians, which we symbolize by H', and of the lattice Hamiltonian H_L, namely

$$H_R = H_Z + H' + H_L \tag{164}$$

with

$$H_Z = \sum_i \hbar\gamma_i H_o I_{iz} \tag{165}$$

and

$$H' = \sum_{i>j} H_d^{ij} + \sum_i H_Q^i + \sum_i H_{SR}^i + \ldots \tag{166}$$

Finally, the interaction Hamiltonian H_c is merely the coupling between the spins and the magnetic field $\underset{\sim}{H}$ of the e.m. field, and we can write:

$$H_c = -\sum_i \int_V d\underset{\sim}{r}\, \underset{\sim}{H}(\underset{\sim}{r})\, \underset{\sim}{\mu}_i\, \delta(\underset{\sim}{r} - \underset{\sim}{r}_i) \tag{167}$$

This expression is formally identical to Eq. (73) with $\underset{\sim}{E}$ replaced by $\underset{\sim}{H}$. The calculation of $\bar{H}_c$ between possible final states proceeds in exactly the same manner as in section 4.3.1 and the result of the interaction is a net absorption of photons.

The ratio of the absorbed power to the incident power is given by the same Eq. (83) where now $C_{\mu\mu}(\omega)$ is the Fourier transform of

$C_{\mu\mu}(t)$ given by

$$C_{\mu\mu}(t) = \frac{1}{N} < \sum_{i,j} \hbar^2 \gamma_i \gamma_j \; (\underset{\sim}{\varepsilon}_I \cdot \underset{\sim}{I}_i^+(o))(\underset{\sim}{\varepsilon}_I \cdot \underset{\sim}{I}_j(t))> \qquad (168)$$

Here, $\underset{\sim}{\varepsilon}_I$ is the unit vector along the incident magnetic field and the brackets necessarily mean a quantum average since spins are purely quantum objects. Let us define $\underset{\sim}{M}$ as the total nuclear magnetization:

$$\underset{\sim}{M} = \sum_i \hbar \gamma_i \underset{\sim}{I}_i \qquad (169)$$

Using Eqs. (15), (163) and the properties of the trace, and assuming that $\underset{\sim}{\varepsilon}_I$ is along the Ox axis, Eq. (178) can be written written in a more usual form:

$$C_{\mu\mu}(t) = \frac{1}{N} \mathrm{Tr}\{\rho_L \cdot \mathrm{Tr}\; (\rho_S \; M_x^+(o) \; M_x(t))\} \qquad (170)$$

5.3.2 The three regimes of molecular motions. Looking at Eqs. (168) or (170) it is seen that the information on the molecular motions contained in the magnetic absorption spectrum is not very direct since $C_{\mu\mu}(t)$ reflects the relative motions of spin variables only. However the result does depend on the lattice variables through H' since we have from Eq. (14) and using the fact that H_L and $\underset{\sim}{I}_j$ commute:

$$\underset{\sim}{I}_j(t) = \exp[-i(H_Z + H')t] \; \underset{\sim}{I}_j(o) \; \exp[i(H_Z + H')t] \qquad (171)$$

This relation suggests the existence of three regimes of molecular motions in which the nature of the absorption will be different. The criterion is the correlation time τ_c for the motion compared to the average frequencies ω_D and ω_L between the levels of H' and H_Z respectively. We consider successively these three cases.

a) $\tau_c^{-1} \ll \omega_D \ll \omega_L$. This case corresponds to very slow motions and is called the rigid lattice regime. The absorption spectrum merely reflects the distribution of levels of $H_Z + H'$. In a high field $|H_Z| \gg |H'|$ and for identical spins, the absorption curve is a line centered around the Larmor frequency ω_L and whose width $\Delta\omega_{RL}$ is of the order of $\gamma|h|$, where h is the mean value of the (random) magnetic field created at the place of a spin by all the neighbours and whose value is roughly such that

$$\Delta\omega_{RL} \approx \gamma|h| \approx \omega_D \qquad (172)$$

This situation is usually found in molecular solids at low

temperatures. With protons, ω_D is typically in the range of kHz and consequently $\tau_c > 10^{-4}$ sec. For details, see chapter IV of ref. [85].

b) $\omega_D << \tau_c^{-1} << \omega_L$. This case corresponds to a situation where the motions are sufficiently rapid to average somewhat the local magnetic field, but too slow to induce transitions between the Zeeman levels. This is reflected on the absorption spectrum by a line which is narrower than in the preceding case. The new width $\Delta\omega_N$ of the spectrum is roughly given by

$$\Delta\omega_N \approx \tau_c (\Delta\omega_{RL})^2 \qquad (173)$$

This case corresponds to the so-called motional narrowing in the adiabatic regime. It is found in solids at high temperature, in mesomorphic phases (liquid crystals, plastic crystals), polymers .. and corresponds typically to 10^{-5} sec $> \tau_c > 10^{-8}$ sec. For details see chapter X of ref. [85].

c) $\omega_D \ll \omega_L \ll \tau_c^{-1}$. In this case, the motions are sufficiently rapid as to induce transition between the levels of H_Z. This is typically what occurs in liquids. In isotropic liquids, H' is zero in average and the absorption line is generally very sharp, in principle a delta peak at ω_L. In fact, the width is non-zero and limited by the relaxation time of the Zeeman levels. This case corresponds to the so-called motional narrowing in the non adiabatic regime or extreme narrowing. Typically we have $\tau_c < 10^{-8}$ sec. For details see chapter XIII of ref. [85].

5.3.3 Spin-spin relaxation time and measurement of self-diffusion coefficients.

The above discussion has shown that the width of the magnetic absorption spectrum $C_{\mu\mu}(\omega)$ is a function of the rate of molecular motions and consequently that its measurement versus temperature can give valuable qualitative information on the time scale of such motions. Instead of measuring the absorption, another equivalent manner of operating is to observe directly the decay of $C_{\mu\mu}(t)$. When this decay is exponential (i.e. the spectrum is a Lorentzian line), the corresponding time constant T_2 is called the spin-spin relaxation time. It can be pictured as the mean time for the x components of the spins to get out of phase due to the existence of different local random magnetic fields. If these fields vary rapidly and their average value is zero (as in an isotropic liquid), then the effective local field is very weak and consequently T_2 is rather long. Suppose now that the measurement is made in the presence of a static field gradient oriented along Ox. During the time scale of the experiment which is $\sim T_2$, the particles with spin diffuse in the sample. This means that the static field "seen" by the spins is not the same during the experiment. This causes an extra dephasing and $C_{\mu\mu}(t)$

decays more rapidly. If the molecular diffusion can be described by a single diffusion coefficient D_t [Eq. (54)], it can be shown [85] that, in the presence of the gradient g, we have:

$$C_{\mu\mu}(t) = C_{\mu\mu}(o) \exp\left[-\frac{t}{T_2} - \frac{\gamma^2 g^2 Dt^3}{12}\right] \quad (174)$$

This equation shows that by performing two measurements, in the presence and in the absence of a field gradient, one has a way of reaching the value of the self-diffusion coefficient. These measurements are usually performed using the spin-echo method. Describing the principle of this method is outside the scope of this paper, but the reader can refer to the standard textbooks [85,86] or papers [87] for details.

In principle, this spin-echo method using a field gradient is the best one, together with the neutron method (see section 3.6.2) for measuring self-diffusion coefficients in pure systems. The advantage of spin echo is that smaller values of D_t can be reached ($\sim 10^{-9}$ cm^2/sec, compared to $\sim 10^{-7}$ cm^2/sec with neutrons). Moreover it tests the diffusion mechanisms over a much longer time ($\sim T_2 \sim 10^{-4}$ sec) than the neutron does ($\sim 10^{-9}$ sec), and is much less time consuming. However it has a practical drawback, namely that T_2 must not be too short. This reduces possible experiments to systems with spins 1/2 (for $I > 1/2$, T_2 is very short due to strong quadrupolar relaxation - see below), namely protons, fluorines, phosphorus. But the neutron method is also limited to protons since the proton is practically the only nucleus whose scattering is almost purely incoherent. However, it is still unique for anisotropic liquids such as liquid crystals, since here, T_2 is rather short (the dipolar interactions are not completely averaged out to zero) and the spin-echo method is difficult to apply [88]. In normal liquids, the spin-echo method is the most widely used for economic reasons, but it should be noted that in the few cases where they have been applied to the same system, both methods have led to results in very good agreement with one another, as is the case for liquid cyclopropane [46,47] - see Fig. 14.

5.4 Nuclear Spin-Lattice Relaxation in Molecular Liquids

5.4.1 General.

The discussion of section 5.3.2 has shown that for usual liquids the average local field is very weak. This means that the effective interaction between spins is also very weak. Since, moreover, the motions are sufficiently rapid to induce transitions between the Zeeman levels, one can say that in liquids, the interaction of the spins with the lattice is dominant. We can

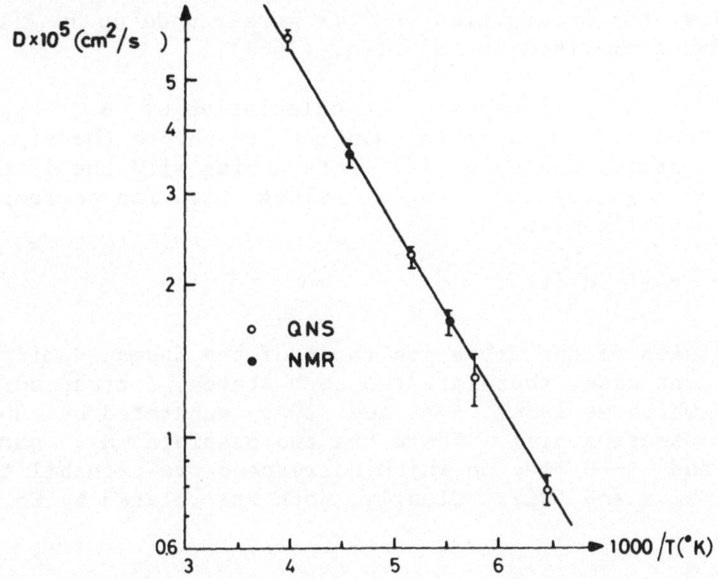

Figure 14: Neutron (QNS) and spin-echo (NMR) measurements of the self-diffusion coefficient of liquid cyclopropane as a function of temperature. From ref. [46].

thus reduce the problem to one or a few spins in a strong magnetic field in interaction with the lattice. If we could study the rate at which the molecular motions induce transitions between the Zeeman levels, this would give a more direct method than absorption for studying these motions. This is in fact possible, and corresponds to spin-lattice relaxation measurements. The principle of these measurements is as follows. First we detect the absorption signal amplitude, then perturb the amplitude of this signal in some way (e.g. using a strong radiofrequency pulse at ω_L) and look at the return to equilibrium of the absorption signal in the absence of any radiofrequency field. When this return is exponential, the corresponding constant time T_1 is called the spin lattice relaxation time.

5 4.2 Spin-lattice relaxation as a spectroscopic method.

Considered as a spectroscopic method, spin lattice relaxation should be described according to the general scheme of Chapter 2. The physical system of study is a liquid made up of N particles i possessing a spin $\underset{\sim}{I}_i$, these spins being weakly interacting between themselves, and put in a strong magnetic field $\underset{\sim}{H}_o$. Clearly in this case, the probe is made up by the spins in the field. The corresponding Hamiltonian is H_Z given by Eq. (165). The reservoir is necessarily the "lattice" whose Hamiltonian is

H_L. Finally, the interaction H_c is merely made up by all the Hamiltonians summarized in H' (Eq. (166)).

In order to continue with the calculation of $\bar{H}_c$, W_{nm}, and its connection to a measurable quantity, we choose the simplest ideal case, namely one spin 1/2 interacting with the lattice via a random magnetic field $H_x(t)$ along Ox. The corresponding interaction Hamiltonian H_c is:

$$H_c = -\hbar\gamma I_x H_x(t) \tag{175}$$

The eigenstates of the probe are those of the Zeeman Hamiltonian. In the present case, there are two such states, corresponding to $m = \pm 1/2$, which we label $|+\rangle$ and $|-\rangle$, separated by $\Delta = \hbar\gamma H_o = \hbar\omega_L$ (cf. section 5.1). There are two possible $\bar{H}_c$, namely $\langle +|H_c|-\rangle$ and $\langle -|H_c|+\rangle$ to which correspond two probabilities per unit time W_{+-} and W_{-+}. Clearly, both are related by Eq. (18), namely

$$W_{-+} = \exp(-\beta\hbar\omega_L)\, W_{+-} \tag{176}$$

Let us calculate for example W_{+-}. According to Chapter 2, we have successively

$$\bar{H}_c = \langle +|H_c|-\rangle = -\hbar\gamma\langle +|I_x|-\rangle H_x(t) = -\frac{\hbar\gamma}{2} H_x(t) \tag{177}$$

where the matrix element of I_x has been calculated using Eqs. (148) and (149), and consequently

$$W_{+-} = \frac{\pi\gamma^2}{2} C_{H_xH_x}(\omega_L) \tag{178}$$

where $C_{H_xH_x}(\omega_L)$ is the value of the Fourier transform of the auto correlation function of $H_x(t)$ taken at $\omega = \omega_L$. If we assume an exponential decay with a constant time τ_c, we have:

$$C_{H_xH_x}(t) = H_x^2 \exp(-t/\tau_c) \tag{179}$$

where H_x^2 is the mean square amplitude of $H_x(t)$. Inserting Eq. (179) into (178) we finally obtain

$$W_{+-} = \gamma^2 H_x^2 \frac{\tau_c}{1 + \omega_L^2\tau_c^2} \tag{180}$$

We should now relate W_{+-} and W_{-+} to a measurable quantity. This quantity is the longitudinal magnetization $M_Z(t)$. Let n_+ and n_- be the number of spins in states $|+\rangle$ and $|-\rangle$. This concept is meaningful since the spins are independent. The net longitudinal magnetization is:

$$M_Z(t) = \hbar\gamma \left[\frac{1}{2} n_+(t) - \frac{1}{2} n_-(t)\right] \tag{181}$$

In the linear approximation, the rate equations governing the evolution of n_+ and n_- are:

$$\left.\begin{aligned} \frac{dn_+}{dt} &= -W_{-+}n_+ + W_{+-}n_- \\ \frac{dn_-}{dt} &= +W_{-+}n_+ - W_{+-}n_- \end{aligned}\right\} \tag{182}$$

constrained to the condition that

$$n_+ + n_- = N = \text{constant} \tag{183}$$

The solution of this system is straightforward and we have:

$$M_Z(t) = A + B \exp\left(-\frac{t}{T_1}\right) \tag{184}$$

where

$$T_1^{-1} = W_{+-} + W_{-+} = \left[1 + \exp(-\beta\hbar\omega_L)\right] W_{+-} \tag{185}$$

and A and B are constants which depend on the initial and equilibrium conditions. It is seen that for this particularly simple case, the decay of the longitudinal magnetization is exponential and the corresponding constant time T_1, called the spin lattice relaxation time, is given in the high temperature approximation ($\beta\hbar\omega_L \ll 1$), and extreme narrowing regime ($\omega_L\tau_c \ll 1$), by

$$T_1^{-1} \approx 2W_{+-} = 2\gamma^2 H_x^2 \frac{\tau_c}{1+\omega_L^2\tau_c^2} \approx 2\gamma^2 H_x^2 \tau_c \tag{186}$$

For spins greater than 1, there are $2I + 1$ levels and consequently $2I$ independent linear equations such as Eqs. (182). The most general solution of such a system, and consequently the longitudinal magnetization, is a sum of exponentials. However, it turns out that in practice, the relaxation is often a single exponential.

Looking at Eqs. (178) and (186), it is seen that the measurement of T_1 simply yields the value of the spectral density of $C_{\bar{H}_C \bar{H}_C}(t)$ at the Larmor frequency. In order to obtain the total spectrum (and thus the model) we should in principle perform measurements as a function of ω_L (i.e. H_o). This is not easy in practice at least in a wide range, since we are dealing with resonance phenomena. Alternatively, if one assumes a model for the motion, one can extract the parameters of this model (e.g. the correlation time τ_c) from a few measurements only. These measurements can be repeated as a function of temperature, and quantities such as the activation energy for the motion can be deduced. In real liquids, the relaxation is caused by Hamiltonians such as H_d, H_Q, H_{SR}, etc. ... and the calculation of T_1^{-1} is more complicated than above. In what follows, we give the formulae for these three cases and illustrate them by results taken from the literature.

5.4.3 Relaxation via the dipole-dipole interaction.

In this case the system to be considered is two identical spins 1/2 in a static magnetic field, coupled to the lattice via H_d given by Eq. (154). The eigenstates of the Zeeman Hamiltonian are labelled $|++\rangle$, $|+-\rangle$, $|-+\rangle$ and $|--\rangle$, the signs referring to the value of m for each spin. The distances between the levels are $\hbar\omega_L$ and $2\hbar\omega_L$) (note that the levels $|+-\rangle$ and $|-+\rangle$ are degenerate). Inspection of Eqs. (154) shows that H_d has non-zero matrix elements between all these levels. Consequently, the relaxation time T_1 will involve values of spectral densities at ω_L and $2\omega_L$, and is necessarily given by a formula of the form [85]

$$T_{1d}^{-1} = \sum_m A_m J_m(\omega_L) + B_m J_m(2\omega_L) \qquad (187)$$

where the A_m and B_m are constants and the $J_m(\omega)$ are the spectral densities of the correlation function of $r_{ij}^{-3} Y_2^m(\theta_{ij}, \phi_{ij})$. This correlation function has two contributions, an intramolecular contribution where the spins i and j belong to the same molecule, and an intermolecular contribution. The intra-contribution depends on the rotational motions only (variation of θ and ϕ) while the inter- contribution depends both on rotational (variation of θ and ϕ) and translational (variation of r) motions. The problem is thus rather complicated for a polyatomic liquid. As an example of such a study, we mention the work on liquid trimethylamine $(CH_3)_3N$ [89]. The authors have performed a series of measurements of the proton spin lattice relaxation time versus temperature. They used solutions of $(CH_3)_3N$ diluted in

$(CD_3)_3N$ in order to separate the intra and inter contributions and measured separately the diffusion coefficient D_t using the spin-echo method. With these data, they calculated the theoretical intra and inter contributions to T_1 within the quasi lattice random flight model, but this model could account for about half of the total relaxation only. These results are summarized in Fig. 15.

In summary, relaxation via the dipole-dipole interaction can give information on both translational and rotational motions, but the separation between them is not simple, at least when the former contribute significantly, as is the case for protons. With C^{13} nuclei (I = 1/2), only the rotational motions contribute, but the measurements are difficult due to the very low natural abundance of this isotope [99].

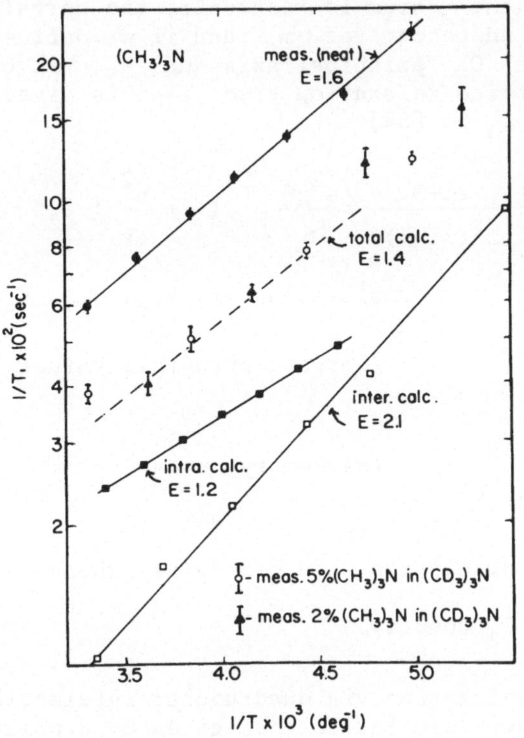

Figure 15: Proton spin-lattice relaxation rate T_1^{-1} of $(CH_3)_3N$ versus temperature (●). Also given are values of T_1^{-1} for protons in dilute solutions of $(CH_3)_3N$ in $(CD_3)_3N$ (Δ and O), and the calculated values for the inter- (□) and intra- (■) molecular contributions to T_1^{-1}. The sum of the calculated T_1^{-1} (intra) and T_1^{-1} (inter) is shown by the dashed line. The activation energies, E, are indicated near the lines. From ref. [89].

5.4.4 Relaxation via the quadrupole interaction.

In this case, the system to be considered is one spin $I > 1/2$ in a strong magnetic field coupled to the lattice via H_Q given by Eq. (156). The Zeeman Hamiltonian has $2I + 1$ levels. Inspection of eqs. (160) shows that H_Q has non-zero matrix elements between these levels. On the other hand, the electric field gradient V_{rs} is mainly produced by the electric charges within the molecule, and consequently the principal axes of this tensor are attached to the molecule. When the molecule rotates, these axes rotate also and this causes the relaxation. Relaxation via the quadrupolar interaction is thus adequate to study rotational motions. Let us call V_{xx}, V_{yy} and V_{zz} ($|V_{xx}| \leqslant |V_{yy}| \leqslant |V_{zz}|$) the principal values of V_{rs} in the molecular frame. Expressed in the laboratory frame ($\underset{\sim}{O}z \equiv \underset{\sim}{H}_o$), the new components V_{rs} will be a combination of these principal values weighted by the second order spherical harmonics $Y_2^m(\Omega)$ where Ω symbolizes the orientations of the molecular principal axes with respect to the laboratory frame. If the medium is isotropic, the correlation functions of the Y_2^m are independent of m, and if we define θ as the angle between the OZ principal axis and $\underset{\sim}{H}_o$, it can be shown that the spin lattice relaxation time T_{1Q} is given, in the extreme narrowing case, by [85]

$$T_{1Q}^{-1} = \frac{3}{40}\frac{2I+3}{I^2(2I-1)}\left(\frac{e^2qQ}{h}\right)^2\left(1+\frac{\eta^2}{3}\right)\tau_2 \tag{188}$$

where eq and η are combinations of V_{xx}, V_{yy}, V_{zz} such that

$$eq = V_{zz} \qquad \text{(largest principal value)}$$

$$\eta = \frac{V_{yy} - V_{xx}}{V_{zz}} \qquad \text{(asymmetry parameter)} \tag{189}$$

and τ_2 is the correlation time of $F_2(t)$ where

$$F_2(t) = \langle P_2(\cos\theta(t))\rangle \tag{190}$$

It is seen that relaxation via quadrupolar interaction gives information which is very similar to that given by depolarized Raman scattering of light (see section 4.4.3). Comparison between the results obtained by these two methods can thus be very fruitful. As an example of such a study, we mention that on liquid deuterated acetonitrile $CD_3\text{-}C \equiv N$ [91]. In this paper, the author analyzes, amongst other things, both Raman and deuterium spin lattice relaxation data of acetonitrile in terms of the anisotropic rotational diffusion model, which is characterized by two diffusion constants

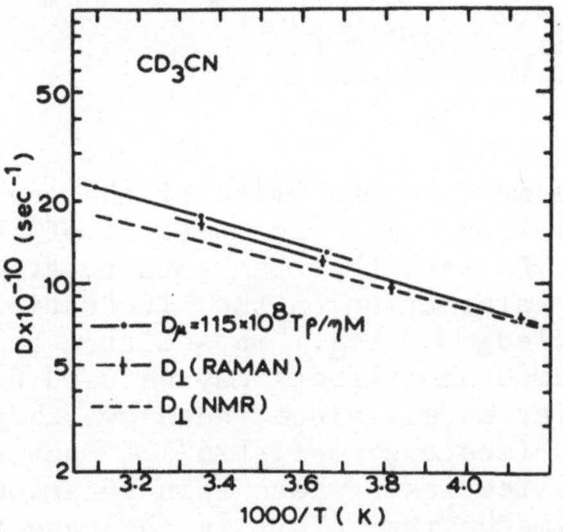

Figure 16: Variation of the perpendicular diffusion constant for reorientation $D_\perp$ of liquid CD_3CN, versus temperature, obtained from quadrupolar relaxation of ^2D (NMR), Raman scattering of light and hydrodynamic calculations. From ref. [91].

$D_{//}$ and $D_\perp$ (diffusion around the symmetry axis and diffusion of the symmetry axis). Fig. 16 summarizes the results for $D_\perp$. It is seen that both methods, as well as a calculation based on hydrodynamic theory, give results which are essentially in agreement with one another.

In conclusion, nuclear spin lattice relaxation via the electric quadrupole interaction is sensitive to rotational motions only and is a priori the best way, when possible, of obtaining the correlation time τ_2 of $F_2(t)$. However, a serious inconvenience of this method is that the quantities eq and η are not always known with sufficient accuracy for this purpose.

5.4.5 Relaxation via the spin-rotation interaction.

As in the dipolar case, in practice this relaxation mechanism is relevant mainly for spins 1/2. The structure of the Hamiltonian (Eqs. (161) and (162)) shows that H_{SR} has non-zero matrix elements between the Zeeman levels. On the other hand, the molecular angular momentum $\tilde{J}$ fluctuates due to the random collisions and this causes the relaxation. Relaxation via spin-rotation is thus an alternative way of studying rotational motions. For a symmetric top molecule, the spin rotation tensor is characterized by two independent components $C_{//}$ and $C_\perp$. If moreover the medium is isotropic, it can be shown that the spin lattice relaxation time T_{1SR} is, in the extreme narrowing case, given by [28]

$$T_{1SR}^{-1} = \frac{2Ik_BT}{\hbar^2} \langle J^2 \rangle \frac{2C_\perp^2 + C_{//}^2}{3} \tau_J \tag{191}$$

where $\langle J^2 \rangle$ is the mean square value of the angular momentum $\underset{\sim}{J}$ and τ_J its correlation time. For details of the calculation see e.g. ref. [94]. It is seen that the spin-rotation mechanism is unique in yielding information on the fluctuations of the angular momentum. The knowledge of τ_J, in addition to the knowledge of τ_2 (e.g. from a Raman experiment) may be used as a rather severe test of the molecular models since these two characteristic times are not independent (see e.g. ref. [26]). Such a study is possible however only in the few cases where spin-rotation is the main spin-lattice relaxation mechanism. This is the case in liquids composed of rather spherical molecules which rotate relatively freely, yielding a value for τ_J, and consequently of T_{1SR}^{-1}, which is not too small. As an example of such a study we mention that of liquid hexafluorides [93] where τ_J was determined by fluorine spin lattice relaxation and τ_2 by depolarized Raman scattering of light. Fig. 17 summarizes the results for a few hexafluorides (SF_6, SeF_6, MoF_6, WF_6) in a τ_2-τ_J plot. It is concluded that the results are fairly consistent with the J model of McClung [15].

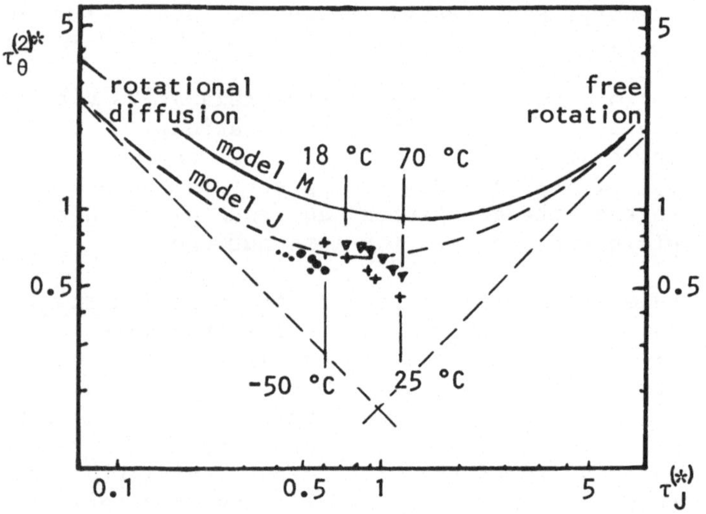

Figure 17: τ_2 - τ_J plot for a number of liquid hexafluorides. τ_2 is deduced from Raman experiments, and τ_J from fluorine spin lattice relaxation measurements. Theoretical curves are also shown. + SF_6, ● SeF_6, ∇ MoF_6, o WF_6. From ref. [93].

In conclusion, spin lattice relaxation via the spin-rotation mechanism is a very important source of information on rotational motions. However, it seems that for ^{13}C, it is relatively more efficient than for other spins [99].

5.6 Conclusion

In conclusion, it appears that although the nuclear magnetic absorption spectrum is not very useful, on the contrary, nuclear spin-lattice relaxation may be an interesting tool for studying the molecular motions in liquids. A problem is that the experiments are usually made at a fixed frequency ω_L and it is thus difficult to have the complete spectral density from T_1 measurements. However, in liquids we are always in the extreme narrowing case, and T_1^{-1} is always proportional to a correlation time. Measurements of T_1 as a function of temperature can thus easily give information on the activation energy of the relevant mechanism of motion. On the contrary, to test models of the motion, relaxation measurements on one nucleus only are generally not sufficient and one should combine results obtained by other methods (light, neutrons, relaxation of other nuclei). However, spin lattice relaxation is unique when (i) relaxation via spin rotation is efficient, and (ii) when the motions are too slow to be detected by the other methods. In this context, it should be noted that spin lattice relaxation and magnetic resonance absorption are practically the only methods for studying slow motions in condensed matter, especially in solids; but for liquids, the other methods are competitive. Finally, spin-spin relaxation measurements in the presence of a field gradient are, together with high resolution neutron scattering, the only way to measure translational self diffusion coefficients in pure liquids. For papers concerning spin lattice relaxation in molecular liquids, in addition to those already cited in this chapter, we mention refs. [95-98] taken from ref. [1].

CHAPTER 6

OTHER METHODS

In the preceding sections, we have described in some detail the main spectroscopic methods which are used to study molecular motions in liquids. There are however a few other ones, which are less popular for various reasons, but which may be very useful in some specific cases. These other methods are generally linked in some way to the main ones, but are sufficiently specific to merit a separate description. In this section, we present briefly these methods, their principles, their range of applicability and give the main references for detailed information on them.

6.1 Fluorescence and Resonant Raman Scattering

The principle of this method is the same as that of the usual Raman scattering method (section 4.4) except that the incident frequency is chosen to excite the molecules in specific rotational-vibrational levels of an excited electronic state. The light is then re-emitted according to the structure of these levels. This re-emission can occur in two ways: (i) one incident photon is absorbed and simultaneously another photon is emitted: this is the usual Raman process (resonant in this case), and (ii) one incident photon is absorbed and after a delay, one photon is emitted: this is the so-called resonance fluorescence. If the excited light is polarized and if molecular motions are present, it is easy to imagine that both the Raman lineshapes and the fluorescence depolarization phenomenon are functions of the rotational and vibrational properties of the molecules. The main advantage of this method is its very large senstivity due to the resonance phenomenon, compared to the usual Raman effect and it can thus be used for very dilute solutions of active molecules. The drawback is that it is limited to suitable molecules and to the existence of non radiative processes, which superimpose on the radiative ones, and greatly complicate the analysis. For details, we refer the reader to refs. [27],[100],[101] and [102].

6.2 Mössbauer Effect

6.2.1 General.

The Mössbauer effect is the following: in some conditions (e.g. when diluted in a crystal) some radioactive nuclei decay by emitting a γ ray of well defined energy. For example, radioactive cobalt 57, imbedded in a solid matrix of platinum or rhenium decays according to the following scheme:

$$^{57}Co \rightarrow {}^{57}Fe^{*} \rightarrow {}^{57}Fe + \gamma \quad (14.1\ keV) \qquad (192)$$

where the linewidth of the γ emission is about 10^{-3} μeV. This corresponds to an energy definition of $\Delta E/E \sim 10^{-13}$. These γ rays can thus be considered as a monochromatic source of electromagnetic waves of very short wavelength (~ 0.8 Å) which can be used for studying molecular motions in liquids. Two kinds of experiments can be made: absorption and scattering experiments.

6.2.2 Absorption experiments.

Absorption experiments may be useful for studying translational diffusion of suitable particles immersed in a liquid. For example, suppose we send γ rays from a ^{57}Co source onto a sample containing ^{57}Fe ions in solution. According to Eq. (192), the incident γ rays can be absorbed to give excited $^{57}Fe^{*}$ nuclei which, in turn, decay by emitting the same γ rays in any direction. A detector placed behind the sample will thus detect an absorption line. If the solution is frozen, the width of this absorption line will be the same as that of the incident line. If the solution is liquid, the ^{57}Fe particles move and the absorption line will be broader because the molecular motions cause a Doppler-shift of the re-emitted γ ray. This broadening is thus directly linked to the motion of the Fe particles. As an example of such a study, we mention that of colloidal cobaltous hydroxy stannate in solutions containing glycerol [103] where the authors were able to test models for self-diffusion of the colloidal particles. Clearly this method can be useful only if the broadening of the Mössbauer line is neither too small nor too large. This limits the values of measurable diffusion coefficients D_t between 10^{-9} and 10^{-7} cm^2/s a range which is typical for supercooled liquids.

6.2.3 Scattering experiments.

Like any electromagnetic waves, monochromatic γ rays can also be used for scattering experiments. However, as opposed to the infrared or light cases, the long wavelength approximation is no longer valid since λ is of the order of interatomic distances ($\lambda \sim 1$ Å). Consequently, the formalism of section 4.4 does not apply. In fact, this wavelength is comparable to that of thermal neutrons (cf. section 3.1), and therefore, one expects both methods to give the same kind of information. This is the case and it can be shown that the relevant correlation function for γ-ray scattering has the same general form as Eq. (39) for neutrons, except that the scattering lengths b_i are replaced by atomic form factors $f_i(\underset{\sim}{Q})$, namely:

$$\sigma(\underset{\sim}{Q},t) = \frac{1}{N} \sum_{i,j} f_i(\underset{\sim}{Q})\, f_j(\underset{\sim}{Q}) \langle \exp i\underset{\sim}{Q}[r_i(t) - r_j(o)] \rangle \qquad (193)$$

The complication related to the Q dependence of the f_i comes from the fact that for γ rays the wavelength is of the same order as the dimensions of the atomic electron clouds, while for neutrons, the comparison should be made with the dimensions of the nuclei. The other difference is that for γ rays, the scattering is purely coherent (since the electron clouds are the same for all the isotopes of the same atom) and thus the interpretation of a γ-ray scattering experiment on a molecular liquid will be at least as difficult as that of the corresponding (coherent) neutron scattering experiment. The advantage of the γ-ray scattering experiment is its much better energy resolution ($\sim 10^{-3}$ μeV compared to 2×10^{-1} μeV in the best case for neutrons), but with little possibility to worsen it. As for the absorption case, in practice this limits the use of this method to liquids with very slow molecular motions, such as supercooled liquids. The other inconvenience of the method is its low intensity and the fact that only high Q values (typically Q from 0.6 to 4 $Å^{-1}$) can be reached. As an example of such a study, we mention that on supercooled glycerol [104] from which some quantitative information about the centre of mass motions and rotational motions has been extracted and compared with results obtained by other methods.

6.3 Electron Paramagnetic Resonance (EPR) [86]

6.3.1 General.

This method applies to systems which exhibit electronic paramagnetism. It is very similar to the nuclear magnetic resonance (NMR), and all the concepts of chapter 5 apply here. However there are a few important practical differences:

(i) in NMR, the nuclei are well protected by their electron clouds and their properties (I, γ_I, eQ) are weakly sensitive to the actual state of the matter. In EPR, on the contrary, the electrons which are responsible for the paramagnetism, being generally in the outer shells, are very sensitive to their surroundings.

(ii) the nuclear magnetic moments (and consequently the typical frequencies ω_D, ω_L) are about 10^3 times smaller than for electrons ($\gamma_e = 660\ \gamma_{proton}$). Consequently, for EPR, the time scales corresponding to the three regimes of molecular motions (section 5.3.2) are shifted towards higher frequencies by a factor $\sim 10^3$.

The most common paramagnetic species which can be observed by EPR are the following: (i) transition metal ions with incomplete electronic shells, (ii) defects in solids (F centers ...), (iii) electrons in metals, (iv) free radicals .... Concerning our problem of molecular motions, the 'usual way to operate' is to dissolve in the liquid one of these species (usually a free radical), and study the EPR absorption spectrum. The corresponding line-shape contains the information on the motion of the radicals.

6.3.2 The relevant Hamiltonians and connection with molecular motions.

Similarly to nuclear spins, electronic spins couple to a static magnetic field (Zeeman interaction) and to other spins (dipolar interaction). However the form of the Hamiltonians is slightly different. For the Zeeman interaction, the external field couples not only with the electronic spin $\underset{\sim}{S}$, but also with the electronic orbital momentum $\underset{\sim}{L}$, which is itself coupled to $\underset{\sim}{S}$ via the spin-orbit interaction. Although the orbital momentum is quenched, second order perturbation theory involving the higher electronic states shows that these various couplings can be taken into account in the form of an anisotropic Zeeman Hamiltonian:

$$H_Z = \beta \underset{\sim}{H}_o \underset{\approx}{g} \underset{\sim}{S} \qquad (194)$$

where $\underset{\approx}{g}$ is the so-called g tensor. The isotropic part of $\underset{\approx}{g}$ defines the position of the EPR line and its anisotropic part $\Delta \underset{\approx}{g}$ is a source of electronic spin relaxation when the radicals rotate. For the dipolar interaction between an electronic spin and another spin (electronic or nuclear), the corresponding tensor $\underset{\approx}{D}$ (Eq.(154)) is generally no longer traceless, but has an isotropic part, of purely quantum origin, corresponding to the so-called contact interaction. The contact interaction between an electronic spin 1/2 and a nuclear spin I splits the EPR spectrum into 2I + 1 components (hyperfine splitting), each of these components being characterized by one value of the quantum number m between -I and +I.

Now, it can be shown that in certain regimes of molecular motions, and in dilute solutions, the width $\Delta H(m)$ of these components is a polynomial in m. For example for I = 1, we have

$$\Delta H(m) = \alpha + \beta m + \gamma m^2 \qquad (195)$$

where the coefficients α, β, γ are functions of the values of elements of the anisotropic parts of the tensors $\underset{\approx}{g}$ and $\underset{\approx}{D}$, and of the rotational correlation times τ_2 of $F_2(t) = \langle P_2(\cos\theta(t))\rangle$ where the θ are the angles between the tensors principal axes and the magnetic field. In practice, the values of τ_2 which can be measured with this method lie between 10^{-8} and 10^{-10} sec. Examples of EPR studies of molecular liquids are given in refs. [105] and [106].

The EPR method is thus sensitive to rotational motions (of the radical) and the information obtained is similar to that obtained by nuclear spin-lattice relaxation via quadrupole coupling and depolarized Raman scattering of light. However, as for the other methods, the accuracy of the value of τ_2 is strongly dependent on the accuracy of the values of the tensors elements. Finally, it is important to note that the EPR method is in fact more often used for the study of molecular order and dynamics in anisotropic systems (liquid crystals, biological media) [107] than in ordinary liquids.

6.4 Dynamic Nuclear Polarization (DNP)

The above section has shown that the EPR method gives information on the motions of the (dilute) radicals only, but not on the relative motions of the radicals and the host molecules. If these host molecules contain nuclear spins, the fluctuating dipolar (and eventually contact) interaction with the electronic spins can be used to obtain information on these relative motions. The principle is the following: consider one electronic spin $S = 1/2$ and one nuclear spin $I = 1/2$. The Zeeman Hamiltonian of this system has four levels $|\alpha\rangle$ whose distances are $\hbar\omega_L{}^S$, $\hbar\omega_L{}^I$ and $\hbar(\omega_L{}^S \pm \omega_L^I)$, with $\omega_L{}^S >> \omega_L{}^I$ (typically GHz and MHz, respectively). Suppose first that we detect the NMR signal of the spins I. Its amplitude I_o is proportional to $\hbar\omega_L{}^I/k_BT$ (cf. section 5.3). Suppose now that during this detection we send a strong magnetic field at the electronic Zeeman frequency $\omega_L{}^S$ (we pump at $\omega_L{}^S$. Then the electronic Zeeman levels tend to be equalized. But the interaction between $\underset{\sim}{I}$ and $\underset{\sim}{S}$, which has non-zero matrix elements between all the levels $|\alpha\rangle$ and, being in a liquid, fluctuate with a correlation time τ_c of the order of $(\omega_L{}^S)^{-1}$ of less, can induce transitions between these levels and tend to re-establish the Boltzmann distribution. A steady state regime is obtained where the various populations are different than at thermal equilibrium. The NMR signal amplitude I is thus different from that without pumping and it can be shown that the corresponding enhancement factor η is given by

$$\eta = \frac{I-I_o}{I_o} = \frac{\gamma_s}{\gamma_I} \, s \, f \, \xi \qquad (196)$$

where s is proportional to the pumping power at $\omega_L{}^S$, f is the leakage factor given by $(1 - T_1/T_{10}) - T_1$ and T_{10} are the nuclear spin lattice relaxation times with and without the radicals and ξ is a function of the probabilities of transition $W_{\alpha\beta}$ induced by the fluctuating interaction, and which contains the information on the molecular motions. This phenomenon is called dynamic nuclear polarization by the Overhauser effect. If the interaction is purely dipolar, then η is negative (the NMR signal is reversed) and it is sometimes called by extension the "Underhauser" effect.

Since the $W_{\alpha\beta}$ contain spectral densities of the motion at $\omega_L{}^S \pm \omega_L{}^I \approx \omega_L{}^S$ and since $(\omega_L{}^S)^{-1}$ is of the order of τ_c, experiments at various magnetic fields are useful for scanning the spectral densities in their interesting regions, i.e. around τ_c^{-1}. This has in fact been used, and in particular cases it could be deduced if the radicals are completely free in the liquid (then the spectral density is that of translational self-diffusion) or solvated by the host molecules (in which case the spectral density is rather that of rotational diffusion). Details on this method can be found in the review paper ref. [108]. Finally, it should be noted that the phenomenon of dynamic nuclear polarization in solids is of a different nature [85].

6.5 γ-γ Differential Perturbed Angular Correlation (DPAC)

This method is similar to EPR except that the probe used is radioactive rather than paramagnetic. The principle of angular correlation is as follows. Consider, for example, a radioactive nucleus of hafnium. It decays according to the scheme

$$^{181}\mathrm{Hf} \rightarrow {}^{181}\mathrm{Ta}^{*} \rightarrow {}^{181}\mathrm{Ta} + \gamma_1 + \gamma_2 \tag{197}$$

The de-excitation of the tantalum nucleus occurs by emission of two <u>successive</u> γ rays. Correlation means that if the γ_1 is emitted in one direction, the direction of the emitted γ_2 is not uniform in space because the spin state I of the intermediate level is determined by the direction of γ_1. If now $I > 1/2$ and if we are in condensed matter, the quadrupolar interaction can <u>perturb</u> this correlation.

The result is that if we detect the γ_2 in one direction at time t knowing that γ_1 has been detected at zero time in another direction at angle θ, then the corresponding probability $W(\theta,t)$ of detecting the γ_2 is no longer the simple law $\exp(-t/T_N)$ where T_N is the lifetime of the intermediate level. Two typical situations occur: (i) the solution is frozen: then $W(\theta,t) \propto \exp(-t/T_N)$ $F_f(\theta,t)$ where F_f is a periodic or pseudo periodic function of t whose period is defined by ω_Q where $\hbar\omega_Q$ is the distance between the levels of the quadrupolar Hamiltonian, and (ii) the solution is liquid with $\tau_2 \ll \omega_Q^{-1}$; in this case quadrupolar relaxation takes place (cf. section 5.4.4) and we have $W(\theta,t) \propto$ $\exp(-t/T_N) \exp(-\lambda_2 t)$ with $\lambda_2 \sim \omega_Q^2 \tau_2$. In practice, the values of τ_2 which can be measured by this method lie between 10^{-9} and 10^{-10} sec. It is clear that the method is limited to suitable nuclei such that T_N is not too short compared to λ_2^{-1}. Another limitation is that the electronics may not be sufficiently rapid. As an example of such a study, we mention that of a hafnium chelate in chloroform [109-110]. A general bibliography on these radioactive methods can be found in ref. [111].

6.6 Sound Velocity and Absorption

Last, but not least, we want to mention sound waves as a means of studying molecular motions in liquids. The probe here is different from those in all the other methods and corresponds to "phonons" rather than photons or neutrons. The velocity and attenuation of the sound waves are associated with the collisions between molecules. The local increase of translational energy at the peak of the wave is converted into molecular energy vibrations, rotations, electronic excitations ... Thus, the attenuation is related to molecular motions. However, other processes also dissipate energy: shear viscosity, heat conductivity ... and therefore extracting molecular information from such data is difficult. Since the frequencies used are in the same range as those of dielectric absorption, there is some analogy between the results obtained by these two methods. A useful review paper on this subject can be found in ref. [112].

7 ACKNOWLEDGEMENTS

The author is indebted to Drs. P. Averbuch, M. Besnard, C. Brot, F. Derrida, A.J. Dianoux, R.E. Lechner, S. Lovesey, C. Marti, A.F. Martins and R. Pynn for numerous and helpful discussions.

REFERENCES

[1] "Molecular Motions in Liquids", ed. J. Lascombe, D.Reidel Publishing Co. (1974).
[2] S. Chandrasekar, Rev. Mod. Phys. 15, 1 (1943).
[3] M.C. Wang and G.E. Uhlenbeck, Rev. Mod. Phys. 17, 323 (1954).
[4] D.W. Condiffand, J.S. Dahler, J. Chem. Phys. 44, 3988 (1966).
[5] F. Perrin, J. Phys. Radium 5, 497 (1934) ; 7, 1 (1936).
[6] W.H. Furry, Phys. Rev. 107, 7 (1957).
[7] L. Pauling, Phys. Rev. 36, 430 (1930).
[8] L.D. Favro, Phys. Rev. 119, 53 (1960).
[9] W.T. Huntress, Jr., J. Chem. Phys. 48, 3524 (1968).
[10] E.N. Ivanov, Sov. Phys. JETP, 18, 1041 (1964).
[11] W.A. Steele, J. Chem. Phys. 38, 2404 (1963).
[12] J.E. Anderson, J. Chem. Phys. 47, 4879 (1967).
[13] M. Fixman and K. Rider, J. Chem. Phys. 51, 2425 (1969).
[14] R.G. Gordon, J. Chem. Phys. 44, 1830 (1966).
[15] R.E.D. McClung, J. Chem. Phys. 51, 3842 (1969); 57, 5478 (1972); Adv. Molec. Relax. 10, 23 (1977).
[16] P.S. Hubbard, J. Chem. Phys. 52, 563 (1970).
[17] P.S. Hubbard, Phys. Rev. A, 6, 2421 (1972).
[18] D. Frenkel, G.H. Wegdam and J. Van der Elsken, J. Chem. Phys. 57, 2691 (1972).
[19] D.E. O'Reilly, J. Chem. Phys. 55, 2876 (1971).
[20] A.G. St. Pierre and W.A. Steele, J. Chem. Phys. 57,4638 (1972).
[21] D. Kivelson and T. Keyes, J. Chem. Phys. 57, 4599 (1972).
[22] R.I. Cukier, J. Chem. Phys. 60, 734 (1974).
[23] K.E. Larsson, J. Chem. Phys. 59, 4612 (1973).
[24] C. Thibaudier and F. Volino, Molec. Phys. 26, 1281 (1973); 30, 1159 (1975).
[25] C. Brot and B. Lassier-Govers, Ber. Ges. Phys. Chem. 80, 31 (1976).
[26] J.G. Powles and G. Rickayzen, Molec. Phys. 33, 1207 (1977).
[27] R.G. Gordon, Adv. in Mag. Resonance, 3, 1 (1968).
[28] C. Brot, in "Molecular Fluids", 37 (1976) ed. by J. Balian and G. Weill, Gordon and Breach.
[29] J.G. Powles, Ber. Bunsenges. Phys. Chem. 80, 259 (1976).
[30] W.A. Steele, in Advances in Chemical Physics, XXXIV, ed. by I. Prigogine and S.T. Rice, J. Wiley (1976).
[31] I.I. Gurevitch and L.V. Tarasov,"Low Energy Neutron Physics", North-Holland (1968).
[32] "Thermal Neutron Scattering", Academic Press (1965) ed. by P.A. Egelstaff.
[33] W. Marshall and S.W. Lovesey,"Theory of Neutron Scattering", Oxford University Press (1971).
[34] T. Springer, Springer Tracts in Modern Physics, 64 (1972); ref.[1]p. 411.
[35] T. Springer, Topics in Current Physics, 3, 255 (1977).

[36] F. Volino and A.J. Dianoux, Proceedings of the EUCHEM Conference on Liquids FRG (April 1976), J. Wiley (1978).
[37] A.J. Leadbetter and R.E. Lechner in "The Plastic Crystalline State",ed. J.N. Sherwood, J. Wiley (1978).
[38] V.F. Sears, Can. J. of Phys. 44, 1279 (1966); 44, 1299 (1966).
[39] A.J. Dianoux, F. Volino and H. Hervet, Molec. Phys. 30, 1181 (1975).
[40] A.J. Dianoux and F. Volino, Molec. Phys. 34, 1263 (1977).
[41] F. Volino and A.J. Dianoux, Molec. Phys. (1978), in print.
[42] A.J. Dianoux, H. Hervet and F. Volino, J. Physique, 38, 809 (1977) and references therein.
[43] A.J. Dianoux and F. Volino, Proceedings of the IAEA Conference Vienna (1977).
[44] H. Hervet, A.J. Dianoux, R.E. Lechner and F. Volino, J. Physique 37, 587 (1976).
[45] F. Volino. A.J. Dianoux and H. Hervet, J. Physique Colloques 37,C3-55(1976).
[46] M.E. Besnard, A.J. Dianoux, P. Lalanne and J.C. Lassègues, J. Physique, 38, 1417 (1977).
[47] M.E. Besnard, A.J. Dianoux, J. Lascombe, J.C. Lassègues and P. Lalanne, Proceedings of the IAEA Conference, Vienna (1977).
[48] Théorie Quantique Relativiste, L. Landau and E. Lifshitz IV, Chap. 1, Ed. Mir, Moscow.
[49] B.J. Berne and R. Pecora, "Dynamic Light Scattering" John Wiley (1976).
[50] R.H. Cole, ref. [1], p. 97.
[51] H. Kilp, G. Klages and W. Noerpel, ref. [1], p. 123.
[52] P. Desplanques, E. Constant, R. Fauquembergue, ref.[1], p.133.
[53] J.L. Greffe,J. Goulon, J. Brondeau, J.L.Rivail, ref.[1],p.151.
[54] J. Goulon; J.L. Rivail, J. Chamberlain and G.W. Cantry, ref. [1], p. 163.
[55] R.S. Wilson, ref. [1], p. 172.
[56] J.P. Badiali, H. Cachet, A. Cyrot and J.C. Lestrade, ref. [1], p. 179.
[57] "Infrared Physics", J.T. Houghton and S.D. Smit, Oxford Clarendon Press (1966).
[58] S. Bratos, Y. Guissani and J.C. Leicknam, ref. [1], p. 187.
[59] J. Lascombe, M. Besnard, P.B. Caloine, J. Devaure and M. Perrot, ref. [1], p. 197.
[60] P.C. Van Woerkom, J. de Bleyser and J.C. Leyte, ref. [1], p. 233.
[61] P.J. Perchard, C. Perchard and D. Legay, ref. [1], p. 235.
[62] W.G. Rotschild, ref. [1], p. 247.
[63] J. Vincent-Geisse and C. Dreyfus, ref. [1], p. 294.
[64] J. Soussen-Jacob, J. Vincent-Geisse, C. Alliott, A.M. Bize, J.C. Briquet,E. Dervil, J. Loisel and J.P. Pinan-Lucarre, ref. [1], p. 301.
[65] P. Dorval and P. Saumagne, ref. [1], p. 319.

[66] S. Bratos, J. Rios and Y. Guissani, J. Chem. Phys. 52, 439 (1970).
[67] M. Davies, ref. [1], p. 615.
[68] K. Müller, P. Etique and F. Kneubühl, ref. [1], p. 265.
[69] I. Laulicht and S. Meirman, ref. [1], p. 213.
[70] R. Arndt, R. Moorman and A. Schäffer, ref. [1], p. 217.
[71] G. Döge, ref. [1], p. 225.
[72] S. Sundler and R.E.D. McClung, ref. [1], p. 273.
[73] M. Gilbert and M. Drifford, ref. [1], p. 279.
[74] N.I. Rezaev, ref. [1], p. 309.
[75] J.E. Griffiths, ref. [1], p. 327.
[76] S. Bratos and E. Maréchal, Phys. Rev. A 4, 1078 (1971).
[77] P. Lallemand, ref. [1], p. 517.
[78] J.P. Munch and S. Candau, ref. [1], p. 535.
[79] N.D. Gershon and I. Oppenheim, ref. [1], p. 553.
[80] C. Demoulin, C.J. Montrose and N. Ostrowsky, ref. [1], p. 575.
[81] J.V. Champion and D.A. Jackson, ref. [1], p. 585.
[82] J. Dill and T.A. Litovitz, ref. [1], p. 605.
[83] P.A. Fleury and J.P. Boon, Advances in Chemical Physics (Prigogine and Rice) XXIV, 1, 1973.
[84] W.M. Gelbart, Advances in Chemical Physics (Prigogine and Rice) XXVI, 1 (1974).
[85] A. Abragam, "The Principles of Magnetic Resonance", Oxford University Press (1960).
[86] C.P. Schlichter, Principles of Magnetic Resonance, Harper and Row (1963).
[87] H.G. Hertz, ref. [1], p. 337.
[88] A.J. Dianoux, A. Heidemann, F. Volino and H. Hervet, Molec. Phys. 32, 1521 (1976).
[89] A. Loewenstein, E. Glaser and R. Ader, ref. [1], p. 403.
[90] W.T. Huntress, Jr., Advances in Magnetic Resonance, 4, 2 (1970).
[91] J.E. Griffith, ref. [1], p. 327.
[92] P. Rigny and J. Virlet, J. Chem. Phys. 47, 4645 (1967).
[93] M. Gilbert and M. Drifford, ref. [1], p. 279.
[94] P.S. Hubbard, Phys. Rev. 131, 1155 (1963).
[95] J. Jonas, J. Dezwaan and J.H. Campbell, ref. [1], p. 359.
[96] R. Eckert, G. Loos and H. Sillescu, ref. [1], p. 385.
[97] R. Mills, ref. [1], p. 391.
[98] M.D. Zeidler, ref. [1], p. 421.
[99] E. Von Goldammer, H.D. Ludemann and A. Müller, J. Chem. Phys. 60, 4590 (1974).
[100] C.N.R. Rao. "UV and Visible Spectra", Butterwords (1967).
[101] T.J. Chuang and K.B. Eisenthal, J. Chem. Phys. 57, 5094 (1972).
[102] P.A. Madden and H. Wennerström, Molec. Phys. 31, 1103 (1976).
[103] K.P. Singh and J.G. Mullen, Phys. Rev. A, 6, 2354 (1972).
[104] M. Soltwisch, M. Elwenspoek and D. Quitmann, Molec. Phys. 34, 33 (1977).

[105] J.H. Wang, D. Kivelson and W. Plachy, J. Chem. Phys. 58, 1753 (1973).
[106] G. Martini, M. Romanelli and L. Burlamacchi, ref. [1], p. 371.
[107] "Spin Labeling : Theory and Applications". Ed. L.J. Berliner, Academic Press (1976).
[108] K.H. Hausser and D. Stehlik, Advances in Magnetic Resonance 3, 79 (1968).
[109] P. Boyer, A. Tissier, J.I. Vargas and P. Vulliet, Chem. Phys. Lett. 14, 601 (1972).
[110] P. Boyer and P. Vulliet, Proceedings of the EUCHEM Conference on Liquids, FRG April (1976) - J. Wiley (1978).
[111] R.M. Stephen and H. Fauenfelder, "Alpha, beta and gamma spectroscopy". North-Holland (1965). Chap. XIX, ed. Sieghbahn.
[112] J. Lamb, ref. [1], p. 29.

THE STRUCTURE OF IONIC LIQUIDS

J.E. Enderby

H.H. Wills Physics Laboratory, University of Bristol
Tyndall Avenue, Bristol BS8 1TL, England

1. THE BASIC STRUCTURE PROBLEM

Since we are dealing with liquids which though electrically neutral overall, are made up of ions, such systems must contain at least two ion types or components. The simplest molten salt contains two components while the simplest electrolyte solution contains four.

Let us label each component by the dummy suffices α and β which may take values 1,2, ..., j for a liquid containing j ion-types. The atomic concentration of the α ion-type is denoted by c_α and is subject to the sum rule

$$\sum_\alpha c_\alpha = 1 \tag{1}$$

If any type of radiation is incident on a mixed assembly of ions, a measure of the amplitude of the scattered waves is given by

$$\sum_\alpha f_\alpha \sum_{i(\alpha)} \exp\left(-i\, \underset{\sim}{k}.\underset{\sim}{r}_{i(\alpha)}\right) \tag{2}$$

where f_α is an appropriate scattering factor and $\underset{\sim}{r}_{i(\alpha)}$ denotes the position of the ith molecule of α-type. The mean intensity, which we denote $\frac{d\sigma}{d\Omega}$, becomes

$$\sum_\alpha \sum_\beta f_\alpha f_\beta^* \overline{\sum_{i(\alpha)} \sum_{j(\beta)} \exp\left(i\, \underset{\sim}{k}.(\underset{\sim}{r}_j(\beta) - \underset{\sim}{r}_i(\alpha))\right)} \tag{3}$$

$$=\sum_{\alpha}\sum_{\beta} f_{\alpha}f_{\beta}^{*}Nc_{\alpha}\ (\delta_{\alpha\beta} + \overline{\sum_{i \neq j} \cos \underset{\sim}{k}.\underset{\sim}{R}_{ji}}) \qquad (4)$$

where N is the total number of ions in the sample.

Let us now introduce partial structure factors $S_{\alpha\beta}(k)$ defined by

$$S_{\alpha\beta}(k) = 1 + \frac{4\pi N}{k\,V}\int_{0}^{\infty} dr(g_{\alpha\beta}\ (r) - 1)\ r \sin kr \qquad (5)$$

where V is the volume of the sample. In (5) $g_{\alpha\beta}(r)$ is the pair distribution function which measures the average distribution of type β ion observed from an α atom at the origin. $g_{\alpha\beta}(r)$ tends, like the analogous quantity for a one-component system, to unity at large values of r. If there is an α ion at r = o, the probability of finding a β ion at the same instant with its centre in a small element of volume $d\underset{\sim}{r}$ is simply

$$\frac{Nc_{\alpha}}{V}\ g_{\alpha\beta}(r)\ d\underset{\sim}{r} \qquad (6)$$

In terms of $S_{\alpha\beta}$, we can rewrite (4) as

$$\frac{d\sigma}{d\Omega} = \sum_{\alpha} Nc_{\alpha}\ f_{\alpha}f_{\alpha}^{*} + \sum_{\alpha}\sum_{\beta} Nc_{\alpha}c_{\beta}f_{\alpha}f_{\beta}^{*}\ (S_{\alpha\beta} - 1) \qquad (7)$$

and for those cases where f is real, more simply as

$$\frac{d\sigma}{d\Omega} = N[\sum c_{\alpha}f_{\alpha}^{2} + F(k)] \qquad (8)$$

$$\text{where } F(k) = \sum_{\alpha}\sum_{\beta} c_{\alpha}c_{\beta}f_{\alpha}f_{\beta}(S_{\alpha\beta} - 1) \qquad (9)$$

If we perform scattering experiments on ionic systems, the quantity which can be extracted - and even then not directly - is F(k). In practice, we observe an intensity I of either neutron, X-rays of

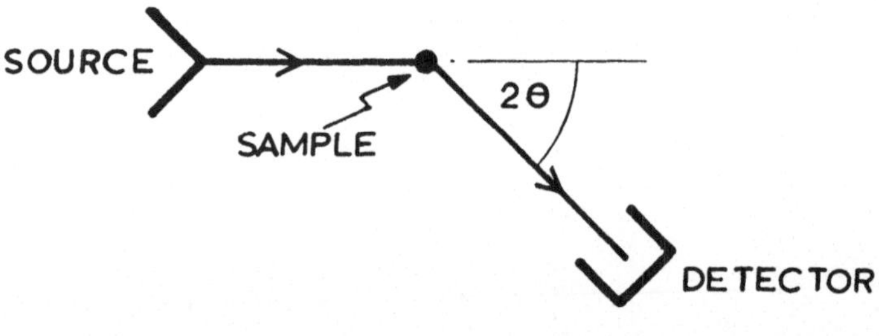

Fig. 1 Conventional lay-out for diffraction experiment.

electrons as a function of a scattering angle Θ in the geometry shown in figure 1.

Quite generally,

$$I(\Theta) = \alpha(\Theta)\left[\frac{d\sigma}{d\Omega} + \delta(\Theta)\right] \qquad (10)$$

and the challenge which faces experimentalists is the determination by theory, experiment or both, the calibration parameters $\alpha(\Theta)$ and $\delta(\Theta)$.

For the moment, let us suppose that α and δ have been satisfactorily determined - a point we shall return to for the neutron case in more detail in 3. We must now ask what information is to be gained from a knowledge of F(k). To do this, we shall review and comment on selected X-ray studies made over the past twenty or so years.

2. X-RAY STUDIES OF THE TOTAL SCATTERING ($F_X(k)$)

2.1. Introduction

X-ray diffraction arises from the scattering of X-rays by all the electrons associated with a given ion. Thus f(k), which measures the scattering from an isolated atom, is a product of the so-called 'form factor' and the 'polarisation correction'. The essential point is that f(k) depends strongly on k and in particular falls to very low values at high k. If we consider the data expressed in r-space, a total $G_X(r)$ can be obtained through

$$G_X(r) = \frac{1}{2\pi^2\rho r}\int F_X(k)\sin kr\, dk \qquad (11)$$

Thus because $F_X(k)$ contains both $S_{\alpha\beta}(k)$ and $f_\alpha(k)$, $G_X(r)$ is not a linear combination $g_{\alpha\beta}(r)$. For this reason, some authors prefer to divide F(k) by

$$M(k) = [\sum_\alpha c_\alpha f_\alpha(k)]^2 \qquad (12)$$

on the grounds that $f_\alpha(k)\, f_\beta(k)\, M(k)^{-1}$ will be more or less independent of k. The modified distribution function obtained in this way is related to $g_{\alpha\beta}(r)$ through a convolution function thus:

$$G_X^M(r) = \sum_\alpha\sum_\beta G_{\alpha\beta}(r) \qquad (13)$$

$$\text{with } G_{\alpha\beta} = \frac{1}{r}\int_{-\infty}^{\infty} u\, g_{\alpha\beta}(u)\, T_{\alpha\beta}(u-r)\, du \qquad (14)$$

$$\text{where } T_{\alpha\beta}(r) = \frac{1}{M}\int_0^{\infty} c_\alpha f_\alpha c_\beta f_\beta M(k)^{-1}\cos(kr)\, dk \qquad (15)$$

2.2 Molten Salts

For simple salts consisting of two types of ion α,β = 1 or 2 so that there are three pair correlation functions corresponding to cation-cation, anion-anion and cation-anion distributions. Even so, the amount of real structural information which have been obtained from X-ray studies carried out so far is most definitely limited. We shall see later on why this is so but meanwhile we refer to the review by Rhodes (1972) and give some examples of apparent distances and coordination numbers (CN) for the first shell. The numbers in brackets indicate the solid state value.

Liquid Salt	r_{+-} Å	CN
LiCl	2·47 (2·66)	4·0 (6)
KCl	3·10 (3·26)	3·7 (6)
NaI	3·15 (3·35)	4·0 (6)

2.3 Aqueous Solutions

The simplest salt solution contains ten partial structure factors. This makes for formidable difficulties in interpretation but in spite of this fact there have been several X-ray studies of aqueous solutions reported in the literature (Brady 1958; Lawrence & Kruh 1967; Wertz & Kruh 1969; Licheri, Piccalgua & Pinna 1975, 1976; Narten, Vaslow & Levy 1973). The advantage of the X-ray method is that there is no need to make the (difficult) Placzek corrections to the experimental data. Thus $d\sigma/d\Omega$ is much more readily related to the intensity data than is the case for neutrons. A major disadvantage is, however, that $d\sigma/d\Omega$ contains all ten partial structure factors and there is no clear way to separate them out. Accordingly, rather qualitative conclusions - some of them not altogether convincing - have, in the past, tended to be drawn from the data. Most of these have been concerned with what can be loosely referred to as hydration phenomena.

Wertz & Kruh (1969), for example, considered a 3.75M solution of $CoCl_2$ in H_2O. They evaluated the Fourier transform (equation 11) of $F_X(k)$ to give an average radial distribution function $G_X(r)$. The first peak in $G_X(r)$ at 2.1 Å was identified as due entirely to Co-O correlations. If this statement is accepted as its face value, it implies that Co II is, on the average, octahedrally coordinated by six waters. Cristini et al. (1974) investigated a $(Cr(H_2O)_6)Cl_3$ aqueous solution at 0.25 M. They also considered $G_X(r)$ and concluded that a peak at 1.90 Å could be identified with the Cr^{3+} -H_2O distance. The weakness of approaches based on the Fourier transform of $F_X(k)$ is that it is impossible to identify unambiguously all the structural combinations present in

$G_X(r)$. This fact has been recognized by Narten et al. (1973) who combined X-ray and neutron measurements on solutions of LiCl in D_2O to clarify the nature of hydration around the Li and the Cl ions.

A systematic study of a variety of electrolytes is at present underway in the group based at Cagliari (see for example Cristini et al (1974)). A useful survey of the results available to date has been given by Soper (1977). The general view which has emerged from these studies is that cations are definitely hydrated by 4 - 6 water molecules; the nature of the hydration around the anions is not so clear. Hydration numbers around six are usually quoted for the Cl^- ion, for example, with Cl-O distances around 3·20 A. The detailed conformation of the ion-water system is not known either for cations or anions except for those cases where the local structure of the hydrated salt can be expected to be retained in the liquid.

3. NEUTRON STUDIES OF THE TOTAL SCATTERING ($F_N(k)$)

3.1 Introduction

Neutron diffraction arises from the scattering of neutrons by atomic nuclei; the f-factor (usually referred to as the scattering length) does not depend on k. It does, however, vary with the isotopic state of the nucleus as the table below shows. A list of recommended scattering lengths has been published by the Neutron Diffraction Commission (1969).

Examples of Coherent Scattering Lengths (10^{-12} cm)

Element or Isotope	f	Element or Isotope	f
H	-0·372	Fe	0·951
D	0·670	^{54}Fe	0·42
^{6}Li	0·18	^{56}Fe	1·01
^{7}Li	0·21	^{57}Fe	0·23
^{35}Cl	1·18		
^{37}Cl	0·26	Ni	1·03
		^{58}Ni	1·44
Ca	0·49	^{60}Ni	0·282
^{40}Ca	0·49	^{62}Ni	-0·87
^{44}Ca	0·18	^{64}Ni	-0·037

It is necessary for the neutron case to distinguish between coherent and incoherent scattering. If all the nuclei in the sample are identical and have zero spin the scattering is said to be coherent. If the scattering length varies in a random way from nucleus to nucleus, there will be a structure independent contribution to the scattering known as incoherent. The origin of this fluctuation in scattering length can either be isotopic or arise from nuclear spin.

For simplicity we consider an element in which there are two stable isotopes. Let x_1 and x_2 be the concentration of the isotopes with scattering length f_1 and f_2. The coherent scattering length is $f_{coh} = x_1 f_1 + x_2 f_2$ and defines a coherent cross section through

$$\sigma_{coh} = 4\Pi(x_1 f_1 + x_2 f_2)^2 \tag{16}$$

The total cross section is $4\Pi[x_1 f_1^2 + x_2 f_2^2]$ so that the incoherent cross section σ_{inc} is simply

$$\sigma_{inc} = 4\Pi[x_1 f_1^2 + x_2 f_2^2] - 4\Pi[x_1 f_1 + x_2 f_2]^2 \tag{17}$$

$$= 4\Pi x_1 x_2 (f_1 - f_2)^2$$

If the nucleus has a spin I, the interaction with a neutron may give rise to a compound nucleus$(I \pm \tfrac{1}{2})$ each with a characteristic scattering length b_+ and b_-. By direct analogy with (16) and (17) we obtain σ_{coh} and σ_{inc} by noting that

$$x_1 \rightarrow (I + 1)/(2I + 1)$$

$$x_2 \rightarrow I/(2I + 1)$$

$$b_1 \rightarrow b_+$$

$$b_2 \rightarrow b_-$$

Another important distinction between X-ray work and neutron work involves the so-called Placzek corrections (Powles, 1973). These arise because when a neutron interacts with a nucleus, a transfer of both momentum and energy takes place. Thus under normal experimental conditions in diffraction studies, an effective rather than a true cross-section is measured. For elements whose nuclear mass is ~ 10 neutron masses, the correction terms are reasonably well understood. However, for aqueous solutions the presence of D (or H) gives rise to correction terms which so far have proved to be incalculable.

3.2 Molten Salts

We refer again to the review by Rhodes (1972); the reliability of the coordination numbers is not high because $F_N(k)$ contains three partial structure factors although the total real space function for neutron $G_N(r)$ is now a linear combination of the form

$$G_N(r) = c_1{}^2 f_1{}^2 g_{11} + c_2{}^2 f_2{}^2 g_{22} + 2c_1 c_1 f_1 f_2 g_{12}$$

Examples of reported distances and coordination numbers

Liquid Salt	r_{+-}(A)	CN
LiCl	2·45 (2·66)	3·5 (6)
KCl	3·53 (3·26)	3·5 (6)
CsBr	3·55 (3·86)	4·7 (8)

3.3 Aqueous Solutions

Apart from the study of LiCl in D_2O by Narten et al(1973)referred to earlier (section 2.3)and our own work, essentially no other data exists for $F_N(k)$

4. THE METHOD OF ISOTOPIC ENRICHMENT AS APPLIED TO MOLTEN SALTS

4.1 Introduction

As we have already seen the quantity which can be extracted from a single diffraction experiment on a liquid containing two species,a and b (Enderby et al 1966), is the total structure factor F(k)*defined by

$$F(k) = c_a^2 f_a^2 (S_{aa} - 1) + c_b^2 f_b^2 (S_{bb} - 1) + 2c_a c_b f_a f_b (S_{ab} - 1), \tag{18}$$

It is now generally accepted that the aim of structural analyses on two-component systems should be to extract S_{ab} rather than F(k). This can be done by varying f_a or f_b to give values for F_1, F_2 and F_3 which are sufficiently different to enable three linear equations to be solved for S_{aa}, S_{bb} and S_{ab}. In the experiments reported below the variation in f was achieved by isotopic substitution.

4.2 Molten NaCl

For the purpose of exposition we shall first consider in detail the work of Edwards, Enderby, Howe and Page (1975)on molten NaCl; we shall then review the current situation with regard to other molten salts in the light of this study.

*From now on we shall drop the subscript N

Total structure factors F(k) for $Na^{35}Cl$ $Na^{37}Cl$ and $Na^{N}Cl$* shown in (figure 2) were determined by the methods described by North et al (1968).

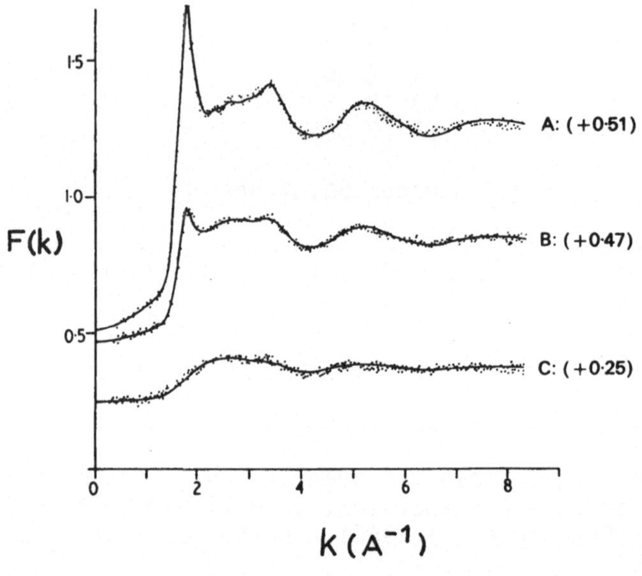

Fig. 2 $F(k) + c_a f_a^2 + c_b f_b^2$ for liquid $Na^{35}Cl$ (curve A) NaCl (curve B) and $Na^{35}Cl$ (curve C)

Edwards et al then considered the numerical method required to extract $S_{ab}(k)$.

The equations to be solved were written in the matrix form:

$$[A].[X(k)] = [F(k)] \tag{19}$$

* A mixture of ^{35}Cl and ^{37}Cl was used by Edwards et al (1975)

where

$$[A] = \begin{bmatrix} c_a^2 f_a^2 & c_b^2 f_b^2 & 2c_a c_b f_a f_b \\ c_a^2 (f_a')^2 & c_b^2 f_b^2 & 2c_a c_b (f_a') f_b \\ c_a^2 (f_a'')^2 & c_b^2 f_b^2 & 2c_a c_b (f_a'') f_b \end{bmatrix} = \begin{bmatrix} a_{11} & a_{12} & a_{13} \\ a_{21} & a_{22} & a_{23} \\ a_{31} & a_{32} & a_{33} \end{bmatrix}$$

$$[X] = \begin{bmatrix} (S_{aa} - 1) \\ (S_{bb} - 1) \\ (S_{ab} - 1) \end{bmatrix} = \begin{bmatrix} X_1 \\ X_2 \\ X_3 \end{bmatrix}$$

and

$$[F] = \begin{bmatrix} F_1 \\ F_2 \\ F_3 \end{bmatrix}$$

In this notation, f_a, f_a' and f_a'' represent the coherent neutron scattering lengths of the chlorine nucleus appropriate to the degree of enrichment, f_b is the coherent neutron scattering length for sodium, c_a and c_b are both 0·5 and S_{aa}, S_{bb} and S_{ab} represent the partial structure factors for Cl-Cl, Na-Na and Na-Cl correlations, respectively.

The formal solution of (19) is

$$[X] = [A]^{-1}[F]$$

and is unique in principle because the determinant of [A] is different from zero. However, it is essential to consider the conditioning of (19) since relatively unimportant experimental errors in [F] may produce considerable uncertainties in [X]. Although ill-conditioning occurs when [A] is nearly singular, the use of |A| as a measure of condition is not valid, since the equations can be multiplied by a constant to obtain any value of |A|. When the equations are normalized by dividing the ith row by

$$\left(\sum_{j=1}^{3} a_{ij}^2\right)^{1/2},$$

the normalized determinant, $|A|_n$, may be used as a test for ill-conditioning (Westlake 1968). Well conditioned equations yield $|A|_n$ of order ± 1. By the choice of a ^{37}Cl-^{35}Cl mixture rather than natural chlorine, Edwards et al achieved $|A|_n$ =-0·03, which, though sufficiently high to allow the algorithm described below to converge, does not enable a direct solution of equation (19) to be undertaken.

The algorithm developed for the extraction of S_{ab} was as follows:

(i) Solutions were found for the equation

$$[x] = [A]^{-1}[f] \tag{20}$$

where [f] is the experimental error in [F] and [X] is the corresponding uncertainty in [X]. The components of [f] were allowed to vary between experimentally defined limits subject to the condition that no component of $[X \pm x]$ can fall outside a defined range, or violate the conditions:

$$c_a + c_a^2(S_{aa} - 1) > 0,$$

$$c_b + c_b^2(S_{bb} - 1) - \frac{c_a^2 c_b^2 (S_{ab} - 1)^2}{c_a + c_a^2(S_{aa} - 1)} > 0$$

(see Enderby et al 1966). This gave S_{aa}, S_{bb}, S_{ab} each with a characteristic but non-independent error band.

(ii) A smooth curve was then drawn through the S_{aa} band such that the sum rule (Enderby et al 1966)

$$\int (S_{aa} - 1)\, k^2 dk = -\, 2\pi^2 n \tag{21}$$

is satisfied. Values of S_{aa} derived from this curve were taken as fixed. The numerical problem is then to solve a pair of simultaneous equations in which there are only two unknowns, S_{ab} and S_{bb}. When use was made of F(k) for $Na^{35}Cl$ and $Na^{37}Cl$, $|A|_n$ = 0·33 and the uncertainty in S_{ab} was only $\pm$ 0·2. It was therefore possible to test whether an S_{ab} can be found which satisfies, as it must, the same sum rule as S_{aa}.

(iii) If it did, S_{bb} was evaluated. If it did not S_{aa} was adjusted and the process started again at step (ii).

The final S_{aa}, S_{ab} and S_{bb} obtained by Edwards et al are shown in figure 3. Numerical inversion of the broken curves through S_{ab} yield the two-body radial distribution functions shown in figure 4.

The experimental work reported on molten NaCl enables some definite conclusions to be drawn about the structure of molten NaCl and to make some general comparisons with molecular dynamics.

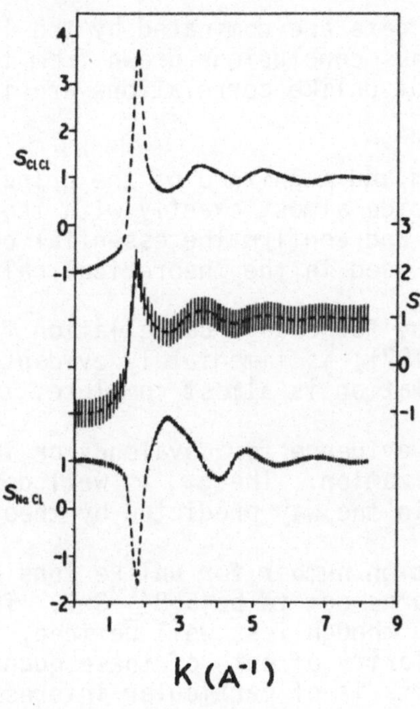

Fig. 3 The partial structure factors for liquid NaCl

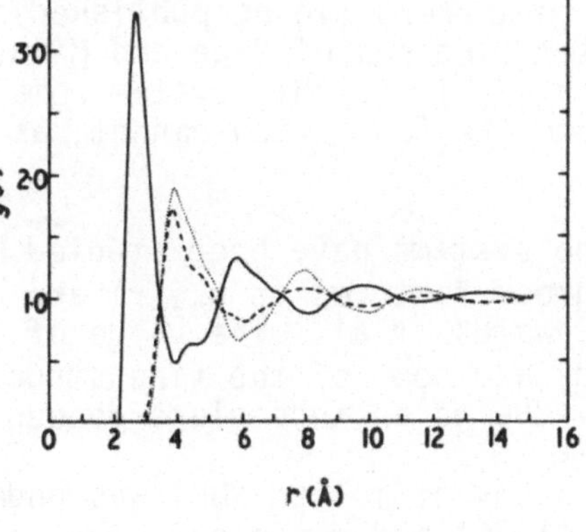

Fig. 4 The radial distribution functions for liquid NaCl
—— $g_{+-}(r)$ ...... $g_{--}(r)$ ---- $g_{++}(r)$

(i) The k-space data are dominated by the like partial structure factors. Thus conclusions drawn from total diffraction data about the nature of unlike correlations are invalid (Levy et al 1960).

(ii) The position and magnitude of the principal peaks in g_{++}, g_{+-} and g_{--} coincide almost exactly with those predicted by Lantelme et al (1974) and confirm the essential correctness of the interionic potentials used in the theoretical calculations.

(iii) The tendency for charge cancellation first noted by Woodcock and Singer (1971) is immediately evident from our data. Beyond 5A, the cancellation is almost complete.

(iv) There is no evidence for covalency or indeed for effects associated with polarization. The g_{+-} is well defined and is not reduced in magnitude in the way predicted by theory.

(v) The coordination number for unlike ions can be determined with confidence and turns out to be $5{\cdot}8 \pm 0{\cdot}1$. The coordination number for like atoms, though less well defined, turns out to be $13{\cdot}0 \pm 0{\cdot}5$. The similarity of both of these quantities with their solid-state counterparts is of particular interest.

4.3 Other systems

(a) CuCl In a pioneering study, Page and Mika (1975) investigated CuCl by changing both the Cl and the Cu isotope. The g_{++} g_{+-} and g_{++} found in this work (fig 5) are quite different from those expected for simple ionic potentials. Attempts to explain the comparative absence of structure in $g_{++}(r)$ by supposing that CuCl is predominately molecular (Powles 1975) are not consistent with other experimental data (Boyce and Mikkelson 1977). A new study by Dupuy and co-workers (to be published) has essentially confirmed the earlier findings of Page and Mika. Understanding the particular form of $g_{\alpha\beta}$ (r) for this system remains a challenge; the possibility of incomplete ionisation cannot, as this stage, be ruled out.

b) KCl and CsCl The systems have been studied by Derrien and Dupuy (1975); the gross features in $g_{\alpha\beta}(r)$ are consistent with these reported by Edwards et al. The value of $|A|_n$ used in the separation was 0·009 and some of the fine structure found by these workers is probably due to an incomplete decoupling of $S_{\alpha\beta}(k)$.

c) RbCl A careful study of molten RbCl was undertaken by Mitchell, Poncet and Stewart (1976) and represents the best structural study of a molten salt made so far. The sample was held in vanadium and so the need to correct the observed data for the coherent scattering

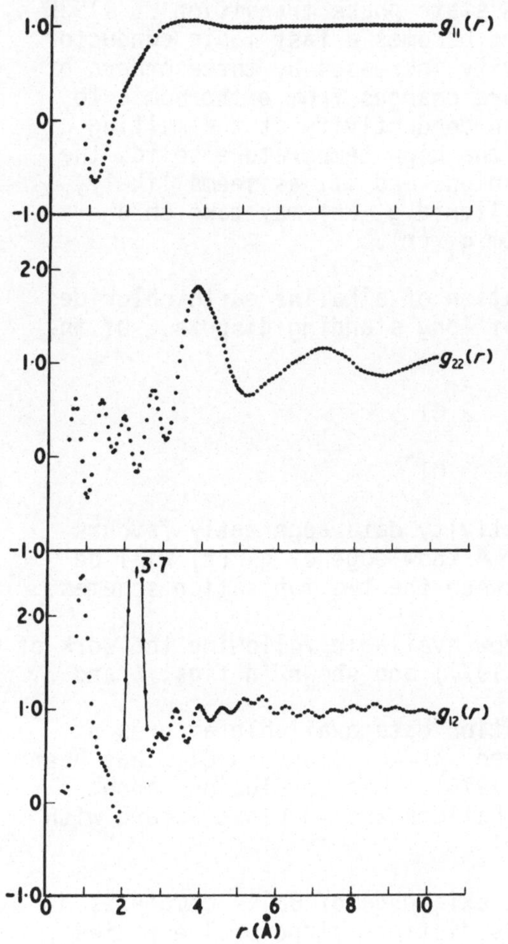

Fig. 5 The radial distribution functions for liquid CnCl (Page and Mika; 1975) 1 ≡ Cu; 2 ≡ Cl

from the container was avoided. The general conclusions of this study are in excellent agreement with those of Edwards et al.

d) $BaCl_2$ Molten $BaCl_2$ is of interest from many points of view. First as a 2 - 1 system it represents a class of molten salt for which little is known experimentally about g_{++}, g_{+-} and g_{--} ; in this connection, it should be noted that computer simulation studies of ionic metals have been restricted to 1 - 1 systems except for the cases of $SrCl_2$ and CaF_2 to which we shall refer in due course.

Secondly, $BaCl_2$ undergoes a solid state phase transition at 1193K (Derrington and O'Keeffe 1974) and becomes a fast ionic conductor. At this transition, the conductivity increases by three orders of magnitude and the crystal structure changes from orthorhomic to cubic. By contrast, the change in conductivity at the melting point (1233K) is negligible. In the high temperature solid, the mobile ions are believed to the anions and if, as seems likely, this continues to be true in the liquid $g_{--}(r)$ may take on a form which is quite different from $g_{++}(r)$.

Finally, the state of ionisation of alkaline earth chlorides in the liquid state is a matter of long standing dispute. Of the two models proposed,

$$MCl_2 \longrightarrow M^{2+} + 2\,Cl^-$$

$$MCl_2 \longrightarrow MCl^+ + Cl^-$$

the indirect evidence from conductivity data apparently favours the latter (Bockris et al 1960). A knowledge of $g_{+-}(r)$ will be considerable help in deciding between the two ionisation schemes.

Data for S(k) and g(r) are now available following the work of Edwards, Enderby, Howe and Page (1977) and shown in figs. 6 and 7.

There are no computer simulation data available at present for liquid $BaCl_2$ although a related system, liquid $SrCl_2$, has been extensively studied by de Leeuw (1976). His conclusions about the nature of the alkaline earth halides are in close accord with the experimental findings.

There is no evidence for the existence of $BaCl^+$ complexes of life long enough to be regarded as distinct structural entities. This conclusion is based on three facts: (a) g_{+-} does not contain any feature attributable to a well defined Ba - Cl distance with a coordination number of unity (b) g_{++} does not bear the expected phase relationship to g_{--} and g_{+-} if the melt were a pseudo 1 - 1 system (c) there is excellent overall agreement between the rigid ion simulation study based on Sr^{2+} and Cl^- and the results of the present investigation. A re-evaluation of the conductivity data for liquid $BaCl_2$ and its interpretation is now urgently required in view of the disagreement between our conclusions and those of Bockris *et al.* (1960).

Unlike the situation 1 - 1 systems, g_{++} and g_{--} are quite different. This is partially accounted for by the charge difference (de Leeuw 1976); however, the comparative absence of structure in g_{--} when compared with g_{++} beyond 6A must be linked with the fact that the anions are the more mobile species in both the solid and (presumably) the liquid phase.

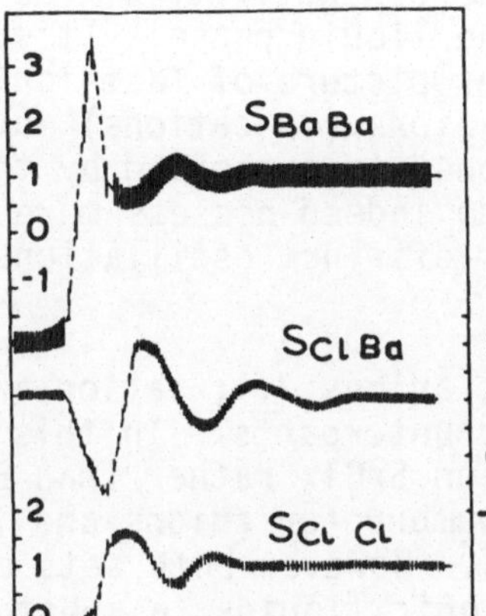

Fig. 6 The partial structure for liquid $BaCl_2$

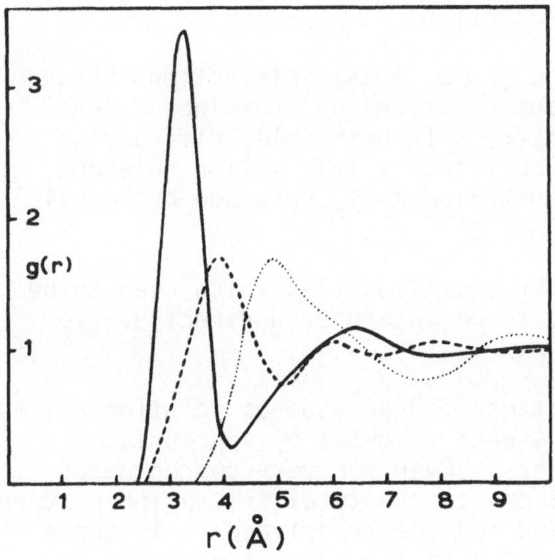

Fig. 7 (Below) The radial distribution function for liquid $BaCl_2$: ——— g_{BaCl} ····· g_{BaBa} ----- g_{ClCl}

The ability of the anions to squeeze past the cations may reflect their great polarisibility. Polarisation effects are known to change the form of $g_{--}(r)$ (Dixon and Sangster, 1975) and must be incorporated in discussions of the conductivity of fast ionic conductors in both the solid and the liquid phase. It should be remarked, however, that the first-order picture of fast ionic conductors as a liquid-like assembly of anions (or cations) moving in a lattice composed of cations (or anions) is supported by the present study. The less mobile species do indeed possess more long range order as evidenced by g_{++} which has distinct oscillations out to at least 10A.

Coordination numbers (cations: 14$\pm$2; anions 7$\pm$1; cation-anion: 7.7$\pm$0.2) are close to their solid-state counterparts. In this respect the simulation studies, admittedly on $SrCl_2$ rather than $BaCl_2$, appear to overestimate the coordination number for anions and to under estimate the number for unlike ions. However both sets of data emphasize the crucial fact that in ionic liquids, a substantial amount of penetration by the anions into the first coordination shell occurs. For this reason most published values for coordination numbers for ionic liquids are not reliable since they were deduced from F(k) and are therefore averages over at least two or the three correlation functions.

5. THE METHOD OF ISOTOPIC ENRICHMENT AS APPLIED TO AQUEOUS SOLUTIONS I

5.1 Introduction

The nature of the short-range (5Å or less) interactions between water molecules and ions in aqueous solutions has remained a long-standing problem in electro-chemistry. In principle, diffraction experiments can provide valuable qualitative information relevant to the problem, as was first demonstrated many years ago by Bernal and Fowler (1933).

We have earlier identified three difficulties which need to be resolved if such measurements are to be interpreted directly. To summarize the earlier remarks:

(i) The total scattering pattern for an aqueous solution arises from ten partial structure factors each weighted by appropriate scattering and concentration factors. Even for very concentrated solutions the ion-water contributions to the total scattering pattern are typically only 30% (x-rays) and 10% (neutrons) and it is therefore exceedingly difficult to disentangle the ion-water unambiguously. As the electrolyte concentration is reduced, the situation becomes progessively less favourable.

(ii) x-ray methods suffer from the disadvantage that the

contribution to the total pattern from ion-hydrogen correlations is essentially zero.

(iii) The effect of the complex energy - momentum relationships which obtain when neutrons are scattered by light elements in the condensed state leads to a characteristic, but so far incalculable, 'droop' in the total neutron scattering at high values of momentum transfer, $\hbar k$. (The 'Placzek' corrections.)

The object of this section is to describe a relatively simple method to overcome these difficulties. In order to help the subsequent discussion, we first of all write down an explicit expression for F (k) for the case when a salt MX_n is dissolved in D_2O (equation 9), c representing the atomic concentration of M

$$\begin{aligned} F(k) = {} & \tfrac{1}{3}(1 - c - nc)\, f_o^{\,2}\, (S_{OO} - 1) + \tfrac{4}{3}(1 - c - nc) f_D^{\,2}(S_{DD} - 1) \\ & + \tfrac{4}{3}(1 - c - nc)\, f_o f_D(S_{OD} - 1) + c^2 f_M^2\, (S_{MM} - 1) \\ & + n^2c^2 f_X^2(S_{XX} - 1) + 2nc^2\, f_X\, f_M(S_{XM} - 1) \\ & + \tfrac{2}{3}\, c(1 - c - nc)\, f_M\, f_o\, (S_{MO} - 1) \\ & + c\,(1 - c - nc)\, f_M\, f_D\, (S_{MD} - 1) \\ & + \tfrac{4}{3}\, nc(1 - c - nc)\, f_X\, f_D\, (S_{XD} - 1) \\ & + \tfrac{2}{3}\, nc(1 - c - nc)\, f_X f_o(S_{Xo} - 1) \end{aligned} \qquad (22)$$

5.2 First-order difference spectroscopy

Let $\Delta_M^0(k)$ represent the algebraic difference between the corrected neutron scattering cross-section in absolute units between two solutions which are identical in all respects except that the isotopic state of M has been changed. It follows from equation (22) that

$$\Delta_M^0(k) = {}_M(k) + \text{correction terms} \qquad (23)$$

where

$$\begin{aligned} \Delta_M(k) = {} & \tfrac{2}{3}c(1 - c - nc)\, f_o(f_M - f_M')(S_{MO}(k) - 1) \\ & + \tfrac{4}{3}\, c(1 - c - nc)\, f_D\, (f_M - f_M')\, (S_{MD}(k) - 1) \\ & + 2nc^2\, f_X\, (f_M - f_M')(S_{XM}(k) - 1) \\ & + c^2\, (f_M^{\,2} - (f_M')^2)(S_{MM}(k) - 1) \end{aligned}$$

$$= A_1(S_{MO}(k) - 1) + B_1(S_{MD}(k) - 1) + C_1(S_{MX}(k) - 1)$$
$$+ D_1(S_{MM}(k) - 1) \qquad (24)$$

and the correction terms arise from kinematic effects for both the self and the interference scattering processes (Powles 1973) and from the change in incoherent scattering.

In (23) f_M and f'_M are the coherent scattering lengths for the two isotopic state of M used in the experiment.

Equ (23) differs from equ (22) in two significant ways. First, the relevant corrections are sufficiently small to allow $\Delta_M(k)$ to be derived from $\Delta_M(k)$ (see 5·3). Secondly, the water terms, which dominate the total scattering and therefore mask the solute-water interactions, no longer appear in (23). Thus the principal obstacles to progress referred to in the introduction have been overcome. The fact that the 'droop' disappears from $\Delta_M(k)$ is of particular significance when considering the errors in normalizing the experimental data. There is, however, a third advantage of working with $\Delta_M(k)$. For a wide range of solutions C_1 and D_1 are much smaller than A_1 and B_1; hence, for practical purpose, $\Delta_M(k)$ is determined by S_{MO} and S_{MD}, that is to say, the short-range correlations between the solvent and M.

We conclude this section by writing down an expression for $\Delta_X(k)$, the first-order difference obtained by isotopically substituting the anion X. By direct analogy with (23) it follows that

$$\Delta_X(k) = A_2(S_{XO}(k) - 1) + B_2(S_{XD}(k) - 1) + C_2(S_{XM}(k) - 1)$$
$$+ D_2(S_{XX}(k) - 1) \qquad (24)$$

where

$$A_2 = \tfrac{2}{3}nc\,(1 - c - nc)\, f_O(f_X - f'_X)$$
$$B_2 = \tfrac{4}{3}nc\,(1 - c - nc)\, f_D(f_X - f'_X)$$
$$C_2 = 2nc^2 f_M\,(f_X - f'_X)$$
$$D_2 = n^2c^2(f_X^2 - (f'_X)^2)$$

5.3 Corrections to the first-order difference $\Delta_M^0(k)$

It is convenient to express the effective differential scattering cross-section for a multi-component system as the sum of two terms, an interference term and a self term, namely

$$\left.\frac{d\sigma}{d\Omega}\right|_{\text{effective}} = \left.\frac{d\sigma}{d\Omega}\right|_{\text{effective}}^{\text{int}} + \left.\frac{d\sigma}{d\Omega}\right|_{\text{effective}}^{\text{self}} \tag{25}$$

If the isotopic state of the nucleus M is changed, the difference in $d\sigma/d\Omega\,|_{\text{effective}}$ is $\Delta_M^0(k)$ and may be written as

$$\Delta_M^0\,(k) = \Delta_M(k) + P_1(k) + \Delta_M^0(k)\,\Big|^{\text{self}} \tag{26}$$

where $P_1(k)$ is the Placzek correction to the interference terms. We therefore wish to calculate $P_1(k)$ and $\Delta_M(k)|^{\text{self}}$, the 'correction terms' which appeared in equation (23).

The Placzek corrections to the self term in equation (26) can be obtained in terms of the moments of the energy transfer (Placzek 1952) provided that the detector law is known. On frequent assumption made is that the detector is black (i.e. constant efficiency of 100%) in which case is can be shown (Powles 1973) that

$$\left.\frac{d\sigma}{d\Omega}\right|_{\text{effective}}^{\text{self}} = \sum_j c_j <f_j^2> \left(1 + \frac{m}{\mu_j}\right)^{-2} \Bigg\{ 1 + \frac{m}{\mu_j}\ [2\cos\theta + \tfrac{1}{3}(\overline{K}_j/E_0)]$$

$$+ \tfrac{1}{2}(m/\mu_j)^2(3\cos^2\theta - 1)\,[1 + \tfrac{2}{3}\overline{K}_j/E_0] + \text{terms in } (m/\mu_j)^3 \text{and higher} \Bigg\} \tag{27}$$

where $\overline{K}_j$ and μ_j are the mean kinetic energy and mass of the jth nucleus, m is the neutron mass of energy E_0 and θ is the scattering angle.

For a single isotopic substitution of the nucleus M, resulting in a change of scattering length f_M to f_M' and a change of mass μ_M to μ_M', the corresponding change in $d\sigma/d\Omega|^{\text{self}} = \Delta_M^0(k)^{\text{self}}$ arises from those terms in (27) that contain the subscript j = M.

Thus,

$$\Delta_M^0(k)\Big|^{\text{self}} = c(<f_M^2> = <f_M'^2>)\ \ 1 - \frac{2m}{\mu^2_M}\,\mu_M'\,(1 - \cos\theta) \tag{28}$$

where $\overline{K}/E_0$ and terms in $(m/\mu_M)^2$ are neglected. Equation (28) demonstrates that the correction to the 'bound' cross-sections is of order (m/μ_M) i.e. 3% for the case of nickel substitutions.

The Placzek corrections for the interference term of equation

(25) are not so straight-forward and involve the momentum correlation between the two nuclei involved in any particular pair component of the interference term. Analogous to equation (27) we write

$$\left.\frac{d\sigma}{d\Omega}\right|_{\text{effective}}^{\text{int}} = \sum_{j,k} c_j c_k f_j f_K \left\{ \exp(i\underset{\sim}{k}.\underset{\sim}{r}_{jk}) + \frac{\hbar^2}{\mu_j \mu_k} \left\langle \exp(i\underset{\sim}{k}.\underset{\sim}{r}_{jk}) \left[\frac{\hbar^2 k^2}{4} + \text{terms involving momentum correlation} \right] \right\rangle + \ldots \right\} \quad (29)$$

$$= \sum_{j,k} c_j c_k f_j f_k \left\{ \left[S_{jk}(k) - 1 \right] + \frac{m^2}{\mu_j \mu_k} \hat{F}_{jk}(k) + \ldots \right\} \quad (30)$$

where $\hat{F}_{jk}(k)$ is, in general, a function of $S_{jk}(k)$ and its derivates.

From (30) $P_I(Q)$ is a sum of terms arising from every pair of nuclei involving M. The largest contribution to P_1 is

$$-\tfrac{4}{3}c(1 - c - nc)\, f_D(f_M - f_M) \frac{m^2}{\mu_D \mu_M} \hat{F}_{MD}(k) \quad (31)$$

and there are no contributions from the troublesome terms involving $\hat{F}_{DD}(k)$ and $\hat{F}_{DO}(k)$. Equation (31) shows that the Placzek correction to the interference term is of order $m^2/\mu_D\mu_M$ and is, therefore even smaller than that for the self term.

5.4 Experimental procedures: the long-wave limit of $S_{\alpha\beta}(k)$

The diffraction data can be obtained by conventional means (e.g. on the D4 spectrometer at the ILL, Grenoble). The results are put on an absolute scale by the vanadium technique (North et al 1968) Since $\Delta(k)$ does not 'droop' at high k further normalization, which allows for inaccuracies in the absorption corrections and the published values of the scattering lengths as well as the systematic error due to the slight variations in the light water content of the samples, was achieved by reference to the high and low k limits of $\Delta(k)$, both of which are known from macroscopic parameters (Beeby 1973).

(a) at high k, $S_{\alpha\beta} \to 1$ so that $\Delta(k) \to 0$

(b) at low k, the work of Beeby (1973) allows $S_{\alpha\beta}$ to be determined in terms of the fluctuations in N_a and N_b, the number respectively of solvent and solute molecules disolved in the volume V. These fluctuations can be determined from a knowledge of the compressibility, the osmotic coefficient and the partial molar volume.

5.5 Results

So far three solutions have been investigated (Soper et al 1977; Cummings et al 1978). The three solutions chosen were $NiCl_2$ (isotopically changing the Ni) and $CaCl_2$ and NaCl (isotopically changing the Cl) both dissolved in D_2O. The concentrations and the corresponding coefficients A_1, B_1 ... D_1 are given in the table and the statement made above, above that C and D are small compared with A and B is verified.

Great care must be taken with the preparation of the samples as the technique relies on the exact cancellation of the water terms. Accordingly, the difference in light water content of each pair of samples was held to within $\pm$ 0·1% by preparative techniques involving the use of an infrared spectrometer. Activity measurements ensured that the c scale was reproducible to within $\pm$ 1%. Finally, mass spectrographic techniques were employed to determine to within 0·1% the abundances of the various isotopes characteristic of the Ni and Cl used in the preparation of the samples.

In order to study hydration effects in real space it is necessary to construct the Fourier transform of $\Delta(k)$. The quantity

$$G(r) = \frac{1}{2\pi^2 \rho r} \int \Delta(k)\, k \sin kr\, dk, \qquad (32)$$

where ρ is the total number of atoms per volume, can be obtained by standard numerical quadrature. It then follows from the definition of $\Delta(k)$ that

$$G_{Ni}(r) = A_1 g_{NiO} + B_1 g_{NiD} + C_1 g_{NiCl} + D_1 g_{NiNi} + E_1 \qquad (33)$$

and

$$G_{Cl}(r) = A_2 g_{ClO} + B_2 g_{ClO} + C_2 g_{ClNa} + D_2 g_{ClCl} + E_2 \qquad (34)$$

where

$$E_1 = -(A_1 + B_1 + C_1 + D_1) \text{ and } E_2 = -(A_2 + B_2 + C_2 + D_2)$$

Two examples of $\Delta(k)$ are shown in figs.8&9. The points represent unsmoothed differences and the broken curve the one used to obtain G(r).

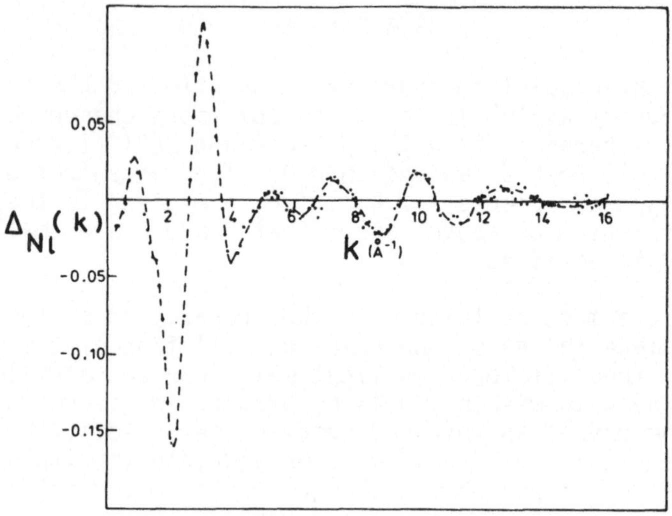

Fig. 8 $\Delta_{Ni}(k)$ 4·41 molal $NiCl_2$ in D_2O full circles, unsmoothed data points; broken curve, smoothed curve used for the Fourier transform

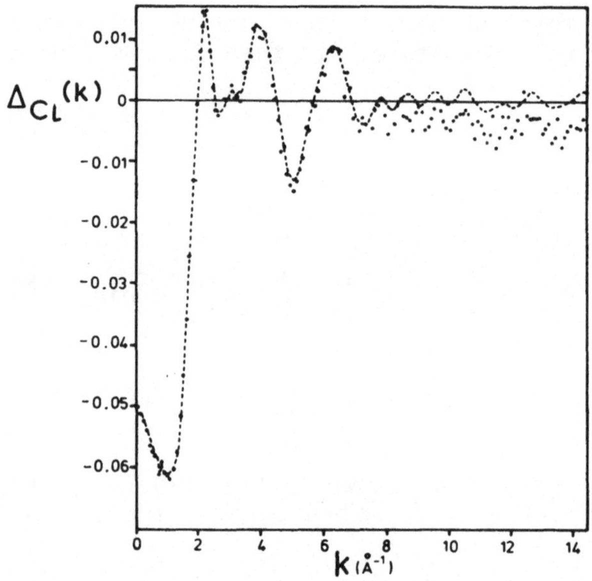

Fig. 9 $\Delta_{Cl}(k)$ for 5·32 molal NaCl in D_2O full circles, unsmoothed data points; broke curve smoothed curve for the Fourier transform

Table

Scattering Length and Sample Parameters

Electrolyte solution	Isotopes	Abundance	Scattering lengths (10^{-12} cm)	c	Molality	A (barns) 10^{-2}	B (barns) 10^{-2}	C (barns) 10^{-3}	D (barns) 10^{-3}
$NiCl_2.D_2O$	$Ni^{natural}$	-	1·03	0·0275	4·41	1·74	4·00	5·05	0·32
	^{62}Ni	94·9	-0·79						
$NaCl.\ D_2O$	^{35}Cl	99·35	1·17	0·0331	5·32	0·99	2·27	0·655	0·14
	^{37}Cl	90·4	0·35						
$CaCl_2.D_2O$	^{35}Cl	99·35	1·17	0·0275	4·49	1·62	3·72	1·20	0·38
	^{37}Cl	90·4	0·35						

(a) The $NiCl_2$ Solution The form of G(r) in figure 10 shows the well-defined nature of the hydration atmosphere surrounding the Ni ion. It is particularly significant to note that at $r \sim 3{\cdot}0$Å, $G(r) \sim - E_1$; thus g_{NiO} and g_{NiD} are both small at this value or r which reflects the stability of the first hydration shell. The two peaks located at 2·05 Å and 2·65Å we identify with Ni-O and Ni-D correlations respectively on the grounds that the ratio of the areas beneath them,

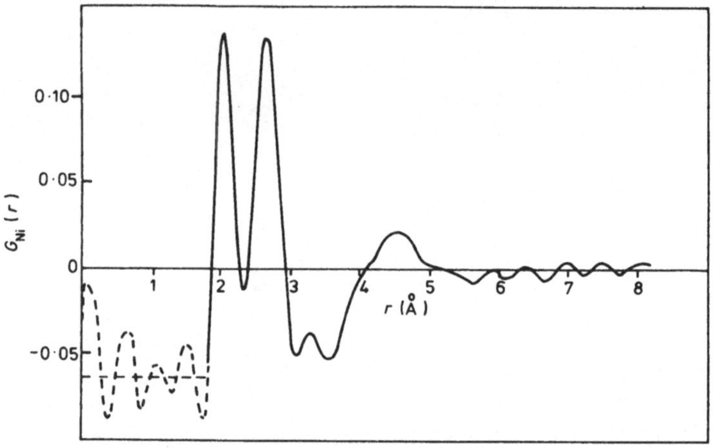

Fig. 10 $G_{Ni}(r)$ for 4·41 molal $NiCl_2$ in D_2O

when weighted by r^2 are almost exactly $A_1 : 2B_1$. An integral over $4\pi r^2 G(r)$ for $1{\cdot}8 < r < 3{\cdot}0$Å yields a total of $5{\cdot}8 \pm 0{\cdot}2$ water molecules in the first coordination shell. The relative peak positions show that each water molecule is at angle of 30° to the Ni-O axis (see 12a). The nickel ion, together with the six or so water molecules attached to it can therefore be thought of as an entity to use in statistical mechanical calculations of solution properties (Quirke and Soper 1977).

A substitution of chlorine isotopes in the same system will enable the hydration state of the anions to be determined and a detailed structure picture of this particular solution will, for the first time, be within out grasp. In passing, we should remark that the feature of 3·25Å appears to reflect real structure in that it survives when a variety of window functions are applied to $\Delta_M(k)$. We tentatively associate this peak with g_{NiCl} and note, for this particular electrolyte, the apparent absence of Cl in the first coordination shell.

(b) <u>The NaCl solution</u> The form of G(r) (fig.11) indicates well-defined hydration about the Cl ions; again it should be noted that $G(r) \sim -E_2$ for $r \sim 2{\cdot}8$Å.

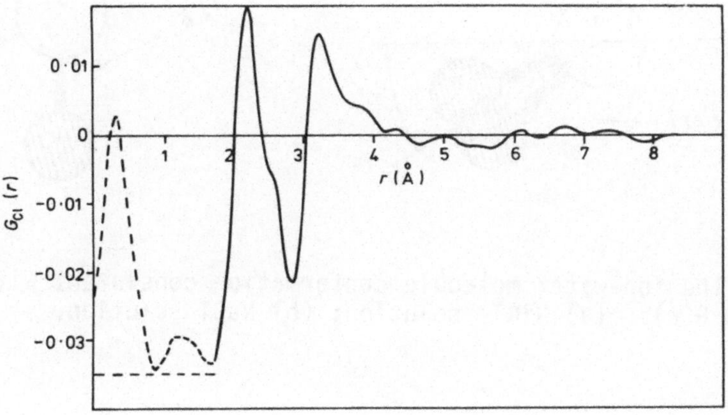

Fig. 11 $G_{Cl(r)}$ for 5·32 molal solution in D_2O

One interpretation of these results is to suppose that each of the two principal peaks contain only contributions for g_{ClO} and g_{ClD} in a ratio corresponding to D_2O molecules. This yields a value for the number of water molecules in the first shell of 2·1 + 0·2. On the other hand, it suggests a Cl-O distance of 2·7Å which seems, on general chemical grounds to be highly implausible.

We therefore favour a second model (fig.12b) which is based on the assumption that the first peak in G(r) arises solely from g_{ClD} and that there is a single hydration shell or radius 3·20Å. Integration of G(r) yields a coordination number of 5.5 $\pm$ 0.2 and it will be of interest to see whether this value increases towards six as predicted by a molecular dynamics study (Vogel and Heinzinger 1976) as the concentration of NaCl is reduced.

(c) <u>The $CaCl_2$ solutions</u> The $G_{ce}(r)$ for this system (fig.13) is of interest because, when scaled for the new values of A_2 and B_2, it strongly resembles that found for the NaCl solution. Since, however, the coefficient A_2 and B_2 and the experimental statistics were both rather more favourable, we are able to state that the conformation of Cl - D_2O is consistent with the second of the two models proposed for NaCl (fig.12b) The absence of structure beyond 5Å shows that only the first shell of water molecules can be said to be co-ordinated.

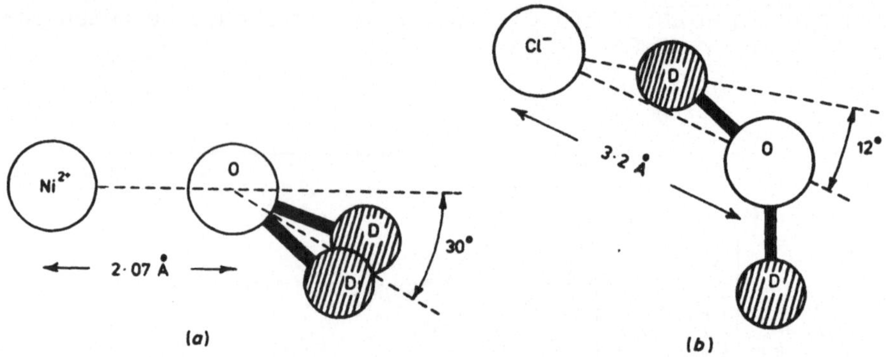

Fig. 12 The ion-water molecule conformation consistent with the observed G(r): (a) $NiCl_2$ solution; (b) NaCl solution.

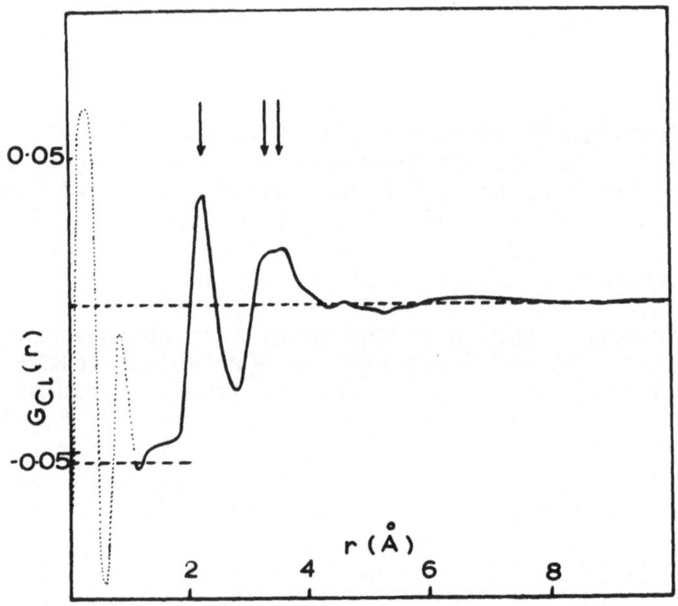

Fig. 13 $G_{Cl}(r)$ for 4·49 molal $CaCl_2$ in D_2O. The three vertical arrows refer, from the left, to Cl-D, Cl-O and Cl-D distances.

6. THE METHOD OF ISOTOPIC ENRICHMENT APPLIED TO AQUEOUS SOLUTIONS II

6.1 Second - Order Differences

Let us consider again a salt MX_n dissolved in D_2O and let $\Delta_{M_1}(k)$ and Δ_{M_2} represent two first-order differences for three solutions[1] with M in the isotopic state M, ′M and ″M. Similarly let $\Delta_{X_1}(k)$ and $\Delta_{X_2}(k)$ be the corresponding quantities for isotopic substitutions of the anion X. It follows from equation (23) and (24) that

$$S_{MM}(k) = \frac{\Delta_{M_1}(k)}{A_M^{(2)}} - \frac{\Delta_{M_2}(k)}{B_M^{(2)}}$$

$$S_{XX}(k) = \frac{\Delta_{X_1}(k)}{A_X^{(2)}} - \frac{\Delta_{X_2}(k)}{B_X^{(2)}}$$

with coefficients A_M ..... B_X given by

$$A_M^{(2)} = c^2 (f_M - f_M') (f_M' - f_M'')$$

$$B_M^{(2)} = c^2 (f_M - F_M'') (f_M' - f_M'')$$

$$A_X^{(2)} = n^2 c^2 (f_X - f_X') (f_X' - f_X'')$$

$$B_X^{(2)} = n^2 c^2 (f_X - f_X'') (f_X' - f_X'')$$

It therefore follows that a <u>three pattern experiment</u> enables individual <u>ion-ion correlation functions to be isolated</u>.

In order to obtain S_{MX}, we must use four samples whose isotopic state can be represented by

$$MX_n, \quad {}'MX_n \quad M'X \quad {}'MX'_n$$

Let Δ_M^X represent the algrebraic difference in intensity between the scattering from the first and the second samples and Δ'^X_M the difference between the third and the fourth sample It follows that

$$S_{MX}(k) = \frac{\Delta_M^X - \Delta_M'^X}{2nc^2 \; (f_M - f_M')(f_X - f_X')} + 1$$

6.2 Application

This second-order difference technique has been applied so far to one solution, a 4.41 Molal solution of $NiCl_2$ in D_2O (Enderby

Solution	Isotopes	Scattering Length (10^{12}cm)	$A_M^{(2)}$ (Barns)	$B_M^{(2)}$ (Barns)
$NiCl_2$-D_2O	$Ni^{natural}$	1·03	$6{\cdot}15 \times 10^{-4}$	$1{\cdot}09 \times 10^{-3}$
	^0Ni(62-60 mixture)	0·00		
	^{62}Ni	-0·79		

et al. 1975). The isotopic state of the nickel was changed and the sample parameters and the values of $A_M^{(2)}$ and $B_M^{(2)}$ are shown in the table.

The data for $S_{NiNi}(k)$ (Fig. 14) are characterised by a well-defined first peak centred at k value of ~ $1A^{-1}$. Originally, it was believed that these data supported the idea that ions in concentrated ionic solutions were charge ordered. Quirke and Soper (1977), however, attempted to fit the data assuming simply that the repulsive interactions between like ions could be neglected and that the dominant contribution to the structure came from the random packing of the spherical hydrated ions.

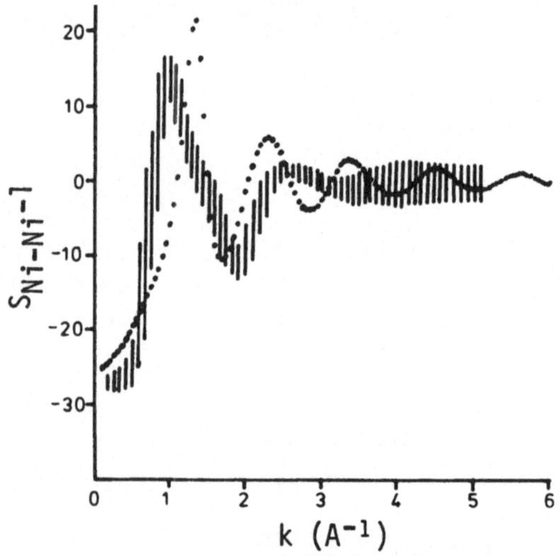

Fig. 14 S_{NiNi} for 4 41 molal $NiCl_2$ in D_2O. Vertical Lines experiment; full circles, a hard sphere simulation (Quirke and Soper 1977)

Neilson et al (1975) had earlier shown that if the prepeak at $\sim 1A^{-1}$ in F(k) could be identified with the main peak in $S_{NiNi}(k)$ the concentration dependence of the position of this peak (k_0) supported the ordered model of liquid.

In figs.14 and 15, we compare the theoretical behaviour of $S_{NiNi}(k)$ and of k_0 with that observed. The hard sphere model, though give roughly the right form for $S_{NiNi}(k)$ fails to predict the variation of k_0 with molarity at 3 molar and less; Until high quality experiments are performed so that $S_{NiNi}(k)$ can be derived with confidence as a function of concentration, a resolution of these two points of view may not be possible. There is, however,

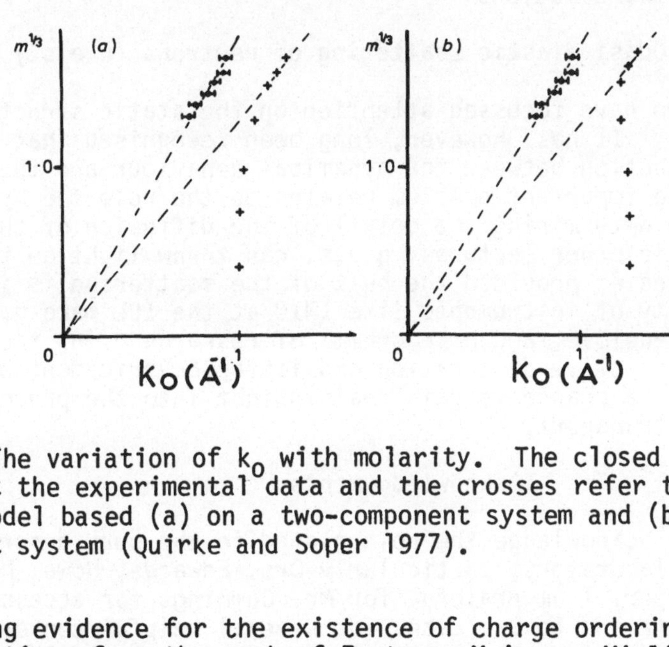

Fig. 15 The variation of k_0 with molarity. The closed circles represent the experimental data and the crosses refer to a hard sphere model based (a) on a two-component system and (b) a three-component system (Quirke and Soper 1977).

supporting evidence for the existence of charge ordering in these NiCl solutions from the work of Fontana, Maisano, Migliardo and Wanderlingh (1977).

7. CONCLUSIONS AND FUTURE PROSPECTS

It is now, we believe, quite clear that the neutron method allied to the technique of isotopic replacement has allowed new insight to be gained into the structure of the ionic systems. Apart from the need to apply the technique systematically to a wide variety of molten salts and aqueous solutions (at various concentrations) two new experiments seem to us to be worth emphasizing.

7.1 High pressure studies of solutions

Already preliminary work in our laboratory at Bristol has shown that the pressure dependence of F(k), which we will denote $F'(k)$ depends on the nature of the solute. A simple scaling argument assuming that all distances go like $(\text{density})^{\frac{1}{3}}$ allows $F'(k)$ to be predicted from a knowledge of F(k) and F(o) (Egelstaff et al 1972). This model works well for dilute solutions but fails for those cases where a high degree of water co-ordination is known to occur (Neilson, to be published). With structure breakers, however, the simple scaling rule seems to apply even at high values of c. These exciting data are still under analysis but seem to offer a real way to distinguish at a molecular level the role of the ion-type in determining the bulk dynamical properties (e.g. the viscosity) of the solutions.

7.2 Quasi-elastic scattering of neutrons (q.e.s.)

So far, we have focussed attention on the static structure of ionic liquids. It has, however, long been recognised that there is a deep connection between the dynamical behaviour and the static structure. One important problem relates to the role the hydration cloud plays in determining the detail of the diffusion of the ions. As dealt with in other lectures, q.e.s. can throw light on the diffusion processes provided the bulk of the scattering is incoherent. The availability of instruments like IN10 at the ILL make q.e.s. work of this sort feasible, and a programme of research using the technique of selective deuteration and isotopic enrichment offers, for the future, a chance to gain real insight into the processes governing ion transport.

Acknowledgements

I wish to acknowledge the role played in the work described here by my collaborators, particularly Drs. Edwards, Howe, Howells, Neilson and Soper. I am grateful for Mr. Cummings for access to unpublished data and to Dr. Evans for several helpful conversations. The work carried out at the ILL would not have been possible without the assistance of Drs. Chieux and Knoll.

References

Beeby, J.L., 1973, J. Phys. C. 6, 2262

Bernal, J.D. and Fowler, R.H., 1933, J. Chem. Phys. 1, 55

Brady, G.W., 1958, J. Chem. Phys. 28, 464, Proc. Roy. Soc. A, 255, 558

Bockris, J.O.M., Crook, E.H., Bloom, H. and Richards, N.E., 1960 Proc. Roy. Soc. A 255, 558

Boyce, J.B. and Mikkelson, J.C., 1977, J. Phys. C. 10, L41

Cristini, A. Lichen, G. Piccaluga, G. and Pinna, G., 1974, Chem. Phys. Lett. 24, 289

Cummings, S. and Enderby, J.E. 1978, to be published

de Leeuw, S.W., 1976, Thesis: Computer Simulation of the Alkaline Earth Halides (University of Amsterdam)

Derrien, Y. and Dupuy, J., 1975, J. Phys. Paris 36, 191

Derrington, C.E. O'Keeffe, M. 1974, Solid State Comm. 15, 1175

Dixon, M. and Sangster, M.J.L. 1975, J. Phys. C. 8 L8

Edwards. F.G., Enderby, J.E., Howe, R.A. and Page D.I., 1975 J. Phys. C. 8, 3483

Edwards, F.G., Enderby, J.E., Howe, R.A. and Page D.I., 1977, in the press

Egelstaff, P.A. Page, D.I. and Heard, C.R.T., 1971, J. Phys. C. 4 1453

Enderby, J.E. North, D.M. and Egelstaff, P.A., 1966, Phil. Mag. 14 961

Fontana, M.P. Maisano, G. Migliardo, P. and Wanderlingh, F., 1977 Solid State Comm. 23, 489

Lantelme, F., Turq, P. Quentric, B. and Lewis, J.W.E., 1974, Mol. Phys. 28, 1357

Lawrence, R.J. and Kruh, R.F., 1967, J. Chem. Phys. 47, 4758

Levy, H.A., Agron, P.A., Bredig, M.A. and Danford, M.D. 1960, Ann. Acad. Sci. N.Y., 79, 762

Licheri, G., Piccaluga, G. and Pinna, G., 1975, J. Chem. Phys. 63., 4412

Licheri, G., Piccaluga, G. and Pinna, G., 1976, J. Chem. Phys. 64., 2437

Mitchell, E.W.J. Poncet, P.F.J. and Stewart, R.J., 1976, Phil. Mag., 34, 721

Narten, A.H., Vaslow, F. and Levy, H.A., 1973, J. Chem. Phys. 5B, 5017

Neilson G.W., Howe, R.A. and Enderby, J.E., 1975, Chem. Phys. Letts. 33, 284

Neutron Diffraction Commission, 1969, Acta Cryst A25, 391

North, D.M., Enderby, J.E. and Egelstaff, P.A., J. Phys. C. 1. 784

Page, D.I. and Mika, K., 1971, J. Phys. C. 4, 3034

Placzek, G., 1952, Phys. Lett., 86, 377

Powles, J.G., 1973, Adv. Phys. 22, 1

Powles, J.G., 1975, J. Phys. C. 8, 895

Quirke, N. and Soper, A.K., 1977, J. Phys. C. 10, 1802

Rhodes, E., 1972, in Water and Aqueous Solutions: Structures Thermodynamics and Transport Processes Ed: R.A. Howe (New York: Wiley)

Soper, A.K., Ph.D. Thesis, University of Leicester

Soper, A.K., Neilson, G.W. Enderby, J.E. and Howe, R.A. 1977, J. Phys. C 10, 1793

Vogel, P.C. and Heinzinger, K., 1976, Z. Naturf. 31a, 476

Wertz, D.L. and Kruh, R.F., 1969, J. Chem. Phys. 50, 4313

Westlake, J.R., 1968, A Handbook of Numerical Matrix Inversion and Solution of Linear Equation (New York: Wiley)

Woodcock,L.V., and Singer, K., 1971, Trans. Faraday, Soc., 67, 12

CRITICAL PHENOMENA IN FLUIDS

P. C. Hohenberg

Physik Department, TU München, 8046 Garching, W. Germany

and Bell Laboratories, Murray Hill, N.J. 07974, USA

OUTLINE

I. INTRODUCTION

The present lectures describe the modern theory of critical phenomena, with emphasis on applications to real fluids. The aim is not to treat the subject exhaustively, but rather to explain the principal ideas, and to provide a guide to the vast literature on critical phenomena for the noninitiate. The original lectures consisted of three parts: (i) An account of static critical phenomena in fluids (theory and experiment), (ii) a brief introduction to the mathematics of the renormalization group, and (iii) a discussion of dynamic critical behavior. Only the first part is written up in any detail in the present notes, since there exist many recent reviews of renormalization group theory on both an elementary and an advanced level. In particular, the author has just completed a review article with B. I. Halperin on critical dynamics, to which the reader is referred for the contents of item (iii) above (Hohenberg and Halperin, 1977).

II. STATIC CRITICAL PHENOMENA IN REAL FLUIDS

The field of critical phenomena has witnessed striking progress in the last 15 years, by both theorists and experimentalists. The most dramatic achievement is the elucidation of the concept of universality. This is the idea that the detailed quantitative behavior of a system near its critical point depends only on very general geometrical properties, and is largely independent of the microscopic nature of the constituent particles and their interactions. The gas-liquid critical point of a fluid and the critical consolute point of a binary mixture are prominent examples of critical behavior which have been studied for over 100 years (see Levelt Sengers, 1974, 1976). The modern approach explains critical phenomena in these systems in terms of a simple statistical model, the lattice-gas or Ising model, which also serves to describe the ferro- or antiferromagnetic transitions in uniaxial magnets, as well as critical points in other systems. Experimental investigations of fluids have played an important role in the history of critical phenomena, since these systems may be studied with a high degree of precision. Moreover, there exist many different substances with critical points in convenient ranges of temperature and pressure, which permits detailed study of the similarities and differences between various systems.

The aim of these lectures is to describe the modern theory in elementary terms, with emphasis on connections between theory and experiment. For the reasons stated above, we confine our discussion to static behavior, namely thermodynamic properties and instantaneous spatial correlations in a system in thermal equilibrium. We begin by reviewing the Van der Waals and mean-field (or Landau) theories of the critical point, and then discuss the more accurate treatments in terms of scaling, universality, and the renormalization group.

A. Mean-field Theories

We include under this heading all the theories which neglect fluctuations (more properly, theories in which the interaction between fluctuations are neglected, i.e. in which fluctuations are only taken into account insofar as they determine the values of parameters such as T_c, P_c, ρ_c, etc.).

1. Van der Waals and Landau Theories. The Van der Waals theory of the critical point of a pure fluid follows from the equation of state (Landau and Lifshitz, 1969; Stanley, 1971)

$$\left(P + \frac{a}{V^2}\right)(V - b) = RT \quad , \tag{1}$$

which gives the pressure P as a function of the volume V and the temperature T, with two parameters a and b characterizing each fluid (the gas constant R merely fixes the temperature scale). It is easy to see from Eq. (1) that the isotherms, shown in Fig. 1, are monotonically decreasing for

$$T > T_c \equiv 8a/27bR \quad , \tag{2}$$

and nonmonotonic for $T < T_c$. In the latter case certain points in the P-V plane are excluded by means of a "Maxwell construction". States of the system corresponding to these values of P and V consist of two coexisting phases, one with a high density (the liquid) and the other with a low density (the gas). The <u>coexistence curve</u> ends at the critical point, where the density difference between gas and liquid disappears, and where

$$\left(\frac{\partial P}{\partial V}\right)_T = 0 \quad , \tag{3}$$

$$\left(\frac{\partial^2 P}{\partial V^2}\right)_T = 0 \quad . \tag{4}$$

Applying Eqs. (3) and (4) to Eq. (1) we find

$$P_c = a/27b^2 \quad , \quad V_c = 3b \quad , \tag{5}$$

with T_c given by (2), i.e.

$$P_c V_c / RT_c = 3/8 \quad . \tag{6}$$

The thermodynamic properties of the system in the vicinity of the critical point are determined by expanding the equation of state (1) about the point P_c, V_c, T_c. We note first that the isothermal compressibility

$$K_T = -\frac{1}{V}\left(\frac{\partial V}{\partial P}\right)_T \tag{7}$$

diverges at the critical point, because of (3). The full thermodynamic behavior depends on the free energy A(V,T) given by

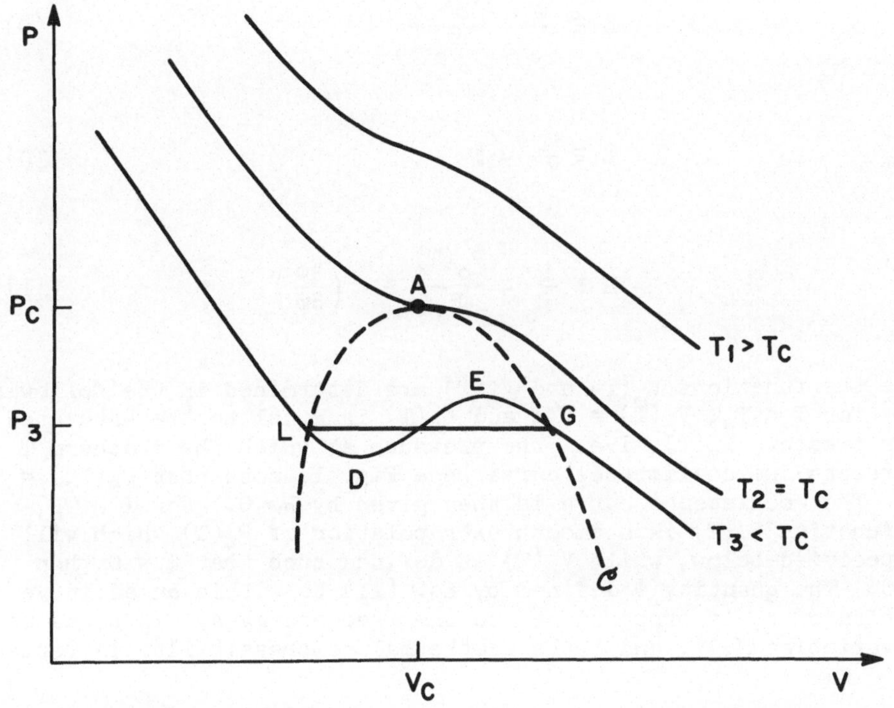

Fig. 1 Schematic phase diagram for the Van der Waals equation of state (1), showing isotherms (solid lines) and the coexistence curve C (dashed line). On the isotherm $T_3 < T_c$, the part LDEG inside the coexistence curve, does not correspond to physical states, and is replaced by the superposition of gas (point G) and liquid (point L) phases, via the "Maxwell construction" (line LG). The pressure P_3 is the common saturated vapor pressure of the points L and G corresponding to the temperature T_3, i.e. $P_3 = P_s(T_3)$. The critical point $[P_c = P_s(T_c), V_c]$ is denoted A in the figure.

$$A(V,T) = A_o(T) - \int P\, dV \quad . \tag{8}$$

Let us define the reduced variables

$$\psi \equiv \frac{V}{V_c} - \frac{V_o(T)}{V_c} \quad , \tag{9}$$

$$t \equiv \frac{T}{T_c} - 1 \quad , \tag{10}$$

$$-H \equiv \frac{P}{P_c} - \frac{P_o(T)}{P_c} \equiv -\left(\frac{\partial \Phi}{\partial \psi}\right)_t \tag{11}$$

where the functions $P_o(T)$ and $V_o(T)$ are determined in the following way: for $T < T_c$, $V_o(T) = V_c$, and $P_o(T)$ is equal to the saturated vapor pressure $P_s(T)$, i.e. the pressure at which the isotherm T intersects the coexistence curve [see Fig. 1; note that $P_s(T_c) = P_c$]. The coexistence curve is then given by $H = 0$. For $T > T_c$, the function $P_o(T)$ is a smooth extrapolation of $P_s(T)$ which will be specified below, while $V_o(T)$ is defined such that $\psi = 0$ when $H = 0$. The quantity Φ defined by Eq. (11) to within an additive function of t, is proportional to the free energy A. In terms of the variables H, ψ, and t the isothermal compressibility is replaced by

$$\chi_\psi \equiv \left(\frac{\partial \psi}{\partial H}\right)_t \quad . \tag{12}$$

The equation of state (1) may be differentiated and expanded near the critical point in terms of the small quantities t and ψ; the compressibility is written as

$$\chi_\psi^{-1} = 6t + \frac{9}{2}\psi^2 + \ldots \quad . \tag{13}$$

Integrating Eq. (13) we find the equation of state

$$H = 6t\,\psi + \frac{3}{2}\psi^3 + \ldots \quad . \tag{14}$$

The function $P_o(T)$ in (11) is now determined above T_c by the condition that the constant of integration in (14) should vanish. The free energy is obtained by integrating once again, i.e.

$$\Phi = 3t\,\psi^2 + \frac{3}{8}\psi^4 + \ldots + \Phi_o(t) \quad , \tag{15}$$

where the constant of integration $\Phi_o(t)$ is assumed to be a smooth function of t.

We shall discuss the critical behavior which follows from (15), but first we wish to generalize the treatment somewhat, by discussing the <u>Landau</u> or <u>mean-field</u> theory of the critical point. This approach does not start with a phenomenological equation of state as in (1), but rather from the <u>assumption</u> (see Landau and Lifshitz, 1969) that in the vicinity of T_c the free energy is an analytic function of the "order parameter" ψ,

$$\Phi = \Phi_o(t) + \frac{1}{2}\bar{r}t\,\psi^2 + u\,\psi^4 + \ldots \quad , \tag{16}$$

where the constants $\bar{r}$ and u are parameters of the theory, as are the quantities $P_o(T)$, $V_o(T)$, and T_c which enter the definitions of H, ψ, and t, Eqs. (9-11). The pressure is obtained from Eq. (11),

$$H = \bar{r}t\,\psi + 4u\,\psi^3 + \ldots \quad , \tag{17}$$

and the coexistence curve is again given by the condition H = 0. The Van der Waals theory thus appears as a special case of the Landau theory, in which the parameters P_c, V_c and T_c are expressible in terms of a and b [Eqs. (2) and (5)], and the constants $\bar{r}$ and u have the values

$$\bar{r} = 6 \;, \quad u = \frac{3}{8} \;. \tag{18}$$

Note, however, that the correspondence between the Van der Waals and Landau theories only holds in the vicinity of the critical point, and for low-order thermodynamic derivatives, since the expansions in (14-16) only retain two terms. Singularities in $(\partial^2P/\partial T^2)_V$, for instance, which occur in the Van der Waals theory, are not correctly represented by (11) and (16), unless the function $P_o(T)$ has singularities.

The critical behavior which follows from the Landau assumption (16) may be obtained by an elementary calculation (see Stanley, 1971). Let us enumerate the principal results, in the standard exponent notation:

(i) Coexistence curve. The coexistence curve is given by $H = 0$, for $t < 0$ ($T < T_c$), namely

$$\psi = B(-t)^{\beta} = (\bar{r}/4u)^{1/2}(-t)^{1/2} \quad . \tag{19}$$

(ii) Compressibility. The compressibility χ_ψ is given by

$$\chi_\psi^{-1} = \left(\frac{\partial H}{\partial \psi}\right)_t = \bar{r}t + 12u\,\psi^2 \quad . \tag{20}$$

Along the critical isochore (the line $H = 0$, $t > 0$) we have

$$\chi_\psi \equiv \Gamma\, t^{-\gamma} = (\bar{r})^{-1}\, t^{-1} \quad , \tag{21}$$

and on the coexistence curve ($H = 0$, $t < 0$),

$$\chi_\psi \equiv \Gamma'(-t)^{-\gamma'} = (2\bar{r})^{-1}(-t)^{-1} \quad . \tag{22}$$

(iii) Specific heat. The constant volume specific heat is obtained by differentiating the free energy

$$C_v = -\left(\frac{\partial^2 \Phi}{\partial t^2}\right)_\psi \quad . \tag{23}$$

On the critical isochore only the background term contributes,

$$C_v = -\frac{d^2\Phi_o}{dt^2} = C_v^o \quad , \quad t > 0 \tag{24}$$

and on the coexistence curve we have

$$\Phi = \Phi_o - \frac{\bar{r}^2 t^2}{16u} \quad , \tag{25}$$

and

$$C_v - C_v^o = \bar{r}^2/8u \equiv \Delta C_v \quad , \quad t < 0 \quad . \tag{26}$$

Table I. Definitions of Exponents and Amplitudes and Their Mean-Field Values

Quantity	Thermodynamic Path	Equation	Mean-field exponent	Mean-field amplitude
Coexistence curve	$H = 0, \; t < 0$	$\psi = B(-t)^{\beta}$	$\beta = \frac{1}{2}$	$B = (\bar{r}/4u)^{\frac{1}{2}}$
Susceptibility	$H = 0, \; t > 0$	$\chi_{\psi} = \Gamma \, t^{-\gamma}$	$\gamma = 1$	$\Gamma = 1/\bar{r}$
	$H = 0, \; t < 0$	$\chi_{\psi} = \Gamma' (-t)^{-\gamma'}$	$\gamma' = 1$	$\Gamma' = 1/2\bar{r}$
Specific heat	$H = 0, \; t > 0$	$C_v = \frac{A}{\alpha} t^{-\alpha} + B_o$	$C_v = C_o$	—
	$H = 0, \; t < 0$	$C_v = \frac{A'}{\alpha'} (-t)^{-\alpha'} + B_o$	$C_v = C_o + \Delta C_v$	$\Delta C_v = \bar{r}^2/8u$
Critical isotherm	$t = 0$	$H = D\psi^{\delta}$	$\delta = 3$	$D = 4u$
Correlations at T_c	$H = 0, \; t = 0$	$\chi_{\psi}(k) = 1/c_1 k^{2-\eta}$	$\eta = 0$	c_1
Correlation length	$H = 0, \; t > 0$	$\xi = \xi_o^+ \, t^{-\nu}$	$\nu = \frac{1}{2}$	$\xi_o^+ = (c_1\Gamma)^{\frac{1}{2}}$
	$H = 0, \; t < 0$	$\xi = \xi_o^- (-t)^{-\nu'}$	$\nu' = \frac{1}{2}$	$\xi_o^- = (c_1\Gamma')^{\frac{1}{2}}$

The above calculation shows that the specific heat experiences a jump as the temperature is lowered below T_c. More generally, the singularity in C_v is defined by power laws,

$$C_v = (A/\alpha)t^{-\alpha} + B_o \quad , \qquad t > 0 \qquad\qquad (27a)$$

$$C_v = (A'/\alpha')(-t)^{-\alpha'} + B_o \quad , \quad t < 0 \quad . \qquad (27b)$$

The mean-field result (26) corresponds to $\alpha = \alpha' = 0$.

The specific heat at constant pressure, on the other hand, is proportional to the compressibility χ_ψ, as can be seen from the relation (Landau and Lifshitz, 1969)

$$C_p - C_v = - T\left(\frac{\partial P}{\partial T}\right)_V^2 \Big/ \left(\frac{\partial P}{\partial V}\right)_T \propto - \left(\frac{\partial V}{\partial P}\right)_T \quad , \qquad (28)$$

since $(\partial P/\partial T)_V$ goes to a constant at the critical point.

(iv) <u>The critical isotherm</u>. The critical isotherm is obtained by setting $t = 0$ in (17),

$$H \equiv D\psi^\delta = 4u\,\psi^3 \quad . \qquad (29)$$

The different critical exponents and amplitudes are summarized in Table I.

2. <u>Ornstein-Zernike theory</u>. For spatial correlations the equivalent of the Landau – Van der Waals approach is the Ornstein-Zernike theory (see Stanley, 1971). It is well known that thermodynamic derivatives such as the compressibility are related to fluctuations of the variables about their mean values in equilibrium (Landau and Lifshitz, 1969), e.g.

$$\left\langle (V - \langle V\rangle)^2 \right\rangle = k_B T V K_T = - k_B T \left(\frac{\partial V}{\partial P}\right)_T \quad . \qquad (30)$$

Let us generalize the above relation to states of the system which are spatially nonuniform and time dependent, by defining ψ in terms of the local volume per particle $v(x,\tau)$, at a spatial point x and a time τ,

$$\psi(x,\tau) = \frac{v(x,\tau)}{v_c} - 1 \quad , \tag{31}$$

with $v_c = V_c/N$. The correlation function is

$$C_\psi(x,\tau) \equiv \left\langle (\psi(x,\tau) - \langle\psi\rangle)(\psi(0,0) - \langle\psi\rangle) \right\rangle \quad , \tag{32}$$

whose Fourier transform in d-dimensions is given by

$$C_\psi(x,\tau) = \int \frac{d^d k}{(2\pi)^d}\, e^{i\vec{k}\cdot\vec{x}}\, C_\psi(k,\tau) \quad . \tag{33}$$

The average in (32) is still taken over the equilibrium ensemble of the system. The study of correlation functions with finite time differences ($\tau \neq 0$) near T_c, is the subject of critical <u>dynamics</u> [Sec. III]. In the present discussion we confine ourselves to static properties, represented by equal-time correlation functions. Let us define

$$k_B T\, \chi_\psi(k) \equiv C_\psi(k,\tau=0) = C_\psi(k) \quad , \tag{34}$$

in terms of which Eq. (30) may be written as

$$\chi_\psi = \chi_\psi(k=0) = (k_B T)^{-1} \int d^d x\, C_\psi(x,0) \quad . \tag{35}$$

The Ornstein-Zernike theory is a generalization of (35) to finite k, which follows from the assumption, analogous to (16), that the small quantity $\chi_\psi^{-1}(k)$ may be expanded in k^2 near T_c, for $k \to 0$,

$$\chi_\psi^{-1}(k) = \chi_\psi^{-1} + k^2 c_1 + \ldots \quad . \tag{36}$$

Here c_1 is phenomenological parameter assumed to remain finite at T_c. Equation (36) may be rewritten

$$\chi_\psi^{-1}(k) = \chi_\psi^{-1}[1 + k^2\xi^2 + \ldots] \quad , \tag{37}$$

where the quantity

$$\xi = [c_1/\chi_\psi]^{1/2} = \xi_o^{\pm}|t|^{-\nu} \tag{38}$$

has the dimensions of a length, which diverges at T_c, with an exponent

$$\nu = \gamma/2 = 1/2 \quad , \tag{39}$$

and amplitudes

$$\xi_o^+ = (c_1\Gamma)^{1/2} \quad , \quad \xi_o^- = (c_1\Gamma')^{1/2} \quad , \tag{40}$$

above and below T_c, respectively. The small-k behavior of $\chi_\psi(k)$ is reflected in the large-x form of the Fourier transform in d-dimensions

$$C_\psi(x) \sim \frac{c_1' \; e^{-x/\xi}}{x^{d-2}} \quad , \tag{41}$$

which shows that ξ measures the range of correlations of ψ in the fluid. At T_c, the length ξ is infinite and the correlations fall off as a power of the distance

$$C_\psi(x) \sim \frac{c_1'}{x^{d-2+\eta}} \quad , \quad x \to \infty \quad , \quad T = T_c \tag{42}$$

where according to (41) we have

$$\eta = 0 \tag{43}$$

in the Ornstein-Zernike theory. Alternatively, the Fourier transform at T_c is written as

$$\chi_\psi(k) \sim 1/c_1 k^{2-\eta} \quad , \quad k \to 0 \quad , \quad T = T_c \quad . \tag{44}$$

As is well known, the equal-time density-correlation function of a fluid may be measured by quasielastic scattering of light or neutrons (see Heller, 1967). At the critical point the growth in scattering intensity associated with the divergence of χ_ψ is known as critical opalescence.

3. Law of corresponding states vs. universality. In the Van der Waals theory, each fluid is characterized by only 2 parameters, a and b in Eq. (1), which determine the critical constants P_c, V_c, T_c. If one defines the reduced variables ψ, H, and t as in Eqs. (9-11), then the equation of state is the same for all fluids. This property is a consequence of the "law of corresponding states", which holds exactly in the Van der Waals theory. In particular, the exponents and critical amplitudes are the same for all fluids.

The Landau theory is a generalization of the Van der Waals theory, in which P_c, V_c, and T_c are independent phenomenological parameters, as are the constants $\bar{r}$ and u in Eq. (16). The critical amplitudes are functions of $\bar{r}$ and u, and are therefore not in general the same for all systems. The critical exponents, on the other hand, are pure numbers which do not depend on $\bar{r}$ and u. Moreover, there exist certain combinations of critical amplitudes which are also independent of $\bar{r}$ and u, and are therefore the same for all fluids in this theory. Examples of such universal amplitude ratios are

$$\Gamma/\Gamma' = 2 \quad , \quad \Gamma D B^2 = 1 \quad , \quad \Delta C_v \Gamma B^{-2} = \frac{1}{2} \quad . \tag{45}$$

The foregoing discussion, which is summarized in Tables I and II, presents the simple mean-field picture of static critical phenomena in a language appropriate to fluids. As is well known, the theory may be applied to a variety of other critical points, such as those of binary mixtures, superfluids, magnets, alloys, etc., but in no case does it describe the observed singular behavior in the vicinity of T_c in a quantitative way (see Fisher, 1967). In the next section we shall discuss a model of a fluid which does appear to be correct asymptotically close to T_c.

B. The Lattice Gas and Ising Models

1. Lattice-gas approximation. The lattice gas yields a rather crude approximation to the partition function of a fluid, but one which retains the important statistical fluctuations in the system, which dominate the critical behavior (Yang and Lee, 1952; Fisher, 1967). Let us consider a fluid with N particles in a volume V, interacting via pairwise potentials

$$U_N(x_1, \ldots x_N) = \sum_{\langle x,x' \rangle} \hat{\phi}(|x - x'|) \quad , \tag{46}$$

where the sum in (46) is over all pairs of particles. The partition function is given by

$$Z_{f\ell}(T,z,V) = \sum_{N=0}^{\infty} z^N Q(T,N,V) \quad , \tag{47}$$

$$Q(T,N,V) = (N!)^{-1} \int dx_1 \ldots dx_N \exp(-U_N/k_B T) \quad , \tag{48}$$

where $z = \exp(\mu/k_B T)$ is the activity, and μ the chemical potential. Let us now divide the system into $\mathcal{N}$ cells and replace the density by an occupation number

$$\sigma_i = 1 \quad , \quad \text{if the } i^{th} \text{ cell is occupied,} \tag{49a}$$

$$\sigma_i = -1 \quad , \quad \text{if the } i^{th} \text{ cell is empty.} \tag{49b}$$

The potential $\hat{\phi}$ is approximated by

$$-J_{ij} = \hat{\phi}(|x_i - x_j|)\Big|_{av} \quad , \quad i \neq j \tag{50a}$$

$$-J_{ii} = \infty \quad , \tag{50b}$$

where the average in (50a) is taken with x_i in the ith cell and x_j in the jth cell. Under these approximations it is easy to show that the partition function becomes that of a ferromagnetic Ising model (Fisher, 1967)

$$Z_I(T,H) = \sum_{\{\sigma_i=\pm 1\}} \exp(k_B T)^{-1} \left[\sum_i H\sigma_i + \sum_{<ij>} J_{ij}\sigma_i\sigma_j \right] \quad , \tag{51}$$

where the magnetic field H is identified with the chemical potential μ (to within an additive constant). The order parameter ψ, which is proportional to the volume or density in the original fluid, is the magnetization M in the Ising model, whereas the compressibility of the fluid becomes the susceptibility of the magnetic system.

The lattice-gas approximation leads to a partition function and coexistence curve which are symmetric under reversal of the sign of ψ, a property which is shared by the mean-field expression (16). If we write the sum of the coexisting densities as

$$\rho_L + \rho_G = 2\rho_c[1 + a(t)] \quad , \tag{52}$$

then $a(t) = 0$ for the lattice gas. More generally, the coexistence curve is asymmetric, but according to the empirical "law of rectilinear diameter", $a(t)$ is an analytic function

$$a(t) = at + \ldots \quad , \tag{53}$$

where a is a constant. For real fluids there is some experimental evidence, which is corroborated by model calculations, for a violation of this law, with

$$a(t) = at^{1-\alpha} \quad , \tag{54}$$

where α is the specific heat exponent (Weiner et al., 1974; Rehr and Mermin, 1973; Ley-Koo and Green, 1977).

1. Lattice-gas assumption. Having made the approximation of replacing the exact partition function $Z_{f\ell}$ by Z_I [Eq. (51)], we formulate the hypothesis that the leading singularities are identical for both systems, so that a study of the critical behavior of the Ising model should give exact information on fluids near the critical point. This hypothesis seems a priori rather unlikely to be verified exactly, but we shall see that it is a special cases of the much more general universality hypothesis, whose verification is a central achievement of the modern theory of critical phenomena.

2. Critical behavior of the Ising model. Let us note briefly what is known concerning the critical behavior of the Ising model, which is one of the simplest models displaying nontrivial critical behavior, and has been studied extensively for many years (see Fisher, 1967; Stanley 1971; Domb, 1974).

(i) The mean-field approximation has precisely the form given in (16), where the parameters T_c, $\bar{r}$ and u can be expressed in terms of the interaction J_{ij}.

(ii) In 2 dimensions, the Ising model in zero field was solved exactly by Onsager (1944), with results which differed dramatically from those of mean-field theory. The specific heat, for example, diverges logarithmically at T_c. The ensuing exponents and amplitude ratios are listed in Table II.

(iii) In 3 dimensions the model is not soluble, but there exist a number of sophisticated calculational schemes, based on series extrapolations, which yield results for exponents and amplitudes, thought to be accurate to within a few percent (see Domb, 1974). An important check on the various methods is an evaluation in 2 dimensions, where the exact answer is known, and the results are generally quite impressive. Exponents and amplitude ratios obtained in 3 dimensions by various methods (high and low temperature series, renormalization group) are also listed in Table II, along with measured values for a number of fluids. It is clear from the numbers given there, that the 3-dimensional Ising magnet or lattice gas is a much better model for the critical behavior of a fluid than the Van der Waals theory or the 2-dimensional Ising model. We shall examine the validity of the lattice-gas hypothesis in more detail below.

3. Phenomenological scaling theory. A phenomenological theory, known as scaling, was developed some years ago, in an effort to understand, or at least systematize, the information contained in the exponent and amplitude values calculated in the various models, and measured on real systems (see Fisher, 1967; Griffiths, 1967). The scaling approach may be formulated (Widom, 1965) as an attempt to generalize the equation of state of mean-field theory (17), which may be written as

$$H = \psi^3[4u + \bar{r}t/\psi^2] = \psi^\delta[4u + \bar{r}t/\psi^{1/\beta}] \quad . \tag{55}$$

The generalization is

$$H = \psi|\psi|^{\delta-1}\, h(t/|\psi|^{1/\beta}) \quad , \tag{56}$$

where the function $h(x)$ can have a more complicated form than the simple linear function in (55), and the exponents δ and β can have values other than 3 and 1/2. It is easy to show by differentiation of (56) that the susceptibility takes the form

$$\chi_\psi = t^{-\beta(\delta-1)}\, h_\chi(t/|\psi|^{1/\beta}) \quad , \tag{57}$$

where the function $h_\chi(x)$ is simply related to $h(x)$. It follows from (57) that on the critical isochore, χ_ψ has exponents and amplitudes

$$\gamma = \beta(\delta-1) \quad , \tag{58}$$

Table II. Exponents and Universal Amplitude Ratios for Various Systems

Quantity	Mean-field theory	2d Ising	3d Ising	Renormalization group d = 3, Ising symmetry	Typical values for pure fluid
Ref.	a	b,c	b,c	d,e	f
α	O (jump)	O'(log)	0.10 ± 0.02	0.11 ± 0.003	0.10 ± 0.04
β	1/2	1/8	0.312 ± 0.005	0.325 ± 0.001	0.355 ± 0.007
γ	1	7/4	$1.250 \pm {}^{0.003}_{0.007}$	1.240 ± 0.001	1.19 ± 0.03
δ	3	15	-	4.815 ± 0.004	4.35 ± 0.10
ν	1/2	1	$0.638 \pm {}^{0.002}_{0.008}$	0.630 ± 0.001	0.63 ± 0.02
η	0	1/4	$0.041 \pm {}^{0.006}_{0.003}$	0.031 ± 0.002	0.07 ± 0.04
Δ_1	1	-	0.5	0.493 ± 0.007	-
A/A'	ΔC_v=const.	1	0.51	-	0.5
Γ/Γ'	2	37.69	5.07	-	4.0
$AB^{-2}\Gamma$	$\Delta C_v B^{-2} \Gamma = \frac{1}{2}$	0.319	0.059	-	0.05
$\Gamma DB^{\delta-1}$	1	6.78	1.75	-	1.6
$\xi_o^+ A^{1/d}$	-	0.398	0.26	-	0.25
ξ_o^+/ξ_o^-	1.41	3.23	1.96	-	2.0

a. Stanley (1971)
b. Domb (1974)
c. Aharony and Hohenberg (1976)
d. Baker et al. (1976)
e. Le Guillou and Zinn-Justin (1977)
f. Levelt Sengers and Sengers (1975)

$$\Gamma = h_\chi(\infty) \quad , \tag{59}$$

above T_c, and

$$\gamma' = \gamma \tag{60}$$

$$\Gamma' = h_\chi(-B^{-1/\beta}) , \tag{61}$$

below T_c. Similarly, it may be shown that the form (56) implies the additional <u>scaling relations</u>

$$\alpha = \alpha' \tag{62}$$

$$2-\alpha = \beta(\delta+1) = 2\beta + \gamma \quad . \tag{63}$$

Thus all thermodynamic exponents are expressed in terms of two of them, β and δ, say.

The Ornstein-Zernike expression for the correlation function (37),

$$\chi_\psi(k) = \frac{\chi_\psi}{1 + k^2\xi^2} \quad ,$$

may also be generalized to

$$\chi_\psi(k) = \chi_\psi \; g(k\xi) \quad , \tag{64}$$

which implies the scaling relations (Widom, 1965; Kadanoff, 1966)

$$\gamma = \nu(2 - \eta) \quad , \tag{65}$$

$$d\nu = 2 - \alpha \quad . \tag{66}$$

The evidence in favor of the scaling hypothesis for models and real systems is quite impressive, and the reader is referred to the literature for further information (see Fisher, 1967; Heller, 1967; Stanley, 1971). It should be remarked, however, that the best series values for the 3-dimensional Ising model, shown in Table II, seem to violate the relation $d\nu = 2\beta + \gamma$ obtained from (63) and (66). On the other hand, scaling is exactly obeyed in 2 dimensions, and it is also a consequence of the renormalization group theory, which we shall now briefly discuss.

C. Universality and the Renormalization Group

1. Universality. We have seen earlier that mean-field theories lead to exponents and amplitude ratios which are the same for all critical points (magnetic, gas-liquid, superfluid), regardless of dimensionality, symmetry, etc. This "maximal universality" is in fact not fully obeyed either experimentally or in more accurate theories, although the critical behavior is observed to be insensitive to many microscopic details. One of the main achievements of the modern theory has been to elucidate and refine the concept of universality as it applies to real systems.

Let us begin by listing some examples of the insensitivity of critical behavior to microscopic properties:

(i) For the 2-dimensional Ising model the exact calculation by Onsager (1944) already showed that the critical exponents do not depend on the lattice, or on the ratio J_x/J_y of interaction constants in the x and y directions (for a square lattice), even though T_c, for instance, does depend on these properties.

(ii) In the 3-dimensional Ising model, series extrapolation results are quite consistent with exponents and amplitude ratios which are insensitive to lattice structure, spin value, and interaction range (as long as it remains finite) (see Domb, 1974).

(iii) Measured values of exponents and amplitude ratios for fluids have been known for some time to be universal (see Table II), although small systematic differences appeared to exist between the critical point of pure fluids on the one hand, and the consolute point of binary mixtures on the other (see Levelt Sengers, 1974).

(iv) Similarly, a comparison of fluids with the Ising model shows a certain similarity, but small differences between the two systems which are apparent in Table II, had been noticed (Levelt Sengers, 1974). We shall comment on these differences in Sec. D below.

2. Renormalization group. A precise formulation of universality follows from the renormalization group theory of critical behavior (see references in Sec. IV, below). In this approach,

the critical point is described in terms of the fixed point of a special type of scale change R, known as a renormalization group transformation, applied to the Hamiltonian of the system. A full discussion of these ideas is far beyond the scope of these notes, but we would like to illustrate the basic concepts schematically, in order to explain the notion of universality. Let us consider an abstract space, each point of which is a Hamiltonian, either for a model or for a real system (more precisely, each point also specifies the temperature, pressure, external field, etc., i.e., it corresponds to a state of the system). The transformation R takes one point in this space into another. Let us start from a system in a particular state which we denote by the point H_o of our space (see Fig. 2). Repeated application of R will generate a trajectory in the space (also known as a Hamiltonian flow). Each point along a trajectory provides an equivalent description of the original system, but with each application of R the length scale is expanded and the "resolution" is coarsened. According to the Wilson renormalization group hypothesis, if H_o is at the critical point of the system, ($T = T_c$, $P = P_c$, etc.) then the flow emanating from H_o will end at a fixed point H^* of R, i.e. at a point such that

$$RH^* = H^* \quad . \tag{67}$$

Moreover, the critical exponents pertinent to H_o can be obtained by linearizing the transformation R in the vicinity of H^*

$$R(H - H^*) \approx R_L(H - H^*) \quad , \tag{68}$$

and diagonalizing the linear transformation R_L. Indeed, if the system H_o is not originally chosen to be at its critical point, then the flow will not reach the fixed point, but the rate at which it "passes" H^* determines the critical exponents, and this rate can be calculated by diagonalizing R_L. Thus the critical exponents are characteristic of the fixed point reached.

The set of systems in our space whose trajectories reach a particular fixed point H^* is called the universality class of H^*. All systems in a given universality class have the "same" critical behavior, in a sense which will be made more precise below. A relevant variable (or field) is one which takes a system outside of its universality class, whereas an irrelevant variable changes the state without removing it from its universality class. For example, the addition of a next-nearest-neighbor interaction to a nearest-neighbor Ising model will be an irrelevant change, since both systems will reach the same "Ising-like" fixed point, if the temperature in each system is equal to its critical value. By

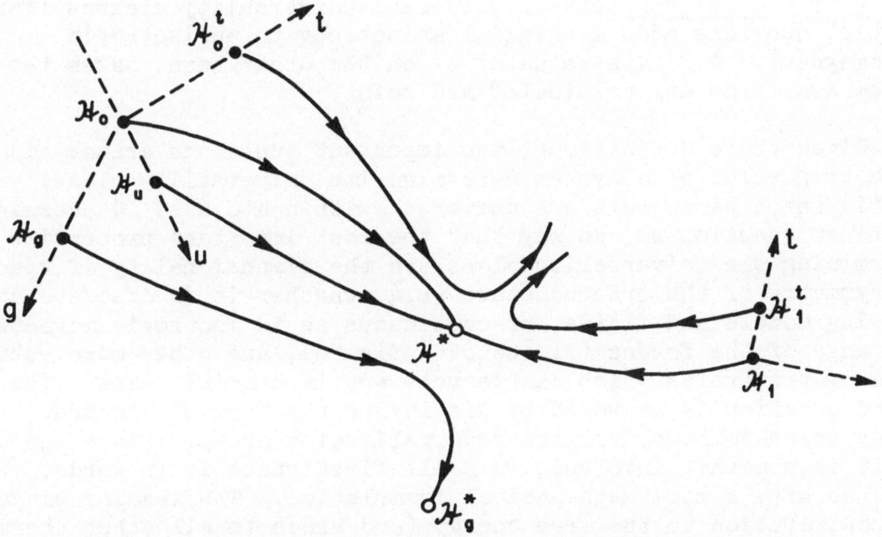

Fig. 2 Abstract space of states and trajectories generated by the renormalization group transformation $\mathcal{R}$. The system H_o is at its critical point (t = 0) and its trajectory passes through the <u>fixed point</u> H^*. When T is not equal to T_c ($t \neq 0$), the trajectory starting at H_o^t does not reach a finite fixed point, but the rate at which it passes H^* determines the critical exponents for that fixed point. The field u is an <u>irrelevant field</u>, since the system H_u can be chosen at its critical temperature, and it reaches the same fixed point H^*. The field g is a <u>relevant field</u>, since for finite g the system H_g (when chosen at its critical temperature) <u>crosses over</u> to a different fixed point H_g^*. The point H_1 is another example of a system reaching H^*, i.e. belonging to the same <u>universality class</u> as H_o and H_u.

extension of the above definitions, we can say that the reduced temperature $t = (T - T_c)/T_c$ is also a relevant field, since a finite value of t prevents the system from reaching its fixed point. In the immediate vicinity of H^*, there is no essential difference between t and any other relevant field. Globally, however, a distinction can be made, since certain fields will change the fixed point from one type to another, in which case one speaks of a crossover between different universality classes (for example, when one adds a uniaxial anisotropy to an isotropic ferromagnet). A finite value of t, on the other hand, takes the system away from any critical fixed point.

Given these definitions, two important questions arise: i) which properties of a system determine its universality class? and ii) which parameters are universal within a class? Concerning the first question we can say that the most important properties determining the universality class are the dimensionality of space, the symmetry of the ordered state (e.g. whether it is discrete as in Ising models and fluids, or continuous as in isotropic magnets), the range of the forces (finite or infinite), and other more subtle symmetry properties which can be relevant in special cases. The second question is answered by displaying the form of the free energy which follows from the renormalization group. Since the result is somewhat involved, we shall first state it in words, and then give a more mathematical formulation. The leading singular contribution to the free energy (and hence to all other thermodynamic functions) has a scaling form analogous to (56), in which the exponents and scaling functions are universal, i.e. the same for all members of a class. The only system-dependent (nonuniversal) quantities are two constants which fix the thermodynamic scales, as well as the values of the critical parameters T_c, V_c, $P_s(T)$, which are used to define the variables t, ψ, and H. In order to describe data a finite distance from the critical point, singular corrections to scaling must be added to the leading terms (Greywall and Ahlers, 1972; Saul et al., 1975; Wegner, 1972a). Each such correction involves an additional nonuniversal constant, and a universal exponent and scaling function. Only a finite number of correction terms contribute if one remains close to the critical point (typically one or two, for $|t| < 10^{-2}$). The correlation function also has a scaling form, as in (64), with no new nonuniversal constants. This last result is known as "two-scale-factor universality", since the two nonuniversal thermodynamic scale factors suffice to determine the leading singularity in the correlation function (Stauffer et al., 1972; Hohenberg et al., 1976).

The significance of the fixed point concept was spelled out dramatically by the papers of Wilson and Fisher (1972), and Wilson (1972), who showed that above 4 dimensions a Gaussian fixed point (with mean-field exponents) is stable, whereas below 4 dimensions the Gaussian fixed point is unstable, and is replaced by the "nontrivial" or "Wilson" fixed point. A great deal of qualitative information on fixed point stability can be obtained by studying the renormalization group transformation $\mathcal{R}$ in the vicinity of four dimensions. A generalization of the dimensionality d to a continuous variable leads to the famous ε-expansion in the small parameter

$$\varepsilon = 4 - d \quad . \tag{69}$$

This expansion yields surprisingly accurate quantitative information on the critical exponents when extrapolated to three dimensions ($\varepsilon = 1$). The most important contribution of the renormalization group method, however, has undoubtedly been the more qualitative aspect of the theory, i.e. its ability to predict which variables will be relevant, and in general to identify a small number of fixed points, which characterize most of the physically important universality classes (see Fisher, 1974).

An interesting application of the ε-expansion is to the question of the validity of neglecting the asymmetry in the coexistence curve, as is done in the lattice-gas model. This approximation is tested by introducing asymmetric terms into the Hamiltonian of a symmetric system, and studying the stability of the symmetric fixed point under such a perturbation, near four dimensions. For example Hubbard and Schofield (1972) showed that a term proportional to ψ^3 in (16) can be eliminated by adjusting the chemical potential μ. There has been a recent suggestion by Valls and Hertz (1977), that a term in ψ^5 becomes relevant at $d \approx 3.33$, and leads to a fixed point in three dimensions which is distinct from that of the Ising model. This claim seems unlikely to us to be correct, however, in view of the result of Wegner (1972b), who showed that for the "nontrivial" Ising fixed point, a ψ^5 term becomes more irrelevant as the dimensionality is lowered below $d = 4$. It is for the Gaussian fixed point that a ψ^5 term becomes relevant at $d = 3.33$, but this fixed point is already destabilized by the ψ^4 term just below $d = 4$.

In order to give a more precise formulation of the results of renormalization group theory, it is convenient to change variables from P and V to the chemical potential μ and the density ρ. The reduced variables are

$$H = [\mu - \mu_o(T)]/\mu_c \quad , \tag{70}$$

$$\psi = [\rho - \rho_o(T)]/\rho_c \quad , \tag{71}$$

$$t = (T - T_c)/T_c \quad , \tag{72}$$

where for $T < T_c$, $\mu = \mu_o(T)$ is the equation for the coexistence curve $[\mu_c = \mu_o(T_c)]$, and for $T > T_c$, $\mu_o(T)$ is any smooth extrapolation of the coexistence curve; the function $\rho_o(T)$ is defined such that $\psi = 0$ for $H = 0$ above T_c, and $\rho_o(T) = \rho_c$ below T_c. Reduced variables analogous to the above can be defined for binary fluids, superfluids, magnets, alloys, etc. The free energy (per unit volume) then takes the form (Wegner, 1972a)

$$F(\mu,T) = F_{reg}(\mu,T) + F_{sing}(H,t) \quad , \tag{73}$$

where $F_{reg}(\mu,T)$ is an analytic function near the critical point, and

$$F_{sing}(H,t) = a_o|t|^{2-\alpha} f(t/b_o|H|^{1/\Delta})\Big[1 + a_1|t|^{\Delta_1} f_1(t/b_o|H|^{1/\Delta}) + a_2|t|^{\Delta_2} f_2(t/b_o|H|^{1/\Delta}) + \ldots\Big] , \tag{74}$$

with

$$0 < \Delta_1 < \Delta_2 < \ldots \quad , \tag{75}$$

and a_o, a_1, a_2,...b_o are constants. The function $F(H,t)$ is the Legendre transform of the free energy $\Phi(\psi,t)$ considered above. When t is small, the terms in a_i with $i \geq 1$ are small <u>corrections</u> to the leading asymptotic behavior, because of (75). Let us impose the conditions

$$f(0) = 1 \quad , \quad f(-1) = 1 \quad , \tag{76}$$

on the scaling function f. This can be done by a proper choice of the constants a_o and b_o, i.e. by fixing the scales of the free energy and the field H, respectively. Similarly, we impose the conditions

$$f_i(0) = 1 \quad , \quad i \geq 1 \quad , \tag{77}$$

by a proper choice of the a_i. It is easy to calculate the thermodynamic functions by differentiating F(H,t), e.g. the equation of state is

$$\psi = a_o|t|^{\beta}\, m(t/b_o|H|^{1/\Delta})\left[1 + a_1|t|^{\Delta_1}\, m_1(t/b_o|H|^{1/\Delta}) + \ldots\right] \quad (78)$$

with

$$\beta = 2 - \alpha - \Delta \quad , \quad (79)$$

which implies

$$\Delta = \beta\delta \quad . \quad (80)$$

The functions m, m_i are related to the corresponding functions f, f_i, and Eqs. (76-77) for the f's imply similar, more complicated conditions for the m's. Note that the leading term in the equation of state (78) has the scaling form (56) discussed earlier.

We are now in a position to state the universality properties of thermodynamic functions: all members of a class have the same exponents α, β, Δ_1, Δ_2, ... , and the same scaling functions f, f_1, f_2 ... . [This is why the exponents and scaling functions are referred to as universal]. It may be shown that certain amplitude ratios can be expressed in terms of the universal function f(x) and the exponents, so that these ratios are also predicted to be universal. Only the scale constants a_o, a_1, a_2, ... and b_o are different from system to system (as well as T_c, ρ_c, and $\mu_o(T)$, which serve to define the variables). Note that if the law of corresponding states were valid, then the constants a_o, b_o, a_i, would also be universal, since in that case only the critical constants ρ_c, T_c, and $\mu_o(T)$ would differ from system to system. As mentioned earlier, however, this law is already violated in the Landau theory.

The foregoing discussion is still somewhat oversimplified. In a more correct formulation the free energy has the form (Wegner, 1972a)

$$F_{sing}(H,t) = a_o|g_t|^{2-\alpha}\, f(g_t/|g_H|^{1/\Delta})\left[1 + a_1|g_t|^{\Delta_1}\, f_1(g_t/|g_H|^{1/\Delta}) + \ldots\right] \quad (81)$$

where the scaling fields g_t and g_H are analytic functions of t and H, which reduce to $g_t = t$ and $g_H = b_o H$ to lowest order in t and H. For a symmetric system g_H is linear in H to all orders, and g_t is a (nonlinear) function of t alone. More generally, the coefficients in the expansion of $g_t(t,H)$ and $g_H(t,H)$ are further nonuniversal parameters, which will appear as correction terms in an expansion in t and H of the form (74), and complicate the universality properties of the higher-order corrections f_2, f_3, etc., and of the exponents Δ_1, Δ_2, ... The inclusion of correction terms as in (78) is sometimes referred to as "extended" scaling, whereas the mixing of t and H in g_t and g_H is called "revised" scaling (see Sengers and Levelt Sengers, 1977).

The correlation function at small k takes on a scaling form similar to (64) near the critical point, namely

$$\chi_\psi(k) = c_o k^{-2+\eta} g(k\xi)\left[1 + a_1 k^{\Delta_1/\nu} g_1(k\xi) + a_2 k^{\Delta_2/\nu} g_2(k\xi) + \dots\right]. \tag{82}$$

Let us define the scaling function g such that

$$g(0) = 1 \quad . \tag{83}$$

Then the constant c_o is related to the thermodynamic scales by condition (35). The length scale defined by ξ may be chosen such that

$$g(1) = 1 \quad . \tag{84}$$

Under these conditions the functions g(z) and $g_1(z)$ are universal, and the correlation function contains only one new constant, the length ξ_o. It turns out, however, that this length is related to the thermodynamic scales by a universal ratio (Stauffer et al., 1972; Hohenberg et al., 1976)

$$R_\xi = \xi_o A^{1/d} \quad , \tag{85}$$

where A is the specific heat amplitude which is simply related to the coefficient a_o appearing in the free energy (per unit volume) (74). Thus the correlation function contains no new nonuniversal constants besides those present in the thermodynamic functions.

D. Summary of Experiments on Static Properties

1. The situation up to 1973. As mentioned above, thermodynamic and correlation data in fluids were shown to obey scaling quite accurately, with measured values of the critical exponents and amplitudes which were reasonably close to series results for the Ising model. Upon closer examination, however, it appeared that the consistency of these quantities in pure fluids, binary mixtures, and Ising models separately, was much greater than the similarities between these different systems (Levelt Sengers, 1974). Indeed, Table II shows systematic discrepancies between fluids and Ising models which are outside the errors quoted for either case. This represented a breakdown of the lattice-gas hypothesis discussed in Section IIB, and it was important to clarify the situation.

2. More recent developments. The development of renormalization group theory has led to a revision of quantitative estimates, both experimental and theoretical, of the critical exponents, with results which are in much better agreement with the universality hypothesis.

The theoretical exponents were calculated from perturbation theory (Nickel, 1977) for a three-dimensional continuum field theory with Ising symmetry, using sophisticated extrapolation techniques (Baker et al., 1976; Le Guillou and Zinn-Justin, 1977). The results are given in Table II, with error bars which reflect the scatter of the different estimates, but not possible systematic errors. The differences between these values and those obtained from series expansions on lattice Ising systems, also shown in Table II, are at the moment unexplained.

Experimental estimates of the exponents have been revised by more careful attention to the possibility of singular correction terms, as in (78). Indeed, with a dimensionless nonuniversal amplitude a_1 of order unity, and a correction exponent $\Delta_1 \simeq 0.5$ [see Table II], the true asymptotic exponent differs by an amount of order 0.03, from the effective exponent, obtained by fitting (78) to a simple power law in the range $10^{-4} \lesssim |t| \lesssim 10^{-3}$. It is these effective exponents which appear in the earlier analyses listed in the last column of Table II. More recent experiments have often focussed on a region closer to T_c, $|t| < 5\times10^{-5}$, where correction terms are much less important. Alternatively, data over a wider range have been analyzed with a finite correction amplitude a_1, and the results were generally shown to be consistent with Ising-like values obtained from renormalization group theory (Balfour et al., 1978; Ley-Koo and Green, 1977). A number of exceptions have been reported, however, such as the coexistence curve of Xe measured by Garland and Thoen (1976), and thermodynamic

Table III. Some Recent Experimental Results

System	Quantity	Ref	Value	Remarks
Xe	β	a	0.329	$2\times10^{-6} < \|t\| < 1.5\times10^{-5}$ no correction term
		b	0.328	$3\times10^{-6} < \|t\| < 2\times10^{-4}$
SF_6	β	a	0.321	$2\times10^{-6} < \|t\| < 2\times10^{-5}$
		f	0.327	$2\times10^{-5} < \|t\| < 10^{-2}$ Correction term with $\Delta_1 \gtrsim 0.5$
H_2O	A/A'	c	0.52	$10^{-4} < \|t\| < 5\times10^{-3}$ correction term used with leading exponents fixed at renormalization group values
	Γ/Γ'		4.96	
	$\Gamma DB^{\delta-1}$		1.7	
Binary mixture of isobutyric acid and water	β	d	0.328	$10^{-4} < \|t\| < 6\times10^{-3}$, correction term not needed
Binary mixture of 3-Methylpentane and nitroethane	ν	e	0.625 ± 0.005	$0.18 < k\xi < 26$, correction term not needed
	η		0.016 ± 0.014	

a. Hocken and Moldover (1975)
b. Hayes and Carr (1977)
c. Balfour et al. (1978)
d. Greer (1976)
e. Chang et al. (1976)
f. Ley-Koo and Green (1977)

data on the critical point of the quantum fluid ^{3}He, by Doiron et al. (1976), in which correction terms do not bring the exponents into agreement with Ising values. A representative set of recent results on both exponents and amplitude ratios are listed in Table III.

The correlation function (82), which can be measured by quasi-elastic scattering experiments, contains the exponents η and ν in addition to the scaling function g(z). Attempts to determine η either directly or via the scaling relation (65) led to conflicting results (see Sengers and Levelt Sengers, 1977). A recent experiment in the binary fluid 3-methylpentane-nitroethane (Chang et al., 1976), spanning the range $0.18 < k\xi < 26$, has yielded exponents in excellent agreement with renormalization group values (see Tables II and III). Moreover, the data were also shown to be consistent with the universal scaling function g(z) calculated in the Ising model, and inconsistent with the Ornstein-Zernike result (37).

3. Discussion and conclusions. Although many of the apparent violations of the lattice-gas hypothesis reported earlier have now been removed, either through more careful analyses or by more accurate measurements and calculations, it cannot be said that universality has been established conclusively in fluids. Apart from the specific deviations mentioned earlier, the number of cases in which universality has been verified quantitatively is still rather small, especially as regards the scaling functions and the universal amplitude ratios. In addition, very little is known on the systematics of the correction terms, except for the value of the exponent Δ_1. A number of models containing an infinite series of correction terms have been studied (Nelson and Rudnick, 1975; Nicoll et al., 1976; Chang et al., 1977), but these models are not realistic enough to yield quantitative information on corrections to scaling in real fluids (see also Green, 1977). Empirically, it is known that in binary fluids correction terms tend to have a smaller effect than in pure fluids (Greer, 1976). It would be quite interesting to systematize available data on corrections to scaling, in order to see if empirical trends become apparent. A convenient method seems to be the parametrization in terms of effective exponents $\beta^*(t)$, $\gamma^*(t)$, as originally suggested by Kouvel and Fisher (1964), and more recently employed by Riedel and Wegner (1974), and by Hayes and Carr (1977). Such a parametrization does not remove the necessity for direct least-squares fits, but it is a useful additional method of displaying data, which removes some of the arbitrariness in the initial choice of fitting function.

In conclusion, we may say that if further work corroborates the lattice-gas hypothesis, then the modern theory will have explained the critical point of a fluid quantitatively, using a model which depends purely on the symmetry of the transition and the dimensionality of space. The chemical and structural information which distinguishes between various fluids, only enters into the results via two numerical scale factors.

III. DYNAMIC PHENOMENA (SEE OUTLINE AND SEC. IV)

IV. RECOMMENDED ADDITIONAL REFERENCES

The following is intended as a partial guide to the literature, and not as a complete bibliography.

Static critical phenomena in fluids

∿ general introductions to critical phenomena (written before 1970) with substantial discussions of scaling in pure fluids and binary mixtures: Stanley (1971), Fisher (1974), Heller (1967), Griffiths and Wheeler (1970).

∿ discussions for nonspecialists of recent theory and experiments confirming universality: Levelt-Sengers et al. (1977).

∿ reviews of experiments and modern concepts: Voronel (1976), Anisimov (1974); Sengers and Levelt Sengers (1977).

∿ superfluid transition in ^{4}He: Ahlers (1977).

Renormalization group

∿ elementary discussions: Wilson (1973), Nelson (1977).

∿ textbooks: Ma (1976), Toulouse and Pfeuty (1975).

∿ lecture notes: Mazenko (1977), Widom (1975).

∿ review articles: Wilson and Kogut (1974), Fisher (1974), Barber (1977), Domb and Green (1976), Patashinskii and Pokrovskii (1977).

Critical dynamics

∿ Swinney and Henry (1973), Sengers (1973), Hohenberg and Halperin (1977 , and references therein), Calmettes (1977), Mazenko (1977).

Acknowledgements

The author is indebted to J. M. H. Levelt Sengers, G. Ahlers, and D. R. Nelson for informative discussions.

REFERENCES

Aharony, A. and P. C. Hohenberg, 1976, Phys. Rev. B13, 3081.

Ahlers, G., 1977, in Quantum Liquids, ed. by J. Ruvalds and T. Regge, North-Holland, Amsterdam.

Anisimov, M. A., 1974, Usp. Fiz Nauk 114, 249 [Sov. Phys.-Usp. 17, 722 (1975)].

Baker, G. A., B. G. Nickel, M. S. Green, and P. I. Meiron, 1976, Phys. Rev. Lett. 36, 1351.

Balfour, F. W., J. V. Sengers, M. Moldover, and J. M. H. Levelt Sengers, 1978, in Proceedings, 7th Symposium ASME, ed. by A. Cezairliyan and J. V. Sengers (to be published).

Barber, M. N., 1977, Phys. Rep. 29C, 1.

Calmettes, P., 1977, Phys. Rev. Lett. 39, 1151.

Chang, R. F., H. Burstyn, J. V. Sengers and A. J. Bray, 1976, Phys. Rev. Lett. 37, 1481.

Chang T. S., C. W. Garland, and J. Thoen, 1977, Phys. Rev. A16, 446.

Doiron, T., R. P. Behringer, and H. Meyer, 1976, J. Low Temp. Phys. 24, 345.

Domb, C., 1974, in Phase Transitions and Critical Phenomena, ed. by C. Domb and M. S. Green, Academic, N.Y., Vol. 3.

Domb, C. and M. S. Green 1976, eds. Phase Transitions and Critical Phenomena, Academic, N.Y., Vol. 6.

Fisher, M. E., 1967, Rep. Prog. Phys. 30, 615.

Fisher, M. E., 1974, Rev. Mod. Phys. 46, 597.

Green, M. S. 1977, in Statistical Mechanics and Statistical Methods in Theory and Application, ed. by U. Landman, Plenum, N.Y.

Greer, S. C., 1976, Phys. Rev. A14, 1770.

Greywall, D. S.,and G. Ahlers, 1972, Phys. Rev. Lett. 28, 1251.

Griffiths, R. B., 1967, Phys. Rev. 158, 176.

Griffiths, R. B., and J. C. Wheeler, 1970, Phys. Rev. A2, 1047.

Hayes, C. E. and H. Y. Carr, 1977, Phys. Rev. Lett. 39, 1558.

Heller, P., 1967, Rep. Prog. Phys. 30, 731.

Hocken, R., and M. R. Moldover, 1975, Phys. Rev. Lett. 37, 29.

Hohenberg, P. C., A. Aharony, B. I. Halperin, and E. D. Siggia, 1976, Phys. Rev. B13, 2986.

Hohenberg, P. C., and B. I. Halperin, 1977, Rev. Mod. Phys. 49, 435.

Hubbard, J., and P. Schofield, 1972, Phys. Lett. 40A, 245.

Kadanoff, L. P., 1966, Physics 2, 263.

Kouvel, J. S., and M. E. Fisher, 1964, Phys. Rev. 136, A1626.

Landau, L. D., and E. M. Lifshitz, 1969, Statistical Physics, Addison-Wesley, Reading, Mass. 2nd edition.

Le Guillou, J. C., and J. Zinn Justin, 1977, Phys. Rev. Lett. 39, 95.

Levelt Sengers, J. M. H., 1974, Physica 73, 73.

Levelt Sengers, J. M. H., 1976, Physica 82A, 319.

Levelt Sengers, J. M. H., R. Hocken, and J. V. Sengers, 1977, Physics Today (to be published).

Levelt Sengers, J. M. H., and J. V. Sengers, 1975, Phys. Rev. A12, 2622.

Ley-Koo, M., and M. S. Green, 1977, Phys. Rev. A (Dec. issue).

Lunacek, J. H., and D. S. Cannell, 1971, Phys. Rev. Lett. 27, 841.

Ma, S., 1976, Modern Theory of Critical Phenomena, Benjamin, N.Y.

Mazenko, G., 1978, in Correlation Functions and Quasiparticle Interactions, ed. by J. W. Halley, Plenum, N.Y.

Nelson, D. R., 1977, Nature 269, 379.

Nelson, D. R., and J. Rudnick, 1975, Phys. Rev. Lett. 35, 178.

Nickel, B. G., 1977 (to be published).

Nicoll, J. F., T. S. Chang, and H. E. Stanley, 1976, Phys. Rev. Lett. 36, 113.

Onsager, L., 1944, Phys. Rev. 65, 117.

Patashinskii, A. Z., and V. L. Pokrovskii, 1977, Usp. Fiz. Nauk. 121, 55 [Sov. Phys.-Usp. (to be published)].

Rehr, J. J. and N. D. Mermin, 1973, Phys. Rev. A7, 379.

Riedel, E. K., and F. J. Wegner, 1974, Phys. Rev. B9, 294.

Saul, D. M., M. Wortis, and D. Jasnow, 1975, Phys. Rev. B11, 2571.

Sengers, J. V., 1973, AIP Conference Proceedings No. 11, (AIP, N.Y.).

Sengers, J. V., and J. M. H. Levelt Sengers, 1977, in Progress in Liquid Physics, C. A. Croxton, ed., Wiley, N.Y., Ch. 4.

Stanley, H. E., 1971, Introduction to Phase Transitions and Critical Phenomena, Oxford, N.Y.

Stauffer, D., M. Ferer, and M. Wortis, 1972, Phys. Rev. Lett. 29, 345.

Swinney, H. L., and D. L. Henry, 1973, Phys. Rev. A8, 2586.

Toulouse, G., and P. Pfeuty, 1975, Introduction au Groupe de Renormalisation et à ses Applications, Presses Universitaires, Grenoble.

Valls, O. T., and J. A. Hertz, 1977, Phys. Rev. B (to be published).

Voronel, A. V., 1976, in Phase Transitions and Critical Phenomena, ed. by C. Domb and M. S. Green, Academic, N.Y., Vol. 5b.

Wegner, F. J., 1972a, Phys. Rev. B5, 4529.

Wegner, F. J., 1972b, Phys. Rev. B6, 1891.

Weiner, J., K. H. Langley, and N. C. Ford, Jr., 1974, Phys. Rev. Lett. 32, 879.

Widom, B., 1965, J. Chem. Phys. 43, 3898, 3892.

Widom, B., 1975, in Fundamental Problems in Statistical Mechanics, edited by E. G. D. Cohen, North-Holland, Amsterdam.

Wilson, K. G., 1972, Phys. Rev. Lett. 28, 548.

Wilson, K. G., 1973, in AIP Conference Proceedings, No. 10: Magnetism and Magnetic Materials - 1972, ed. by C. D. Graham and J. J. Rhyne (AIP, N.Y.), p. 843.

Wilson, K. G., and M. E. Fisher, 1972, Phys. Rev. Lett. 28, 240.

Wilson, K. G., and J. Kogut, 1974, Phys. Rep. 12C, 75.

Yang, C. N., and T. D. Lee, 1952, Phys. Rev. 85, 808.

Seminars

PICOSECOND LASER TECHNIQUES

A. Laubereau and W. Kaiser

Physik Department der Technischen Universität München
München, Germany

In recent years, experimental investigations of vibrational relaxation processes have received increasing attention. Interesting information is obtained from the band shapes observed in infrared and Raman spectroscopy [1, 2]. The interpretation of the measured lines is difficult, however, since several physical processes contribute in general to the observed band contours. Line broadening factors are : rotational motion, vibrational dephasing, energy relaxation, and inhomogeneous broadening due to a distribution of vibrational frequencies, e.g. isotopic line splitting. Under certain assumptions it is possible to separate the rotational contribution. The rest is sometimes called the "intrinsic vibrational part" and contains the other line-broadening factors. At the present time, one cannot isolate the different contributions by spectroscopic methods. For instance, the population lifetime of an excited vibrational state was unknown until very recently for any vibrational mode in the liquid state. Similarly, the inhomogeneous part of a spectroscopic line is not known in many cases. As a result, the time constants deduced from spectroscopic band contours are not well understood.

In this summary we briefly discuss new experimental techniques to study the dynamics of vibrational modes in liquids. Picosecond light pulses are used for direct, time-resolved investigations of ultrafast relaxation processes yielding detailed dynamical information [3]. A number of recent experiments will be presented to illustrate the state of the art and the potential of our new techniques.

The experimental methods consist of two steps. First, the excitation of molecules is achieved by an intense short laser pulse via stimulated Raman scattering or by resonant absorption of an infrared pulse. After the excitation process free relaxation e.g. energy transfer of the excited mode occurs. A second, weak probe pulse properly delayed with respect to the first pulse monitors the instantaneous state of the excited vibrational system. Three probing processes were used to obtain different dynamical information :

(i) Coherent Raman scattering occurs from the interaction of the probe pulse with the coherently excited system [4]. This scattering process is sensitive to the phase relation between the vibrating molecules and requires a carefully adjusted geometry of the relevant wave vectors. For homogeneously broadened modes (with essentially one vibrational frequency) the vibrational dephasing time is directly connected with the loss of phase correlation of the excited molecules and deduced from the decay of the probe scattering signal. The situation is more complex for molecular vibrations with isotopic substructure or with a distribution of vibrational frequencies (inhomogeneous lines). For these cases the experimental information obtained by coherent probe scattering depends on the k-matching situation.

We have developed a highly selective k-vector geometry in order to isolate one frequency component [5]. In this way we are able to measure a homogeneous dephasing time of a molecular subensemble selected out of a slowly fluctuating distribution of vibrational transition frequencies.

(ii) Spontaneous anti-Stokes Raman scattering allows the study of the instantaneous occupation of a vibrational energy level. With this technique it was possible, for the first time, to observe population lifetimes, energy transfer, and energy redistribution [6-8]. Time constants between 1 psec and several tens of psec were measured for different relaxation processes in a number of polyatomic molecules.

(iii) We have also used a fluorescence probing technique where a first infrared pulse excited a well-defined vibrational mode and a second probe pulse promotes the molecules close to the fluorescent first singlet state. The degree of fluorescence is a direct measure of the instantaneous vibrational excitation when the probe pulse is present in the sample. This technique was found very useful for investigations of highly diluted systems [9].

In this brief report we wish to discuss in some detail the following examples : (a) the dephasing time of a homogeneously broadened transition; (b) the dephasing time of a single component of a complex vibrational system and the collective beating due to different vibrational species; (c) the homogeneous dephasing of a molecular subgroup selected out of an inhomogeneously broadened vibrational band by k-vector spectroscopy; (d) the population lifetime and the energy redistribution after direct infrared excitation.

In our examples (a) to (c) transient stimulated Raman scattering is used to excite a specific vibrational mode of a polyatomic liquid. This scattering mechanism generates a coherent vibrational excitation represented by the amplitude $\langle q \rangle$, the expectation value of the normal mode operator [10]. When the excitation process has terminated, $\langle q \rangle$ decays according to the dephasing time T_2 : $\langle q \rangle \propto \exp(-t/T_2)$. The excess population n of the excited vibrational level, on the other hand, relaxes with the population lifetime T_1, $n \propto \exp(-t/T_1)$. Treating the vibrational transition as a two-level system (allowed on account of the anharmonic shifts of the higher lying levels), T_2 and T_1 are fully analogous to the time constants of a spin system. The coherent vibrational excitation is measured by coherent Raman scattering of a delayed probe pulse. $\langle q \rangle$ gives rise to a macroscopic polarization $P = N \frac{\partial\alpha}{\partial q} \langle q \rangle E$ via the Raman polarizability $\partial\alpha/\partial q$ (number density N; electromagnetic field amplitude E of the probe pulse) [5, 10]. The induced polarization generates the scattering emission, which is shifted by the vibrational frequency to the larger (anti-Stokes) or smaller (Stokes) frequencies. Since the molecules in the excited volume vibrate with a defined spatial phase relation, described by a wave vector $\vec{k}_o$ the system behaves like an oscillating three-dimensional phase grating. Scattering off this grating occurs when the k-matching condition for the relevant wave vectors of the incident probe pulse, the scattered light and the vibrational system,is fulfilled.

The experimental set-up is shown schematically in Fig. 1. A well defined single picosecond pulse is first generated by a laser system which consists of a mode-locked Nd:glass laser oscillator followed by an electro-optic switch and an optical amplifier. The pulse is subsequently converted to the second harmonic at 18910 cm^{-1} (0.53 μm) in a potassium dihydrogen phosphate crystal and traverses the liquid cell containing the sample.

Stimulated Raman scattering is effectively generated in samples of several cm of length. A beam splitter in the path of the incident pulse provides a second pulse of small intensity, approximately 10^{-2} of the exciting pulse. The weak pulse serves as a probing pulse; it passes a variable delay set-up and a second beam splitter, so that the probe pulse travels parallel to the excitation pulse (collinear geometry). In a number of experiments, non-collinear systems (not shown here) have been used. In Fig. 1, the scattering signal is highly collimated and observed with a spectrometer and photomultiplier as a function of time delay between excitation and probing pulse.

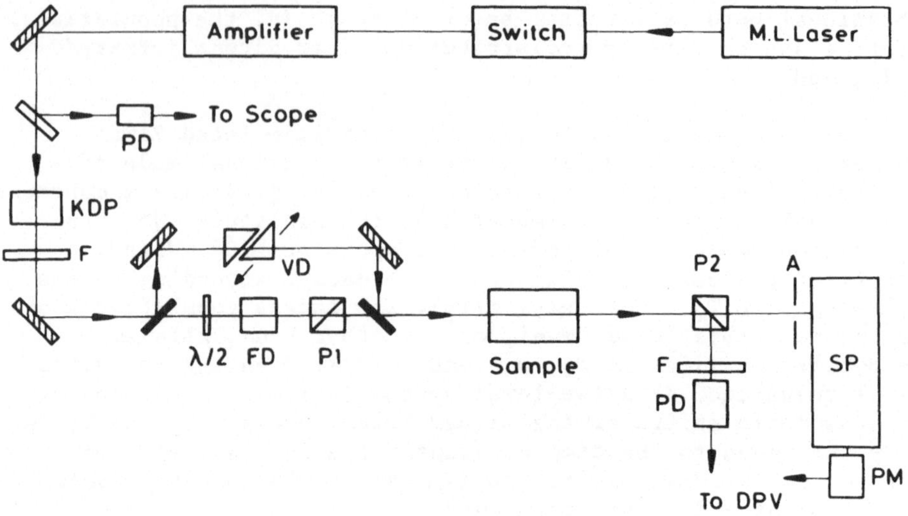

Fig. 1

Experimental set-up to measure coherent probe scattering in a collinear geometry. A beam splitter generates the probe pulse which is properly delayed before travelling collinearly with the exciting pulse. The scattered signals of the two pulses are separated by two polarizers P1 and P2.

a) As a first example we discuss the symmetrical CH_3-stretching mode of $CH_3CC\ell_3$. Fig. 2 presents our results on the dephasing time of the molecular vibration. The coherent probe scattering signal is plotted as a function of delay time between excitation and probe pulse. From the exponential decay of the signal curve we deduce a dephasing time of $T_2/2 = 1.2$ psec, which corresponds to a homogeneous line broadening of $\sim 5\ cm^{-1}$. This number is in excellent agreement with the linewidth of the isotropic scattering component observed in sponaneous Raman spectroscopy; i.e. the dynamic processes described by the dephasing time T_2 (pure dephasing and population relaxation) fully account for the spectroscopic line broadening. We have also investigated the population lifetime T_1 of the first excited level of the same vibrational mode [6, 8] . A value of $T_1 = 5.2$ psec was found for the neat liquid. The difference between the time constants $T_2/2$ and T_1 indicates that the phase correlation between the $CH_3CC\ell_3$ molecules may be lost under conservation of the excited state population (pure dephasing). Theoretical estimates of T_2 from a quasi-elastic collision model satisfactorily agree with the experimental results within a factor of approximately two [11].

b) For a homogeneously broadened mode, e.g. our example $CH_3CC\ell_3$, the time constant T_2 is directly deduced from the decay of the probe scattering signal. The situation is more complex

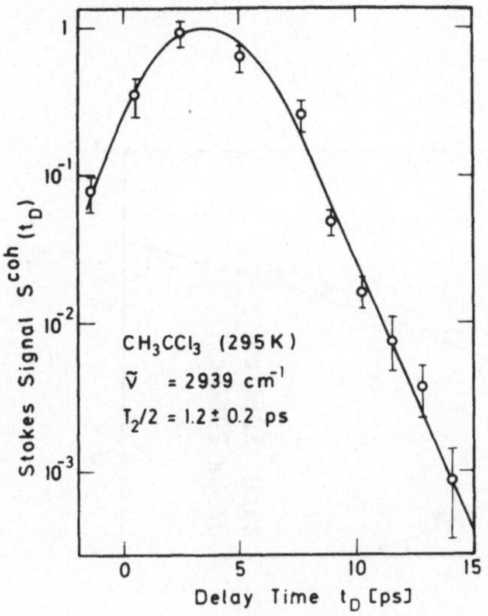

Fig. 2

Coherent Raman probe scattering signal $S^{coh}(t_D)$ versus delay time between pump and probe pulse for the symmetric CH_3-stretching mode of $CH_3CC\ell_3$.

for molecular vibrations with isotopic substructure or with a distribution of vibrational frequencies (inhomogeneously broadened lines). For these cases the observed time dependence of the coherent probe scattering depends on the k-matching situation. We have developed a highly selective k-vector geometry in order to select a molecular subensemble with one transition frequency [5]. In this way, we are able, for the first time, to measure a dephasing time of an inhomogeneously broadened vibrational band. For non-selective k-matching of the probe scattering we observe coherent superposition of neighbouring frequency components with interesting beating effects.

The tetrahalides $CC\ell_4$ and $SnBr_4$ are discussed here as examples. The isotopes of $C\ell$ and Br give rise to vibrational multiplicity of the totally symmetric tetrahedron vibration around 460 cm^{-1} and 220 cm^{-1}, respectively. In $CC\ell_4$, the isotropic structure is clearly seen in the spectrum with a frequency spacing of approximately 3 cm^{-1} of the various lines (see inset of Fig. 3a). The results for coherent probe scattering with a selective k-geometry (measuring one isotropic species) are depicted in Fig. 3a [5]. The decaying part of the signal curve represents loss of phase correlation between the coherently excited molecules with time constants $T_2/2$ = 3.6 psec. The measured dephasing time fully

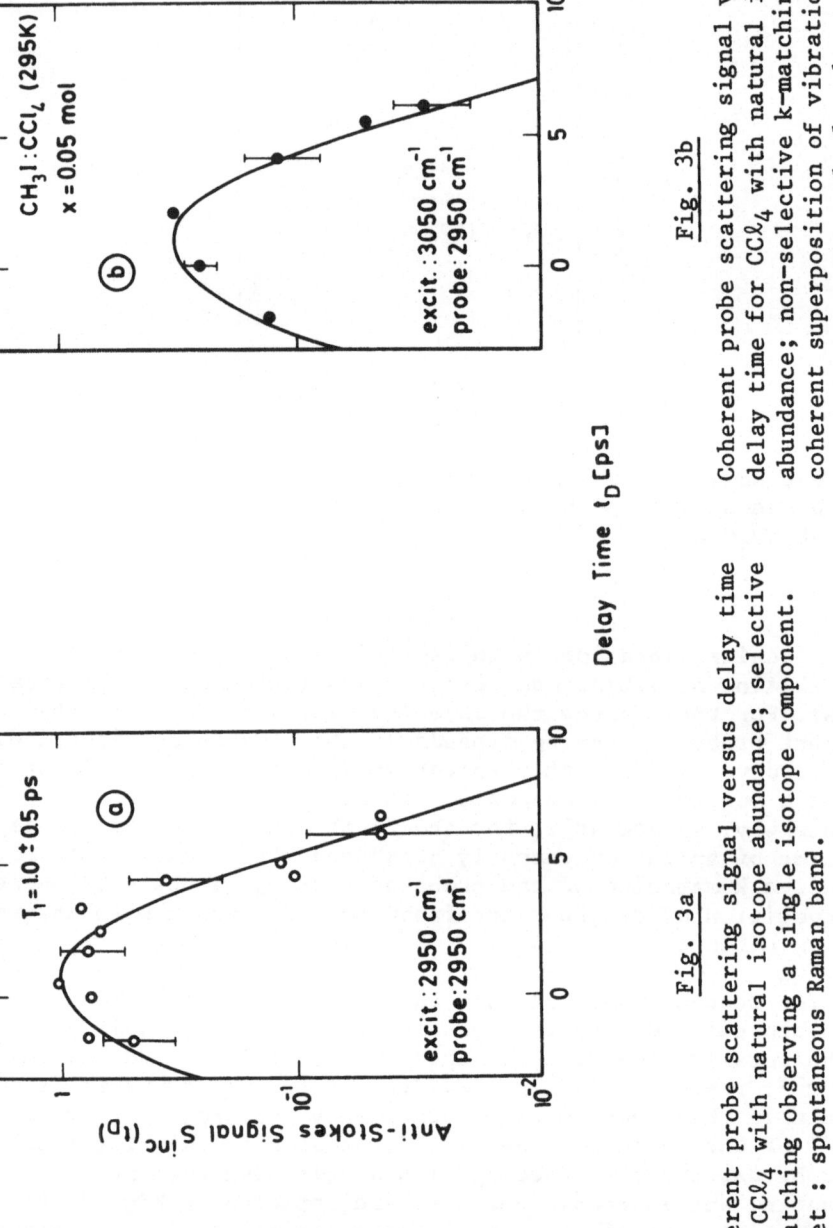

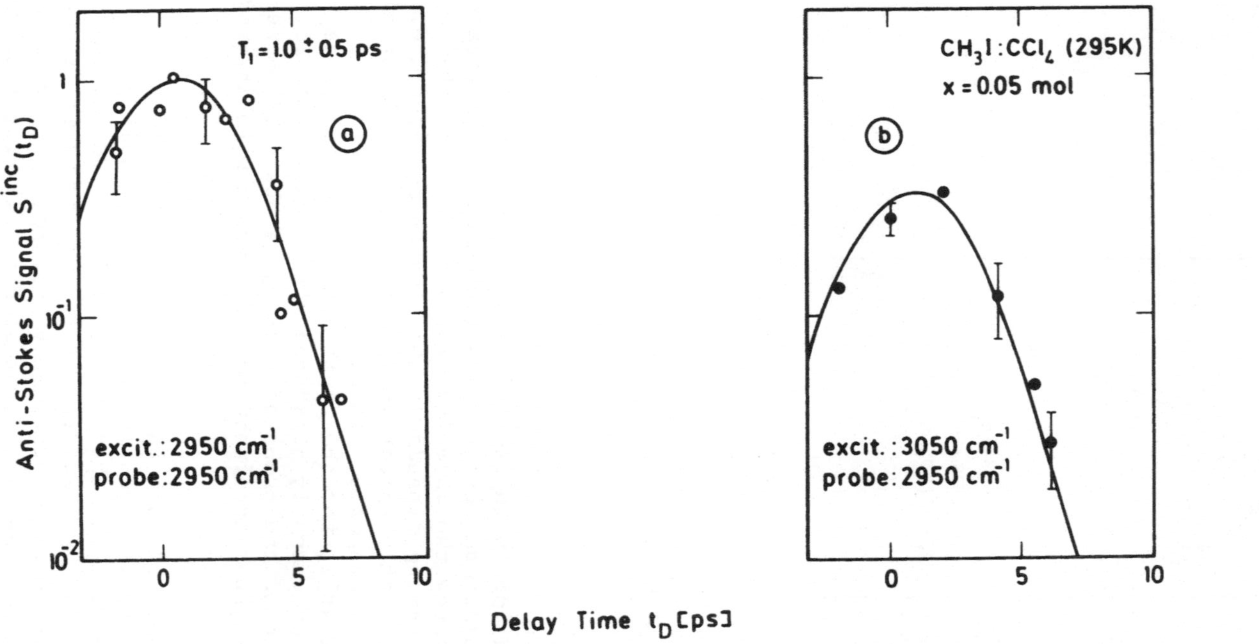

Fig. 3a

Coherent probe scattering signal versus delay time for CCl_4 with natural isotope abundance; selective k-matching observing a single isotope component. Inset : spontaneous Raman band.

Fig. 3b

Coherent probe scattering signal versus delay time for CCl_4 with natural isotope abundance; non-selective k-matching with coherent superposition of vibrational states; curves are calculated.

accounts for the Raman linewidth of one isotope component of 1.4 cm^{-1}. Additional information is obtained by coherent probing with non-selective k matching. In these experiments, a coherent superposition of the different isotopic species is observed (Fig. 3b) [5]. The various isotope components are first excited with approximately equal phases (rising part of the signal curves). The excitation process then terminates and free relaxation of the collective excitation is observed. The frequency differences of the individual species lead to a striking interference phenomenon with a beating period of ∿ 12 psec.

Similar results were obtained for $SnBr_4$, where a smooth Raman band of ∿ 3.2 cm^{-1} is measured in the spontaneous scattering not resolving the isotopic substructure with a frequency spacing of ∿ 0.7 cm^{-1}. In spite of the overlap of the individual components our coherent probing technique can select a single isotope species. We measure $T_2/2 = 3.0$ psec. This value corresponds to a homogeneous line broadening of 1.8 cm^{-1}, which is significantly smaller than the observed total bandwidth of 3.2 cm^{-1} [5].

c) There are vibrational modes where experimental evidence exists of a distribution of transition frequencies. For instance, liquids with strong hydrogen bonding show extended inhomogeneously broadened OH-bands. Spontaneous spectroscopic techniques provide the total Raman or infrared band, but do not allow a separation between homogeneous and inhomogeneous contributions to the total line shape. As a result, it is impossible,in general, to decide to what extent a normal mode is inhomogeneously broadened. The investigation discussed now is aimed at tackling this problem. The vibrational system we have investigated is the CH-stretching mode of pure $(CH_2OH)_2$ at 2935 cm^{-1}. The spontaneous Raman band of this mode is broad with a linewidth of ∿ 60 cm^{-1}.

Similar to the investigations discussed above for $CC\ell_4$ we have carried out measurements with our coherent probing technique with different k-matching geometries varying the sample length and the divergence of the detected Stokes beam. We present experimental data of two widely different experimental situations [12]. In Fig. 4a, a highly selective k-matching situation is achieved with a sample length of 10 cm and with a small divergence of the Stoke beam of 3 mrad. In Fig. 4b, on the other hand, we devised a less selective k-vector geometry by using a shorter cell of 1 cm and a larger Stokes divergence of 10 mrad.

The experimental results of Fig. 4a are of special interest. After the maximum of the probe signal we find an exponential decay with a time constant of $T_2/2 = 3.0 \pm 0.5$ psec. For an interpretation of this time constant we estimate the frequency spread of the molecules monitored by the selective k-vector geometry to be smaller

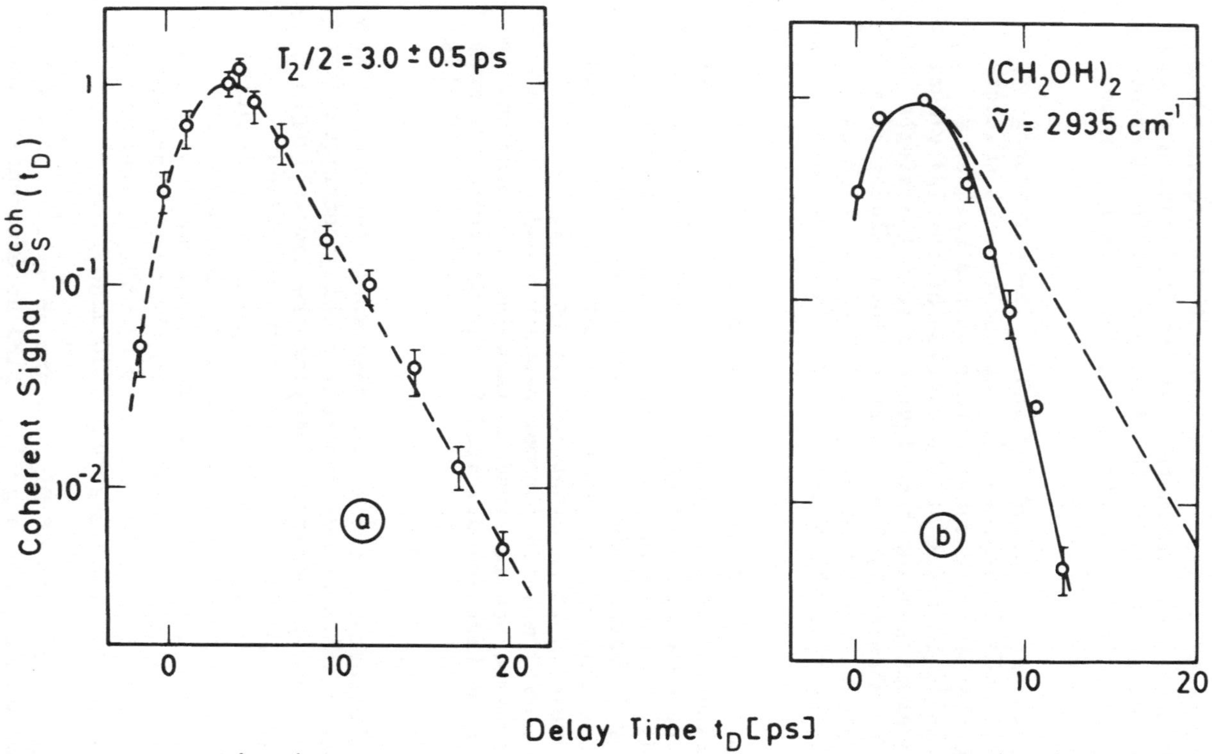

Fig. 4a

Coherent Stokes scattered probe signal versus delay time measured in a highly selective collinear k-matching geometry in $(CH_2OH)_2$. We observe a dephasing time $T_2/2 = 3.0$ psec. corresponding to a homogeneous linewidth of 1.8 cm^{-1}.

Fig. 4b

Coherent Stokes signal for a less selective k-matching geometry. The signal decays rapidly on account of destructive interference of neighbouring molecules oscillating with different transition frequencies.

than 0.5 cm^{-1}; i.e. dephasing by a distribution of frequencies is negligible. The time constant of 3 psec represents the homogeneous dephasing of a small group of molecules, the frequency of which is close to the center frequency of the broad Raman band at 2935 cm^{-1}. Our value of T_2 corresponds to a homogeneous linewidth of 1.8 cm^{-1} which is smaller by a factor of approximately 30 than the spontaneous Raman band.

The time dependence of $S^{coh}(t_D)$ is completely different for less selective phase-matching (Fig. 4b). Under these experimental conditions the coherent scattering signal disappears rapidly. The measured time dependence represents the destructive interference of molecules which vibrate with a wide distribution of frequencies. Our picosecond data of Fig. 4 provide direct evidence that vibrational bands in liquids are inhomogeneously broadened [12].

Our findings of Figs. 4a and b point to (at least) two different mechanisms determining the dynamics of the investigated CH_2-mode. The first mechanism occurs on a rapid time scale < 1 psec, leading to the exponential decay with the time constant T_2 = 6 psec (close to the limit of motional narrowing). The second mechanism, on the other hand, is responsible for the inhomogeneous broadening. This interaction obviously proceeds on a comparatively long time scale > 10^{11} psec, longer than the duration of our picosecond experiment allowing us to observe the distribution of transition frequencies.

Similar results were obtained for the CH-mode of CH_3OH at 2835 cm^{-1}. The coherent probing technique offers the possibility to distinguish between broadening due to short dynamic time constants and frequency distributions resulting from different molecular surroundings. In general, there are several physical processes which contribute to the loss of phase of the coherently excited molecules. Pure dephasing may occur by quasi-elastic nearest neighbour interaction, by resonant transfer of vibrational quanta and by exchange coupling to lower frequency modes [11-14]. In addition, the dephasing time is affected by rapid energy (T_1) relaxation, e.g. population decay or energy redistribution processes with neighbouring vibrational levels. At the present time, the dominant mechanism determining T_2 of the specific vibrational mode in $(CH_2OH)_2$ cannot be stated definitely.

We believe that the observed inhomogeneously broadened CH-bands in alcohols are connected to the local order in the liquids, in particular with the hydrogen bonding of the individual molecules [15]. The molecular association also determines the long time constant of the fluctuating transition frequencies of individual molecules. In fact, it is known from other techniques that the liquid structure in the alcohols changes at moderate speed. For example, studies of dielectric relaxation yield relaxation

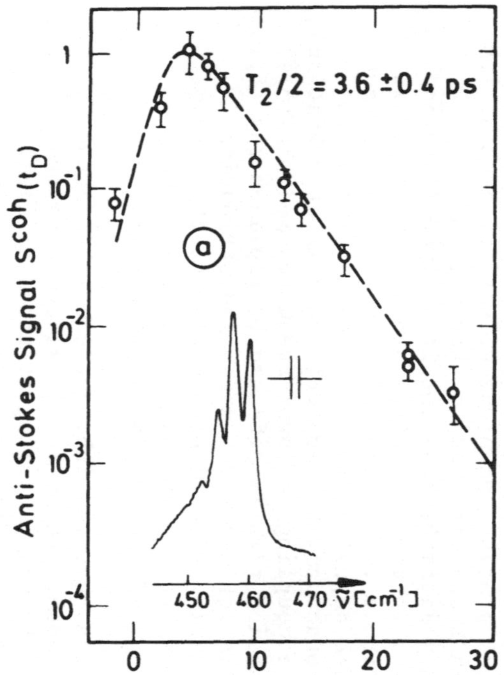

Fig. 5a
Incoherent anti-Stokes probe scattering $S^{inc}(t_D)$ versus delay time of the symmetric CH_3-stretching mode ($\tilde{\nu}$ = 2950 cm^{-1}) of CH_3I dissolved in $CC\ell_4$ (mole fraction 0.05); the mode at 2950 cm^{-1} is directly populated by the resonant infrared excitation pulse.

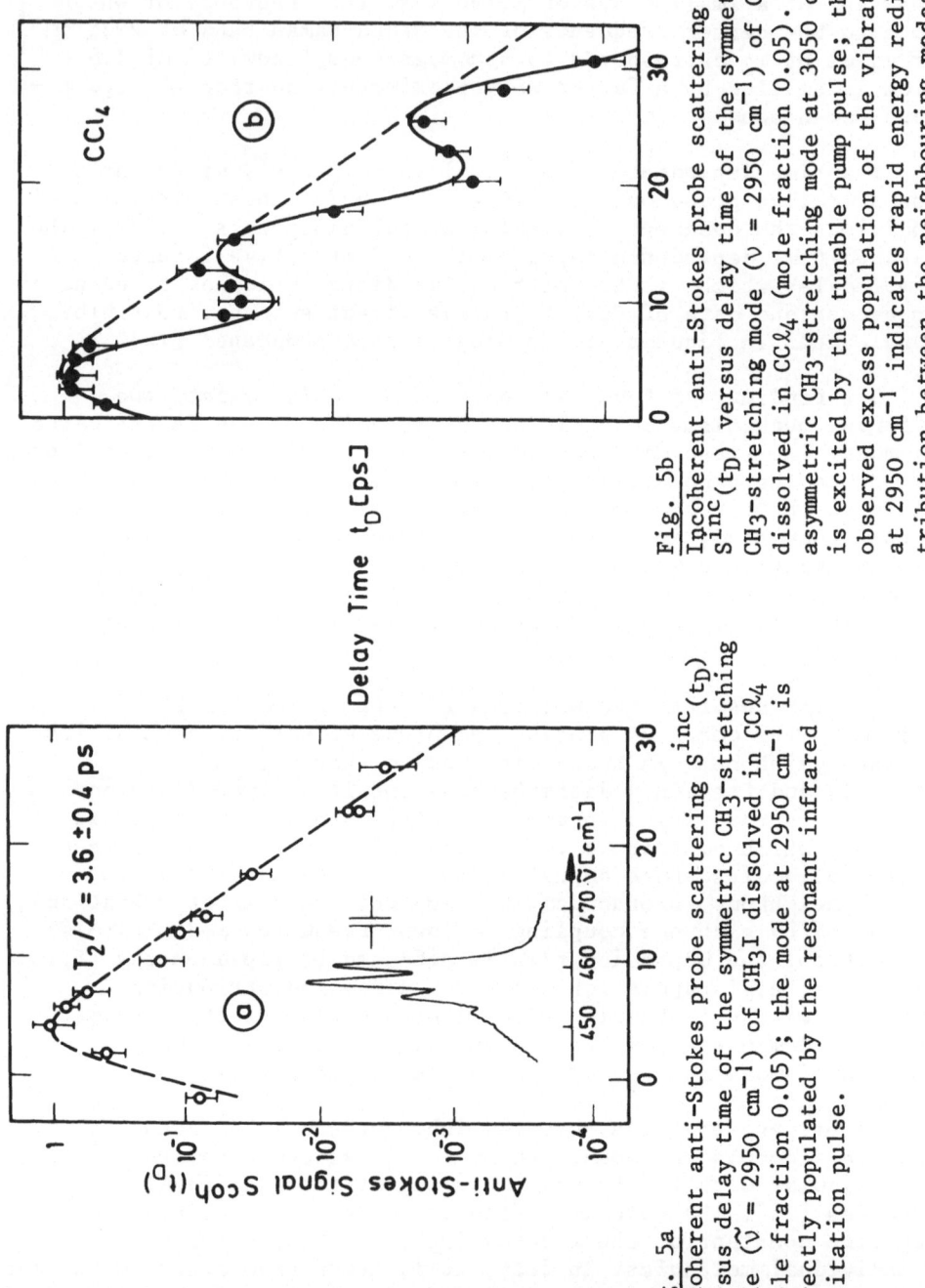

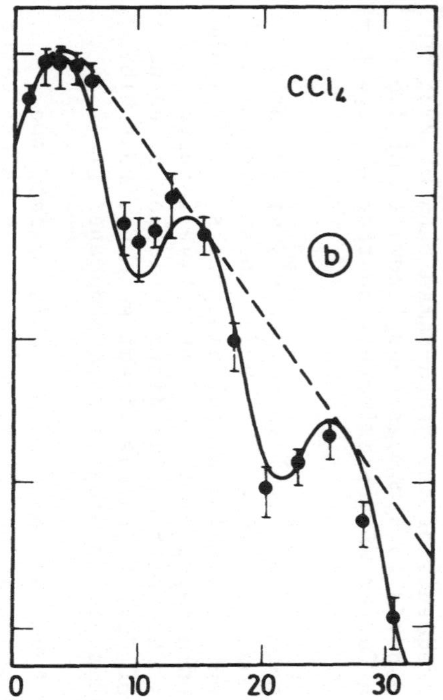

Fig. 5b
Incoherent anti-Stokes probe scattering $S^{inc}(t_D)$ versus delay time of the symmetric CH_3-stretching mode (ν = 2950 cm^{-1}) of CH_3I dissolved in $CC\ell_4$ (mole fraction 0.05). The asymmetric CH_3-stretching mode at 3050 cm^{-1} is excited by the tunable pump pulse; the observed excess population of the vibration at 2950 cm^{-1} indicates rapid energy redistribution between the neighbouring modes.

times of $\sim$ 70 psec and $\sim$ 130 psec for CH_3OH and $(CH_2OH)_2$, respectively [16].

d) It is well known that the lifetimes of energy levels contribute to the broadening of spectral lines. Population decay and vibrational energy transfer in molecules has been directly studied in polyatomic molecules with picosecond techniques. For several systems the energy relaxation was found to give only a small contribution to the measured Raman band shapes. For the example to be discussed here the situation is remarkably different.

We have investigated the CH_3-stretching modes of CH_3I in solution of $CC\ell_4$ in a concentration range of several mol per cent. [17]. In Fig. 5a and b, the incoherent anti-Stokes Raman signal, which is a direct measure of the excited state population, is plotted as a function of delay time between an infrared excitation pulse and a green (0.53 μm) interrogating pulse. The probe scattering signal was measured at a frequency shift of 2950 cm^{-1} which corresponds to the symmetric CH_3-stretching ν_1-vibration. The different experimental conditions of Fig. 5a and 5b should be noted : In one experiment we excited the ν_1-mode by an infrared pulse at a frequency of 2950 cm^{-1} (Fig. 5a) and observed the population lifetime T_1 of the same vibration. A value of T_1 = 1.0 ± 0.5 psec is directly indicated by the data. In the second experiment we tuned our infrared pulse to 3050 cm^{-1} in order to excite the asymmetric CH_3-mode ν_4. We monitor again the build-up and decay of the lower lying ν_1-vibration (Fig. 5b). The measured excess population of the ν_1-mode directly indicates rapid energy exchange between ν_1 and ν_4. Our investigations of Fig. 5 show that the population relaxation proceeds very fast in CH_3I; the observed energy exchange between the two neighbouring CH_3-stretching vibrations is accounted for by a time constant of approximately 1.5 psec.

In summary, it is pointed out that investigations with picosecond light pulses provide detailed information on a series of molecular processes. The results should advance the deeper understanding of spectral line shapes and of molecular dynamics in liquids.

REFERENCES

[1] R.G. Gordon, J. Chem. Phys. 40, 1973 (1964); 42, 3658 (1965); 43, 1302 (1965).

[2] S. Bratos and E. Maréchal, Phys. Rev. A4, 1078 (1971); F.J. Bartoli and T.A. Litovitz, J. Chem. Phys. 56, 404, 413 (1972); G. Döge, Z. Naturforsch. 28a, 919 (1973). H.S. Goldberg and P.S. Pershan, J. Chem. Phys. 58, 3816 (1973); W.G. Rotschild, G.J. Rosasco and R.C. Livingston, J. Chem. Phys. 59, 5310 (1973).

[3] For a review see A. Laubereau and W. Kaiser, Ann. Rev. Phys. Chem. 26, 83 (1975).

[4] D. von der Linde, A. Laubereau and W. Kaiser, Phys. Rev. Lett. 26, 955 (1971); A. Laubereau, Chem. Phys. Lett. 27, 700 (1974).

[5] A. Laubereau, G. Wochner and W. Kaiser, Phys. Rev. A13, 2212 (1976).

[6] A. Laubereau, D. von der Linde and W. Kaiser, Phys. Rev. Lett. 28, 1162 (1972).

[7] R.R. Alfano and L.L. Shapiro, Phys. Rev. Lett. 29,1655 (1972).

[8] A. Laubereau, K. Kirschner, and W. Kaiser, Opt. Commun. 9, 182 (1973); A. Laubereau, G. Kehl, W. Kaiser, Opt. Commun. 11, 74 (1974).

[9] A. Laubereau, A. Seilmeier and W. Kaiser, Chem. Phys. Lett. 36, 232 (1975).

[10] For a more detailed discussion see A. Laubereau and W. Kaiser in "Chemical and Biochemical Applications of Lasers", ed. C.B. Moore, vol. 2, Academic Press (New York 1977).

[11] S.F. Fischer and A. Laubereau, Chem. Phys. Lett. 35, 6 (1975); P.A. Madden and R.M. Lynden-Bell, Chem. Phys. Lett. 38, 163 (1976).

[12] A. Laubereau, G. Wochner and W. Kaiser, Chem. Phys. in press,

[13] S. Bratos, J. Chem. Phys. 63, 3499 (1975).

[14] W.G. Rotschild, J. Chem. Phys. 65, 455 (1976); ibid 65, 1958 (1976).

[15] G.C. Pimentel and A.L. McClellan, "The Hydrogen Bond", Freeman (San Francisco, 1960); Ann. Rev. Phys. Chem. 22, 347 (1971).

[16] N. Koizumi, J. Chem. Phys. 27, 625; H. Fellner-Feldegg, J. Phys. Chem. 73, 616 (1969).

[17] K. Spanner, A. Laubereau and W. Kaiser, Chem. Phys. Lett. 44, 88 (1976).

ON DIPOLAR FLUIDS

J.J. Weis

Laboratoire de Physique Théorique et Hautes Energies[+]
Université de Paris-Sud, 91405 Orsay, France
[+]Laboratoire associé au C.N.R.S.

ABSTRACT

We give a brief summary of recent progress in the equilibrium statistical mechanics of a simple model of polar fluids consisting of hard-spheres with a permanent point dipole at the centre. Various theoretical approaches for the structural, dielectric and thermodynamic properties are examined. One of these, the linearized hypernetted-chain equation (LHNC), turns out to be remarkably accurate. This conclusion is reached by comparing the Monte Carlo calculations for the pair distribution function of a system with a spherically truncated potential with the results given by the LHNC integral equation solved with a similar truncation of the potential. The knowledge of this accurate solution enables us then to show that all presently available methods fail to adequately treat the long-range part of the dipole-dipole interaction in a computer simulation. The effect of polarizability is investigated.

1. INTRODUCTION

The purpose of this paper is to give a brief survey of recent progress in the equilibrium statistical mechanics of simple models for polar fluids. The molecules are assumed to be spherical with a permanent point dipole $\vec{\mu}$ at the centre. The interaction potential can therefore be written

$$v(12) = v_S(r_{12}) + v_D(12) \quad (1)$$

where the symbol $1 \equiv (\vec{r}_1, \vec{\Omega}_1)$ indicates both positional and orientational coordinates and r_{12} is the distance between the centres of molecules 1 and 2. The dipole-dipole interaction is given by

$$v_D(12) = -\mu^2 D(12)/r_{12}^3 \tag{2}$$

with

$$D(12) = 3r_{12}^{-2}(\vec{s}_1.\vec{r}_{12})(\vec{s}_2.\vec{r}_{12}) - \vec{s}_1.\vec{s}_2 \tag{3}$$

$$= 2\cos\theta_1\cos\theta_2 - \sin\theta_1\sin\theta_2\cos(\phi_1 - \phi_2)$$

(Here μ is the magnitude of the dipole moment, θ_1 and θ_2 are the angles between the dipole vectors $\vec{s}_i=\vec{\mu}_i/\mu$ and $\vec{r}_{12}$ and ϕ_1,ϕ_2 are the azimuthal angles about $\vec{r}_{12}$).

The spherically symmetric part vs of the interaction is generally chosen to be either the hard-sphere or the Lennard-Jones (LJ) potential. In the former case the model is referred to as dipolar hard-spheres (and this will be our choice throughout this paper); the latter case is the so-called Stockmayer potential.

Most real polar molecules are polarizable, have a non-spherical shape and have extended charge distributions not represented by point dipoles or possibly higher point multipoles. Therefore the model can hardly be regarded as adequate for a quantitative interpretation of the behaviour of polar molecules. However, it certainly contains qualitatively the essential features of dielectric polarization while remaining sufficiently simple to be handled both by theory and computer simulations.

Dipolar hard-spheres can be characterized by two dimensionless parameters, the packing fraction $\eta = \pi\rho d^3/6$ (where ρ is the number density and d the hard-sphere diameter) and the reduced dipole moment $\mu^{*2} = \beta\mu^2/d^3$ ($\beta = 1/kT$, k Boltzman constant, T temperature). For Stockmayer systems we need to specify in addition the temperature:

$$\rho^* = \rho\sigma^3 \;, \quad T^* = kT/\epsilon \text{ and } \quad \mu^{*2} = \mu^2/kT\sigma^3$$

where ϵ and σ are the parameters of the LJ potential. Typical values for the reduced dipole moments are given in Table 1.

Sections 2 - 4 will be concerned with the structural, dielectric and thermodynamic properties of dipolar hard-spheres, section 5 with generalizations of this model to include polarizability and higher order multipoles.

2. STRUCTURAL PROPERTIES

Information on the structure of dipolar hard-spheres can be

	μ (debyes)	ε/k (°K)	σ (Å)	μ^*	μ^{*2}
CO	0.12	110	3.585	0.14	0.02
H Cℓ	1.03	360	3.305	0.77	0.6
Cℓ CH_3	1.89	380	3.43	1.30	1.70
N H_3	1.47	320	2.60	1.68	2.83
H_2 0	1.83	380	2.65	1.84	3.39

Table 1. Typical values for the reduced dipole moment $\mu^* = (\mu^2/\epsilon\sigma^3)^{1/2}$ (from Hirschfelder et al.[39]).

obtained from the pair distribution function (p.d.f.) defined by

$$g(r_{12}, \vec{\Omega}_1, \vec{\Omega}_2) = \frac{\Omega^2 N(N-1)}{\rho^2} \frac{\int e^{-\beta U} d(3)\ldots d(N)}{\int e^{-\beta U} d(1)\ldots d(N)} \tag{4}$$

(N number of molecules, U total configurational energy, $\Omega = 4\pi$ for linear molecules, $d(i) = \int d\vec{r}_i d\vec{\Omega}_i$).

It is convenient to expand the p.d.f. on a basis set of rotational invariants[1]

$$g(12) = g_S(r) + h_\Delta(r)\Delta(12) + h_D(r)D(12) + \ldots \tag{5}$$

where

$$\Delta(12) = \vec{s}_1 \cdot \vec{s}_2 \tag{6}$$

and D(12) defined by (3) are linear combinations of products of spherical harmonics associated with molecules 1 and 2. Projections on higher order spherical harmonics also exist but will not be considered here. The thermodynamic and dielectric properties of the system are completely determined by h_Δ and h_D.

Several approximate theories have been proposed for the p.d.f.:

A) Mean Spherical Approximation (MSA)

The MSA consists in solving the Ornstein-Zernike (OZ) equation (which defines the direct correlation functions c(12))

$$h(12) = c(12) + \frac{\rho}{\Omega} \int c(13)h(32)\, d(3) \tag{7}$$

subject to the boundary conditions

$$h(12) = -1 \qquad r < d \tag{8}$$

$$c(12) = -\beta v_D(12) \qquad r > d \tag{9}$$

(with $h(12) = g(12) - 1$).

The first condition is clearly an exact condition for hard spheres, the second one holds only asymptotically ($r \rightarrow \infty$) and defines the MSA. Wertheim [2] succeeded in giving a completely analytical solution of this integral equation. His main results are that the pair correlation function (p.c.f.) h(12) contains no other components than those on Δ and D, i.e.

$$h(12) = h_S(r) + h_\Delta(r)\Delta(12) + h_D(r)D(12) \tag{10}$$

(and no other terms). A similar relation is obtained for c(12). The angle-averaged component h_S decouples from the two other components and is equal to the MSA for a pure hard-sphere system (which coincides with the Percus-Yevick (PY) approximation [3] and is available in closed form [4]). Finally h_Δ and h_D can be expressed in terms of the PY solution for pure hard-spheres only. As will be shown below, the MSA for the p.c.f. turns out to be extremely poor.

B) Beyond MSA. Cluster Series Expansion

A remarkable feature of the MSA p.c.f. for dipolar hard-spheres is that it appears [5-7] to be the basic ingredient entering the successive approximations in a cluster series expansion of the p.c.f. [8, 9].

The first approximation is simply [5,7]

$$h(12) = h_{HS}(r) + C(12) \tag{11}$$

where **C**(12) is defined as

$$C(12) = h^{MSA}(12) - h_S^{MSA}(r) = h_\Delta^{MSA}(r)\Delta(12) + h_D^{MSA}(r)D(12) \tag{12}$$

It is seen to differ from the MSA only by the fact that the angle-averaged component is equal to the exact hard sphere p.c.f. h_{HS}

and no longer to the PY p.c.f.. This approximation is identical to the optimized random phase approximation of Andersen and Chandler [8,6,7] and is also obtained as the lowest order approximation (LOGA) in the γ-expansion scheme of Lebowitz, Stell and Baer [10].

Two higher-order approximations, EXP and N3, have been worked out [9, 8, 6, 11]. They read

$$\text{(EXP)} \qquad g(12) = g_{HS}(r) \exp C(12) \tag{13}$$

and

$$\text{(N3)} \qquad g(12) = g_{HS}(r) \exp C(12) \left(1 + \rho\int d(3)\; S(13)S(23) + 2\rho\int d(3)\; S(13)H(23) \right) \tag{14}$$

with

$$S = g_{HS}(\exp C - 1 - C) + Ch_{HS} \tag{15}$$

$$H = h_{HS} + C$$

$$g_{HS} = h_{HS} + 1$$

As they stand the EXP and N3 approximations turn out to be hopelessly poor [6]. Verlet and Weis [6] showed that linearizing the exponential terms with respect to Δ and D ($\exp \mathbf{C} \to 1 + C$) leads to much more acceptable results. The accuracy of the linearized versions of the EXP and N3 approximations (which we shall call LIN and L3) will be discussed below.

C) Linearized and Quadratic Hypernetted-Chain Equations (LHNC and QHNC)

An extremely successful integral equation approach has been proposed recently by Patey [12]. One starts with the OZ equation (7) and assumes that no terms other than the Δ and D terms need to be retained in the p.c.f., i.e.

$$h(12) = h_S(r) + h_\Delta(r)\Delta(12) + h_D(r)D(12) \tag{16}$$

(and a similar expression for c(12)). The OZ equation is complemented by condition (8) for $r < d$ and an HNC-closure for $r > d$

$$c(12) = h(12) - \ln g(12) - \beta v_D(12) \tag{17}$$

To be consistent with eq. (16) one further keeps in lng only terms linear in Δ and D (this gives the linearized hypernetted-chain equation (LHNC)). As for the MSA the angle-averaged p.c.f. h_S decouples from the other components and is equal to the p.c.f. of pure hard-spheres in the HNC approximation and thus independent of μ. For high densities and not too high dipole moments this is a fair approximation provided one uses the exact p.c.f. for hard-spheres (cf. ref.[6, 11, 13]).Having made this choice for h_S, the projections h_Δ and h_D are obtained by the numerical solution of 2 coupled equations [12].

At low density and high dipole moments h_S differs considerably from the pure hard-sphere result [14]. In this case an almost perfect approximation for h_S can be obtained by retaining in the expansion of the logarithm in (12) also terms quadratic in Δ and D [14]. This leads to the so-called quadratic hypernetted-chain equation (QHNC) [14, 38]. The advantage of this approximation is that the equation giving h_S no longer decouples so that h_S depends now on the dipolar interaction. However, contrary to LHNC, the QHNC approximation is inconsistent with assumption (16) and at present its only justification lies in the remarkably accurate results which are obtained [14].

D) Monte Carlo (MC)

In fact we have no idea so far how accurate the preceding approximations are and to elucidate this question one would like to resort to computer simulations. However, one is then faced with the rather difficult problem of how to treat the long-range part of the dipole-dipole interaction. If, as is customarily done in computer simulations one simply truncates the potential (e.g. spherically or at the surface of a cube having the same dimensions as the fundamental MC cell ("minimum image" - MI - method [15])), one cannot expect to get the infinite system behaviour (which one is interested in). Indeed, as can be seen from Figs. 1 and 2, the p.c.f. depends dramatically on the cut-off radius of the potential. In particular, for a spherical cut-off of the potential, h_D is discontinuous at the cut-off distance instead of decreasing as $1/r^3$. However, meaningful comparison between the computer simulation and integral equation results can still be made by solving the integral equation for a truncated potential. In Figs. 3 and 4 we compare MC calculations for h_Δ and h_D using cut-off radii of the dipole-dipole interaction of 3.4d and 5.1d with LHNC results for a similar truncation of the potential. The agreement is remarkably good and it is likely that similar accuracy is also obtained when the LHNC equation is solved for an untruncated potential. The good agreement between the MC and LHNC results implies further that the influence of the periodic boundary conditions (used in the MC calculation but not in the integral equation) must be quite small. This conclusion is further supported by varying the size of the MC system but keeping the cut-off radius of the

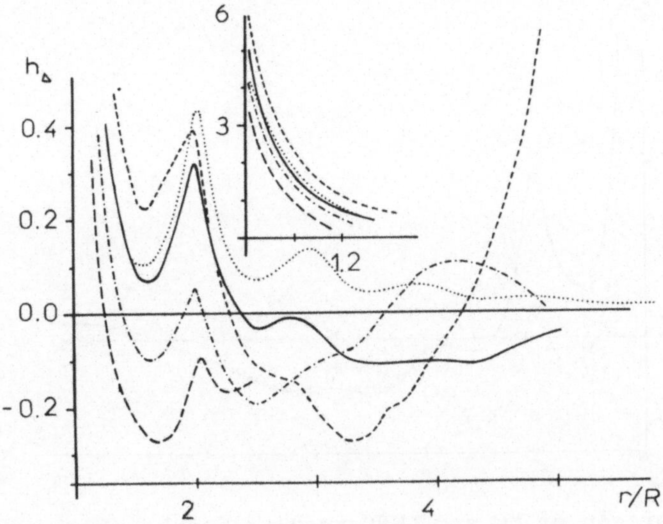

Fig. 1. Monte Carlo results for h_Δ at $\rho^* = 0.8$ and $\mu^{*2} = 2.75$. The symbols represent the following: — —, h_Δ^{SC}, N = 108, $R_c/d = 2.5$; -•-•-, h_Δ^{SC}, N = 256, $R_c/d = 3.4$; ———, h_Δ^{SC}, N = 864, $R_c/d = 5.1$; -----, h_Δ^{MI}, N = 256; •••••, LHNC for an infinite system with untruncated potential. SC denotes a spherical cut-off of the potential at R_c and MI the minimum image method. [Reproduced by permission from Mol. Phys. 34, 1077 (1977).]

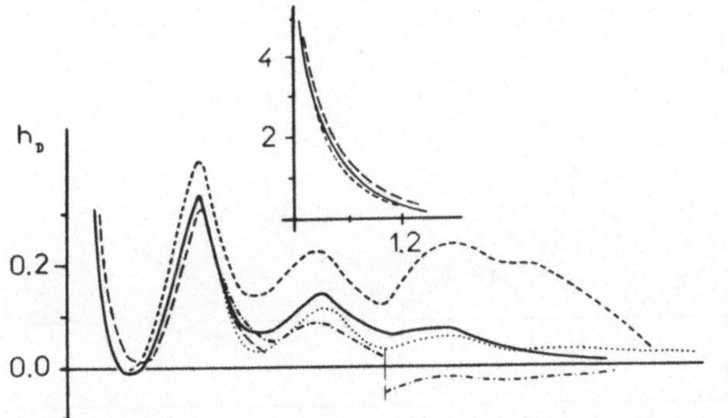

Fig. 2. Monte Carlo results for h_D at $\rho^* = 0.8$ and $\mu^{*2} = 2.75$. The symbols have the same meaning as in Fig. 1. [Reproduced by permission from Mol. Phys. 34, 1077 (1977).]

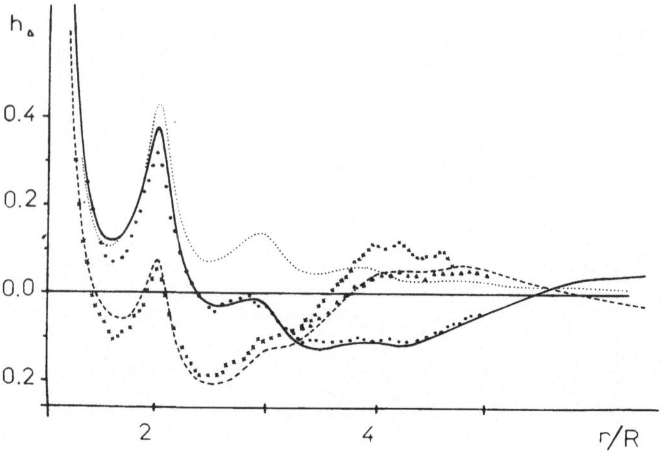

Fig. 3. Comparison of MC and LHNC results for h_Δ at $\rho^* = 0.8$ and $\mu^{*2} = 2.75$. The symbols represent the following: ----, LHNC for an infinite system with a potential cut-off at $R_c/d = 3.4$; ××××, MC result for h_Δ^{SC}, N = 256, R_c/d =3.4; ▲▲▲▲, MC result for h_Δ^{SC}, N = 864, $R_c/d = 3.4$, up to $r = R_c$ both MC results coincide within statistical error; ——, LHNC for an infinite system with a potential cut-off at $R_c/d = 5.1$; ••••, MC result for h_Δ^{SC}, N = 864, $R_c/d = 5.1$; ····, LHNC for an infinite system with an untruncated potential; SC denotes a spherical cut-off of the potential at R_c. [Reproduced by permission from Mol. Phys. 34, 1077 (1977).]

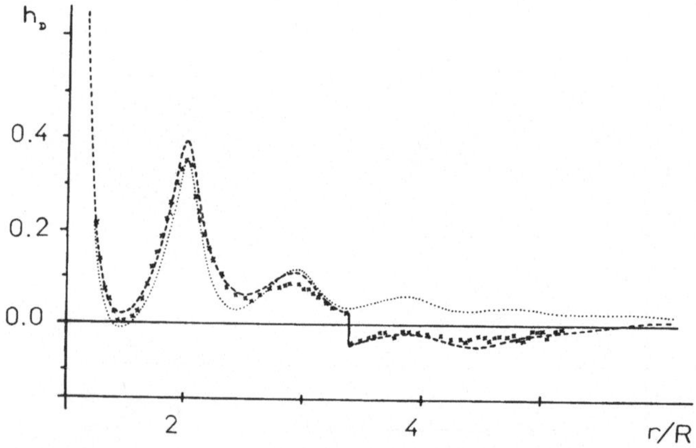

Fig. 4. Comparison of MC and LHNC results for h_D at $\rho^* = 0.8$ and $\mu^{*2} = 2.75$. The symbols have the same meaning as in Fig. 3. [Reproduced by permission from Mol. Phys. 34, 1077 (1977).]

potential fixed [18]. Fig. 5 demonstrates that the QHNC equation is equally successful at low density and high dipole moment.

Now that we have an accurate p.c.f. (given by LHNC and QHNC), we are in position to compare it with other approximations introduced in this Section. The MSA, LIN, L3 and LHNC approximations for h_Δ and h_D are compared in Figs. 6 and 7.It is seen that the LIN and L3 approximations considerably improve upon the MSA in the region near core, but whereas the former completely underestimates the structure especially in the second-nearest-neighbour region, the latter exaggerates this structure (see also ref.[11]). The cluster series expansion for the p.c.f. appears to be slowly convergent.

The knowledge of an accurate p.c.f. further allows us to examine the adequacy of methods devised to take into account the long-range part of the dipole-dipole interaction in a computer simulation. The uniform reaction field method of Barker and Watts [16, 17], applied by these authors to fluids of "water-like" molecules, consists essentially of surrounding the cut-off sphere by a dielectric continuum having the same dielectric constant as the sample. The long-range interactions are then approximated by allowing each dipole to interact with the reaction field arising from the polarization of the continuum. Roughly speaking this means that molecular fluctuations are ignored in the region beyond the spherical cut-off (typically of the order of 3-4 hard-sphere diameters). By comparison with LHNC results Levesque et al. [18] come to the conclusion that this approximation does not adequately correct for the truncation of the potential (cf. Fig. 9 of ref.[18]).

An alternative method is to view the MC system as truly periodic. The Ewald method, adapted to dipolar systems by Jansoone [19], Smith and Perram [20] and Adams and McDonald [21] enables one to calculate the contribution to the energy from all periodic images. The physical realism of this approximation has been discussed thoroughly by Valleau and Whittington [15]. Only one computer calculation has been reported so far in the fluid phase [19]. However, the small size of the system (32 particles) and the insufficient data reported do not allow conclusions to be drawn. Similar in spirit is a method recently proposed by Ladd [22]. The interactions between the particles in the fundamental MC cell are evaluated exactly whereas the interactions between the particles in the fundamental cell and the particles in the neighbouring cells are evaluated by means of a multipole expansion. Although MC calculations for a dense dipolar hard-sphere liquid have been performed recently using this method [23], the results have not yet been compared with LHNC predictions.

The large dependence of the p.c.f. on the boundary conditions when the potential is truncated (even for a cut-off radius of 5

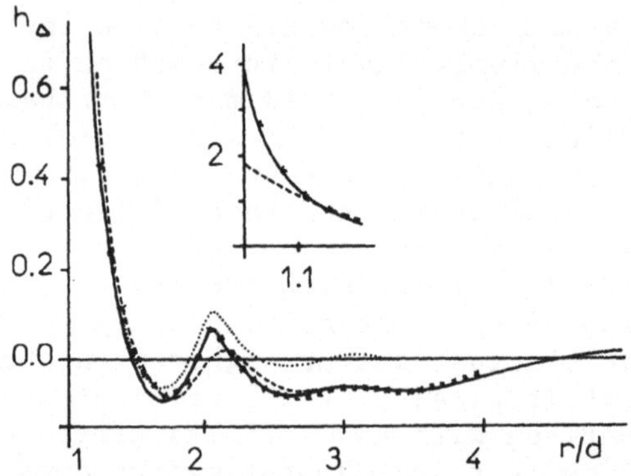

Fig. 5. h_Δ for $\rho^* = 0.4$ and $\mu^{*2} = 2.75$. The symbols represent the following: ——, QHNC for an infinite system with a potential cut-off at $R_c/d = 4.2$; ----, LHNC for an infinite system with a potential cut-off at $R_c/d = 4.2$; ××××, MC result for a potential truncated at $R_c/d = 4.2$; ••••, QHNC for an infinite system with an untruncated potential.

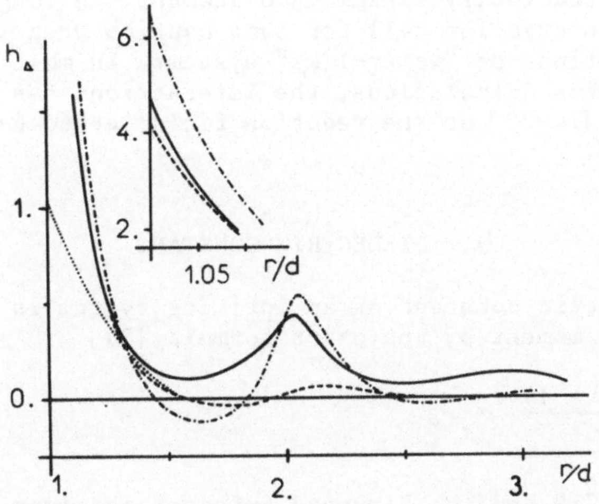

Fig. 6. Comparison of different approximations for h_Δ at $\rho^* = 0.8$ and $\mu^{*2} = 2.75$. ····, MSA; ----, LIN; -·-·, L3; ——, LHNC.

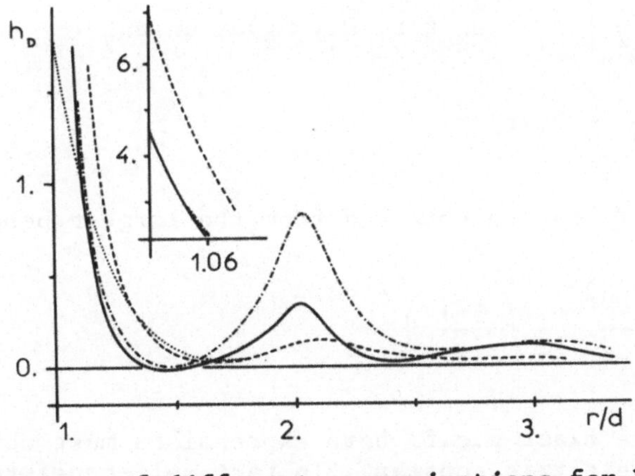

Fig. 7. Comparison of different approximations for h_D at $\rho^* = 0.8$ and $\mu^{*2} = 2.75$. ····, MSA; ----, LIN; -·-·, L3; ——, LHNC.

hard-sphere diameters the p.c.f. bears only very qualitative resemblance to the infinite p.c.f.) and the inability of available methods to satisfactorily taking into account the long-range dipole-dipole interaction call for some caution when interpreting computer simulations on "water-like" systems. In most MC and molecular dynamics calculations, the interactions has either been truncated [24-29] or the reaction field method has been applied [16,17].

3. DIELECTRIC CONSTANT

The dielectric constant of an infinite system is related to the mean square moment by the exact formula [30]

$$\frac{(\epsilon - 1)(2\epsilon + 1)}{9\epsilon} = yg \tag{18}$$

where $y = 4\pi\rho\beta\mu^2/9$ and the Kirkwood g-factor is given by

$$g = \frac{\langle M^2 \rangle}{N\mu^2} = \frac{\langle (\sum_i^N \vec{\mu}_i)^2 \rangle}{N\mu^2}$$

$$= 1 + (N-1) \langle \vec{\mu}_1 \cdot \vec{\mu}_2 \rangle$$

$$= 1 + \frac{\rho}{\Omega^2} \int \vec{\mu}_1 \cdot \vec{\mu}_2 \, g(r, \vec{\Omega}_1, \vec{\Omega}_2) d\vec{r} \, d\vec{\Omega}_1 d\vec{\Omega}_2$$

$$= 1 + \frac{\rho}{3} \int h_\Delta(r) d\vec{r} \tag{19}$$

Alternatively ϵ can be obtained from the large-r behaviour of $h_D(r)$ [30].

$$h_D(r) \to \frac{\beta\mu^2}{r^3} \frac{1}{\epsilon} \left(\frac{\epsilon - 1}{3y}\right)^2 \tag{20}$$

If h(12) is the exact p.c.f. both expressions must obviously lead to the same dielectric constant. In fact self-consistency holds under much weaker conditions [12] and is satisfied by the MSA and LHNC theories but not by the other theories considered previously (LIN, L3). Values of ϵ calculated via (19) are compared in Table 2 for various approximations. It is seen that the LHNC equation leads to dielectric constants notably higher than those given by "older" theories particularly that of Onsager (which completely neglects correlations between dipole moments, i.e. $h_{\Delta}^{ONS} = 0$ for all r).

When a spherical cut-off of the potential is used the appropriate relation between ϵ and $<M^2>$ is

$$\frac{\epsilon - 1}{\epsilon + 2} = y\, g' \tag{21}$$

with

$$g' = \frac{< M^2 >_{system}}{N\mu^2} \tag{22}$$

$$= 1 + \frac{\rho}{3} \int_{system} h_{\Delta}(r) d\vec{r} \tag{23}$$

where the mean square moment has to be calculated by including all the particles in the system. In a MC calculation this means that the mean square moment has to be calculated by including all the dipole moments of the fundamental MC cube. This relation is usually derived for a spherical sample but it has been shown [31] to be valid for an isotropic cubic system under periodic boundary conditions. Furthermore it holds equally well when a spherical cut-off of the potential is applied or the minimum image convention is used.

$\frac{\beta\mu^2}{d^3}$	$\epsilon_{ONSAGER}$	ϵ_{MSA}	ϵ_{LIN}	ϵ_{L3}	ϵ_{LHNC}	ϵ_{MC}
0.5	3.17	3.59	3.84	3.76	3.76	3.73 ± 0.1
1.0	5.62	7.80	9.27	9.66	9.66	9.0 ± 0.5
2.0	10.60	20.00	27.06	37.93	50.0	
2.75	14.36	31.86	45.62	79.17	250.	

Table 2. Dielectric constant ϵ for dipolar hard-spheres at the density $\rho d^3 = 0.8$ for the various approximations considered in the text. The Monte Carlo (MC) results are taken from ref. [18].

The mean square moment $\frac{\langle M^2\rangle_{cube}}{N\mu^2}$ (obtained by MC) does not depend strongly on the boundary conditions (cut-off distance or system size) provided the system is not too small (N > 108 particles). (cf. Tables I-III of ref.[18]).This result is particularly remarkable in view of the large differences occurring in h_Δ (cf. Fig. 1). It appears that the system reponds to the change in boundary conditions in such a way as to keep $\langle M^2\rangle_{cube}$ constant (as one would expect).

Dielectric constants calculated by MC via formula (22) compare quite favourably with the LHNC results (cf. Table 2). Unfortunately equation (21) has the disadvantage that the statistical error in g' does not permit the calculation of ϵ for dipole moments much larger than $\mu^* = 1$.

If one includes in the calculation of the mean square moment only the particles inside the cut-off sphere, as has been argued before [17], the results depend strongly on the cut-off radius (cf. Tables I-III of ref. 18) and moreover lead to quite unrealistic dielectric constants.

4. THERMODYNAMIC PROPERTIES

The configurational internal energy per particle depends only upon h_D and is given by

$$\beta u = - \frac{3y}{4\pi} \int \frac{h_D(r)}{r^3} d\vec{r} \tag{24}$$

The free energy per particle can then be obtained by

$$\beta f = \beta f_{HS} + \int_o^{\mu^{*2}} \frac{\beta u(\Gamma)}{\Gamma} d\Gamma \tag{25}$$

where Γ is a dummy variable representing the square of the reduced dipole moment and βf_{HS} is the free energy of a pure hard-sphere system given e.g. by the Carnahan-Starling expression [32].

The pressure can be obtained from the virial expression

$$\frac{P}{\rho kT} = 1 + 4\eta\, g_S(d) + \beta u \tag{26}$$

where $g_S(d)$ is the value of the angle-averaged p.d.f. at contact.

In Table 3 we compare the internal energies, calculated via (24), and the excess free energies (with respect to hard-spheres) for the different approximations introduced in Section 2. The MC internal energies are in excellent agreement with the LHNC results, provided that a spherical cut-off of the potential is used in the MC calculation. The MI methods leads to substantially different internal energies. This point is discussed in more detail in ref. 18. The slow convergence of the cluster series expansion (MSA, LIN, L3) is again apparent.

On the free energy level the highest-order cluster series result (L3) of Verlet and Weis [6] compare quite favourably with the LHNC results. Note that, contrary to βf^{LHNC}, these results have been obtained by a direct cluster series expansion of the free energy and not by integration of the internal energy βu^{L3} obtained through (24) and the use of h_D^{L3}. Somewhat less accurate (cf. Table 3) but vastly more convenient to use is the Padé approximant of Rushbrooke et al. [5] based upon an expansion of the free energy in powers of the dipole moment (which itself is slowly convergent [6]).

$$\beta\Delta f^{Padé} = \beta(f^{Padé} - f_{HS}) = \frac{a_2\ \mu^{*4}}{1 - \frac{a_3}{a_2}\mu^{*2}} \tag{27}$$

where a_2 and a_3, depending only on the density $\rho^* = \rho d^3$, are given with high accuracy by [33,34]

$$a_2 = -\frac{\rho^*}{6}\,(4.1888 + 2.8287\rho^* + 0.8331\rho^{*2} + 0.0317\rho^{*3} + 0.0858\rho^{*4} - 0.0846\rho^{*5}) \tag{28}$$

and

$$a_3 = \frac{\rho^{*2}}{9}\ \frac{2.7097 + 1.68918\rho^* - 0.31570\rho^{*2}}{1 - 0.59056\rho^* + 0.20059\rho^{*2}} \tag{29}$$

Expression (27) for the free energy has been used by Rushbrooke et al. [5] to locate the gas-liquid phase boundary for dipolar hard-spheres. The critical point is found to be $\rho_c^* = 0.159$, $\mu_c^{*2} = 3.584$. The location of the critical point could probably be somewhat refined by use of the QHNC theory (at low density there is an error of about 10 % in the Padé free energy).

μ^{*2}	N	R_c/d	$\Delta\beta f^{LHNC}$	$\Delta\beta f^{L3}$	$\Delta\beta f^{Padé}$	βu^{MC}	βu^{LHNC}	βu^{L3}	βu^{LIN}	βu^{MSA}	$\beta u^{Padé}$
0.5	256	3.4	-0.185	-0.187	-0.191	-0.327	-0.329	-0.341	-0.382	-0.214	-0.347
1.0	864	5.1	-0.612	-0.617	-0.645	-0.999	-1.001	-1.09	-1.287	-0.688	-1.091
1.405			-1.056	-1.08	-1.132		-1.648				-1.828
1.942			-1.734	-1.79	-1.885		-2.603				-2.894
2.287			-2.210	-2.29	-2.415		-3.257				-3.608
2.75	864	5.1	-2.891	-3.01	-3.167	-4.294	-4.18	-5.37	-6.56	-3.15	-4.587

Table 3

Excess free energy (with respect to hard-spheres) and internal energy of dipolar hard-spheres at $\rho d^3 = 0.8$ for the different approximations considered in the text. N and R_c denote the number of particles and the cut-off distance used in the Monte Carlo (MC) [18] calculations. The Padé values for the internal energy have been obtained by differentiation of the free energy Padé.

5. BEYOND DIPOLAR HARD-SPHERES

A) Polarizable Dipolar Hard-Spheres

Patey and Valleau [35] use perturbation theory to estimate the magnitude of the effect of polarizability on the thermodynamic properties of dipolar hard-spheres. In addition to their permanent dipole moment $\vec{\mu}$, these are now characterized by a polarizability tensor $\overleftrightarrow{\alpha}$ which (in a coordinate system fixed to the molecule and with the z-axis taken along the direction of the dipole) has the form

$$\overleftrightarrow{\alpha} = \begin{pmatrix} \alpha_{XX} & 0 & 0 \\ 0 & \alpha_{YY} & 0 \\ 0 & 0 & \alpha_{ZZ} \end{pmatrix}$$

The total configurational energy can be written as the sum

$$U = U_\mu + U_{ind}$$

where the first term represents the contribution from the hard-sphere and permanent dipole interactions and U_{ind} is the contribution from the permanent-induced and induced-induced dipole interactions. Expanding the free energy around that, βf_μ, of pure dipolar hard-spheres ($\overleftrightarrow{\alpha} = 0$), one has, to first order,

$$\beta f - \beta f_\mu = \beta < U_{ind} >_\mu /N + O(\beta^2) \tag{30}$$

where the average $< \; >_\mu$ is understood to be taken over the configurations of a system of pure dipolar hard-spheres. The Gibbs-Bogoliubov inequality tells us that $\beta < U_{ind} >_\mu /N$ is in fact an upper bound on the excess free energy $\beta(f-f_\mu)$ or, as $< U_{ind} >_\mu$ is found to be negative [35], a lower bound on the magnitude of the excess free energy. In order to calculate U_{ind} we need to know the total dipole moment $\vec{p}_i$ associated with particle i. It is the sum of its permanent dipole moment $\vec{\mu}_i$ and the induced moment

$$\vec{m}_i = \overleftrightarrow{\alpha}_i \cdot \vec{E}_i^{loc} \tag{31}$$

where the local field acting on particle i is

$$\vec{E}_i^{loc} = -\sum_{(j\neq i)} \overleftrightarrow{T}_{ij} \cdot \vec{p}_j \tag{32}$$

and

$$\overleftrightarrow{T}_{ij} = \frac{1}{r_{ij}^3}\left(\overleftrightarrow{I} - \frac{3\vec{r}_{ij}\,\vec{r}_{ij}}{r_{ij}^2}\right) \qquad (33)$$

($\overleftrightarrow{I}$ unit tensor).

The total configurational energy is then given by

$$U = U_{HS} - \frac{1}{2}\sum_i \vec{p}_i \cdot \vec{E}_i^{loc} \qquad (34)$$

It is clearly not pairwise additive.

MC results for $< U_{ind} >_\mu$ are presented in Table 4 for a system characterized by $\rho^* = 0.834$, $\mu^* = 1$ and a polarizability tensor such that $\alpha_{XX} = \alpha_{YY}$ and a mean value $\overline{\alpha}^* = 0.03$ ($\alpha^* = \alpha/d^3$, $\overline{\alpha} = (\alpha_{XX} + \alpha_{YY} + \alpha_{ZZ})/3$). For this state the excess free energy of dipolar hard-spheres (with respect to pure hard-spheres) is [36]

$$\beta(f_\mu - f_{HS}) = - 0.638$$

so that equation (30) can be rewritten

$$\beta(f - f_{HS}) = - 0.638 + < U_{ind} >_\mu/NkT + O(\beta^2) \qquad (35)$$

α^*_{XX}	α^*_{ZZ}	α_{ZZ}/α_{YY}	$<U_{ind}>_\mu/NkT$	$\beta f - \beta f_{HS}$	
				Perturbation	MC
0.045	0.0	0.0	−0.170±0.005	−0.808±0.007	
0.03	0.03	1.0	−0.365±0.008	−1.003±0.009	−1.02±0.01
0.0225	0.045	2.0	−0.481±0.012	−1.119±0.013	−1.22±0.02
0.015	0.06	4.0	−0.613±0.019	−1.251±0.019	−1.54±0.04

Table 4

Contribution to the excess free energy (with respect to hard spheres) of polarizable dipolar hard spheres from the induced interactions [35] ($\rho^* = 0.834, \mu^* = 1$, $\overline{\alpha}^* = 0.03$). The perturbation results are obtained from (35); MC denotes "exact" Monte-Carlo results [37].

From Table 4 it is immediately seen that the contribution to $\beta(f - f_{HS})$ from the induced interactions is quite large, ranging from 35% for isotropic particles ($\alpha_{ZZ}/\alpha_{XX} = 1$) to 50% when $\alpha_{ZZ}/\alpha_{XX} = 4$. "Exact" MC calculations [37] (not based upon perturbation theory) show that the contribution of the induced interactions to the thermodynamic properties is still larger (cf. Table 4). Table 4 further shows that the mean polarizability assumption ($\alpha_{ZZ}/\alpha_{XX} \sim 1$) is valid only when particles are very nearly isotropic.

Patey and Valleau [35] also come to the conclusion that the pairwise additivity assumption (i.e. the interaction between a pair of particles depends only upon their permanent and mutually induced dipole moments) is a good approximation when the particles are isotropically polarizable (at least for the system considered and in the perturbation approximation).

B) Extension to Higher-Order Multipoles

All the theories mentioned in Section 2 can be generalized in a straightforward way to higher-order point multipoles. The case of hard-spheres plus point quadrupoles has been examined in great detail by Patey [38] and the QHNC equation is again shown to give excellent agreement with MC computations [31] (which are no longer affected by boundary conditions due to the short-range nature of the quadrupole-quadrupole interaction). These results can easily be extended to systems containing both dipoles and quadrupoles and also higher-order multipoles.

Acknowledgements

I thank G.N. Patey and D. Levesque for fruitful collaboration.

REFERENCES

[1] L. Blum and A.J. Torruella, J. Chem. Phys. 56, 303 (1971).
[2] M.S. Wertheim, J. Chem. Phys. 55, 4291 (1971).
[3] J.K. Percus and G.J. Yevick, Phys. Rev. 110, 1 (1958).
[4] M.S. Wertheim, Phys. Rev. Lett. 10, 321 (1963).
E. Thiele, J. Chem. Phys. 39, 474 (1963).
[5] G.S. Rushbrooke, G. Stell and J.S. Høye, Mol. Phys. 26, 1199 (1973).
[6] L. Verlet and J.J. Weis, Mol. Phys. 28, 665 (1974).
[7] G. Stell in Statistical Mechanics, Vol. V of MODERN THEORETICAL CHEMISTRY, edited by B. Berne (Plenum, New York, 1977).
[8] H.C. Andersen and D. Chandler, J. Chem. Phys. 57, 1918 (1972).
[9] G. Stell in PHASE TRANSITIONS AND CRITICAL PHENOMENA, Vol.5B, edited by C. Domb and M.S. Green (Academic, London, 1976).
[10] J.L. Lebowitz, G. Stell and S. Baer, J. Math. Phys. 6, 1282 (1965).
[11] G. Stell and J.J. Weis, Phys. Rev. A16, 757 (1977).
[12] G.N. Patey, Mol. Phys. 34, 427 (1977).
[13] G.N. Patey and J.P. Valleau, J. Chem. Phys. 61, 534 (1974).
[14] G.N. Patey, D. Levesque and J.J. Weis, Mol. Phys. (1978).
[15] J.P. Valleau and S.G. Whittington in STATISTICAL MECHANICS, Vol. V of MODERN THEORETICAL CHEMISTRY, edited by B. Berne (Plenum, New York, 1977).
[16] J.A. Barker and R.O. Watts, Mol. Phys. 26, 789 (1973).
[17] R.O. Watts, Mol. Phys. 28, 1069 (1974).
[18] D. Levesque, G.N. Patey and J.J. Weis, Mol.Phys. 34,1077(1977).
[19] V.M. Jansoone, Chem. Phys. 3, 79 (1974).
[20] E.R. Smith and J.W. Perram, Mol. Phys. 30, 31 (1975).
[21] D.J. Adams and I.R. McDonald, Mol. Phys. 32, 931 (1976).
[22] A.J.C. Ladd, Mol. Phys. 33, 1039 (1977)
[23] A.J.C. Ladd, preprint and private communication.
[24] J.A. Barker and R.O. Watts, Chem. Phys. Lett. 3, 144 (1969).
[25] H. Popkie, H. Kistenmacher and E. Clementi, J. Chem. Phys. 59, 1325 (1973).
[26] G.C. Lie, E. Clementi , J. Chem. Phys. 62 2195 (1975).
[27] G.C. Lie, E. Clementi and M. Yoshime, J. Chem. Phys. 64, 2314 (1976).
[28] F.H. Stillinger and A. Rahman, J. Chem. Phys. 57, 1281 (1972).
[29] F.H. Stillinger and A. Rahman, J. Chem. Phys. 60, 1545 (1974).
[30] J.S. Høye and G. Stell, J. Chem. Phys. 61, 562 (1974); J. Chem. Phys. 64, 1952 (1976).
[31] G.N. Patey and J.P. Valleau, J. Chem. Phys. 64, 170 (1976).
[32] N.F. Carnahan and K.E. Starling, J. Chem. Phys. 51, 635 (1969).
[33] B. Larsen, J.C. Rasaiah and G. Stell, Mol. Phys. 33, 987 (1977).

[34] J.A. Barker, D. Henderson and W.R. Smith, Phys. Rev. Lett. 21, 134 (1968).
[35] G.N. Patey and J.P. Valleau, Chem. Phys. Lett. 42, 407 (1976).
[36] G.N. Patey and J.P. Valleau, Chem. Phys. Lett. 21, 297 (1973).
[37] G.N. Patey and J.P. Valleau, private communication. I am greatly indebted to G.N. Patey and J.P. Valleau for permission to quote from their unpublished results.
[38] G.N. Patey, Mol. Phys. (1978).
[39] J.O. Hirschfelder, C.F. Curtiss and R.B. Bird, MOLECULAR THEORY OF GASES AND LIQUIDS (Wiley, New York, 1954).

ORIENTATIONAL CORRELATIONS IN MOLECULAR LIQUIDS

M. D. Zeidler

Institut für Physikalische Chemie und Elektrochemie der Universität Karlsruhe
Kaiserstr. 12 D-7500 Karlsruhe

In molecular liquids one needs for the description of the configuration of a pair of molecules 6 variables: 3 components of the center-center vector and 3 Eulerian angles to define the relative orientation. The pair-correlation function $g(\mathbf{R},\Omega)$ is thus a function of these 6 variables; $\Omega=(\alpha,\beta,\gamma)$ denotes here the Eulerian angles. However one prefers the symmetrical definition $g(\mathbf{R}_i,\Omega_i,\mathbf{R}_j,\Omega_j)$ where $\mathbf{R}_i$ denotes the center vector and Ω_i the orientation of the i-th molecule with respect to an arbitrary coordinate system, accepting the disadvantage of dealing now with 12 variables in the pair-correlation function. However in an isotropic fluid the pair-correlation function is independent of translation and rotation of the pair of molecules. Translational invariance requires that

$$g(\mathbf{R}_i,\Omega_i;\ \mathbf{R}_j,\Omega_j) = g(R,\Omega_i,\Omega_j) \tag{1}$$

where $\mathbf{R}_{ij}$ is the difference vector $\mathbf{R}_i-\mathbf{R}_j$ and $R = |\mathbf{R}_{ij}|$ is its length. The condition for rotational invariance is easily derived using group-theoretical methods: We need the projection of the pair-correlation function onto the totally symmetric irreducible representation of the continuous rotation group which is obtained using the projection operator

$$P = h\cdot g^{-1} \int \chi_R \, dR \tag{2}$$

where g is the number of elements in the group, χ_R the character of the irreducible representation under the operation R and h the dimension of the irreducible representation. $\chi_R = 1$ for all R in case of the totally symmetric representation, also h = 1. The rotation operator is used in its matrix representation, and following Wigner the matrix elements are written as $D^{(l)}_{mn}(\Omega)$ explicit expressions of which are given for example by Edmonds[1]. For example

$$D^{(l)}_{mn}(\Omega) = e^{im\alpha}\, d^{(l)}_{mn}(\beta)\, e^{in\gamma} \tag{3}$$

with

$d^{(0)}_{mn}(\beta)$

m \ n	0
0	1

$d^{(1)}_{mn}(\beta)$

m \ n	1	0	-1
1	$\frac{1}{2}(1+\cos\beta)$	$\frac{1}{\sqrt{2}}\sin\beta$	$\frac{1}{2}(1-\cos\beta)$
0	$-\frac{1}{\sqrt{2}}\sin\beta$	$\cos\beta$	$\frac{1}{\sqrt{2}}\sin\beta$
-1	$\frac{1}{2}(1-\cos\beta)$	$-\frac{1}{\sqrt{2}}\sin\beta$	$\frac{1}{2}(1+\cos\beta)$

and $l=0,1,2\ldots\infty$. The projection operator eq. (2) is now allowed to operate on the pair-correlation function, and thus it is convenient to expand the latter in terms of Wigner rotation matrices[2,3]:

$$g(R,\Omega_i,\Omega_j) = \sum_{l_i l_j l_{ij}=0}^{\infty} \sum_{\substack{m_i n_i \\ m_j n_j \\ n_{ij}}} g^{(l_i l_j l_{ij})}_{m_i n_i m_j n_j n_{ij}}(R)\, D^{(l_i)}_{m_i n_i}(\Omega_i) D^{(l_j)}_{m_j n_j}(\Omega_j)\, D^{(l_{ij})}_{o n_{ij}}(\Omega_{ij}) \tag{4}$$

where Ω_{ij} defines the orientation of the vector $\mathbf{R}_{ij}$. Application of eq. (2) on (4) yields the rotationally invariant pair-correlation function

$$g(R,\Omega_i,\Omega_j) = \sum_{l_i l_j l_{ij}} \sum_{\substack{m_i n_i \\ m_j n_j \\ n_{ij}}} g^{(l_i l_j l_{ij})}_{m_i n_i m_j n_j n_{ij}}(R) \sum_{k_i k_j k_{ij}} D^{(l_i)}_{m_i k_i}(\Omega_i)\, D^{(l_j)}_{m_j k_j}(\Omega_j)\, D^{(l_{ij})}_{o k_{ij}}(\Omega_{ij}) \left\{ \frac{1}{8\pi^2} \int d\Omega\, D^{(l_i)}_{k_i n_i}(\Omega)\, D^{(l_j)}_{k_j n_j}(\Omega)\, D^{(l_{ij})}_{k_{ij} n_{ij}}(\Omega) \right\} \tag{5}$$

and after carrying out the integration[1]:

$$g(R,\Omega_i,\Omega_j) = \sum_{l_i l_j l_{ij}} \sum_{\substack{m_i n_i \\ m_j n_j \\ n_{ij}}} g^{(l_i l_j l_{ij})}_{m_i n_i m_j n_j n_{ij}}(R) \sum_{k_i k_j k_{ij}} \begin{pmatrix} l_i & l_j & l_{ij} \\ k_i & k_j & k_{ij} \end{pmatrix} \begin{pmatrix} l_i & l_j & l_{ij} \\ n_i & n_j & n_{ij} \end{pmatrix} D^{(l_i)}_{m_i k_i}(\Omega_i)\, D^{(l_j)}_{m_j k_j}(\Omega_j)\, D^{(l_{ij})}_{o k_{ij}}(\Omega_{ij})$$

$$= \sum_{l_i l_j l_{ij}} \sum_{\substack{m_i \\ m_j}} \{ \sum_{n_i n_j n_{ij}} \begin{pmatrix} l_i & l_j & l_{ij} \\ n_i & n_j & n_{ij} \end{pmatrix} g^{(l_i l_j l_{ij})}_{m_i n_i m_j n_j n_{ij}}(R) \}$$

$$\{ \sum_{k_i k_j k_{ij}} \begin{pmatrix} l_i & l_j & l_{ij} \\ k_i & k_j & k_{ij} \end{pmatrix} D^{(l_i)}_{m_i k_i}(\Omega_i) D^{(l_j)}_{m_j k_j}(\Omega_j) D^{(l_{ij})}_{o k_{ij}}(\Omega_{ij}) \}$$

$$= \sum_{l_i l_j l_{ij}} \sum_{m_i m_j} g^{(l_i l_j l_{ij})}_{m_i m_j}(R)\, \Phi^{(l_i l_j l_{ij})}_{m_i m_j}(\Omega_i, \Omega_j, \Omega_{ij}) \quad (6) \tag{6}$$

This expression is called the invariant expansion of the molecular pair-correlation function(4). Since this expansion can be used for any reference coordinate system we may choose the z axis of the system to coincide with the center-center vector $\mathbf{R}_{ij}$. In this case

$$D^{(l_{ij})}_{o k_{ij}}(0) = \delta_{o k_{ij}} \tag{7}$$

and therefore

$$\Phi^{(l_i l_j l_{ij})}_{m_i m_j}(\Omega_i, \Omega_j, \Omega_{ij}) = \sum_k \begin{pmatrix} l_i & l_j & l_{ij} \\ k & -k & o \end{pmatrix} D^{(l_i)}_{m_i k}(\Omega_i)\, D^{(l_j)}_{m_j -k}(\Omega_j) \tag{8}$$

With eq. (8) introduced into (6) we obtain

$$g(R, \Omega_i, \Omega_j) = \sum_{l_i l_j l_{ij}} \sum_{m_i m_j} \sum_k \begin{pmatrix} l_i & l_j & l_{ij} \\ k & -k & o \end{pmatrix} g^{(l_i l_j l_{ij})}_{m_i m_j}(R)\, D^{(l_i)}_{m_i k}(\Omega_i)\, D^{(l_j)}_{m_j -k}(\Omega_j)$$

$$= \sum_{l_i l_j} \sum_{m_i m_j} \sum_k \{ \sum_{l_{ij}} \begin{pmatrix} l_i & l_j & l_{ij} \\ k & -k & o \end{pmatrix} g^{(l_i l_j l_{ij})}_{m_i m_j}(R) \}\, D^{(l_i)}_{m_i k}(\Omega_i) D^{(l_j)}_{m_j -k}(\Omega_j)$$

$$= \sum_{l_i l_j} \sum_{m_i m_j} \sum_k g^{(l_i l_j)}_{m_i m_j k}(R)\, D^{(l_i)}_{m_i k}(\Omega_i)\, D^{(l_j)}_{m_j -k}(\Omega_j) \tag{9}$$

In comparison to eq. (6) this expression is termed the irreducible expansion of the molecular pair-correlation function(4). In eq. (8) and (9) the property of the 3j-symbol was used that the sum of the terms in the lower row must be zero. We now can relate the coefficients of the invariant and irreducible expansions through

$$g^{(l_i l_j l_{ij})}_{m_i m_j}(R) = (2\, l_{ij}+1) \sum_k \begin{pmatrix} l_i & l_j & l_{ij} \\ k & -k & o \end{pmatrix} g^{(l_i l_j)}_{m_i m_j k}(R) \tag{10}$$

where the orthogonality property of the 3j-symbols was used.

The number of terms appearing in the expansions (6) or (9) is restricted by symmetry considerations. Symmetry operations which leave the pair-correlation function unchanged may be carried out in the configurational space of the molecular pair or the single molecule. With respect to the molecular pair the invariance under rotations has already been dealt with. If the liquid contains just one species of molecules the pair-correlation function must be invariant also against permutations of molecules i and j. We will discuss the effect of the permutation on the irreducible expansion (9). Since the reference system is the center-center vector $\mathbf{R}_{ij}$ and the permutation changes its direction by the angle π, the second and third Euler angles β and γ (the rotations transfer the molecular system to the reference system) change their values to

$$\beta \rightarrow \pi + \beta$$
$$\gamma \rightarrow 2\pi - \gamma$$

whereas the first Euler angle α remains unchanged. The effect of changing these angles on the Wigner rotation matrix elements is the following[1]:

$$\begin{aligned} d^{(l)}_{mn}(\pi+\beta) &= \sum_k d^{(l)}_{mk}(\beta)\cdot d^{(l)}_{kn}(\pi) \\ &= \sum_k d^{(l)}_{mk}(\beta)\cdot(-1)^{l+n}\,\delta_{k-n} \\ &= (-1)^{l+n}\, d^{(l)}_{m-n}(\beta) \end{aligned} \tag{11}$$

Observing eq. (3) and the sign change in γ after the permutation we find for the expansion coefficients in eq. (9)

$$g^{(l_i l_j)}_{m_i m_j k} = (-1)^{l_i+l_j}\, g^{(l_j l_i)}_{m_j m_i k} \tag{12}$$

Finally as far as the molecular symmetry is concerned we may consider a K-fold rotation axis and an inversion center. The rotation merely influences the first Euler angle α, since we choose the molecular z-axis to be the rotation axis. According to eq. (3) invariance of the pair-correlation function against a rotation $2\pi/K$ requires

$$e^{im(\alpha + \frac{2\pi}{K})} = e^{im\alpha} \tag{13}$$

and consequently $m/K = 0,1,2,3...$ Thus the expansion coefficients $g^{(l_i l_j)}_{m_i m_j k}$ in (9) must have indices m_i, m_j which are multiples of the order of the rotation axis. For linear molecules with the rotation axis C_∞, for example, these indices must be zero. If a molecular inversion center exists this symmetry operation leads to a change of the first and second Euler angles α and β according to

$$\alpha \rightarrow 2\pi - \alpha$$

$$\beta \rightarrow \pi + \beta$$

Similar considerations as for the permutation operation of the molecular pair lead to the condition

$$g_{m_i m_j k}^{(l_i l_j)} = (-1)^{l_i + m_i} g_{m_i m_j k}^{(l_i l_j)} \tag{14}$$

which can be fulfilled only for $l_i + m_i$ being an even number.
The expansion coefficients $g_{m_i m_j}^{(l_i l_j l_{ij})}$ or $g_{m_i m_j k}^{(l_i l_j)}$ appear in expressions of thermodynamic functions or other properties of molecular liquids[2,3]. For example a straightforward extension of the energy equation for atomic liquids gives

$$U = \frac{6}{2} N k_B T + \frac{1}{2} N\rho \cdot 4\pi \int dR \cdot R^2 \cdot \sum_{l_i l_j l_{ij}} \sum_{m_i m_j} g_{m_i m_j}^{(l_i l_j l_{ij})}(R) \cdot u_{m_i m_j}^{(l_i l_j l_{ij})*}(R) \{(2l_i+1)(2l_j+1)(2l_{ij}+1)\}^{-1} \tag{15}$$

where $u_{m_i m_j}^{(l_i l_j l_{ij})}$ are coefficients in a corresponding expansion of the pair potential. Another interesting relation is with the Kirkwood correlation factor appearing in his theory for the dielectric constant:

$$\langle \mu \cdot \mu \rangle = \mu^2 \{1 - \frac{4\pi\rho}{3\sqrt{3}} \int dR \cdot R^2 \, g_{00}^{(110)}(R)\} \tag{16}$$

We now will deduce the relation between the expansion coefficients and the scattering cross-section since scattering experiments provide some experimental access to the molecular pair-correlation function. For coherent scattering we have

$$\left(\frac{d\sigma}{d\Omega}\right)_{coh} = \frac{1}{N} \sum_{i,j=1} \sum_{\alpha,\beta=1} f_\alpha f_\beta \langle \exp\{i\mathbf{k}(\mathbf{R}_{i\alpha} - \mathbf{R}_{j\beta})\}\rangle \tag{17}$$

$\mathbf{R}_{i\alpha}$ is the position vector of atom (nucleus) α in molecule i. f_α is the scattering length which for X-ray scattering depends on the wave vector **k** but for neutron scattering is independent of **k**. The sample exists of N molecules with m atoms in each molecule. The position vector is split into

$$\mathbf{R}_{i\alpha} = \mathbf{R}_i + \mathbf{r}_{i\alpha} \tag{18}$$

with $\mathbf{R}_i$ the position of the molecular center and $\mathbf{r}_{i\alpha}$ the position of atom α relative to the molecular center. With (18) substituted into (17) we get

$$\left(\frac{d\sigma}{d\Omega}\right)_{coh} = \frac{1}{N} \sum_i \sum_j \langle \{ \sum_\alpha f_\alpha \exp(i\mathbf{k}\mathbf{r}_{i\alpha})\} \cdot \{ \sum_\beta f_\beta \exp(-i\mathbf{k}\mathbf{r}_{j\beta})\} \exp(i\mathbf{k}\,\mathbf{R}_{ij}) \rangle \qquad (19)$$

We now use the Rayleigh expansion of the exponential functions and distinguish between two coordinate systems: the laboratory system defined by $\mathbf{k}$ and the molecular system fixed to the molecule:

$$\exp(i\mathbf{k}\mathbf{r}_{i\alpha}) = \sum_{l=0}^{\infty} (4\pi(2l+1))^{1/2} \cdot i^l \cdot j_l(kr_\alpha) \sum_m Y_m^{(l)}(\theta_\alpha,\Phi_\alpha) D_{mo}^{(l)}(\Omega_i) \qquad (20)$$

j_l is a spherical Bessel function and $Y_m^{(l)}$ a spherical harmonic. The polar angles θ_α and Φ_α fix the vector r_α within the molecular system; the Euler angles Ω_i fix the orientation of the molecular system relative to the laboratory system. If eq. (20) is introduced into (19) and the molecular scattering factor

$$a_m^{(l)}(k) = \sum_\alpha f_\alpha (4\pi(2l+1))^{1/2} j_l(kr_\alpha)\, Y_m^{(l)}(\theta_\alpha,\Phi_\alpha) \qquad (21)$$

is defined, we obtain

$$\begin{aligned} \left(\frac{d\sigma}{d\Omega}\right)_{coh} = \frac{1}{N} \sum_{ij} \langle & \{ \sum_{l_i m_i} i^{l_i} a_{m_i}^{(l_i)}(k) D_{m_i o}^{(l_i)}(\Omega_i)\} \\ & \{ \sum_{l_j m_j} (-i)^{l_j} a_{m_j}^{(l_j)}(k)^* D_{m_j o}^{(l_j)}(\Omega_j)^* \} \\ & \{ \sum_{l_{ij}} i^{l_{ij}} (2l_{ij}+1)\, j_{l_{ij}}(kR)\, D_{oo}^{(l_{ij})}(\Omega_{ij})\} \rangle \end{aligned} \qquad (22)$$

where a corresponding expansion of exp $(i\mathbf{k}\mathbf{R}_{ij})$ was used.

The coherent differential cross-section separates into an intramolecular and an intermolecular contribution. For the intramolecular part we get from eq. (22) using $l_i = l_j = l$, $m_i = m_j = m$, $R = 0$ and $\Omega_{ij} = 0$ and observing the properties $j_{l_{ij}}(o) = \delta_{l_{ij}o}$ and $D_{oo}^{(l)}(o) = 1$:

$$\begin{aligned} \left(\frac{d\sigma}{d\Omega}\right)_{coh}^{intra} &= \sum_{lm} a_m^{(l)}(k)\, a_m^{(l)}(k)^* \langle D_{mo}^{(l)}(\Omega_i)\, D_{mo}^{(l)}(\Omega_i)^* \rangle \\ &= \sum_{lm} \frac{a_m^{(l)}(k) a_m^{(l)}(k)^*}{2l+1} \\ &= \sum_{\alpha\beta} f_\alpha f_\beta \frac{\sin(kr_{\alpha\beta})}{kr_{\alpha\beta}} \end{aligned} \qquad (23)$$

In the last step of (23) the addition theorem of Bessel

functions was used. This result is the common Debye formula.

Eq. (22) contains N(N-1) intermolecular terms ($i \neq j$). Averaging over the variables Ω_i, Ω_j, Ω_{ij} and R requires the molecular pair-correlation function which may be used in its invariant expansion (6):

$$< D_{m_i o}^{(l_i)}(\Omega_i)\, D_{m_j o}^{(l_j)}(\Omega_j)^*\, j_{l_{ij}}(kR)\, D_{oo}^{(l_{ij})}(\Omega_{ij}) > =$$

$$\frac{\iiiint g(R,\Omega_i,\Omega_j)\, D_{m_i o}^{(l_i)}(\Omega_i)\, D_{m_j o}^{(l_j)}(\Omega_j) j_{l_{ij}}(kR)\, D_{oo}^{(l_{ij})}(\Omega_{ij}) 4\pi R^2 dR d\Omega_i d\Omega_j d\Omega_{ij}}{\iiiint g(R,\Omega_i,\Omega_j)\, 4\pi R^2 dR d\Omega_i d\Omega_j d\Omega_{ij}} \tag{24}$$

$$= \frac{\rho}{N-1}\, \frac{(-1)^{-m_i}}{(2l_i+1)(2l_j+1)(2l_{ij}+1)} \begin{pmatrix} l_i & l_j & l_{ij} \\ o & o & o \end{pmatrix} \int_o^\infty g_{-m_i m_j}^{(l_i l_j l_{ij})}(R)\, j_{l_{ij}}(kR) 4\pi R^2 dR$$

Integration over $g(R,\Omega_i,\Omega_j)$ in the denominator of (24) gave $\frac{N-1}{\rho}\cdot(8\pi^2)^3$. A condition for the 3j-symbol $\begin{pmatrix} l_i & l_j & l_{ij} \\ o & o & o \end{pmatrix}$ to be different from zero is that the sum $l_i+l_j+l_{ij}$ must be even. If eq. (24) is substituted into (22) and the Fourier-Bessel transform

$$h_{-m_i m_j}^{(l_i l_j l_{ij})}(k) = \rho \int_o^\infty g_{-m_i m_j}^{(l_i l_j l_{ij})}(R)\, j_{l_{ij}}(kR)\, 4\pi R^2 dR \tag{25}$$

is employed the final result becomes

$$\left(\frac{d\sigma}{d\Omega}\right)_{coh}^{inter} = \sum_{l_i l_j l_{ij}} \frac{(-1)^{l_j+\frac{1}{2}(l_i+l_j+l_{ij})}}{(2l_i+1)(2l_j+1)} \begin{pmatrix} l_i & l_j & l_{ij} \\ o & o & o \end{pmatrix} \sum_{m_i m_j} (-1)^{-m_i} a_{m_i}^{(l_i)}(k) a_{m_j}^{(l_j)}(k)^* h_{-m_i m_j}^{(l_i l_j l_{ij})}(k) \tag{26}$$

If this result is written out for the specific case of a molecule with 3-fold rotation axis observing all symmetry relations and the triangle condition $|l_i-l_j| \leq l_{ij} \leq l_i+l_j$ for the 3j-symbol one gets

$$\begin{aligned} \left(\frac{d\sigma}{d\Omega}\right)_{coh}^{inter} = {} & a_o^{(o)} a_o^{(o)*} h_{oo}^{(ooo)} + 0.385\, a_o^{(1)} a_o^{(o)*} h_{oo}^{(101)} + 0.179\, a_o^{(2)} a_o^{(o)*} h_{oo}^{(202)} \\ & - a_o^{(1)} a_o^{(1)*} \{ 0.064\, h_{oo}^{(110)} + 0.041\, h_{oo}^{(112)} \} \\ & + a_o^{(1)} a_o^{(2)*} \{ 0.049\, h_{oo}^{(121)} + 0.039\, h_{oo}^{(123)} \} \\ & + a_o^{(2)} a_o^{(2)*} \{ 0.018\, h_{oo}^{(220)} + 0.010\, h_{oo}^{(222)} + 0.010\, h_{oo}^{(224)} \} \\ & + \ldots . \end{aligned} \tag{27}$$

if the expansion is terminated at $l_i = l_j = 2$. This termination is permissable if the angular dependence of the intermolecular potential is not too strong. There exists an interdependence between the number of terms needed in the expansion and the relevant k-region. Higher terms correspond to sharper angular correlations and are important only at small intermolecular distances or large k. This is obvious from the Fourier-Bessel transform, because with increasing order of the Bessel function the region af large k is more heavily weighted. It is difficult to determine the intermolecular differential cross-section experimentally at large k, above 3.5 $Å^{-1}$ the data become rather inaccurate. In the region below 3.5 $Å^{-1}$ however the expansion as in eq. (27) should be adequate.

Calculation of the molecular scattering factors $a_m^{(1)}$ from molecular geometry and scattering lengths according to eq. (21) shows that both for X-ray and neutron scattering certain terms in the expansion dominate because of the factors $a_m^{(1)} a_{m'}^{(1')*}$. As a practical example liquid acetonitrile (a molecule with C_3 rotation axis) may be mentioned for which X-ray data on CH_3CN(5) and neutron data on the isotopes CD_3CN^{14} and CD_3CN^{15} exist(6). In this case only the first three terms in eq. (27) need to be considered. Thus it is possible to determine the three coefficients $h_{oo}^{(000)}$, $h_{oo}^{(101)}$ and $h_{oo}^{(202)}$ from the three diffraction experiments by solving the three simultaneous equations. One has to remember that the factors $a_m^{(1)} a_{m'}^{(1')*}$ assume different values for each experiment, since f_α depends on the experiment and on the isotope used. Finally by the inverse Fourier-Bessel transform the g-coefficients are obtained:

$$g_{m_i m_j}^{(l_i l_j l_{ij})}(R) = \frac{1}{\rho} \frac{1}{2\pi^2} \int_0^\infty h_{m_i m_j}^{(l_i l_j l_{ij})}(k)\, j_{l_{ij}}(kR)\, k^2 dk \qquad (28)$$

Limitations in the experimental k-range, as discussed above, of course lead to restricted resolution in the g-coefficients.

Experimental details and data corrections will not be discussed here. It may be pointed out merely that separation of the coherent differential cross-section into intra- and intermolecular contributions is achieved by calculating the intramolecular term from eq. (23) using known molecular data and including Debye-Waller factors to take care of intramolecular vibrations. Required refinements of molecular data by fitting to the scattering curve at large k where the intermolecular contribution is absent were usually included.

After solving the three simultaneous equations for the $h_{n_i n_j}^{(l_i l_j l_{ij})}(k)$ according to the first three terms in eq.(27) at each value of k the g-coefficients are obtained by application of eq. (28) (see figure next page). Due to the limited k-range of the h-coefficients oszillations and

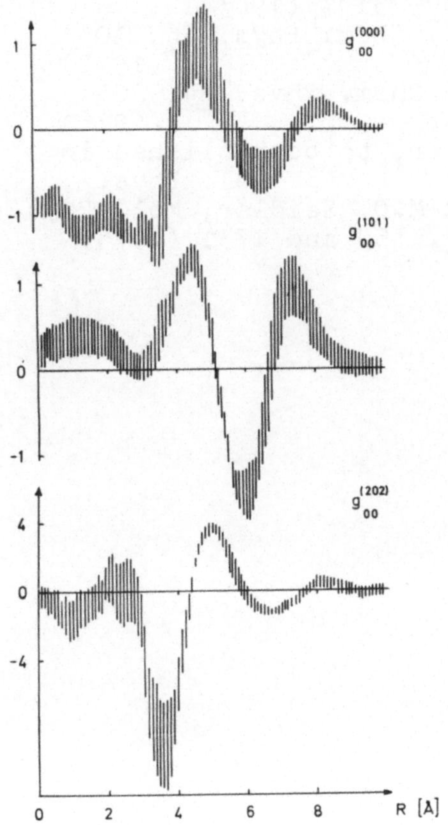

broadening of the results are obvious. The isotropic term $g_{00}^{(000)}$ shows the usual behaviour known from atomic liquids. The orientational terms $g_{00}^{(101)}$ and $g_{00}^{(202)}$ are of special interest since they are determined experimentally for the first time. They only give information about the orientation of a molecule relative to the center-center system independent of the orientation of the second molecule; thus nothing is revealed about the relative orientation of two acetonitrile dipoles. The range of orientational correlations extends up to 9 Å. The change in sign can be discussed on the basis of eq. (6) or (9). Observing eq. (3) and the additional relation for $D_{00}^{(2)}$

$$D_{00}^{(0)}(\Omega_i) = 1$$

$$D_{00}^{(1)}(\Omega_i) = \cos\beta$$

$$D_{00}^{(2)}(\Omega_i) = \frac{3}{2}\cos^2\beta - \frac{1}{2}$$

the following conclusions can be drawn since $g_{00}^{(10)}$ and $D_{00}^{(1)}(\Omega_i)$ must have the same sign: in the range $0^o \leq \beta \leq 90^o$ $g_{00}^{(101)}$ must be negative, but positive in the range $90^o \leq \beta \leq 180^o$; $g_{00}^{(202)}$ must be positive for $0^o \leq \beta \leq 54.7^o$ and $125.3 \leq \beta \leq 180^o$ and negative for $54.7 \leq \beta \leq 125.3^o$. Thus the experimental results tell us that below 4.4 Å preferred orientations of the dipole axis relative to the center-center vector occur in the range $90^o \leq \beta \leq 125.3^o$, whereas above 5.2 Å they are in the range $0^o \leq \beta \leq 54.7^o$. Above 6.8 Å again the first orientational correlation is found.

References

1) A.R. Edmonds, Angular Momentum in Quantum Mechanics, Princeton University Press 1960

2) W.A. Steele, J. Chem. Phys. 39, 3197 (1963)
3) L. Blum and A.J. Torruella, J. Chem. Phys. 56, 303 (1972)
4) L. Blum and A.H. Narten, Adv. Chem. Phys. 34, 203 (1976)
5) H. Bertagnolli and M.D. Zeidler, to be published in Mol. Phys.
6) H. Bertagnolli, P. Chieux and M.D. Zeidler, Mol. Phys. 32, 759 and 1731 (1976)

Round Tables

FAST ION CONDUCTORS - AN INTRODUCTION

K. Funke

Institut für Physikalische Chemie der Universität
Göttingen and Sonderforschungsbereich 126, Germany

1. CURRENT INTEREST AND TRADITION

Tonight we are going to talk about solid electrolytes with high ionic conductivities. In the last few years, these materials have come to be called "super"-ionic conductors. In this introduction however, we will prefer to use the less showy name "fast ion conductors".

The prefix "super" obviously reflects - and possibly even enhances - the high current interest in these materials. This current interest is mainly due to the possibility of using fast ion conductors for the construction of novel battery systems. A cartoon similar to that of Fig. 1 appeared in the TIME-issue of May 16, 1977, illustrating the article "A Look at the Cars of 1985". As you know, battery-driven cars neither go fast nor do they go far. In spite of these shortcomings, much work is now being done in this field, for instance at Ford Motor Company. The battery system having the best chances of being used in future cars seems to be the sodium-sulfur battery which contains the solid electrolyte sodium-β-alumina. In section 4, we will have a brief look at the various technical applications of fast ion conductors.

On the other hand, solid electrolytes, at least in Europe, do have a tradition. This tradition dates back to Faraday[1] who in 1834 recorded the transitions of two solids into highly ion-conducting states. His particular examples were Ag_2S, at 177°C, and PbF_2, near 500°C. While the observed transition was first order in Ag_2S, it was smooth in the case of PbF_2. Last year, O'Keeffe [2] proposed to use the name "Faraday transitions" for those diffuse transitions typically found in several fluoride-ion conductors.

†:

*:

Figure 1

Eighty years later, in 1914, Tubandt and Lorenz reported the discovery of α-AgI [3] which is now known as the prototype of the purely ion-conducting solid electrolytes. When measuring the electrical conductivities of the silver halides, Tubandt and Lorenz were struck by the extraordinary properties of the high-temperature phase of solid AgI. Within this α-phase, which extends over more than 400° C, the electrical conductivity is as high as in a molten salt. It even drops on melting. Fig. 2 is

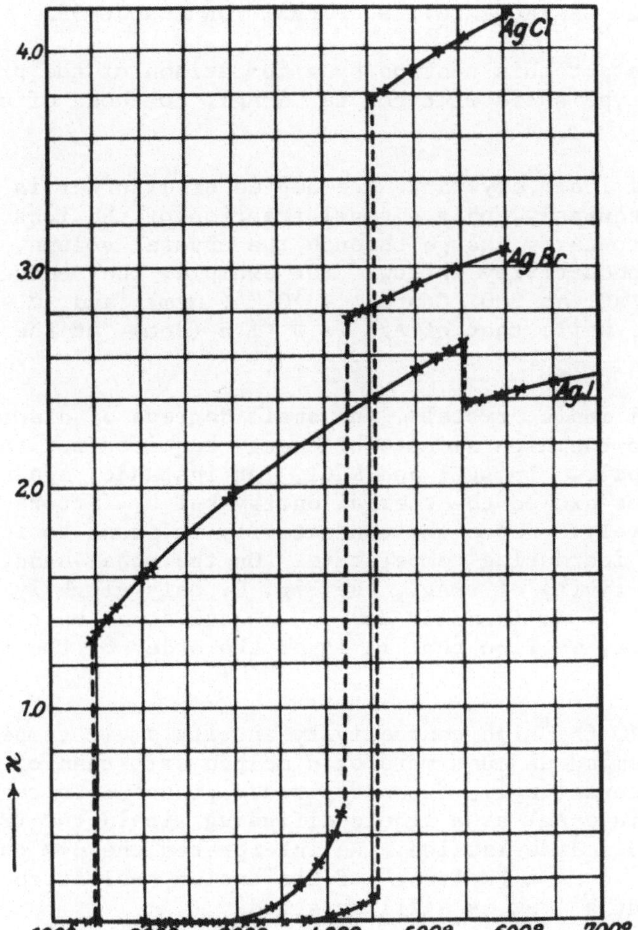

Fig. 2. Conductivity of the silver halides according to Tubandt and Lorenz [3].

Tubandt's and Lorenz' original plot[3]. With the help of transference measurements Tubandt also definitely proved that the unusually high electrical conductivity of α-AgI is entirely due to a rapid diffusive motion of the silver ions[4].

2. CHARACTERISTICS OF FAST ION CONDUCTORS

Let us start this section by a comparison of the properties of our prototype solid electrolyte, α-AgI, to those of usual ionic crystals.

In usual ionic crystals, the degree of disorder is small, $<10^{-2}$. Consequently, only a small fraction of the ions are free to move and to carry charge through the crystal volume. Hence the electrical conductivity is low. For example, the conductivities of AgCl and NaCl at 200° C are $\sigma \approx 10^{-4}$ $(\Omega\text{cm})^{-1}$ and $\sigma \approx 10^{-8} (\Omega\text{cm})^{-1}$, respectively, while that of AgI is $\sigma \simeq 1.6$ $(\Omega\text{cm})^{-1}$ at the same temperature.

In usual ionic crystals, the small degrees of disorder are of course due to the high amounts of energy required for the formation of a defect pair. In AgCl and NaCl, for instance, again at 200° C, these energies exceed the thermal energy k_BT by factors of 35 and 50, respectively. Hence the conductivity is found to increase rapidly with increasing temperature. On the other hand, the electrical conductivity of α-AgI, see Fig. 2, only slightly depends on temperature. If we formally derive an activation energy from an Arrhenius plot, we find that it is of the order of the thermal energy.

In α-AgI, the high conductivity and its small temperature dependence remind us much more of a molten salt than of an ionic crystal. Interestingly, Tubandt already proposed to regard the silver ions in α-AgI as a liquid diffusing within the framework of some solid iodide lattice. He interpreted the β→α phase transition at 147° C as a "melting" of the cation sublattice. Essentially, Tubandt's view is still ours today.

Comparing the silver ions in α-AgI to a diffusing liquid, means considering their dynamical properties. Historically of course, the structure of the ions was analyzed a long time before their microscopic dynamics. The first structure determination performed on a fast ion conductor was done in 1934 on α-AgI[5]. In this investigation, Strock discovered the struccural disorder of the silver ions. By this we mean that the silver ions do not have their own fixed lattice-sites. Rather, they can be found virtually anywhere where there is space for them within the rigid crystal-lattice built by the iodide ions. Strock immediately realized that the structural disorder of the cations could well explain the striking peculiarities of α-AgI. Being structurally disordered, all of the cations are free to move. This results in the high value of the electrical conductivity. Secondly, no defect-formation enthalpy is needed for the thermal activation of the translational motion of the cations. Hence we also understand the small temperature dependence of the conductivity, which seems to be only due to some migration enthalpy.

We will now briefly discuss the structures of α-AgI and of other fast ion conductors. In α-AgI, the iodide ions form a body-centred cubic lattice. On the average, there should be two silver ions within one cube of iodine ions. However, a body-centred cube obviously does not contain any two preferred sites which we might hope to assign to these silver ions. Strock obtained the best fit to his X-ray data by assuming that the two silver ions are distributed at random over 42 different sites. Of course, he did not take these sites too literally, and we should not do so either. They are just an indication of the structural disorder of the silver ions. Later on, Rickert[6] proposed to substitute six "elementary regions" for the 42 Strock sites. These elemenary regions are illustrated in Fig. 3. As you see in the figure, they only provide a convenient subdivision of the space available for the cations within the anion framework. More recent structure work has shown that the probability of finding a silver ion is not at all

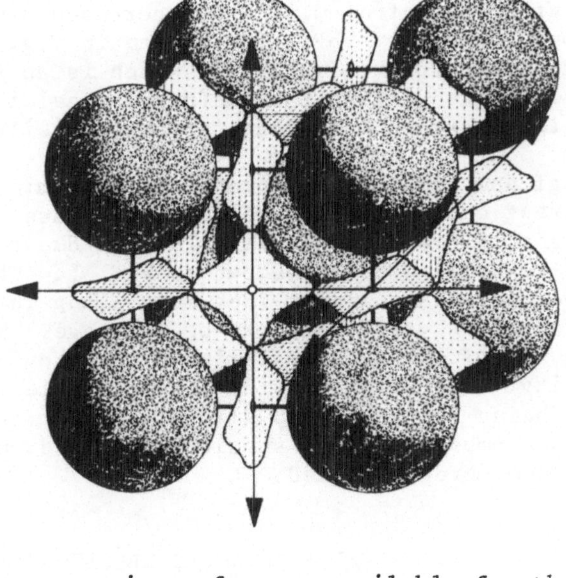

Fig. 3: Elementary regions of space available for the silver ions in α-AgI

homogeneous within these elementary regions. Rather, it is lower at the centre and larger in the outer parts, in particular in the vicinity of the tetrahedrally coordinated sites [7]. For our present discussion, however, another structural aspect is more important because it turns out to be of more general validity; each elementary region can, at least in principle, play two different roles. It can (i) play the role of an actual residence region where a cation might stay for a while and it can (ii) play the role of a portion of a channel-like diffusion path along which a cation is free to move. The situation seems to be similar in other fast ion conductors.

Besides α-AgI, some other solid electrolytes also have b.c.c. structures of their anion lattices. For instance, α-CuBr and α-AgI are isostructural. In both cases there are two cations within a cube with six elementary regions. In α-Ag_3SI and α-Ag_2S, however, there are three and four cations respectively, distributed over the six elementary regions. For translational diffusion to take place, it is of course essential that the number of cations should always be smaller than the number of regions available.

Face-centred cubic arrays of the rigid ion-framework have been found in cation and anion conductors as well. In α-CuI and α-Ag_2HgI_4 the monovalent cations are the mobile species, whereas the fluoride ions are mobile in CaF_2 and PbF_2 which have the fluorite structure. Interestingly, there is a first order phase transition into the highly conducting state in the case of the cation-conductors while in the fluorides the transition is continuous and without change of the rigid cation-framework. The corresponding continuous change of the conductivity of PbF_2 [2] is seen in Fig. 4. Fig. 4 also shows the conductivities of $RbAg_4I_5$, which is an AgI-type solid electrolyte with a difficult structure, of calcia-doped zirconia, which is an oxygen-ion conductor, and of silver-β-alumina.

Besides α-AgI and related compounds, the β-aluminas form the second class of fast cation conductors. In this case, the conduction is however only two-dimensional. This is due to the layer structure of the material which is shown in Fig. 5. The mobile ions can be silver or alkali ions. They move along in different planes, like cars in the various floors of a garage. Within each floor, columns built of oxygen and aluminum ions provide the necessary stability. Two sets of voids are available for the mobile ions, one being energetically slightly preferred. Thus our condition that the number of voids should exceed the number of mobile ions, is once more fulfilled.

3. BRIEF SURVEY ON AgI-TYPE SOLID ELECTROLYTES

In the field of fast ion conductors, there are at present two main directions of interest. Firstly, the investigation of the

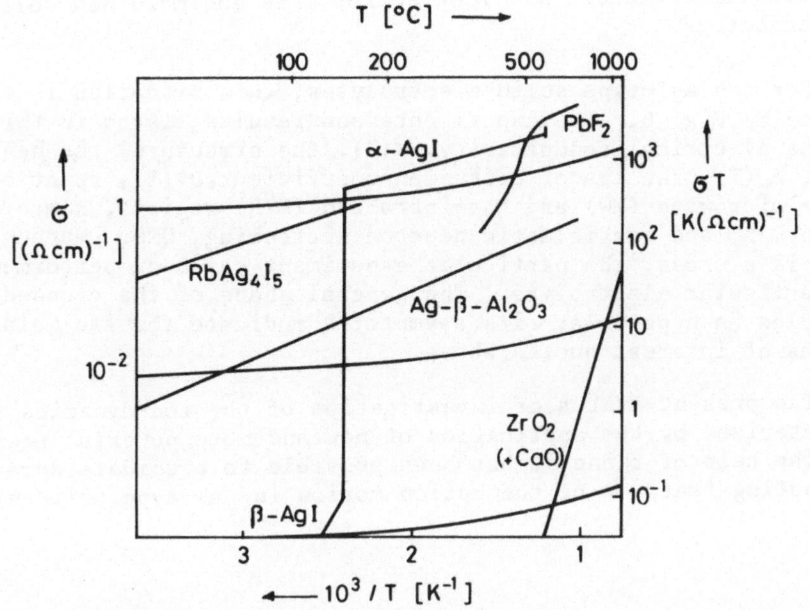

Fig. 4: Electrical conductivity of various solid electrolytes

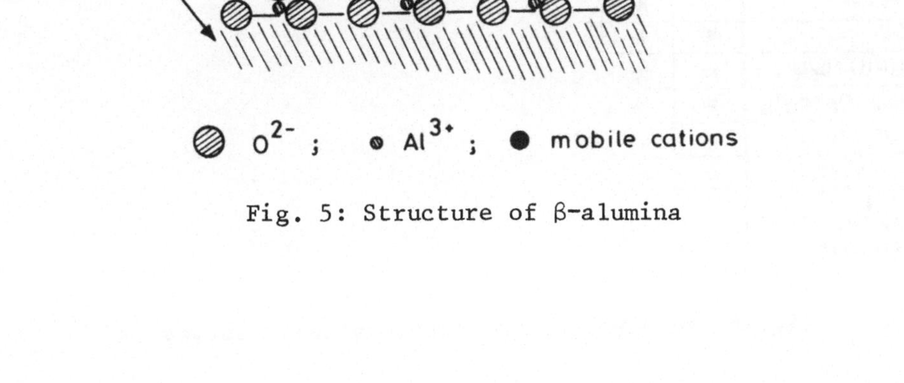

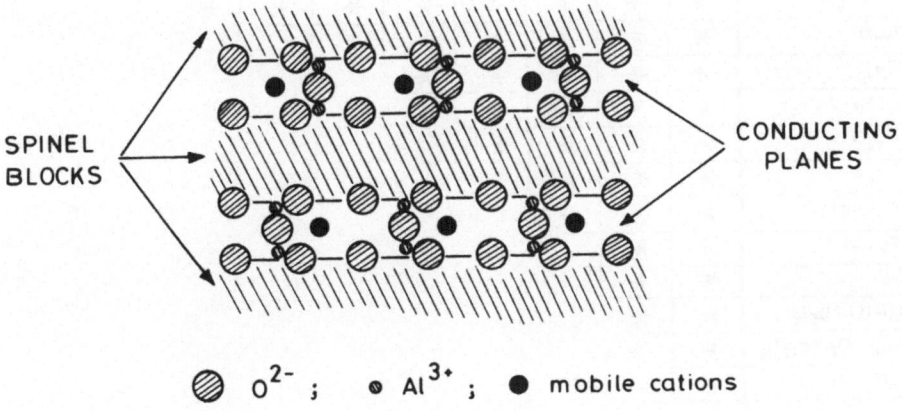

Fig. 5: Structure of β-alumina

dynamical properties of these systems presents a unique and fascinating problem to experimentalists and theoreticians. Secondly, in view of the potential technical applications, scientists and chemical engineers are looking for more and more new solid ion conductors.

For the AgI-type solid electrolytes, this situation is exemplified by Fig. 6. The experiments and results listed in this table are the electrical conductivity, $\sigma(T)$, the structure, the heat capacity, $c_p(T)$, the tracer diffusion coefficient, $D^*(T)$, spectroscopy in the microwave (MW) and far-infra-red (FIR) regions, Raman spectroscopy, R, and quasielastic neutron scattering, QNS. Whenever there is a cross, the particular experiment has been performed on the particular electrolyte. The general shape of the crossed area resembles an hyperbola; its asymptotes indicate the two main directions of interest quoted above.

The present status of investigation of the ion-dynamics is characterized by the application of new and more powerful techniques. With the help of these it has been possible to elucidate some interesting features of the cation motion in AgI-type solid elect-

	$\sigma(T)$	Strukt.	$c_p(T)$	$D^*(T)$	MW	FIR	R	QNS	→ new experim. techniques
α - AgI	*	*	*	*	*	*	*	*	
α - $RbAg_4I_5$	*	*	*	*	*	*	*	*	
β - CuBr	*	*	*	*	*	*			
α - CuI	*	*	*	*	*				
α - CuBr	*	*	*						
α - Ag_2S	*	*	*	*					
α - Cu_2S	*	*	*						
α - Ag_2HgI_4	*	*	*						
α - Ag_3SI	*	*							
Ag_3SBr	*	*							
$(C_5H_5NH)Ag_5I_6$	*	*							
$((CH_3)_4N)_2Ag_{13}I_{15}$	*	*							
$(Ag_2WO_4)Ag_4I_4$	*								
$(Ag_3AsO_4)Ag_4I_4$	*								

↓ new solid electrolytes

Fig. 6: AgI-type solid electrolytes: a survey

rolytes. In Fig. 7, we compare these features to those of mobile particles in other kinds of systems. The two extreme possibilities are shown on the left- and right-hand sides. In a monoatomic liquid, the atoms obey the laws of simple diffusion at least down to one picosecond and one Ångstrøm. On the other hand, hydrogen in metals performs a jump-diffusion from one well-defined site to another. The motion of the cations in AgI-type solid electrolytes appears to be intermediate. As in a liquid, there is a permanent irregular motion of the cations. This motion is however restricted to a system of residence regions and interconnecting passageways provided by the periodic structure of the anion lattice. A typical cation stays within a residence region for a certain residence time, then diffuses away through a channel and is trapped again in another residence region after a certain passage time. Both times are comparable and of the order of 10ps.

We now address ourselves to the other main line of interest, i.e. to materials and applications. The search for new fast ion conductors is dominated by two demands:

(i) high ionic conductivity at low temperature
(ii) low cost.

These demands are not easy to fulfil simultaneously. $RbAg_4I_5$, for instance, is already highly conducting at room temperature, see Fig. 4, but it is not exactly low-priced. Sodium-β-alumina does not cost too much money, but it becomes a really good conductor only at elevated temperatures.

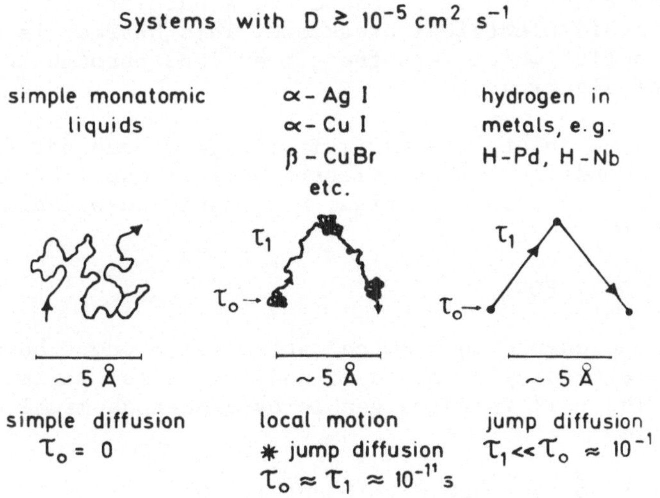

Fig. 7: Diffusion in different systems

4. TECHNICAL APPLICATIONS

In many cases, the use of batteries is preferable to the usual and classical method of energy storage which is the containment of fuel. Using batteries may help saving valuable fossil fuels and reducing the smog problem - provided of course that the energy stored in them was not originally gained from fossil fuels. The field of applications of solid electrolytes in battery systems is wide; it ranges from cardiac pacemakers and satellites to public transportation and utility units which help to meet the fluctuating demands for electrical power. Battery-driven trucks and buses seem to be a likely development in this century, whereas personal electric vehicles do not appear to be very probable because of their well-known shortcomings in performance.

Let us now have a look at electrochemical cells incorporating solid electrolytes. In order to explain the principle, we consider the all-solid-state cell

$$- \, | \, Ag \, | \, \alpha\text{-}AgI \, | \, \alpha\text{-}Ag_2Se \, | \, Se \, | \, C \, | \, +$$

which also serves as an example in our students' laboratory at Göttingen. In this cell we exploit the fact that - at given values of temperature and pressure - the chemical potential of Ag_2Se is lower than that of 2Ag plus 1Se. The tendency of silver and selenium to react to form Ag_2Se leaves electrons behind on the silver side, because of $Ag \rightarrow Ag^+ + e^-$, and requires electrons on the selenium side, because of $Se + 2e^- \rightarrow Se^{2-}$. Thus a potential difference is set up between both ends of the cell. As α-Ag_2Se is a mixed ionic and electric conductor, a barrier for electrons is needed to avoid electrical breakdown. This barrier is provided by the α-AgI pellet which lets the silver ions through and blocks the passage of electrons.

How do we calculate the maximum voltage we can get from this battery? The condition of the electrochemical equilibrium states that the electrochemical potential, η , should be a minimum. We thus have

$$0 = \Delta\eta = \Delta\mu + F\Delta V$$

where $\Delta\mu$ is the change in chemical potential brought about by the reaction. F is Faraday's constant, and ΔV is the maximum voltage obtainable. The cell is rechargeable by electrochemical decomposition.

The most important new develoment in the field of electrochemical energy storage is probably that of the sodium-sulfur battery. This system appears to be best suited to meet the requirements of vehicular and utility applications. It is operated at temperatures between 300° C and 350° C, when sodium and sulfur are both liquid. The elements are separated by a sodium-β-alumina tube which acts as a sodium-ion conductor and electron-barrier.

The sodium-sulfur battery-system is much more efficient than the conventional lead-lead oxide battery: the energy stored per unit volume is roughly doubled (40 Wh/ℓ instead of 20 Wh/ℓ) and the energy stored per unit weight is increased at least by a factor of five (> 150 Wh/kg instead of 30 Wh/kg). At the same time the specific cost of the active materials is reduced from 10 to 0.5 $/kWh [8].

Finally, we mention oxides like calcia-doped zirconia which are good oxygen-ion conductors at temperatures above 1000°C. However, their conductivity vs. temperature curves, cf. Fig. 4, do not qualify them as typical fast ion conductors. The high-temperature applications of these materials are in fuel-cells and lambda-probes.

REFERENCES

[1] M. Faraday, "Experimental Researches in Electricity", Art.1339, Taylor and Francis, London (1839).
[2] M. O'Keeffe, "Phase Transitions and Translational Freedom in Solid Electrolytes", in "Superionic Conductors", G.D. Mahan, W.L. Roth eds., Plenum Press, New York and London (1976).
[3] C. Tubandt and E. Lorenz, Z. physik. Chem. 87, 513, 543 (1914).
[4] C. Tubandt, in "Handbuch der Experimentalphysik XII", W. Wien, F. Harms eds., Akadem. Verlagsges., Leipzig (1932).
[5] L.W. Strock, Z. physik. Chem. B 25, 411 (1934) and B 31, 132 (1936).
[6] H. Rickert, Z. physik. Chem. NF 24, 418 (1960).
[7] R.J. Cava, F. Reidinger, B.J. Wuensch, "Single-Crystal Neutron Diffraction Study of α-AgI between 160° and 300° C", preprint.
[8] J.R. Birk, "Energy Storage, Batteries, and Solid Electrolytes: Prospects and Problems", in "Superionic Conductors", G.D. Mahan, W.L. Roth eds., Plenum Press, New York and London (1976).

GENERAL BIBLIOGRAPHY

A "Superionic Conductors", G.D. Mahan, W.L. Roth eds., Plenum Press, New York and London (1976).
B "Fast Ion Transport in Solids", W. van Gool ed., North-Holland, Amsterdam and London, and American Elsevier, New York (1973).
C "Solid Electrolytes", P. Hagenmuller, W. van Gool eds., Academic Press, New York (1978).
D J.T. Kummer, "β-Alumina Electrolytes", Progr. Solid State Chem. 7, 141 (1972).
E K. Funke, "AgI-type Solid Electrolytes", Progr. Solid State Chem. 11, 345 (1976).

DISCUSSION

De Gennes : A very naïve question I might ask is: Are these systems transparent to the eye or what do they look like?

Funke : Very pure single crystals of α-AgI are red and transparent.

Alder : Did you say that the phase transition is always of first order?

Funke : It is of first order in the simple AgI-type solid electrolytes, whereas the fluorides have smooth transitions, the Faraday transitions.

Hansen : How do you characterize those first-order phase transitions? Is there a volume change or a latent heat?

Funke : Yes, both are observed. AgI contracts by roughly 5% at its $\beta \rightarrow \alpha$ phase transition. The latent heat is roughly 6 kJ/mole.

Angell : I believe a thermodynamic singularity which conforms to lattice gas statistics has been observed in one of the silver halide based superionic conductors, within the high-conducting region. Can you comment on this?

Funke : In Ag_2HgI_4, there is a first-order phase transition into the highly conducting phase. This transition takes place at 51° C, after an initial continuous disordering process of the cations at lower temperatures. This disordering process has been described in terms of order-disorder models. $RbAg_4I_5$ appears to have a second-order phase transition at 209 K, with a continuously decreasing order parameter.

Springer : You mentioned this comparison with hydride systems. It is true that the mean rest time of the hydrogen on its site as compared to the jump time is very long. But it is worthwhile to mention that in vanadium hydride the diffusion is so fast that also there the mean time of the jump itself is comparable to the mean rest time in similarity to these systems.

De Gennes : May I ask you a question about the diffusion coefficients which you listed. Do they usually agree with the conductivity as a single-particle property?

Funke : No, there are collective effects. The Haven ratio is 0.6 in α-AgI and it deviates from unity also in α-$RbAg_4I_5$ and in α-Ag_2S. There are two possibilities for explaining this Haven ratio. The first possibility is the usual one in ionic crystals: the particles jump back and forth because the vacancies, not the particles are moving at random. It is relatively sure that this model does not apply to α-AgI. Rather we believe that there is a forward correlation of the motions of different ions, and we can see that in the microwave spectra.

Springer : You were speaking of these channels where the ions pass through. Are these channels wide enough that the ions can pass through without much deformation of the anion lattice or does it need a large deformation of the anions.

Funke : Considering the ionic radii one should assume that the cations and the anions are deformed. Both kinds of ions are highly polarizable.

De Gennes : Is anything known about the pressure effects on conductivity?

Funke : It is known that in AgI the transition temperature changes with pressure. Of course that should be expected because of the contraction at the $\beta \rightarrow \alpha$ phase transition.

De Gennes : Is there no special sensitivity to application of pressure? I mean, if you have some local strains that are large you might think that some channels get blocked.

Funke : I do not know of any experiment of that kind.

TOWARD A THEORY OF SUPERIONIC CONDUCTORS

L. Pietronero and S. Strässler

Brown Boveri Research Center, CH-5405 Dättwil
Switzerland

ABSTRACT

We describe some theoretical questions about superionic conductors and outline the main open problems.

Superionic conductors are solids characterized by an ionic conductivity comparable to liquid electrolytes [1, 3]. The interest in them is twofold : First they represent a new state of condensed matter intermediate between solid and liquid and second they are of technological importance in electrochemistry.

The remarkable high conductivity of these compounds is due to the low barrier height for diffusion (barriers of the order of k_BT are not uncommon) and to the fact that essentially all ions of one species are mobile within the host lattice provided by the other ionic species. To explain this behaviour is quite difficult and the reasons are clear if one notes that there is no theory even for the melting transition of normal compounds.

The more difficult questions are those connected to the structural principles and the phase diagrams of these materials and little is known about them. Problems of phenomenological nature like the diffusion mechanism, the role of the host lattice etc. are better understood.

In order to say something about the structural principles of superionic conductors [4] it is convenient to recall a few facts about the stability of normal binary compounds. The important parameter is the ionicity f_i. It is known in fact that compounds with an ionicity larger than a critical value $f_i^c = .785$ crystallize into structures with six or eight-fold coordination. This is due to the fact that the bonds are mainly ionic and therefore the electrostatic energy gives the biggest contribution to the binding energy [5]. For ionicities smaller than f_i^c the coordination is four-fold tetrahedral. The bonds are mainly covalent and show sp^3 hybridization [6]. The positions of the ions are very well defined in the two extreme cases $f_i = 0$ and $f_i = 1$. For f_i close to f_i^c the various coordinations are equally possible so that the ions are less strongly localized. This has to be the case if we want to have an ionic species diffusing in the host sublattice provided by the other ionic species. In fact the binary superionic conductors AgI, CuI and CuBr all have an ionicity very close to f_i^c [4]. This competition between ionic and covalent interactions also gives rise in the non superionic phase (low temperature β phase of AgI) to phonons with unusually low frequency [7].

Further information about the stability of the superionic phase for binary compounds comes by considering the melting mechanism of simple ionic compounds. It has been pointed out that certain regularities of the melting temperatures with respect to the ionic radii of alkali halides can be explained starting from the hypothesis that the melting instability occurs when the anions are in "effective" contact between them [8]. This implies a description of the ions in terms of hard spheres with radial (ionic) potentials. Since this criterion is able to describe alkali halides ($f_i \sim .95$) but obviously not superionic conductors like AgI that are stable with the anions in contact between them, it is clear that the superionic phase cannot be explained with purely ionic interactions. Covalency or polarization effects are essential for the stability of these peculiar materials. Of special relevance in this respect, are the molecular dynamics calculations recently reported by Schommers [9]. In fact it was not possible to obtain a stable binary superionic conductor with the interionic interactions simply described by Born-Meyer potentials.

The phase diagram of AgI [3], the prototype of these materials shows the following characteristics : Below 147°C it has a standard (wurzite) structure (β-phase) with tetrahedral coordination. Between 147°C and 555°C the iodines form a b.c.c. lattice while the silvers can occupy several different sites and show a liquid like mobility (α-phase superionic). Above 555°C it is molten.

Let us consider the normal-superionic transition. It is of first order so the transition temperature is given by

$$T_c = \frac{\Delta U}{\Delta S} \tag{1}$$

where the entropy change per ion is $\Delta S \sim 1.8$ and the energy change per ion is $\Delta U = 0.065$ eV. In the superionic phase there are six available sites at lowest energy per each Ag^+ ion. The configurational entropy for a system of N particles and nN sites is

$$W = \binom{nN}{N} \tag{2}$$

$$\Delta S = \frac{1}{N} \ln(W) = n\ln(n) \; (n-1) \; \ln(n-1)$$

For n = 6 we have $\Delta S = 2.7$. But we should notice that some of the sites cannot be occupied simultaneously because the distance between them is smaller than the ionic diameter of Ag^+. The inclusion of this effect gives $\Delta S \simeq 1.7$. The entropy change is therefore of configurational nature.

An estimation of the energy change, as given by the deformation of the unit cell, using the short range force constants derived from the phonon spectrum, can only account for a small fraction of the observed ΔU. The main contribution is therefore given by Madelung and polarization type contributions.

Let us consider now the problem of the ionic dynamics. The relevant quantity is

$$\sigma_{dc} = \frac{n(Ze)^2}{m} \mu \tag{3}$$

where n is the number of diffusing ions, m is their mass and μ their mobility. In order to gain more information about the ionic motion it is useful to consider the full frequency dependent conductivity $\sigma(\omega)$. The mobility shows in general an activated behaviour with very low activation energies $E_A \sim 0.1 \div 0.2$ eV similar to those of molten salts. Since n includes all the ions of a given species it is clear that also σ_{dc} will be of the order of that of molten salts. The large number of equivalent available sites represents the main reason for the large number of diffusing ions and the low activation energies.

The simplest model for the dynamical properties is that of a Brownian particle diffusing in a periodic potential [10]. Assuming the damping to be proportional to the temperature (high temperature limit; no quantum effect is in fact relevant to the physics of superionic conductors) $\Gamma = \Gamma_o k_B T$ and taking for Γ_o a value typical of the ionic interactions we can calculate σ_{dc} as a function of the temperature [11]. The conductivity σ_{dc} shows the following behaviour : for $k_B T < E_A$ it increases exponentially with temperature (activated regime). For $k_B T > E_A$ it decreases linearly with temperature because of the increase of Γ. The maximum is at $k_B T \sim E_A$. The region relevant for superionic conductors is $k_B T \lesssim E_A$. It is still in the activated regime but close to the maximum. This is an interesting point and shows that, unless completely new diffusion mechanisms are present, we cannot expect to find new materials with a conductivity much larger than AgI.

For the single particle Brownian model also the full frequency dependent conductivity can be computed. This topic is the subject of Fulde's lecture [12] so we shall not discuss it here. We only recall that the frequency dependent conductivity obtained from this model can give account of the main features observed in AgI [10,11]. Only in the frequency region from 0 to 20 cm^{-1} the observed structure cannot be explained by a single particle model.

We now consider the problem of many diffusing particles which interact. The first question concerns the equilibrium configuration of such a system. Since the host lattice defines special sites for the diffusing particles it is natural to adopt a lattice gas model [13]. There are more available sites than diffusing particles and in general the sites can be of different types. Two competing effects are present : the occupational entropy of the sites and the interaction between diffusive ions. These two effects can give rise to order disorder transitions of various kinds [14,15] affecting only the diffusing ions. A more complicated situation occurs when the host lattice takes part in an essential way to the phase transition like in the normal-superionic transition of AgI described previously. In this case there is a global structural transformation and not just an order disorder transition.

The standard lattice gas model is generally described as an Ising model in which a spin up corresponds to an occupied site and a spin down to an empty one. Substantial differences occur between the order disorder transitions in superionic conductors and the Ising model if there are different sites with different degeneracies [16].

Of particular interest are the one dimensional models, which can be solved exactly using the method of the transfer matrix [16]. The results can be applied to study those systems in which the ions diffuse only into channels. One of these materials is hollandite for which it was possible to obtain a detailed knowledge of the state of order of the ions including long range Coulomb interactions [17].

The next step is to study how the state of order affects the d.c. conductivity. Some cases where the conductivity can be described by a hopping model have been studied in Refs. [14],[18] and [19]. The microscopic conditions for a hopping model are that the activation energy is larger than k_BT and that the jumps are independent. Within this model it is also possible to obtain the frequency dependent conductivity [19],[20] and some structure is present in the microwave region but it is inconsistent with the AgI data [3]. The observed microwave structure can be instead related to a forward correlation between jumps that enhances the conductivity [21]. The specific mechanism for this forward correlation is not yet clear. A possibility is that it is mediated by the phonons of the lattice. This brings us to the limit of our present understanding and we conclude by summarizing that we have learned together with the main open problems.

From the point of view of structural principles very little is known apart from the empirical although interesting facts described at the beginning. Because of this lack of knowledge it is very difficult to predict whether a material with a given chemical composition will be a good ionic conductor or not. All we can really say is that polarizability plays an important role in the stability of binary superionic conductors.

Some aspects of the ionic dynamics are rather well represented by a Brownian particle in a periodic potential. Strong similarities with the dynamics of molten salts are present [21].

The equilibrium properties of many interacting particles are properly described by lattice gas models. A realistic description of the full dynamics should include explicitly a coupling between hopping and vibrations. This is quite a difficult problem and little is known about it.

Another interesting question is the effect of disorder (due to the diffusing sublattice) on the harmonic modes. In an ordered system only the TO mode at $k = 0$ contributes to $\sigma(\omega)$. In the presence of disorder the k selection rule is broken and some oscillation strength is present also for the other modes.

REFERENCES

[1] "Fast Ion Transport in Solids" ed. by W. van Gool (North-Holland, Amsterdam, 1973).

[2] "Superionic Conductors", ed. by G.D. Mahan and W.L. Roth (Plenum, New York, 1976).

[3] K. Funke, Progr. Solid State Chem. 11, 345 (1976).

[4] J.C. Phillips, J. Electrochem. Soc. 123, 934 (1976).

[5] M.P. Tosi, Solid State Physics, 16, 1 (1964).

[6] J.C. Phillips, "Bonds and Bands in Semiconductors", Academic Press, New York, 1973.

[7] W. Bührer and P. Brüesch, Solid State Commun. 16, 155 (1975).

[8] L. Pietronero, Proceedings of the Int. Conf. on Lattice Dynamics, Paris 1977, in print.

[9] W. Schommers, Phys. Rev. Letters, 38, 1536 (1977).

[10] P. Fulde, L. Pietronero, W. Schneider and S. Strässler, Phys. Rev. Letters, 26, 1776 (1975).

[11] P. Brüesch, L. Pietronero, S. Strässler and H.R. Zeller, Electrochimica Acta, 22, 717 (1977).

[12] P. Fulde, this volume

[13] K. Huang "Statistical Mechanics" J. Wiley, New York (1963).

[14] H. Sato and R. Kikuchi, J. of Chem. Phys. 55, 677 (1971).

[15] K.R. Subbaswamy and G.D. Mahan, Phys. Rev. Letters 37, 642 (1976).

[16] L. Pietronero and S. Strässler, to be published.

[17] H.U. Beyeler, L. Pietronero, S. Strässler and H.J. Wiesmann, Phys. Rev. Letters, 38, 1532 (1977).

[18] W. Dietrich, I. Peschel and W. Schneider, Communications on Physics, in print.

[19] R. Vargas, M.B. Salamon and C.P. Flynn, Phys. Rev. Letters, 37, 1550 (1976).

[20] J.C. Kimball, preprint.

[21] H.R. Zeller, P. Brüesch, L. Pietronero and S. Strässler in Ref. 2, p. 201.

[22] P. Brüesch, L. Pietronero and H.R. Zeller, J. of Phys. C. 9, 3977 (1976).

DISCUSSION

B. Quentrec : It seems to me that this is very similar to a plastic crystal. The motion of the ions has here a linear behaviour; in a plastic crystal it is the same with a rotation.

L. Pietronero : May be that general formulations like the Brownian motion concept can be applied in both cases, but I do not think the specific problems are similar.

J. Smit : Is it known if there is a correlation between the jumping process of the silver ions ?

L. Pietronero : Yes, it is probably responsible of the structure observed in $\sigma(\omega)$ at microwave frequencies.

W. Freyland : Is there any experimental evidence in β-Aℓ crystals from which one can decide which is the order of the transition ?

L. Pietronero : In β-Aℓ the transition is of continuous type. The methods to deduce this are those standard for any phase transition : specific heat and calorimetric measurements.

A. Angell : There is a molecular dynamics study on CaF_2. Is it relevant to compare that study to yours ?

L. Pietronero : CaF_2 is probably better described in terms of extended defects than in terms of a "molten sub-lattice" more appropriate for AgI. The arguments about the structural principles do not apply to CaF_2 because it is not a binary structure but for the ionic dynamics the same models can be applied.

SUPERIONIC CONDUCTORS : THE INDEPENDENT PARTICLE MODEL

Peter Fulde

Max-Planck-Institut für Festkörperforschung
7000 Stuttgart, Germany

1. INTRODUCTION

In parallel with the revived interest of experimental physics in fast ionic conductors[1](superionic conductors) there has been renewed interest in the theoretical aspects of this phenomenon. Basically one is dealing thereby with systems in which one species of atoms is in a liquid like state while the other forms a lattice which acts like a periodic potential on the ionic liquid. An example is AgI. Here a phase transition takes place at T=146°C above which the Ag-lattice is molten while the I-lattice still persists. From a theoretical point of view the dynamical behaviour of the ionic liquid in the potential of the remaining lattice is of particular interest. It shows diffusive as well as oscillatory features and it is of considerable interest to describe both in a unified way. The simplest model which one can apply for such a description is that of independent ions moving through the periodic potential. This constitutes certainly a severe approximation since one expects collective or interaction effects among the ions of the molten lattice to become important for small frequencies e.g. $\omega < 10\ cm^{-1}$. An indication for this is the appearance of considerable structure in the low frequency response of superionic conductors such as AgI in the microwave regime [2]. There is no way to explain such structure within an independent particle model. Nevertheless it is very useful to study and understand the independent particle model since it can describe the coupled diffusive and oscillatory motion. For a possible comparison with experiments one has to calculate certain correlation functions. The frequency dependent conductivity requires a knowledge of the velocity-velocity correlation function $\langle v(t)v(o)\rangle$ while for neutron scattering and Raman scattering the intermediate scattering function $\langle \exp\{iq(x(t)-x(0))\}\rangle$ and the correlation function $\langle \alpha_{\nu\mu}(\tilde{r},t)\alpha_{\nu\mu}(\tilde{0},0)\rangle$ of the Raman tensor $\alpha_{\nu\mu}$ have to be known. In the following we will describe the methods which have been applied to treat the independent particle model and list the results obtained.

2. FORMULATION OF THE PROBLEM

In the independent particle model we consider the motion of a single ion in a periodic potential. The motion is influenced by stochastic forces f(t) acting on the particle. The simplest case is that of a one-dimensional sinusoidal potential (see Fig. 1). In that case the Langevin equation reads

$$m\ddot{x} + m\gamma\dot{x} + m\omega_o^2 \frac{a}{2\pi} \sin\frac{2\pi}{a} x = f(t) \qquad (1)$$

Due to the fluctuation-dissipation theorem this implies a power spectrum of the fluctuating forces of the form $\langle f(t)f(t')\rangle = 2mk_BT\gamma\delta(t-t')$. The aim will be to compute functions such as $\langle v(t)v(o)\rangle$ from which the frequency dependent mobility $\mu(\omega)$ is obtained for example via

$$\mu(\omega) = \frac{1}{k_BT}\int_o^\infty d\omega e^{i\omega t} \langle v(t)v(o)\rangle \qquad (2)$$

The Langevin equation (1) is equivalent to a Fokker-Planck equation for the conditional probability $W(xvt|x_ov_o)$ which is written as

$$\begin{aligned}\frac{\partial W}{\partial t} &= L\,W \\ &= \left\{\gamma\left(1 + v\frac{\partial}{\partial v} + \frac{k_BT}{m}\frac{\partial^2}{\partial v^2}\right)\right. \\ &\qquad \left. - v\frac{\partial}{\partial x} - \frac{K(x)}{m}\frac{\partial}{\partial v}\right\} W \qquad (3)\end{aligned}$$

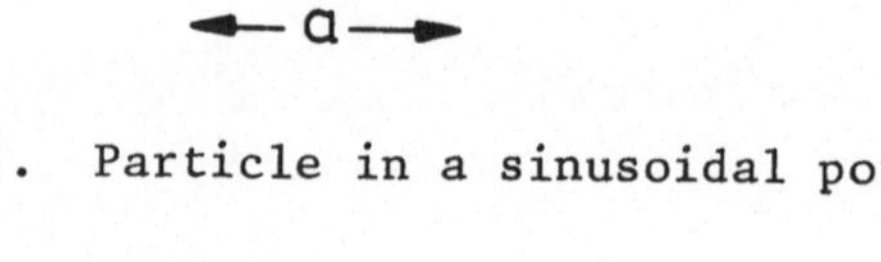

Figure 1. Particle in a sinusoidal potential.

where $K(x) = -m\omega_0^2\frac{a}{2\pi}\sin\frac{2\pi}{a}x$. For $t\to\infty$ the solution of this equation yields the equilibrium distribution $W_{st}(xv) = C\exp\{-\frac{1}{k_BT}\{\frac{m}{2}v^2+V(x)\}\}$ where $V(x)$ is the potential to $K(x)$. In terms of W, the correlation function $\langle v(t)v(o)\rangle$ is written as

$$\langle v(t)v(o)\rangle = \int dxdx'dvdv'vW(xvt|x'v')v'W_{st}(x'v') \qquad (4)$$

Thus calculating the correlation function is equivalent to solving the Fokker-Planck equation. The problem of a particle in a sinusoidal potential is equivalent to that of a pendulum which is hit by Brownian particles. It has also a bearing on a series of other problems among which we mention only the noise in Josephson junctions.

3. THE VELOCITY-VELOCITY CORRELATION FUNCTION

In the limiting cases $k_BT \gg A$ ($A = \frac{m\omega_0^2a^2}{2\pi^2}$ is the barrier height) and $k_BT \ll A$ the correlation function is well known.

α) $k_BT \gg A$. Here Eq.(1) reduces to $m\ddot{x}+m\gamma\dot{x}=f(t)$ and $\langle v(t)v(o)\rangle=\langle v^2\rangle\exp\{-\gamma t\}$. Hence

$$\mu(\omega) = \frac{1}{m}\,\frac{1}{\gamma-i\omega} \qquad (5)$$

β) $k_BT \ll A$. In this case $m\ddot{x}+m\gamma\dot{x}+m\omega_0^2x=f(t)$ and $\langle v(t)v(o)\rangle=\langle v^2\rangle\exp\{-it(\omega_0-i\gamma)\}$. Thus

$$\mu(\omega) = \frac{1}{m}\,\frac{i\omega}{\omega^2-\omega_0{}^2+i\omega\gamma} \qquad (6)$$

Earlier treatments of the combined oscillatory and diffusive motion assumed that for times shorter than a characteristic jump time τ_0 the particle is described by a damped harmonic oscillator while for $t>\tau_0$ it is described by pure diffusion [3]. The mobility is then obtained as a weighted sum of Eqs.(5) and (6). Although this result contains some of the essential features of the combined oscillatory and diffusive motion a more accurate description is clearly desirable. The problem can be solved by the following methods:

a) Continued Fraction Method.

Here one makes an ansatz for $\mu(\omega)$ in the form of a J-continued fraction [4]

$$\mu(\omega) = \frac{1}{m} \cfrac{1}{-i\omega + \gamma_o + \cfrac{\Delta_1}{-i\omega + \gamma_1 + \cfrac{\Delta_2}{\ddots \cfrac{\Delta_n}{-i\omega + G_n(\omega,T)}}}} \tag{7}$$

This form is particularly suitable since the limiting results (5,6) are obtained in a simple fashion, namely by setting $\Delta_1 = 0$ and $\gamma_1 = \Delta_2 = 0$, respectively. The coefficients γ_ν, Δ_ν are calculated from a small time expansion of the correlation function. It is

$$\langle v(t)v(o)\rangle = \sum_{n=o}^{\infty} \frac{c_n}{n!} t^n \tag{8}$$

with $c_n = \int dxdv\, W_{st}(xv)vL_B^n v$ where L_B denotes the "backward" Fokker-Planck operator

$$L_B = \gamma(-v\frac{\partial}{\partial v} + \frac{k_B T}{m}\frac{\partial^2}{\partial v^2}) + v\frac{\partial}{\partial x} + \frac{K(x)}{m}\frac{\partial}{\partial v} \tag{9}$$

The mathematical theory of continued fractions [5] provides prescriptions how to compute the coefficients γ_ν, Δ_ν from the known c_n which in turn are easily calculated. Of particular interest is the choice of the rest term $G_n(\omega,T)$. There are two ways in which one can proceed: In the case that $\mu(\omega=0,T)$ is known from an independent calculation one can approximate $G_n(\omega,T)$ by a frequency independent function $g_n(T)$ such that the function $\mu(0,T)$ is correctly reproduced. We list briefly the temperature and friction regimes for which $\mu(0,T)$ is known analytically:

α) limit of large friction (Smoluchowski limit)

β) limit of low temperatures (Chandrasekhar, Kramers)

γ) high temperature regime (Dieterich, Peschel and Schneider).

There are also extrapolating results available between the high and low temperature regimes except in the case of small frictions [6].

In case $\mu(0,T)$ is not known one has to continue the fraction until convergence is reached. This is possible for $n<30$ except for small values of γ [6]. This way the continued fraction produces also the static result $\mu(0,T)$. Some results for $\mu(\omega,T)$ as obtained by the continued fraction method are shown in Fig. (2a,b). It should be mentioned that it is difficult to calculate the continued fraction for $n>4$ if the potential is different from a sinusoidal one.

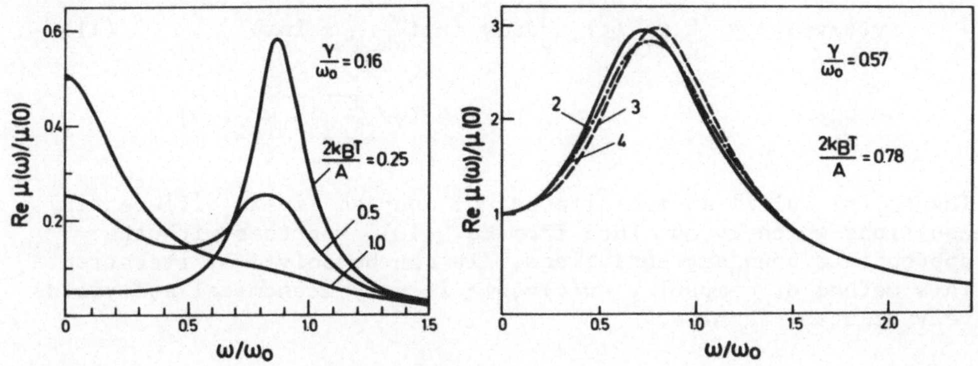

Figure 2

a) Reμ(ω) for given value of μ(o) as calculated from a continued fraction which has been terminated for n=2,3,4 (from Ref.[4]).

b) μ(ω) for different temperatures (for Ref.[6]).

b.) Equation of Motion Method

Although the continued fraction method gives good insight into the dynamical behaviour of the system it has also disadvantages. First it does not work well for some correlation functions such as the intermediate scattering function and second it involves considerable work if one has to go to large values of n. These disadvantages are partially absent in another method which aims at calculating $\langle v(t)v(o)\rangle$ by an equation of motion method[7]. For that purpose one rewrites Eq. (4) in the form

$$\langle v(t)v(o)\rangle = \int dxdv\; W_{st}(xv)e^{L_B t}v \tag{10}$$

where use has been made that $W(xvt|x'v')$ satisfies with respect to the variables x',v' the backward Fokker-Planck equation. It is easy to find the eigenfunctions of the γ-dependent part of L_B, which is $-\gamma v\frac{\partial}{\partial v} + \frac{\gamma k_B T}{m}\frac{\partial^2}{\partial v^2}$. They are given by

$$|\ell\rangle = \exp\left(\frac{mv^2}{4k_BT}\right)\phi_\ell\left(v\sqrt{\frac{m}{2k_BT}}\right)$$

where the $\phi_\ell(x)$ are the harmonic oscillator eigenfunctions. This suggests expanding the r.h.s. of Eq.(10) in terms of $\phi_\ell\left(v\sqrt{\frac{m}{2kT}}\right)$. Furthermore if the potential V(x) is periodic as in our case one can make a Bloch-type expansion and obtains

$$\langle v(t)v(o)\rangle = \sum_n c_{1n}(t) \int dxdv \exp(\frac{mv^2}{4k_BT} + inx) \cdot \phi_1(v\sqrt{\frac{m}{2k_BT}}) v W_{st}(xv) \quad (11)$$

The $c_{\ell n}(t)$ fulfil an infinite set of coupled linear differential equations which is obtained from Eq. (10), together with the appropriate boundary conditions. It can be solved by iteration. This method of computing $\langle v(t)v(o)\rangle$ is very economical and yields very accurate results.

c). The Memory Function Approach

The memory function approach replaces the non-linear equation (1) by the following linear equation

$$m\ddot{x} + m\gamma_0\dot{x} + m\omega_0^2 \int_{t_o}^{t} dt' M(t-t')\dot{x}(t') = F(t) \quad (12)$$

i.e. the external force is replaced by an additional frictional damping term with a memory. Due to this the power spectrum of the stochastic force is rather complicated (fluctuation-dissipation theorem) and can simulate a combined oscillatory and diffusive behaviour [8]. The simplest ansatz is $M(t) = \exp(-t/\tau)$. With $\tau = (\gamma-\gamma_o)\omega_o^{-2}$ this leads to

$$\mu(\omega) = \frac{1}{m} \frac{1}{-i\omega + \gamma_o + \frac{\omega_o^2}{-i\omega + 1/\tau}} \quad (13)$$

and reproduces Eqs. (5) and (6) in the respective limits. This approximation for the memory function corresponds therefore to terminating the continued fraction by $\Delta_2 = 0$. It is a good description of the system for reasonably large damping. However it can not reproduce a behaviour of $\mu(\omega)$ as shown in Fig.(2b) since this requires higher order terms in the continued fraction expansion.

4. THE INTERMEDIATE SCATTERING FUNCTION

The intermediate scattering function $F(q,t)$ is defined as

$$F(q,t) = \langle \exp\{iq(x(t) - x(o))\}\rangle \quad (14)$$

and is connected with the dynamic structure factor $S(q,\omega)$ via

$$S(q,\omega) = \frac{1}{\pi} \mathrm{Re} \int_0^\infty dt \exp\{i\omega t\} F(q,t) \tag{15}$$

Following ref.[6] we discuss several limiting situations.

a) Small q-limit

In this case $F(q,t)$ can be connected with $\mu(\omega)$. In the Gaussian approximation it is $F(q,t) = \exp\{-q^2\rho(t)\}$ with

$$\rho(t) = \int_0^t dt'(t-t') \langle v(t)v(o)\rangle$$

$$= \frac{4k_BT}{\pi} \int_0^\infty d\omega \frac{\sin^2\omega t/2}{\omega^2} \mu(\omega) \tag{16}$$

This leads to [6]

$$F(q,\omega) = \frac{1}{-i\omega + q^2 k_B T\mu(\omega)} \tag{17}$$

The weight of the oscillatory peak in the spectrum I_{osc} is given by $I_{osc} \simeq q^2 k_B T(m\omega_0^2)^{-1}$ and that of the diffusive one by $I_{diff}=1-I_{osc}$.

b) Large Friction Regime

In this case the Fokker-Planck Eq.(3) can be reduced to the Smoluchowski equation. Its stationary solutions $\psi_n(x)$ can be obtained with the ansatz $\psi_n(x) = \exp\left(-\frac{V(x)}{2k_BT}\right) \phi_n(x)$ from a Schrödinger type of equation[6]

$$\left(-\frac{d^2}{dx^2} + U(x)\right) \phi_n(x) = \frac{\lambda_n}{D_o} \phi_n(x) \tag{18}$$

where the "potential" $U(x)$ is given by $U(x)= \frac{K'(x)}{2k_BT} + \left(\frac{K(x)}{2k_BT}\right)^2$ and $D_o=k_BT/m\gamma$ is the diffusion constant for $K(x) \equiv 0$. The eigenvalues λ_n depend on q (reduced to the first Brillouin zone) and a band index n. The problem resembles the computation of energy bands in solid state physics. From the solutions of the Smoluchowski equation one obtains for $S(q,\omega)$ a sum of Lorentzians

$$S(q,\omega) = \frac{1}{\pi} \sum_{\nu=0}^{\infty} |M_\nu(q)|^2 \frac{\lambda_\nu(\tilde{q})}{\omega^2 + (\lambda_\nu(\tilde{q}))^2} \tag{19}$$

where $M_\nu(q) = \int dx \exp[iqx]\phi_{\tilde{q},\nu}(x)\phi_o(x)$ is a "transition" matrix element. $\tilde{q}$ indicates the restriction to the first Brillouin zone. For $T\to 0$ only the term $\nu=o$ contributes to $S(q,\omega)$ and the result reduces to that of jump diffusion[9]. At finite temperatures there are also contributions from $\nu = 1,2$. This leads to a broad background in $S(q,\omega)$ which can be interpreted as coming from the motion of the particle within a potential well, while the narrow peak $\nu = 0$ results from the diffusion of the particle from well to well. The large friction limit has also been treated with the equation of motion method [10] and very accurate results were obtained.

c) Arbitrary q-values

The equation of motion method has been applied to study $S(q,\omega)$ for arbitrary q-values[11]. For not too large frictions an inelastic peak is found near ω_o which results from the oscillations of the particle in a potential well. The weight of this peak increases with increasing values of q until higher harmonics start to appear. The inelastic structure disappears for T becoming larger than the barrier height.

5. POLARIZABILITY CORRELATION FUNCTION

We want to consider the effect of polarizability changes due to the motion of the diffusing ions on the Raman scattering intensity[12]. The latter is expressed by the following correlation function

$$I(\underset{\sim}{q},\omega) \sim \int dt d^3\underset{\sim}{r}\, d^3\underset{\sim}{r}' \exp[i\underset{\sim}{q}(\underset{\sim}{r}-\underset{\sim}{r}')-i\omega t] \times \times \langle\alpha_{is}(\underset{\sim}{r}',0)\alpha_{is}(\underset{\sim}{r},t)\rangle \tag{20}$$

where $\alpha_{is}(\underset{\sim}{r},t)$ is the component of the tensor along the polarization vectors of the incident and scattered light. It can be written as

$$\alpha_{is}(\underset{\sim}{r},t) = \alpha_{is}(\underset{\sim}{r}) \sum_j \delta(\underset{\sim}{r}-\underset{\sim}{r}_j(t)) \tag{21}$$

where as before $\underset{\sim}{r}_j(t)$ denotes the position of the j-th diffusing ion. The function $\alpha_{is}(\underset{\sim}{r})$ describes the polarizability as function of position in the lattice. Clearly the polarizability of a particle will depend in which region of the unit cell it is situated and therefore will change periodically as the particle is moving through the periodic structure.

Taking the Fourier transform of $\alpha_{is}(\underset{\sim}{r}) = \Sigma\alpha_{is}(\underset{\sim}{K})\exp(i\underset{\sim}{K}\underset{\sim}{r})$

leads to

$$I(\omega) \sim N \sum_{\underset{\sim}{K},\underset{\sim}{K'}} \alpha_{is}(\underset{\sim}{K})\, \alpha_{is}(\underset{\sim}{K'})\, S(\underset{\sim}{K'}-\underset{\sim}{q},\underset{\sim}{K}+\underset{\sim}{q},\omega) \qquad (22)$$

with

$$S(\underset{\sim}{Q'},\underset{\sim}{Q},\omega) = \frac{1}{2\pi N} \int dt\, e^{-i\omega t} \sum_{jl} \langle e^{i\underset{\sim}{Q'}\underset{\sim}{r}_{\ell}(o)} e^{i\underset{\sim}{Q}\underset{\sim}{r}_{j}(t)} \rangle$$

We remark that the periodicity of $\alpha_{is}(\underset{\sim}{r})$ is usually not the same as that of the lattice but can be a fraction or a multiple of it. For $\underset{\sim}{Q'}=-\underset{\sim}{Q}$ the function $S(\underset{\sim}{Q'},\underset{\sim}{Q},\omega)$ reduces to the dynamic structure factor $S(\underset{\sim}{Q},\omega)$.

One can compare the Raman scattering due to the diffusion of the ions in a superionic conductor with that in a liquid. In the latter case $\alpha_{is}(\underset{\sim}{r})$ is constant and changes in the polarizability density result only from long wavelength fluctuations in the density i.e. the terms $K,K' \neq 0$ vanish in Eq.(22). $I(\omega)$ contains then a quasielastic Lorentzian which is described by the collective (Hydrodynamic) behaviour of the system. The linewidth Dq^2 is given by the thermal diffusion and is usually smaller than $10^7 sec^{-1}$. The situation is different in a superionic conductor. Here terms $K,K' \neq 0$ can contribute due to a different polarizability of the diffusing ions at different lattice sites and in these cases the independent particle model is expected to be appropriate. That results in a quasielastic line which is roughly by a factor 10^6 broader than in a liquid.

REFERENCES

[1] See for example: Superionic Conductors (edit. by G.D. Mahan and W.L. Roth) Plenum Press, New York, London 1976.

[2] K. Funke, Progr. Solid St. Chem. 11, 345 (1976).

[3] B.A. Huberman and P.N. Shen, Phys. Rev. Letters 33, 1379 (1974).

[4] P. Fulde, L. Pietronero, W.R. Schneider and S. Strässler, Phys. Rev. Letters 35, 1776 (1975).

[5] See for example: H.S. Wall, "Analytic Theory of Continued Fractions", Chelsea Publishing Company, New York, 1948.

[6] W. Dieterich, I. Peschel and W.R. Schneider, Z. Physik B27 177 (1977).

[7] H.D. Vollmer, to be published.

[8] P. Brüesch, S. Strässler, H.R. Zeller, Phys. Stat. Solidi (a) 31, 217 (1975).

[9] C.T. Chudley and R.J. Elliott, Proc. Phys. Soc. 77, 353 (1961).

[10] A.J. Dianoux and F. Volino, this conference and to be published.

[11] W. Dieterich, T. Geisel and I. Peschel, to be published.

[12] T. Geisel, Solid State Comm. 24, 155 (1977).

DISCUSSION

A.J. Dianoux: Have there been experiments on Raman scattering where such an enhancement of the quasielastic linewidth has been found?

P. Fulde: Yes, for example Winterling in Stuttgart has seen the broad quasielastic line in a Raman-scattering experiment on AgI.

A.J. Dianoux: Just another comment on our calculation which corresponds to yours in the large friction limit. There is a pronounced difference to a quasielastic Lorentzian for $q/q_o = 1.25$. But at much lower or higher values of q there is only one Lorentzian. The calculations are not very difficult to perform.

P.C. Hohenberg: Your model is one-dimensional. Do you think that this is important.

P. Fulde: There is no principal difficulty to extend to three dimensional calculations but the numerical work is considerable. I don't think that such an extension would yield new insight. Collective effects are of course expected to be more important in one-dimension than in three dimensions.

P.G. De Gennes: Let me ask a question about S(q,t). I have the impression that you have two limiting regimes if you look at the wave vector q. At low q you have a sort of overall effective diffusion which is very slow if the barriers are high. The diffusion coefficient depends on the activation energy. At large q I have checked only in the weak and strong coupling limits that there is a remarkable sum rule, i.e. that the S(q,t) reduces to the bare diffusion. Do you think that there is an exact sum rule for all values of V/kT such that the large q-limit gives you bare diffusion?

P. Fulde: Dieterich and Peschel have checked for a specific example (V/kT = 1.2) that the linewidth goes as q^2 in the limit of large q indicating ideal gas behaviour.

QUASIELASTIC NEUTRON SCATTERING FROM α-AgI

K. Funke, G. Eckold†
Institut für Physikalische Chemie der Universität
Göttingen and Sonderforschungsbereich 126, Germany

R.E. Lechner
Institut Laue-Langevin, 156X, 38042 Grenoble Cedex, France

1. INTRODUCTION

In this contribution we will present new experimental quasi-elastic neutron scattering data on α-AgI along with an interpretation of these data in terms of a simple model+. As has already been discussed in the introductory part of this session, α-AgI may be visualized to a first approximation as a sea of mobile silver ions within a rigid b.c.c. iodide ion lattice. The question we asked ourselves is the following. How do the silver ions move on microscopic scales of a few Ångstroms and picoseconds ? In particular, how does this microscopic motion depend on temperature ?

Our answer to this problem is essentially illustrated in Fig. 7 of the introductory talk. A typical silver ion stays for a while within a certain region of space defined by the periodic structure of the anions. Within this region, it performs some kind of irregular motion. As soon as it finds an exit out of this region, it starts diffusing along a channel which guides the ion towards another region where it is trapped again, and so on. An important point is that the mean time an ion spends diffusing through a channel, τ_1, is of the same order of magnitude as the mean time the ion spends within a local region of space, τ_0. This is different

† Present address : Institut für Festkörperforschung, KFA Jülich Germany.

\+ In previous papers on this topic [1-3] we reported on measurements at fixed temperature. The experiments have now been extended to different temperatures and the mathematical formulation of our model has been improved.

from the usual treatment of jump-diffusion of hydrogen in metals, where τ_1 is normally assumed to be much shorter than τ_0. Self-diffusion in α-AgI is of course also quite different from self-diffusion in liquids. This is because solid-like features are now introduced by the ordered structure of the anions which provide a system of regions and channels available to the mobile cations. Our present information about the cation motion in α-AgI has been obtained by combined application of various experimental techniques : microwave- and far-infrared spectroscopy [4-6] as well as quasielastic neutron scattering [1-3, 7].

2. EXPERIMENTS

In this paper we will concentrate exclusively on our neutron scattering experiments. These have been performed at the Institut Laue-Langevin, Grenoble, with the multichopper time-of-flight spectrometer IN5. The basic principle is readily explained. Pulses or monochromatic neutrons hit the sample, the neutrons are then scattered by the sample and detected by detectors mounted at several fixed scattering angles ϕ. Measurement of the time of flight of the scattered neutrons yields the time-of-flight spectra which are then corrected and transformed to an energy scale. We thus obtain the total scattering law $S^{tot}(\phi,\omega)$ which includes the incoherent as well as the coherent scattering.

So far, two experiments have been performed, the first of them at constant temperature, 250°C [2,3] , the second at four different temperatures ranging from 160°C to 300°C [7] (The α-phase extends from 147°C to 555°C). In these experiments, the wavelength of the incident neutrons was 5.34 Å and 6.21 Å, respectively. The energy resolution was 0.18 meV and 0.10 meV, respectively.

3. PRESENTATION OF EXPERIMENTAL SPECTRA

In Figs. 1 and 2 we present neutron scattering spectra obtained at fixed temperature, 250°C, but at different scattering angles, ϕ. On the other hand, spectra at fixed ϕ but at different temperatures are displayed in Figs. 3 and 4. The spectrum in Fig. 1, at $\phi = 41.5°$, consists essentially of a narrow quasielastic line. Q being relatively small, $Q_0 \approx 0.5$ Å^{-1} for elastically scattered neutrons, we observe the silver ions on a relatively large spatial scale. Details of the diffusion process cannot be seen in this case. More spatial resolution is achieved if the scattering angle is increased. The spectrum in Fig. 2, at $Q_0 \approx 2.0$ Å^{-1}, already shows a substantial deviation from the Lorentzian line-shape which one would expect for liquid-like diffusion. Rather, we find a relatively narrow quasielastic peak superimposed on a much broader distribution. The narrow peak has

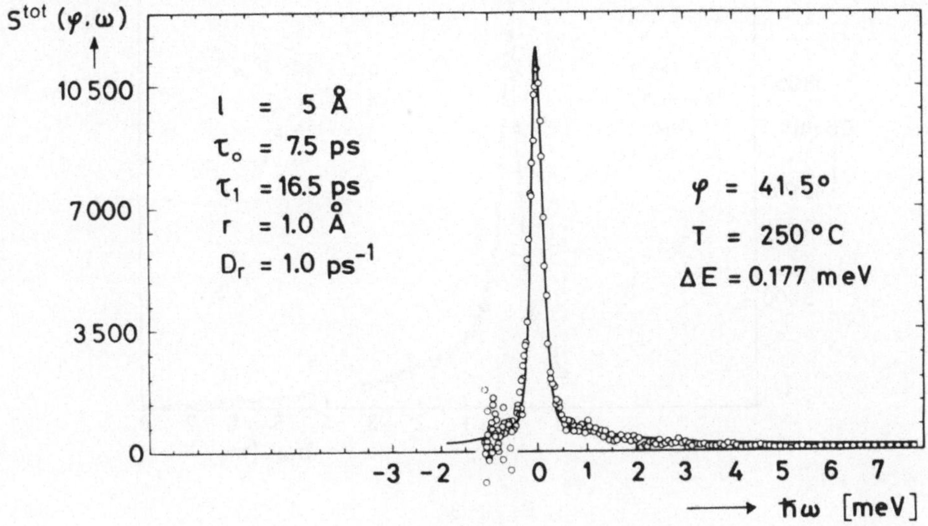

Fig. 1

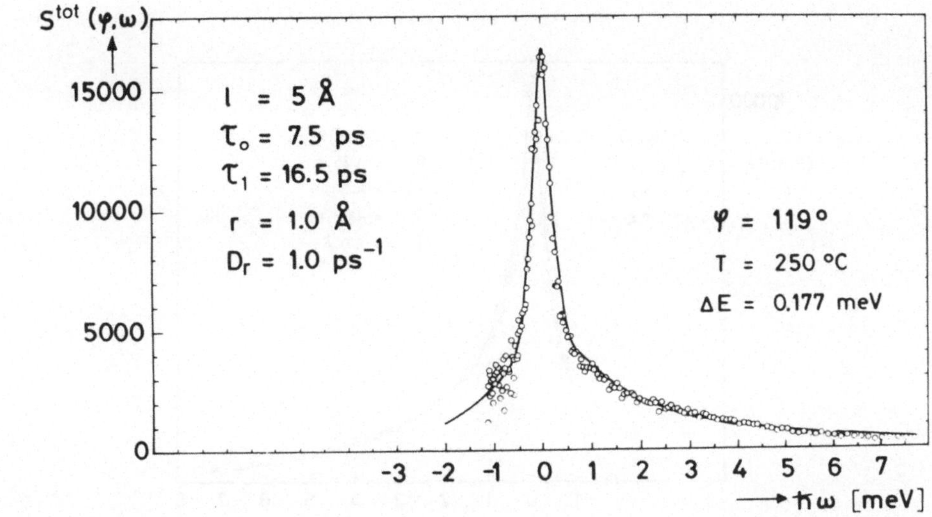

Fig. 2

Experimental scattering function $S^{tot}(\phi, \omega)$ of α-AgI at 250°C as obtained with incident neutrons of $\lambda = 5.34$ Å at scattering angles $\phi = 41.5°$ and $\phi = 119°$; not corrected for resolution, arbitrary units. The solid lines result from a model calculation (parameter values inset, see [3].

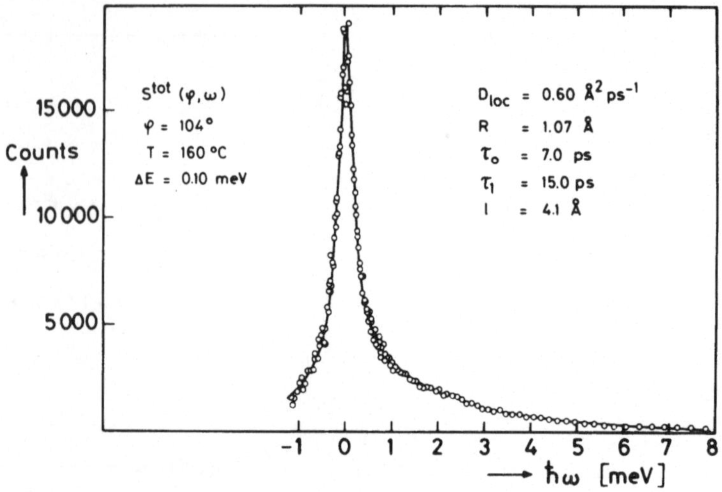

Fig. 3

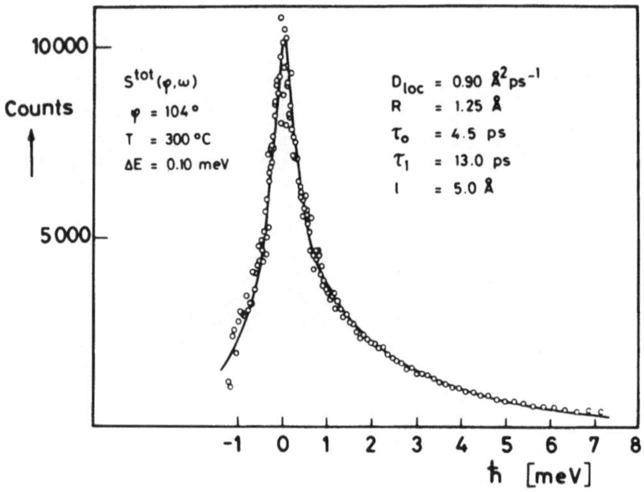

Fig. 4

Experimental scattering function $S^{tot}(\phi,\omega)$ of α-AgI at 160°C and 300°C, as obtained with incident neutrons of $\lambda = 6.21$ Å at $\phi = 104°$; not corrected for resolution, arbitrary units. The solid lines result from the present model calculation (parameter values inset, see [7]).

evolved from the narrow quasielastic line of Fig. 1 as Q has been increased. At relatively large Q a saturation of the width of this line is observed. This saturation directly reflects the physical feature that each silver ions is alternately residing and moving, residing and moving, etc... The mean residence time within a certain region fixed in space should be proportional to the inverse of the width of the narrow line at large Q.

Furthermore, we observe at large Q a second feature, namely the broad distribution extending to relatively high energy transfer. This should correspond to a motion taking place within a local region of space and on a shorter time scale than the resting-moving-resting motion. As the broader distribution is also quasielastic in character, it should also correspond to a diffusive type of motion; thus this might be some irregular motion within a restricted region of space.

What happens to our spectra when the temperature is changed at fixed scattering angle ? The spectrum at 160°C, i.e. 13° above the β-α phase transition, see Fig. 3, very clearly displays the narrow line reflecting the resting-moving-resting process. On the other hand, the spectral shape is considerably changed at 300°C, see Fig. 4. The broad distribution has now become more intense, due to a slight enlargement of the region of space where the local irregular motion is taking place. Moreover, the width of the narrow component is slightly increased since the mean residence time, τ_o, has been decreasing with increasing temperature.

4. QUANTITATIVE INTERPRETATION OF SPECTRA

So far, our discussion has been merely qualitative. In this section, however, we will present a simple mathematical model which we will fit to our experimental spectra. To begin with, two points should be mentioned.

(i) The observed scattering is almost entirely quasielastic.

(ii) It is almost entirely due to the motion of the silver ions.

In order to explain our experimental scattering law, we need models for calculating the incoherent and the coherent scattering. For a calculation of the incoherent distribution, we require a model for the single-particle motion. The coherent contribution will then be estimated by using an empirical Ansatz given by Sköld a few years ago [8]. Application of the Sköld-model however requires a knowledge of the structure factor of the silver ions which is fortunately known [9]. The problem is thus reduced to designing a proper description of the single-particle motion.

At this point we note that the incoherent and the coherent parts of our spectra are connected via the Sköld-model in such a way that they qualitatively have the same kind of shape. We can therefore directly see from our experimental spectra what the incoherent scattering function should look like qualitatively.

Fig. 5 illustrates the procedure how we have been reading in our spectra in order to extract the relevant physical features. The incoherent part of the experimental spectrum is approximately described as a convolution of a relatively narrow line and a broad distribution with a δ-peak on top of it. Considering also the observed Q-dependence of these two different contributions to the spectra, one finds that a jump-diffusion model and a model of a random motion within a limited and locally fixed region of space yield the two different spectral shapes shown in the lower left part of Fig. 5, and the proper Q-dependence. Convolution of the individual scattering laws in energy corresponds to a convolution of the individual self-correlation functions in space. We thus arrive at the simple description of the cation motion already used in [2,3] and presented in the introductory talk of this session.

5. MODEL FOR LOCAL MOTION

The random local motion is treated mathematically by solving the diffusion equation for a particle confined to the interior of a cubic box. This model contains two parameters, the coefficient

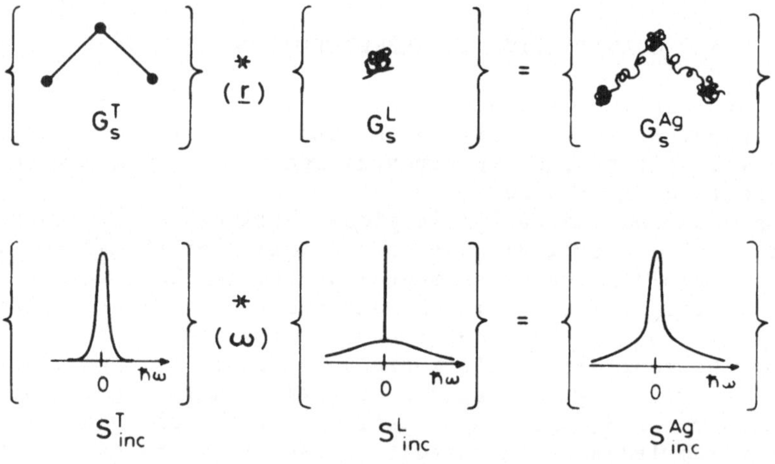

Fig. 5

Incoherent scattering function and self-correlation function of the silver ions in α-AgI represented as convolutions of the individual functions for translational and local motion.

of local self-diffusion, D_{loc}, and the side-length of the box, 2R. A reasonable short-time behaviour originally due to Egelstaff and Schofield [10] has also been included. In the short-time limit, the silver ions are thus treated as ideal-gas particles.

Our local-diffusion model yields good fits to the wings of the experimental spectra. From 160°C to 300°C, D_{loc} increases approximately linearly from 0.60 $Å^2 ps^{-1}$ to 0.90 $Å^2 ps^{-1}$. Note that $Å^2 ps^{-1} = 10^{-4} cm^2 s^{-1} \approx 10\ cm^2/day$. Expressing D_{loc} as a function of temperature by

$$D_{loc}(T) = -a + bT,$$

which is a common representation of self-diffusion coefficients in liquids, we find $a \approx 0.33\ Å^2 ps^{-1}$ and $b \approx 0.21\ Å^2 ps^{-1} K^{-1}$.

The side-length of the cube, 2R, is found to increase from $2R \approx 2.15$ Å at 160°C to $2R \approx 2.50$ Å at 300°C. This increase of 2R explains the observed temperature dependence of the relative intensities of the broad and narrow contributions to our spectra.

Furthermore we want to emphasize that the velocity-auto-correlation function $< v^L(t) \cdot v^L(0) >$ suggested by our simple model agrees qualitatively with the one derived by Schommers [11] from his molecular dynamics study on α-AgI. The behaviour of $< v^L(t) \cdot v^L(0) >$ as obtained by Schommers is rather liquid-like being negative at $t \gtrsim 0.15$ ps and slowly approaching zero in the long-time limit. This kind of behaviour is readily understood in terms of our model. After an initial period of three dimensional diffusion within the box, the silver ion will sooner or later arrive at one of the faces of the cube. It then has to move backwards into the interior of the box, instead of continuing its free diffusion.

One last point should be made concerning the local random motion. We can calculate the local-motion frequency spectrum

$$f^L(\omega) \propto \lim_{Q \to 0} \left(\frac{\omega^2 \cdot S^L_{inc}}{Q^2} \right)$$

from the mathematical expression for the contribution to the incoherent scattering due to the local random motion, $S^L_{inc}(Q,\omega)$ [7]. To a first approximation, different silver ions may be assumed to perform their local motion independently of each other. The frequency spectrum $f^L(\omega)$ should therefore roughly describe the contribution of the local random cation-motion to the frequency spectrum of the electrical conductivity, $\sigma(\omega)$. In

Fig. 6 we compare $f^L(\omega)$ and the far-infrared (FIR) part of $\sigma(\omega)$.

The FIR $\sigma(\omega)$ spectrum consists of two main contributions.

(i) The high maximum of $\sigma(\omega)$ near 3THz = 100 cm^{-1} very much resembles the lattice absorption spectra due to transverse optical phonons one finds in normal ionic crystals. The ordered low temperature phase, β-AgI, also exhibits this $\sigma(\omega)$ peak at 3THz. In α-AgI, this $\sigma(\omega)$-maximum should be interpreted by phonon-like oscillatory motions of the iodide ions against the "cation sea".

(ii) The broad absorption band on the low-frequency side of the FIR spectrum is however quite unusual in ionic crystals. It appears to be a typical feature of AgI-type solid electrolytes. So far, it has been observed in α-AgI, β-CuBr and α-$RbAg_4I_5$.

Comparing the very far infrared part of $\sigma(\omega)$ to the frequency spectrum $f^L(\omega)$ we immediately realize that the shapes, positions and widths of both curves coincide. The unusual FIR absorption band thus results from the same physical feature as the broad distribution in the neutron-scattering spectra. For our present purpose, a reasonable mathematical description of this feature is provided by our local-diffusion model.

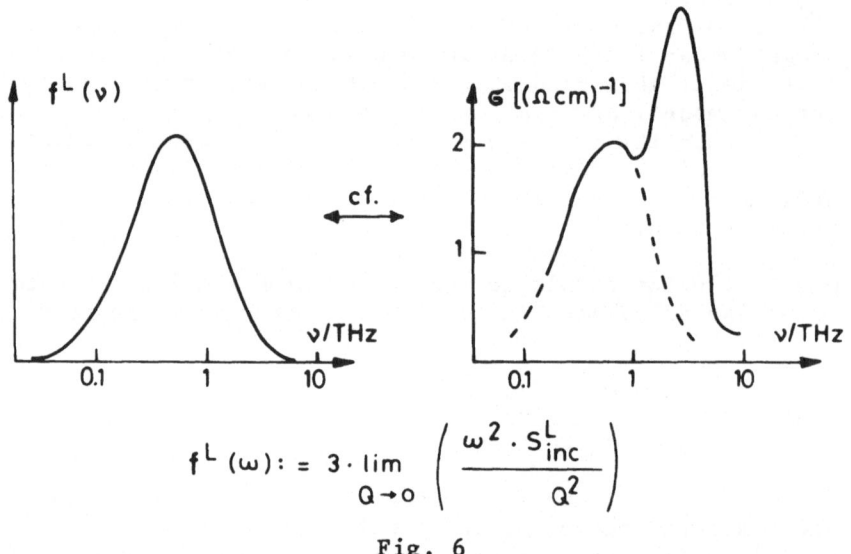

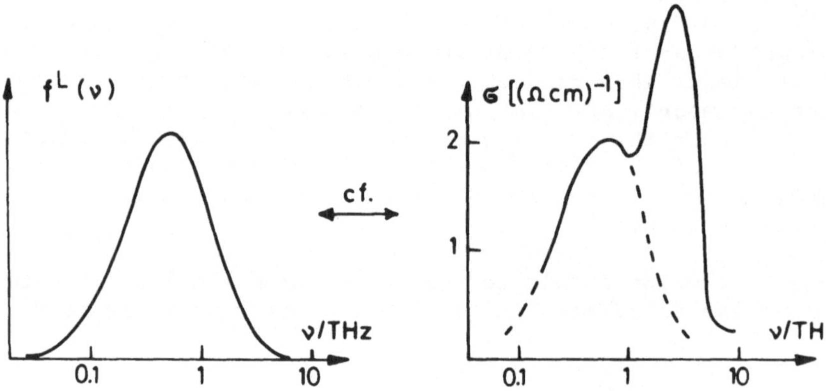

$$f^L(\omega) := 3 \cdot \lim_{Q \to 0} \left(\frac{\omega^2 \cdot S^L_{inc}}{Q^2} \right)$$

Fig. 6

Comparison of the far-infrared part of the electrical conductivity, $\sigma(\omega)$, of α-AgI at 250°C, to the frequency spectrum of the irregular local motion of the silver ions, $f^L(\omega)$, as derived from the neutron spectra.

6. LOCAL PLUS TRANSLATIONAL MOTION

In the actual experimental neutron spectra, the $\delta(\omega)$-peak given by our local-diffusion model is broadened due to the translational motion of the cations. Within our model, cf. Fig. 5, we describe the cation motion in the following way. Each silver ion is permanently confined within its own box performing there its irregular local motion. The box itself rests for an average time, τ_o, moves for an average time τ_1, and then rests again at an average distance, ℓ. Obviously this is a mathematical simplification of a more complicated situation. To be more realistic, we should reformulate the description as follows. During its mean residence time, τ_o, the cation is staying within a cage defined by the neighbouring anions while other silver ions are probably blocking the exits of this cage. If one of these blocking cations moves away, the ion can pass through an exit, enter a channel-like diffusion path, and diffuse within this channel until it is trapped again within one of the cages provided by the anions. The cooperative aspect of this phenomenon will be discussed in more detail in our subsequent paper [12].

In order to reduce the number of independent parameters needed for a description of the resting-moving-resting motion, τ_o, τ_1, and ℓ, we take advantage of two simple equations relating these parameters with one another :

(i) $6D(\tau_o + \tau_1) = \ell^2$

(ii) $2D_{loc} \cdot \tau_1 = \ell^2$.

In writing the second equation we have regarded the motion along the channels as a one dimensional diffusion with the coefficient of local self-diffusion, D_{loc}. For this type of motion we thus have left only one adjustable parameter. We have chosen τ_o to be this parameter. As already mentioned before, the value of τ_o can be estimated independent of any particular model by determining the width of the narrow quasielastic line, ΔE^T, of the experimental spectra at large Q. In the large -Q limit we have

$$\lim_{Q\to\infty} \Delta E^T = \lim_{Q\to\infty} \Delta E^T_{inc} = 2\hbar/\tau_o \quad .$$

One thus estimates τ_o-values of roughly 5 ps. As the temperature is increased from 160°C to 300°C, a slight decrease of τ_o is observed.

The spectra $S^T_{inc}(Q,\omega)$ required for the mathematical simulation of our experimental spectra have been obtained using the Gissler-

Stump model [13]. Its parameters are τ_0, τ_1, and ℓ. Application of this simplifying model together with our model for the local motion finally yields the calculated spectra shown in Figs. 3 and 4 where the parameter values resulting from the fits are also given. Similarly, we find τ_0 = 5 ps, τ_1 = 15 ps, ℓ = 5 Å at 250°C.

Finally two points should be noted :

(i) The values of ℓ obtained are close to one lattice constant which is 5.06 Å at 250°C. A discussion of this result will be given in our second paper [12].

(ii) The mean time a silver ion spends diffusing along one of the channels, τ_1, is roughly 15 ps and appears to be rather independent of temperature. This is in agreement with our findings from microwave experiments which also yields $\tau_1 \approx 15$ ps at various different temperatures within the α-phase.

The main purpose of our second paper will be to relate our model and the values of the fit parameters to the crystal structure of α-AgI, and to emphasize the cooperative aspects of the cation motion.

The calculated spectra presented in Figs. 1 and 2 (also for 250°C) were obtained from a slightly different model used in our previous work [2,3]. The parameter values enclosed in these two figures are quite similar to those obtained for the same temperature in the present experiment.

REFERENCES

[1] K. Funke, J. Kalus, R.E. Lechner, Solid State Commun. 14, 1021 (1974).
[2] G. Eckold, K. Funke, J. Kalus, R.E. Lechner, Phys. Letters 55A, 125 (1975).
[3] G. Eckold, K. Funke, J. Kalus, R.E. Lechner, J. Phys. Chem. Solids, 37, 1097 (1976).
[4] W. Jost, K. Funke, A. Jost, Z. Naturforsch. 25a, 983 (1970).
[5] K. Funke, A. Jost, Ber. Bunsenges. physik. Chemie 75, 436 (1971).
[6] K. Funke, "AgI-Type Solid Electrolytes : Properties at Frequencies between 10^9 and 10^{13} Hz" in "Superionic Conductors", G.D. Mahan and W.L. Roth eds., Plenum, New York 1976.
[7] G. Eckold, K. Funke, R.E. Lechner, to be published in J. Phys. Chem. Solids.
[8] K. Sköld, Phys. Rev. Letters 19, 1023 (1967).
[9] H. Fuess, K. Funke, J. Kalus, phys. stat. sol. (a) 32, 101 (1975).
[10] P.A. Egelstaff, P. Schofield, Nucl. Sci. Engng 12. 260 (1962).
[11] W. Schommers, Phys. Rev. Letters 38, 1536 (1977).
[12] R.E. Lechner, G. Eckold, K. Funke, subsequent paper in these Proceedings.
[13] W. Gissler, N. Stump, physica 65, 109 (1973).

DISCUSSION

De Gennes : Why did you use a cubic box in your model, rather than for instance a harmonic well ?

Funke : We also calculated an irregular local motion within a harmonic well starting with a Langevin equation. In this case the fit to our data turns out to be a little bit worse than that in Figs. 3 and 4. This model calculation will be given in ref. [7].

SINGLE PARTICLE AND COLLECTIVE ASPECTS OF THE SILVER ION MOTION IN α-AgI

R.E. Lechner
Institut Laue-Langevin, 156X, 38042 Grenoble Cedex
France
G. Eckold* and K. Funke
Inst. f. Physikalische Chemie, Univ. Göttingen u.
Sonderforsch. ber. 126, Deutschland

1. INTRODUCTION

As was explained in our previous paper [1] we have derived our model for the diffusive motion of silver ions by starting from qualitative features observed in the neutron spectra of α-AgI. We know from experience that in the here considered Q-range simple monatomic liquids give relatively simple Lorentzian-type quasi-elastic peaks [2]. These are closely related to translational diffusion. Plastic crystals on the other hand yield quasielastic spectra plus a purely elastic contribution [3], because there is a local diffusive motion within a limited volume (see Fig. 5 in ref. [1]). We have therefore chosen a model which allows for both types of effects. A perfect fit to our data has been obtained with the parameter values given in Table 1 for 250°C.

One of the necessary requirements for our model in order to be reasonable is of course that it fits the data. Another requirement is that the parameter values obtained from this fit make sense together with the whole picture provided by the model. In [1] we have already presented several arguments by relating our results to the macroscopic diffusion constant and to microwave and infrared data. Here we will try to give a qualitative insight into the microscopic geometry of this solid electrolyte in order to understand the meaning of the parameter values which we have obtained. From this we will then be able to draw conclusions about single particle and collective aspects of the silver ion motion.

*Now at Inst. f Fesckörperforschung, KFA-Julich, Germany.

Table 1 : Parameter Values for 250°C

Parameter	Symbol	Value
"radius" of local diffusion region	R	1.2 Å
coefficient of diffusion within local region	D_{loc}	0.8 $Å^2 ps^{-1}$
time of residence in local region	τ_o	5.0 ps
time-of-flight through "ld-channel"	τ_1	15.0 ps
average length of "ld-channel"	ℓ	5.0 Å

2. THE STRUCTURE OF α-AgI

From early X-ray structure work [4] it was concluded that in α-AgI the iodide ions form a b.c.c. lattice with the silver ions distributed at random over three kinds of sites, namely 6(b) sites at the centers of distorted iodide octahedra (just in the middle between two 2nd nearest neighbour iodide ions), 12(d) sites at the centers of distorted iodide tetrahedra and 24(h) sites (just at the centers of the lines joining each pair of nearest neighbour "tetrahedral" sites). This is illustrated in Fig. 1 of ref.[14]). More recent neutron diffraction results suggest that the 6(b) and the 24(h) sites should be discarded since the observed Bragg intensities are better described by confining the silver ions to the 12(d) tetrahedral sites. Still better refinements are obtained if the silver ions are assigned to 24(g) sites which are pairs of sites displaced by roughly 0.2 Å symmetrically about tetrahedral sites in the directions towards octahedral sites [6]. However the magnitude of the Debye-Waller factor found in this analysis suggests that such pairs of sites should not be considered as genuine discrete positions of the silver ions in the structure. Therefore, as other authors [7], we prefer to explain the better agreement by the fact that the two sites may just be a better simulation of the real silver ion density distribution than only one tetrahedral site. This distribution probably results from damped motions in a rather anharmonic potential around the tetrahedral sites. Such a picture receives qualitative support from even more recent results of extended X-ray absorption fine structure (EXAFS) measurements [8] as well as neutron and X-ray diffraction data [9]. In [8] it is claimed that the Ag^+-ions are distributed over 48(j) sites which are displaced by ∿0.1 Å from the 12(d) tetrahedral sites along the [110] directions, such that each tetrahedral site is surrounded by a group of four of the 48(j) sites. This is to be compared to an analysis, where the silver ions are distributed at random over the 12(d) sites and allowed to have large thermal vibrations in an anharmonic potential [9]. These authors find the largest amplitudes of motion in the four [110] directions which are along the lines connecting the nearest neighbour 12(d) positions.

Such detailed results should perhaps not be taken too seriously; the absence of measurable high index reflections caused by the large temperature factors should rule out a very detailed analysis of the thermal motion as seen by diffraction experiments [10] , furthermore it must be pointed out that the analysis of EXAFS data is still a subject of controversy [11]. However, what all these recent results have in common, and what should be retained is that the space-averaged silver ion density distribution is composed of some kind of "thermal clouds", which are centered at everyone of the 12(d) tetrahedral sites of the lattice. We consider the precise shape of these "tetrahedral" clouds as unknown but we would like to emphasize that -independently of the shape- our dynamical model [1, 12] is consistent with this general picture [13].

Let us now have a closer look at the arrangement of the abovementioned clouds within the b.c.c. iodide ion lattice. In Fig. 1 four of these clouds located on a cube face with iodide ions at its corners are represented by the (dashed) border lines of the areas which silver ions would occupy if one takes into account the ionic radii ($R_{Ag^+} \approx 0.9$ Å ; $R_{I^-} \approx 1.75$ Å) [14] and the Debye-Waller factors measured at 250°C [7] . One sees immediately that there is a lot of overlap between the four clouds. Let us now think of them as combined to a larger "toroidal" region, which is limited by the hard core potentials of the neighbouring iodide ions on all sides except for four exists. Such a region has a smaller extension perpendicular to the plane of the figure than in the plane and it is centered around an octahedral site, although this site itself may not be favourable to accesss by Ag^+-ions. Every one of the 6(b) octahedral sites has such a toroidal region around it and we note that neighbouring tori are perpendicular to each-other and overlapping (see Fig. 2). They form the system of interconnected regions of space available to the diffusing silver ions, which was represented in Fig.3 of ref.[5] in a slightly different way, namely by drawing schematically the minimum energy paths, the Ag^+-ions might have within the b.c.c. iodide ion lattice.

3. INSTANTANEOUS AND TIME AVERAGED ION-ION POTENTIAL

It must be emphasized that Fig. 3 of ref. [5] is a schematic view of the average density distribution, which gives us an idea of the average potential. The average potential energy of a silver

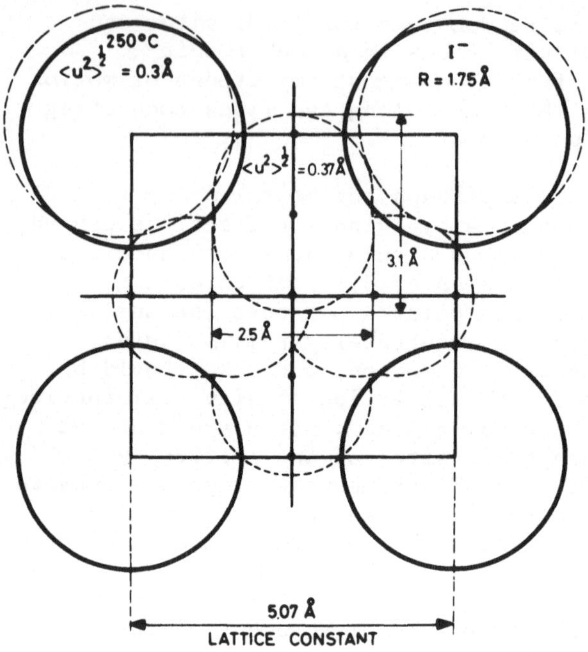

Fig. 1

Silver ion sites and Debye-waller factors : The extension of Ag^+-ion "thermal clouds" corresponding to the measured Debye-Waller factors and taking into account the ionic radii is indicated by dashed closed curves centered at the four "tetrahedral" sites on a cube face of the b.c.c. I^--ion lattice. The I^--ions are represented by circles at the corners of the cube face with ionic radii drawn to scale.

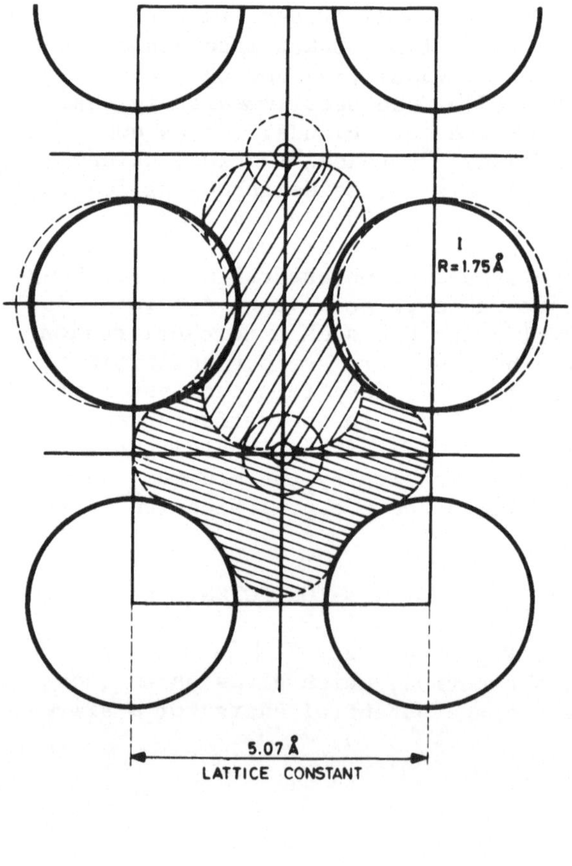

Fig. 2

Two neighbouring overlapping toroidal regions as described in the text are shown as hatched sections between the I^--ions sitting at the corners of the cube faces.

ion as a function of its position in the lattice can be estimated by taking into account the effect of the average positive charge distribution caused by the mobile Ag^+-ions in addition to the interaction with the I^--lattice. If one follows paths in between iodide ions, one obtains a weakly oscillating function with the periodicity of the lattice and potential barriers of the order of kT [14]. It is however clear that the instantaneous potential felt by an individual silver ion does not have the same symmetry and periodicity, mainly because of the disorder in the "molten" cation sublattice. The importance of the instantaneous silver-silver nearest-neighbour interaction is easily seen by considering a numerical example. Let us think of the Ag^+-ions as distributed "uniformly" over the I^--lattice (lattice constant ; 5.07 Å). Then the nearest neighbour Ag^+-Ag^+ distance is about 4 Å. In order to obtain the shape of the instantaneous potential the average positive charge distribution must be replaced by a more realistic distribution of positive charges. A simple estimate can be made by considering the potential energy of a silver ion on a straight line between two Ag^+-neighbours fixed at twice the average Ag^+-Ag^+ distance, 8 Å from each other, see Fig. 3. For this purpose we have used a Born-Mayer repulsion term and a Coulomb term to calculate the potential energy:

$$V^{Ag^+} = \sum_{i=1}^{2} (A \cdot \exp(-r_i/\rho) + e^2/r_i)$$

with $A = 2.34 \times 10^{-11}$ erg and $\rho = 0.33$ Å [15].

It is seen that the potential energy rises relatively fast as the mobile silver ion is displaced from the center in between the two fixed Ag^+-ions. In fact, in order to displace the ion by ∿ 2.5 Å corresponding to the distance between the centers of two of the nearest neighbour toroidal regions mentioned above, an energy of more than 1 eV would be needed. We think this indicates that a simultaneous occupation of nearest neighbour toroidal "cages" by two silver ions will be quite rare. Another conclusion which can be drawn from Fig. 3 is the following : If one adds to the Ag^+-Ag^+ potential a weakly oscillating potential, as obtained in [14] essentially for the Ag^+-I^- interaction, one can easily construct a situation, where the potential well becomes rather flat with a radius of the order of 1 Å at its bottom. Note that this is just what we found for the radius of the local diffusion region of the Ag^+-ion.

4. Ag^+-Ag^+ ION CORRELATIONS

As already mentioned above nearest neighbour toroidal cages will in general not be occupied simultaneously, because this would require too much energy. This means that at any given instant -

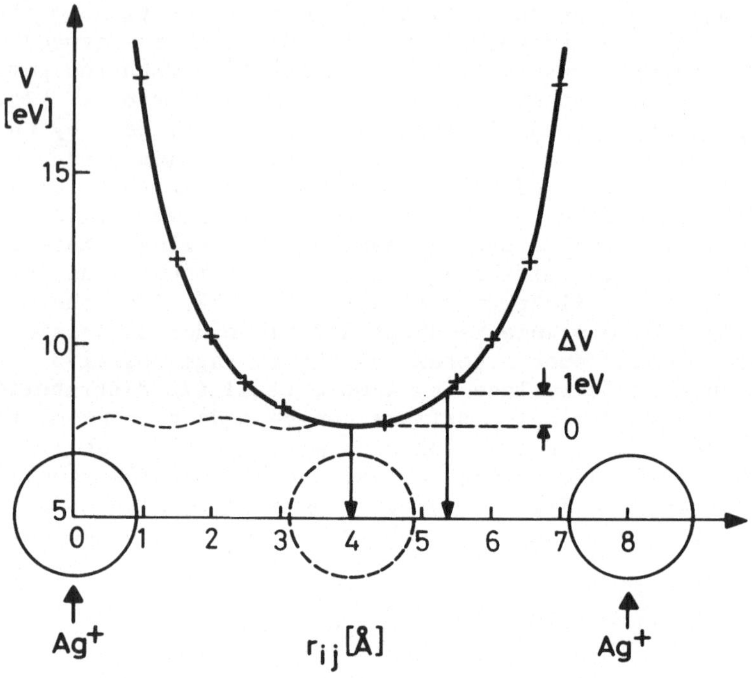

Fig. 3

Potential energy of a silver ion as a function of position on a straight line between two Ag^+-ions fixed at a distance of 8 Å from each other (full line). The estimated oscillatory contribution which the interaction of the Ag^+-ion with the I^--ion lattice would make is shown schematically (dashed line).

due to the presence of other silver ions - the effective number of sites (or cages) which is available to an individual silver ion is reduced as compared to what one would expect from the space-averaged density distribution. In fact, if one distributes the silver ions randomly into toroidal cages with the condition that two ions are never closer to each-other than second nearest neighbour cages, then exactly two thirds of those cages available under such a restriction are filled and one third is empty. This is illustrated (for simplicity in two dimensions) in Fig. 4, where toroidal cages in the plane of the figure are represented as squares, whereas cages perpendicular to the plane are indicated as narrow "channels". Occupied cages are symbolized by large circles, tetrahedral sites by small circles. It is seen that there are situations a) where one Ag^+-ion is completely surrounded by Ag^+-neighbours, so that all exits of the central cage can be considered as closed, and b) where the enclosure is incomplete, so that at least one exit is open. In case a) the central silver ion cannot leave its cage until one of the blocking ions goes away from an exit. This seems to take the average time τ_o, which we obtained as the residence time of the silver ion; τ_o may therefore be considered as a "cage life time". During this time the silver ion has available for its local diffusive motion the space of its cage consisting roughly of four (interconnected) tetrahedral clouds as defined in Section 2.

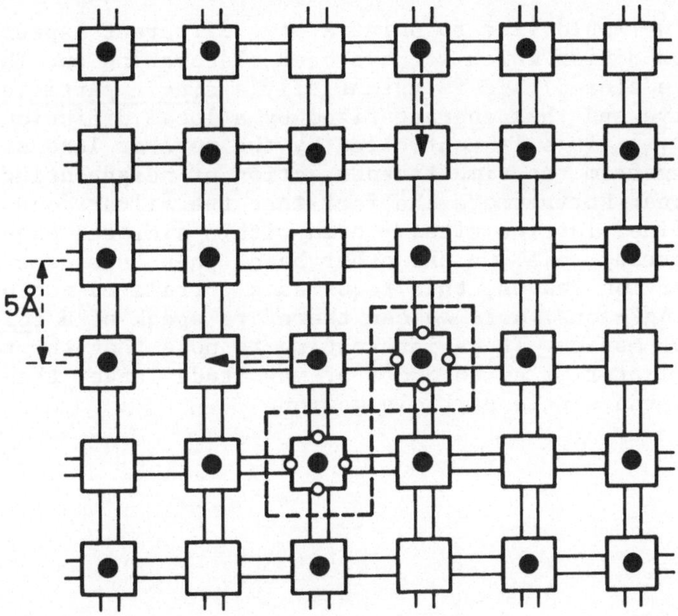

Fig. 4

Two-dimensional representation of a lattice of "toroidal cages" (shown as squares) with random distribution of Ag^+-ions. Occupied cages are indicated by large circles, tetrahedral sites by small circles. The dashed squares correspond to two of the cube faces of the b.c.c. I^--ion lattice.

Now, when a cage opens by the removal of one blocking ion, then this corresponds to case b). Then immediately a "one-dimensional channel" is available for a silver ion, in the sense that a preferential direction is given for a diffusive translational motion of the ion over a limited distance. It is seen from Fig. 4 that such distances are of the order of a lattice constant, i.e. the value which we have found for the "jump length" ℓ. It must however be remembered that the "channel" is composed of a number of tetrahedral cloud regions whose centers are not on a straight line. Thus the ion may move back and forth within and between adjacent clouds without performing a straight flight over a large distance, but nevertheless staying in a general direction during an average time τ_1. After this the ion will again be trapped in a cage. Let us note that the toroidal cages used in the above illustration were meant only as an average local diffusion region with roughly the radius R obtained in our model fit.

5. CONCLUSIONS

Finally we would like to point at two different aspects of the silver ion motion which we have been elaborating on. The motion occurring in a closed cage is essentially a single particle motion. It is diffusive and thus characterized by a local diffusion coefficient D_{loc}. This is controlled by the more or less stochastic forces arising from the simultaneous action of neighbouring iodide and silver ions. Furthermore the fact that the silver ions do not have well-defined lattice sites - even within a closed cage - plays an important role. On the other hand, when this cage opens and the silver ion leaves, this requires a correlated motion of at least two Ag^+-ions. Here we can therefore speak of a collective or cooperative motion. It is interesting to note that the time scale of the latter is by an order of magnitude longer than the time scale of the single particle motion.

REFERENCES

[1] K. Funke, G. Eckold and R.E. Lechner in Proceedings of the Summer School on "Microscopic Structure and Dynamics of Molecular Liquids", Aleria, 28/8 - 10/9 1977, Ed. J. Dupuy (1978).
[2] P.A. Egelstaff, "An Introduction to the Liquid State", Academic Press, London 1967.
[3] A.J. Leadbetter and R.E. Lechner in "The Plastic Crystalline State", Ed. J.N. Sherwood, John Wiley and Sons, in print (1977/78).
[4] L.W. Strock, Z. Phys. Chem. B25, 441 (1934); B31, 132 (1936).
[5] K. Funke in Proceedings of the Summer School on "Microscopic Structure and Dynamics of Molecular Liquids", Aleria, 28/8 - 10/9 1977, Ed. J. Dupuy (1978).
[6] W. Bührer and W. Hälg, Helv. Phys. Acta 47, 27 (1974).
[7] A.F. Wright and B.E.F. Fender, J. Phys. C., Solid State Ph., 10, 2261 (1977).
[8] J.B. Boyce, T.M. Hayes, W. Stutius and J.C. Mikkelsen, Jr., Phys. Rev. Letters 38, 1362 (1977).
[9] S. Hoshino, T. Sakuma and Y. Fujii, Sol. State Comm. 22, 763 (1977).
[10] In fact no more than eleven (powder) reflections have been reported from α-AgI near the α-β transition, and at higher temperatures the number of measurable Bragg peaks was even smaller (9 at 255°C and 6 at 450°C) [7].
[11] J. Goulon (L.U.R.E. - Orsay), private communication.
[12] G. Eckold, K. Funke, J. Kalus and R.E. Lechner, J. Phys. Chem. Solids 37, 1097 (1976).
[13] Contrary to what is said in [9] we have never claimed that the [100]-directions are "channels of the cation diffusion". This was considered as one of the possibilities together with the approximation of isotropic jump diffusion,and it was clearly stated that the two versions of the jump diffusion model could not be distinguished experimentally [12]. It should also be mentioned that the hops between neighbouring tetrahedral sites proposed in [8] occur on a time scale which is covered by our "local random motion". A comparison of the corresponding residence-to-flight-time ratio with the ratio we obtained for the translational "jump" diffusion (which is an order of magnitude slower than the local random motion) is therefore meaningless.
[14] W.H. Flygare and R.A. Huggins, J. Phys. Chem. Solids 34, 1199 (1973).
[15] W. Schommers, Phys. Rev. Letters 38, 1536 (1977).

DISCUSSION

P.G. de Gennes : It was not quite clear to me from the potential diagram, whether the oscillatory part of this diagram was calculated allowing for some displacement of the iodide ions.

R.E. Lechner : This was done for fixed iodide ions.

P.G. de Gennes : So it could give an even flatter potential well if one would include iodide ion displacements (due to vibrations).

R.E. Lechner : Yes. But it is in any case difficult to give an exact shape of the potential, because it depends for example on the sizes and polarizabilities of the ions, which are not well-known. In ref. [14] calculations were made for different ion sizes and this shows qualitatively the possible range of potential wells and barriers.

T. Springer : I was a bit confused by the idea that the motion from cage to cage occurs by diffusion. Should it not (in view of the short distance) rather be like a linear motion with some oscillation ? This is a detail, but it has some consequences on the quantitative features.

R.E. Lechner : I must say that the picture which I showed (Fig. 4) is rather simplified. The "channel" which becomes available for the silver ion, when a cage opens, does not correspond to a straight line, but to a "zigzag" line passing via several time-dependent wells and barriers. This is a channel in the sense that it provides the general direction of the motion during the time τ_1 and over a finite distance ℓ.

J.G. Smit : Do you know something about the time scale of the phonons ?

K. Funke : Yes we do. The conductivity has a maximum at 3 THz ($\approx$ 100 cm^{-1}). This corresponds to the motion of the anions against the cation sea.

L. Pietronero : In your model you state that also along the diffusion paths the ions perform a Brownian motion for which a white noise spectrum of the interactions (with the other ions) is tacitly assumed. Now we know that the motion of a silver ion along a diffusion path requires an in-phase displacement of the neighbouring iodide ions. The iodide ions have a well-defined harmonic motion with a given frequency. How can they provide a white noise spectrum ?

K. Funke : The periods of the iodine vibrations are much shorter than the time the silver ion needs for diffusing along a channel; thus an in-phase motion during all this time is not likely to exist.

R.E. Lechner : The ions may be in phase temporarily, i.e. only during a fraction of the time needed for the displacement over a distance of the order of ℓ. There are several reasons for this : firstly the iodide ions are probably moving in an anharmonic potential; secondly the individual Ag^+-ion motion is also strongly influenced by the irregular motion of neighbouring silver ions; finally it must be remembered that on its way over a distance ℓ the silver ion may pass by several shallow potential minima (near tetrahedral sites) and move back and forth between them.

W. Freyland : What about possible electronic contributions to the electrical conductivity ? From various arguments one would estimate an electronic contribution to the electrical conductivity of the order of 10^{-3} $(\Omega cm)^{-1}$.

K. Funke : The DC-conductivity rises from 1.3 to 2.6 $(\Omega cm)^{-1}$ within the α-phase, in the phonon region we have even 3 $(\Omega cm)^{-1}$, and then at higher frequencies the conductivity decreases (within experimental error) to zero. We cannot say more than that.

W. Freyland : Do precise optical absorption edge measurements exist in α-AgI ?

K. Funke : Yes, absorption measurements by Heitkamp (Göttingen) were performed more than ten years ago, but were not published.

J. Dupuy : From where did you get the values of the ionic radii ?

R.E. Lechner : Our values were taken from ref. [14], and they are probably fairly good. For instance the iodide ion value agrees quite well with the recent result from diffraction displayed in yesterday's poster session (C. Ciaccamo, M. Parrinello et al., University of Messina, "Ionic Radii and Diffraction Patterns of Molten Alkali Halides"), if the values of these authors are extrapolated below the melting point of the molten ionic salt. For the silver ion one can say that its radius can hardly be larger than 0.9 Å, if fast diffusion is to occur.

ROUND TABLE DISCUSSION ON COHERENT NEUTRON SCATTERING

J.R.D. Copley

Institut Laue-Langevin, 156X
38042 Grenoble Cédex, France

I. INTRODUCTORY REMARKS

1. Why Use Neutrons ?

In the context of a discussion about coherent neutron scattering, it is natural to start with a brief description of the properties of the neutron which make it such a useful tool for the microscopic study of motion in liquids.

There are two dominant types of interaction between neutrons and matter. There is the nuclear interaction, characterized by a scattering length b, and there is the magnetic interaction with unpaired electrons. The latter interaction need not concern us here. In general nuclear scattering lengths depend on the isotope, and on the spin of the compound state formed by the nucleus and the neutron. They are of order 10^{-12} cm, so that nuclear cross-sections $4\pi b^2$ are of order 10^{-24} to 10^{-23} cm^2, i.e. 1 to 10 barns. Thermal neutron absorption cross-sections are often, but not always, of the same order, so that linear attenuation coefficients for condensed matter are typically of order cm^{-1}.

It also turns out that thermal neutrons have energies and wavelengths comparable with excitation energies and interatomic distances, respectively, in condensed matter. They are therefore ideally suited to the investigation of many types of structure and excitation in liquids and solids.

2. Phases of Condensed Matter

We may characterize crystalline material by the existence of long range order (l.r.o) or periodicity in both time and space, leading trivially to delta-functions in both $\underset{\sim}{k}$ and ω in the scattering function $S(\underset{\sim}{k},\omega)$. The intermediate amorphous state has lost its l.r.o. in space, so there are no longer any delta-functions in k : on the other hand l.r.o. in time remains, and there is therefore a component of purely elastic scattering for all k. In the liquid state there is no l.r.o. in time or in space, so there are no delta functions in $S(k,\omega)$ (except at $k=0$, $\omega=0$).

An immediate consequence is that in most experiments on non-molecular liquids one is not measuring a well-defined peak (as is often the case in elastic or inelastic experiments on crystals), but just a broad distribution of intensity. In these circumstances it becomes imperative to consider very carefully the various transformations and corrections to data (summarized below), which must in principle be performed no matter what the sample may be. In experiments on solids it is often (and usually rightly) considered unnecessary to perform these manipulations. On the other hand, several of the early neutron experiments on liquids are of little long term significance, since these corrections were not performed : this came about partly because efficient methods of calculation were not available at the time.

I shall confine my discussion to monoisotopic, non-magnetic, isotropic systems. The neutron scattering cross-section is then completely coherent, and it is described by the single function $S(k,\omega)$.

3. Neutron Scattering Experiments

In an ideal neutron scattering experiment neutrons with incident energy E_o and wavevector $\underset{\sim}{k}_o$ are scattered by the sample and leave with scattered energy E_f and wavevector $\underset{\sim}{k}_f$. The relationship between a neutron's energy E_n and its wavevector k_n is $E_n = \hbar^2 k_n^2/2m_n$, where m_n is the mass of the neutron. The energy transfer $\hbar\omega$ and the wavevector transfer $\underset{\sim}{k}$ are just given by

$$\hbar\omega = E_o - E_f \; ; \quad \underset{\sim}{k} = \underset{\sim}{k}_o - \underset{\sim}{k}_f$$

In an inelastic scattering experiment, the intensity $I(\phi, E_f)$ is typically determined as a function of scattering angle ϕ and scattered energy E_f, whereas in a neutron diffraction experiment no energy analysis is performed so that the simpler quantity $I(\phi)$ is measured. For a description of neutron scattering methods as applied to liquids, the reader is referred to Copley and Lovesey [1].

4. Extraction of Correlation Functions

In an ideal inelastic scattering experiment on a very small sample, it is easy to derive the cross-section ($d^2\sigma/d\Omega dE_f$) from the measured intensity $I(\phi, E_f)$. If a time-of-flight experiment has been performed, it is first necessary to make the important "t^3" transformation, in order to take care of the fact that counts are accumulated in channels with a constant width in time. It is then easy to extract $S(k,\omega)$ since

$$\frac{d^2\sigma}{d\Omega dE_f} = \frac{1}{\hbar} b^2 \left(\frac{k_f}{k_o}\right) S(k,\omega).$$

In the analogous ideal diffraction experiment, the intensity $I(\phi)$ is proportional to an experimental cross-section

$$\left.\frac{d\sigma}{d\Omega}\right|_{exp} = \int_0^\infty \frac{d^2\sigma}{d\Omega dE_f} f(E_f) dE_f,$$

where $f(E_f)$ contains the detector efficiency function, and the integral is taken over a locus in k-ω space corresponding to a fixed value of ϕ. On the other hand $S(k)$ is desired, and this is simply the integral of $S(k,\omega)$ over all ω at constant k. The transformation from the experimental $d\sigma/d\Omega$ to $S(k)$ is achieved by applying the Placzek correction procedure. This subject has been discussed in detail by several authors (e.g. Yarnell et al. [2]).

Both diffraction and inelastic scattering measurements must also be corrected for the effects of container scattering, "self-shielding", and multiple scattering. In addition some workers attempt to correct for instrumental resolution. On the whole these corrections are more complicated when applied to inelastic scattering measurements.

5. A Priori Knowledge Concerning S(k) and S(k,ω)

The limiting behaviour of $S(k)$ at small and large k, the detailed balance property of $S(k,\omega)$, the known moments of $S(k,\omega)$ and the limiting behaviour of $S(k,\omega)$ at small k and ω and at large k and ω, are subjects which are discussed elsewhere in these Proceedings. In the present context, this a priori knowledge serves two purposes. It can help in the planning of experiments, and it can be used to make checks on the data, or in some cases to constrain the data by fixing quantities such as the overall normalization.

6. The Kinematically Allowed Region

Given k_o and a k_f, ω is determined. There is also a minimum k and a maximum k, given by $|k_o-k_f|$ and k_o+k_f respectively. Plotting these extreme values of k as a function of ω, for a given k_o (or k_f) gives the kinematically allowed region which is accessible for singly scattered neutrons. Examples for two values of k_o are shown in Fig. 1. Clearly the region is enlarged as k_o is increased. On the other hand the resolution in k and ω rapidly deteriorates as k_o is increased, and a compromise is clearly required.

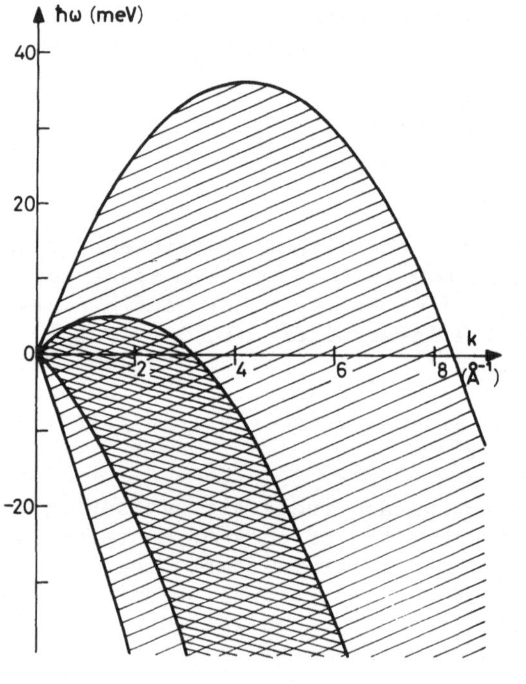

Fig. 1

The kinematically allowed regions for neutrons with wavelengths of 4 Å and 1.5 Å, i.e. k_o = 1.57 and 4.19 Å^{-1}, E_o = 5.1 and 36.4 meV, respectively. The larger region is appropriate to the 1.5 Å neutrons. The maximum positive energy transfer occurs when $E_f = 0$, $\hbar\omega = E_o$, and $k = k_o$. The maximum value of k for elastic scattering ($\hbar\omega = 0$) is $k = 2k_o$.

An important property of the kinematically allowed region is that the limiting slope as $k \to 0$ is simply the neutron velocity, $v_o = \hbar k_o/m_n$. Thus in order to observe a mode at $\omega(k) = c_s k$, the condition $v_o > c_s$ must be satisfied (certainly in neutron energy loss, whereas in energy gain the condition is slightly relaxed). Using present day neutron scattering instruments, it turns out that few systems can be studied in the low k region with a view to examining the "sound wave" peaks, because in most systems the condition $v_o > c_s$ cannot practically be satisfied.

II. DISCUSSION

[The following account is an abridged version of the actual discussion, and speakers' remarks have been paraphrased. The order of the discussion has been significantly rearranged, in order to make it more coherent to the reader.

We start with remarks regarding recent studies of the hard sphere system, since this sytem is a good reference system for the treatment of many properties of liquids. There follow several general remarks regarding the usefulness of neutron scattering experiments, and what we can learn from them.]

B.J. ALDER : I would like to describe some rather extensive hard sphere (HS) molecular dynamics (MD) calculations of $S(k,\omega)$ which have been done in collaboration with S. Yip (M.I.T.) and W.E. Alley (Livermore). We have calculated $F(k,t)$ (the intermediate scattering function) and we have compared our results with both hydrodynamics and the generalized Enskog kinetic theory. A general comment is that $S(k,\omega)$ is extremely insensitive to the collective modes, so that if the neutron scattering people experimentally want to see deviations from theory, they must do very precise work, in the 1% range.

In Fig. 2, we plot the normalized $F(k,t)$ versus t/τ_E, where τ_E is the Enskog mean collision time. The number density is comparable with the critical density, and the wave vector k is a typical value for neutron scattering. The theoretical curve lies within 1 or 2% of the MD results, which are themselves good to 1%. The input data for the theory comes from the MD calculations. Fig. 3 shows the corresponding $S(k,\omega)$. In Fig. 4 we show results for $F(k,t)$ at the triple point density.

We only know two collective modes in fluids. One is the vortex mode described in the lectures of McDonald; the other mode is the viscoelastic mode, and it shows up here. Discrepancies between

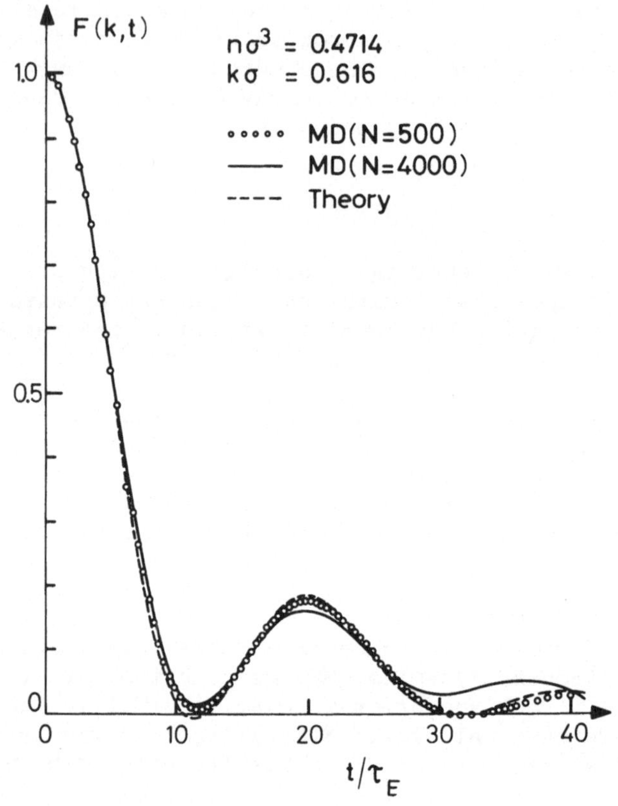

Fig. 2

Intermediate scattering function F(k,t) of a hard-sphere gas at reduced density $n\sigma^3 = 0.4714$ ($V/V_o = 3$) and reduced wavenumber $k\sigma = 0.616$. Time t is given in units of Enskog mean collision time $\tau_E = [4\sqrt{\pi} n V_o \sigma^2 g(\sigma)]^{-1}$. Computer molecular dynamics results are denoted by open circles (500 particle system) and solid line (4000 particle system). The smaller system was run with more collisions per particle and therefore the statistics are better. A kinetic model solution of the generalized Enskog equation is denoted by the dashed curve. This graph shows that kinetic theory describes quite well the density fluctuations in fluids at almost twice the critical density.

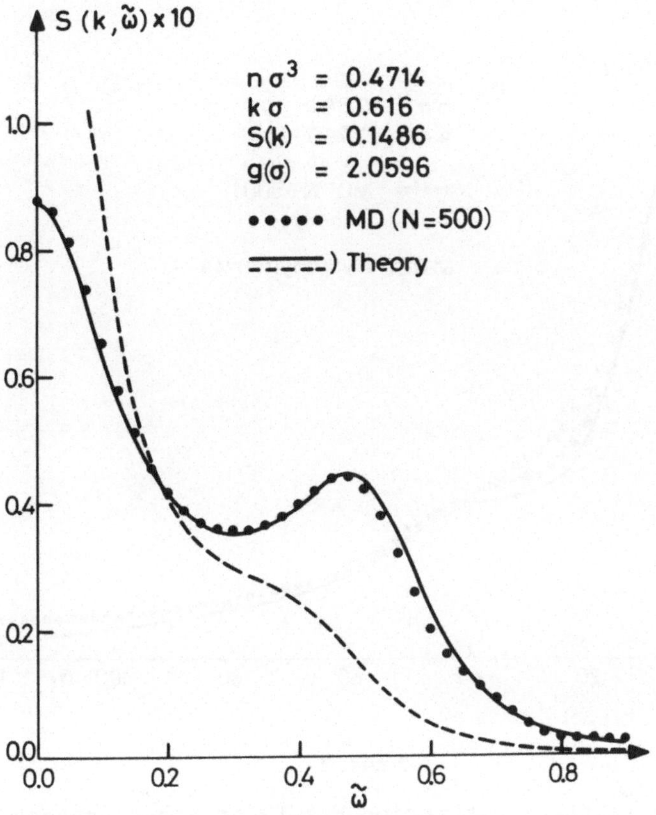

Fig. 3

Dynamic structure factor $S(k,\omega)$ of a hard-sphere gas at reduced density $n\sigma^3 = 0.4714$ and reduced wavenumber $k\sigma = 0.616$. Reduced frequency $\tilde{\omega}$ is $\omega\tau_E/k\sigma$. Computer molecular dynamics results for a 500 particle system are denoted by the closed circles while the kinetic model solution to the generalized Enskog equation is denoted by the solid curve. The dashed curve indicates the kinetic model solution in which the mean field term is obtained from the conventional Enskog equation. This graph shows that a propagating mode in $S(k,\omega)$ is quite well established at intermediate fluid densities and reasonably large wavenumbers. Moreover, this behaviour is quantitatively calculated by kinetic theory. The mean field term in the conventional Enskog equation is known to be incorrect on the basis of sum rules, and from the large difference between the solid and dashed curves one sees that this defect is very important in describing density fluctuations at finite k and ω.

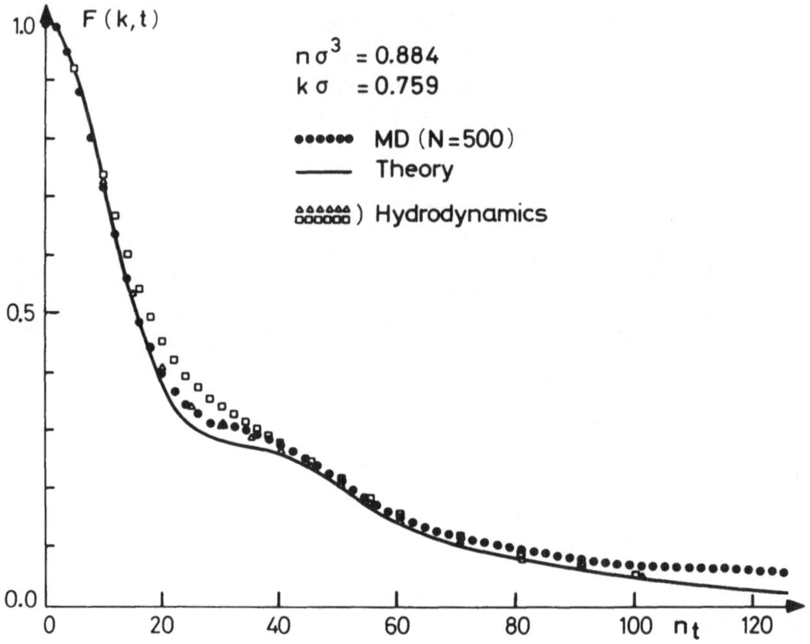

Fig. 4

Intermediate scattering function F(k,t) of a hard-sphere fluid at reduced density $n\sigma^3$ = 0.884 (V/V_o = 1.6) and reduced wavenumber $k\sigma$ = 0.759. Time in units of Enskog mean collision time is given by 0.6163 n_t. Computer molecular dynamics results for a 500-particle system are denoted by closed circles while kinetic model solution of the generalized Enskog equation is denoted by the solid curve. Calculations based on the linearized equations of hydrodynamics with Enskog and renormalized transport coefficients are denoted by triangles and squares respectively. This graph shows that hydrodynamics theory can extend into the time region of $\sim$ 10-20 τ_E if instead of the full (renormalized) transport coefficients one uses the Enskog (short-time) transport coefficients, thus suggesting the need for a frequency-dependent viscosity, or more generally a generalization to include viscoelasticity effects. The graph also shows that kinetic theory is still qualitatively correct even at liquid densities.

the theories and the MD results are of the order of 5% at the triple point. If you modify hydrodynamics by introducing an $\eta(k,\omega)$ you can quantitatively explain $S(k,\omega)$. Near the triple point, η/η_E is considerably larger than one, so that as the system evolves the viscosity changes from a short time (Enskog) value to a long time hydrodynamic value. Introducing such a time-dependent viscosity in the hydrodynamic theory, that is to say undressed transport coefficients at short times and dressed ones at long times, removes any major discrepancies.

M.P. ALLEN : Is it surprising that the results of the Enskog theory, based on a binary collision model, were not very different from the MD results for HS ? After all the HS model is an exclusively binary collision model, representing the ultimate in short range potentials. Would the discrepancy between the theories and the MD be greater if you used a longer range potential ?

B.J. ALDER : Yes, although there exists no analogous kinetic theory for longer range potentials.

J.E. ENDERBY : The structure factor S(k) is rich in information, if you will just look at it very hard. First of all, the damping of S(k) at large k is a very good measure of atomic overlap. A soft potential means a soft g(r), which implies many Fourier components, which largely cancel one another. Thus if S(k) is flat beyond 5 or 6 $Å^{-1}$, you have a very soft potential. If S(k) is oscillatory the potential is hard. Following the work of Alder and others, we have an upper bound for the extent to which S(k) oscillates at large k for hard sphere (HS) systems. The damping in liquid Se, for example, is less than for HS systems, implying the need to include interactions of higher order than pair-wise forces. The quantity $c(k) \equiv [1/S(k) - 1]$ is the Fourier transform of c(r), and at low k it depends on the asymptotic form of the potential. In ionic systems we know now that the charges on the ions place enormous constraints on the phase relationships between the g(r)'s. Thus there is once again a characteristic behaviour in the S(k)'s at large k. In molecular liquids, a well-defined intramolecular distance shows up as a damped oscillation in S(k) of definite frequency.

In conclusion, I would submit that an enormous amount of information about effective potentials is contained in the function S(k). This may of course be measured with neutrons or with X-rays, depending on the system.

J.R.D. COPLEY : Thank you for those remarks. What is your opinion of the usefulness of inelastic scattering measurements ?

J.E. ENDERBY : As you ask more and more difficult questions, you have to do more and more careful experiments, and finally the full information contained in $S(k,\omega)$ will be needed. All I'm saying is "let us not despise $S(k)$".

J.R.D. COPLEY : With regard to diffraction measurements, I have a rather technical comment. In such experiments, one wishes to obtain $S(k)$, i.e. the integral of $S(k,\omega)$ from $\omega = -\infty$ to $\omega = +\infty$, at constant k, but this is of course not possible in practice. At small k, in an experiment where the neutron velocity is less than the sound speed c_s (which is often the case), it is impossible for neutrons to be singly scattered with energy transfer $\hbar\omega$ such that $\omega/k = c_s$. Thus a considerable fraction of the intensity at small k cannot be picked up in the detector. Under such circumstances I find it difficult to believe that the inelasticity correction can be performed with confidence.

J.G. POWLES : There is a common misconception which is that Placzek corrections are small at small k : they are not. There is for instance the kinetic energy term. It is clear that the only way to get $S(k)$ properly is to measure $S(k,\omega)$. The important thing is to measure $S(k)$ and $S(k,\omega)$ to a consistent degree of accuracy.

[Throughout the session, the relationship between interatomic potentials and collective excitations was discussed. Some of the more pertinent remarks are reproduced below].

J.R.D. COPLEY : It seems that the more harmonic the bowl of the potential is, the more likely it is that well-defined collective excitations exist in the liquid at small k. Work (with neutrons, with MD, and with theory) on both rubidium and on various modifications of the Lennard-Jones system, supports this view (see for example the reviews of Copley and Lovesey [1, 3]). Note however that the experimental measurements on neon were performed in the supercritical region ($T > T_C$).

J.B. SUCK : Would you not think that it is the repulsivity of the core, rather than the shape of the bowl in the potential, which determines the existence of collective excitations ?

J.R.D. COPLEY : No, I think the shape of the bowl is the important quantity, i.e. both the repulsive and the attractive parts of the potential.

J.P. HANSEN : You should not look at the pair potential, but at the total potential, which is the sum of all pair potentials involving a given atom. I would think that this total potential is determined primarily by the repulsive cores, rather than by the fact that you have a well which is more or less harmonic. This could be tested by performing MD calculations on a purely repulsive but soft (e.g. r^{-6}) potential, and comparing the results with Alder's HS results.

J.R.D. COPLEY : I agree wholeheartedly with the idea of doing such tests. It would also be interesting to run such potentials through both the generalized Enskog and the hydrodynamic theories for comparison with MD data.

T. SPRINGER : Are the striking differences between the results for Rb and Ne due to the much longer range of interaction in liquid metals ? There is also the fact that the thermal conductivities of Ne and Rb differ by a factor of order 50 or 100.

R. EVANS : There is considerable evidence that the repulsive part of the potential is very soft in alkali metals, much harder in polyvalent metals, and harder still in noble and transition metals. Regarding the attractive part of the potential, we know virtually nothing. If inelastic neutron scattering can help in this respect, that would be splendid. It is clear however, that detailed MD simulations should be carried out in order to determine the extent to which neutron studies will throw light on this question.

O.J. EDER : In our data for liquid aluminium we appear to see collective modes at k-values of order 1.5 $\mathring{A}^{-1}$. Only after further analysis will we be able to make more definite statements.

J.B. SUCK : Aluminium is a fairly good harmonic solid. One can see this from the existence of well-defined phonons nearly up to the melting point, and from the small ratio of melting temperature T_m to Debye temperature θ_D as compared with Rb. Despite this, collective excitations seem to persist less far in liquid Al than in Rb (in terms of reduced k). For this reason I have the feeling that it is the repulsive part of the potential, rather than the "harmonicity" of the well, which is the determining factor.

C.A. ANGELL : The melting temperature is not necessarily a good reference temperature for comparisons of liquid state behaviour in different types of system. In the case of argon, one is also very close to the boiling point, whereas with Rb (with which inert gases were earlier compared) one is very far from the boiling point. The temperature of observation relative to interparticle attractive energies is in fact very different. Maybe an important consideration is how close one is to an effective θ_D. For argon, you can see nice collective modes, even transverse modes, if you computer simulate it under supercooled conditions, which means moving the temperature down relative to the effective θ_D.

T. KEYES : If plots of $\omega(k)$ basically reflect the position of the peak in $S(k)$, is there any extra information to be gained from the harder measurements of $S(k,\omega)$, to obtain $\omega(k)$?

P.C. HOHENBERG : A simple relationship between $S(k)$ and $\omega(k)$ would only seem to me to exist if the mode with frequency $\omega(k)$ dominates $S(k,\omega)$. Since it does not, even in Rb, and of course not in Ne, one does not expect to be able to predict $\omega(k)$ from $S(k)$.

J.R.D. COPLEY : Essentially any viscoelastic theory which satisfies the first few sum rules will, I think, predict peaks in Rb, and no peaks in Ne, for fairly small values of k. On the other hand, the various theories do not give identical line-shapes, nor do they necessarily get $\omega(k)$ quite right, but at least one can obtain a good bit of information from the sum rules.

C.A. ANGELL : Could we push the experimental situation somewhat further, by doing this sort of small k experiment on supercooled gallium, at a temperature of order 0.5 T_m ?

J.R.D. COPLEY : That is an interesting idea, but unfortunately the sound speed c_s is large compared with neutron velocities so we cannot (for the time being) look for collective modes at small k. On the other hand one could hope to learn something by looking at $S(k,\omega)$ either for k in the range from $\sim$60 % to $\sim$90 % of k_p (k_p being the position of the main peak in $S(k)$),or else by looking in the valley between the first and second peaks in $S(k)$. This is substantiated by observations on Rb. Of course the measurements will have to be very accurate.

Furthermore it may be possible to extract information by looking at the "Rayleigh peak" at small k and comparing with $S(k)$. If one can show that the sum rule is not exhausted one can deduce something about the behaviour of $S(k,\omega)$ at larger ω : on the other hand it is not clear to me that $S(k)$ can be satisfactorily determined (see above).

B.J. ALDER : The theoreticians believe that there is a long-time tail in $F(k,t)$. It is expected to be extremely weak : even the MD people have not seen it yet. Could it be observed with neutrons ?

A.J. DIANOUX : We hope to do an experiment at the I.L.L. to look for the long-time tail in $F_s(k,t)$, the incoherent function.

J.R.D. COPLEY : Neutron scattering techniques have advanced tremendously in recent years, and I believe the experiment (on an incoherent scatterer) is worth attempting, but it will be very difficult.

B.J. ALDER : The tail in $F(k,t)$ is probably an order of magnitude weaker than the tail in $F_s(k,t)$.

J.R.D. COPLEY : Well I believe it is time to conclude this discussion. I shall not attempt to summarize what has been said. All I would say is that there is plenty of work remaining to be done, both for the theorists and the experimentalists. It is a difficult field for us all : no advances will be made without a great deal of hard work. Given the difficulties, we must all be careful not to embark on projects without giving very careful consideration to the expected benefits. I believe this discussion has confirmed our belief that one of the most important aspects of this work is the continuing close collaboration between theorists and experimentalists, whether they scatter radiation or do computer simulation work.

REFERENCES

[1] J.R.D. Copley and S.W. Lovesey. Rep. Prog. Phys. 38, 461 (1975).

[2] J.L. Yarnell, M.J. Katz, R.G. Wenzel and S.H. Koenig. Phys. Rev. A 7, 2130 (1973).

[3] J.R.D. Copley and S.W. Lovesey. Liquid Metals 1976 (London : Institute of Physics), p. 575 (1977).

ROUND TABLE ON LIGHT SCATTERING

Animator : J.G. Powles

Canterbury, U.K.

Unfortunately Dr. C. Brot (Nice) was unable to act as Animator owing to illness.

The following is a condensed account of a rather lively discussion reconstructed from the imperfect recollection of the Animator. Where names are mentioned it should be noted that opinions ascribed are as understood by the Animator and may have been expressed in the heat of the argument. The Animator also apologises in advance for not mentioning by name all contributors to the discussion.

The Animator gave a brief introduction to light scattering so as to set the scene, defining some symbols and describing some ideas which would surely arise in the Discussion and which it had not been possible to deal with in detail in Dr. Volino's lectures. He hoped that the Discussion would have some bias towards the experimental aspects of the subject and indeed the persons who had asked to introduce particular topics were mostly concerned with the acquisition and interpretation of experimental results. The general review of light scattering included a discussion of the Brillouin, Rayleigh and Mountain lines in the polarised (VV) spectrum and the information derived therefrom. This includes the thermal conductivity and the velocity and attenuation of thermal phonons at frequencies of order 10^9 Hz, as well as certain molecular vibrational relaxation rates. The observation of partially depolarised spectra due to scattering by anisotropically polarisable molecules gives information about correlated molecular reorientation because the scattering is coherent and also about the rate and nature of the reorientation. In particular, the static orientational correlation factor, g_2, is in principle obtainable from the absolute total intensity. The possibility of observing the effect of translational processes, via its coupling with the reorientation line was discussed. It was shown how 'shear dips' arise in the VH spectrum associated with shearing fluctuations and in the HH spectrum due to longitudinal fluctuations which implicitly involve a shear flow. The various types of 'induced' absorption, due to dipole-induced-dipole interaction or non-additive polarisability of two molecules in close proximity was discussed, especially as regards its contribution in the wings of the spectra, where it may dominate. This, it was noted, makes the experimental determination of moments for the reorientation line very difficult. These moments include the total intensity (see above), the second moment, which gives information also about the dynamics of molecular orientational correlation and the fourth moment, which gives the mean square torque. It was emphasised that all these effects appear together and an important problem is how to separate the various contributions to the observed spectra.

There followed several comments of a general nature which helped to elucidate the physical processes which lead to light scattering. In particular the question of whether one could yet observe effects due to chemical reactions was discussed. It appeared that this has not yet been observed although the Animator felt that increasing experimental accuracy and better understanding of liquids would lead to success eventually. Dr. Boon (Brussels) reported that Berne had suggested that one should start with the simplest case, an isomerisation reaction, $A \rightleftharpoons A^*$, involving charged species in an electric field where the electrophoresis effect may render the scattering spectra easier to interpret, especially for slow reactions.

The relation of fluorescence spectra to light scattering spectra was discussed.

Dr. Keyes (Yale) reported an analysis of light scattering total intensity data due to anisotropy scattering (I_{VH}) with special reference to liquid carbon disulphide. He emphasised that the intensity depends not only on the static correlation function, g_2, but that local field effects mean that the gas phase molecular anisotropy factor, $(\Delta\alpha)^2$, is considerably modified in the liquid. This is well known in cavity field treatments which lead to factors like $(n^2+2/3)^4$ and is related to the dielectric internal field problem. He has attempted to allow for the microscopic local structure and also for the distributed molecular polarisability assuming a three-point polarisability for the CS_2 molecule. Both of these effects are very important and the effective $(\Delta\alpha)^2$ may well be changed by a factor of about two. Thus,apart from its intrinsic interest, this gives a better estimate for g_2. Values of g_2 previously seemed to be less than unity ($g_2 = 1$ for no correlation of orientation) but with the correction, at least for CS_2, g_2 is greater than unity, which seems more reasonable. There followed an interesting discussion of various aspects of the interpretation including a consideration of triplet and higher correlation functions and the relevance of similar local field effects in Raman scattering and for 'induced' scattering by atomic liquids. Professor Powles (Canterbury, U.K.) wondered about the separation of induced scattering in the experimental measurements of absolute intensity (due to Flygare et al.). Professor Buckingham (Cambridge, U.K.) suggested that one might separate correlation and internal field effects by looking at solutions of anisotropic molecules in a solvent of 'spherical' molecules and Dr. Keyes reported that this had been tried but there are still difficulties for solutions of variable concentration. Professor Alder (Livermore) raised the problem of interpreting the absolute intensity for rare-gas liquids which is a closely related problem which seems to require a non-additive polarisability and this ought also to arise in molecular liquids. Dr. Keyes said this would also indeed arise for molecules but may not be as important as for atoms.

The discussion then turned to the question of 'shear dips' etc. and Dr. Searby (Nice) gave an account of these effects and reported some experiments. He considered that the dips could be accounted for by purely viscous effects associated with fluctuations in shear flow rather than with Professor Powles' explanation in terms of vestigial shear rigidity (which was first introduced by Rytov). He pointed out that shear flow fluctuations produce flow birefringence and that a theoretical treatment along those lines leads to a width of the dip which is given by $k^2\eta/\rho$ where η is the shear viscosity. He pointed out that the visibility of the shear dip then depends on the relation of this quantity to the width of the anisotropy line, the effect being easiest to observe when they are comparable in magnitude. He pointed out that in some liquids he had studied, the dip followed the behaviour with temperature as predicted by the above formula. There followed a lively discussion in which, in particular, Dr. Quentrenc also (Nice) felt that one does in fact need some shear rigidity to explain the effects ! The question of how the shear dips change as one goes into the glassy state (if the liquid can be persuaded to do so) and becomes the usual transverse Brillouin peaks in the VH spectrum of an amorphous isotropic solid, aroused considerable interest. The conditions under which one should be able to see these shear dips experimentally was discussed in some detail. Professor de Gennes (Paris) pointed out that it would appear to be desirable to measure at small k but Dr. Searby pointed out that this raised serious experimental difficulties. There was also a brief discussion of the 'dips' in the HH, 90°, anisotropy spectrum which are associated with the shear effects in longitudinal phonons but there is little experimental data available and it appears to be difficult to give a simple physical explanation of the form of the perturbation of the anisotropy spectrum. It is clear however that any experimental indication of possible shear rigidity in a simple liquid is a topic of considerable general interest.

Since long-time tails on velocity autocorrelation functions are apparently now thought to be in some sense visco-elastic there was some discussion as to whether the shear dips were related, but the general feeling was that these are two separate phenomena. This gave rise to a discussion about mode-mode coupling and forces and fluxes and symmetry among the theoreticians, particularly between Professor de Gennes (Paris) and Dr. Lekkerkerker (Brussels), which was clearly important, but the Animator is not competent to summarize it.

Dr. Versmold (Karlsruhe) discussed an alternative method of getting the static correlation of orientation factor, g_2, as proposed originally by Kivelson. This is by using a relation between the correlation time for correlated molecular reorientation which is obtainable from light scattering and the single molecule correlation time obtainable from Raman scattering or magnetic resonance, which involves g_2 and the correlation of angular velocity. If the latter is weak we can get g_2, although it appears that the theory involves the assumption of diffusional reorientation. Professor Powles and Professor Bratos (Paris) both thought this is a rather serious assumption and was probably not true for liquid CS_2 in particular, as indicated by infra-red measurements. Some experimental results for liquid CS_2 were presented and the g_2 values which were found were in reasonable agreement with the values found from intensity measurements. There followed a lively discussion of the relation between intensities and spectra and the local effects of molecules on one another in the liquid to which Dr. Keyes and Professor Buckingham were the main contributors. The discussion then turned to the general problem of measuring g_2 to which Dr. Rouche (Bordeaux) contributed and mentioned the method of the Kerr effect as being valuable.

Dr. Battaglia (Cambridge, U.K. and Australia) reported measurements of the Kerr and the Cotton-Mouton effect in liquids and pointed out that the problems arising in the interpretation are very similar to those for light scattering as discussed by Dr. Keyes, i.e. internal fields and effective anisotropies. He pointed out that one term of many in the measured Kerr effect constant involves g_2 and similarly for the Cotton-Mouton effect. In spite of many terms arising in both expressions he showed how one could in fact get a reasonably accurate estimate of the value of g_2 from the two experiments because in ratioing many unknowns cancel. The value of g_2 seems to be consistent with that obtained from the correctly interpreted light scattering experiments for many molecular liquids including CS_2. It seems that some of the worrying corrections, such as collisional contributions, in light scattering are less important in the Kerr effect but it should be possible to do the correction better by looking at the spectral aspects of the Kerr effect.

Professor Buckingham pointed out the contrast between the virtual independence of the effective mean polarisability on state as shown by the Lorentz-Lorentz relation and the large change in the effective optical anisotropy. This provoked further discussion about internal fields and on this recurring theme the formal discussion ended - to be continued informally at the Beach bar.

The full contribution of Dr. Versmold will appear elsewhere:

H. Versmold - Depolarized Rayleigh Scattering:
Collective Reorientation in Molecular Liquids
Ber. Buns. Gesel. (1978) - to be published

ROUND TABLE ON CRITICAL PHENOMENA

J.G. Powles

Canterbury (U.K.)

The following is a condensed account of the Discussion corresponding to the impressions of the Animator who apologises in advance for any incorrect or imperfect attributions to particular individuals who may be mentioned.

The Animator suggested that the Discussion should be biassed towards the experimental side of the subject, since, as arranged, the lectures by Pr. Hohenberg (Munich and Bell Labs.) had been rather theoretical in nature.

Professor Powles gave a brief introduction to critical phenomena emphasising the difficulties arising in actual experiments - where the dimensionality is exactly three ! He discussed the problem of actually determining a critical temperature, T_c, and a critical exponent, which were more or less taken for granted in theoretical work. He pointed out that all the basic phenomena can be observed in a critical opalescence experiment, namely the increase in magnitude of fluctuations related to the divergence of the compressibility and the correlation length and also the critical slowing-down as one approaches the critical point. He discussed the divergence of the isothermal compressibility as a thermodynamic quantity which is directly measurable, or indirectly by means of the structure factor S(k) for $k \to 0$ and which leads to an evaluation of the critical exponent γ. The k dependence of the structure factor, which near the critical point, is narrow in k, leads to the idea of the long range of the fluctuation, i.e. κ, the width, is small and the correlation length $\xi = 1/\kappa$ is large. Clearly for $k \ll \kappa$ we get essentially thermodynamics but if $k \gg \kappa$ we are looking 'inside' the fluctuation and when ξ is large we see more explicitly critical effects. Then more information may be obtained, for instance we have $S(k) \propto k^{-2}$, or more precisely $k^{-2-\eta}$, thus determining another exponent. It was pointed out that the correlation length plays a central role in all critical phenomena - although that cannot be made apparent in a brief discussion. Looking now at the dynamics we can generalise to the scattering law, $S(k,\omega)$, and since near T_c the fluctuations are slow we expect the spectrum to be narrow in ω and, since it is essentially a Rayleigh line, to be Lorentzian with width Γ. For the hydrodynamic region ($k << \kappa$), $\Gamma = D_T k^2$ and the width 'diverges' to zero since C_p diverges faster than λ. When $k \gtrsim \kappa$ we need to correct hydrodynamics and more elaborate theories for Γ, such as those of Fixman or Kawasaki with mode-mode coupling schemes, are required and more detailed information about the critical situation can be obtained. It was pointed out that critical binary mixtures behave in much the same way where roughly speaking one replaces density fluctuations by concentration fluctuations. Although thermodynamic and scattering methods are the most powerful it is to be expected that almost any physical measurement should show some anomalous behaviour near the critical point. In many cases the effect on the measured quantity may be quite small, as in nuclear magnetic resonance, but continuing improvements in experimental techniques are likely to give more, and more diverse, experimental data concerning the behaviour near the critical point.

It was pointed out that there is a great wealth of more complicated critical phenomena which are becoming of great current interest, such as tricritical and more complex critical points in multiple mixtures. It may be expected that this field will be a very lively one in the next few years, especially in view of the recent dramatic progress in understanding arising from the development of the Renormalisation Group theory.

There was then a general discussion of various topics arising. In particular Dr. Eder (Vienna) asked what happened to $S(k)$ for k much larger than κ and it was pointed out that this was 'normal' in the sense that it reflected the short-range part of $g(r)$. In this connection Pr. Hohenberg pointed out that although to see 'proper' critical effects one is interested in having $k > \kappa$ nevertheless one wants to also have $k < 1/a$ where a is a characteristic microscopic length, e.g. a molecular diameter, in order to see critical effects. Clearly one must be near enough to T_c to have an appreciable 'window' in k available.

Dr. Calmette (Saclay) then discussed some measurements of the light scattering spectrum due to critical concentration fluctuations in binary liquid mixtures where the linewidth is related to a sort of 'mass conductivity' but also involves viscosity. The expression for the linewidth and intensity was discussed at some length since there appeared to be some uncertainty as to what the latest numerical coefficient actually is and what is really meant by the 'viscosity' arising. A number of experimental difficulties such as the effects of absorption (turbidity) and cell scattering were discussed, as well as the difficulty of measuring the critical temperature exactly. This is a case, as pointed out by Pr. Hohenberg, where one is trying to measure an 'amplitude' as opposed to an exponent, and this is always more difficult. This problem prompted Pr. Alder (Livermore) to ask what happens to the viscosity near the critical point. It appears to diverge weakly (exponent ~ 0.04) but, as with all weakly diverging properties, one has problems in allowing for the 'background' viscosity which is also temperature dependent. Dr. Calmette had analysed a lot of data on viscosity and presented the results. After a long discussion the Animator concluded that we do not understand very well the experimental behaviour of viscosity in the critical region or how to use it confidently in analysing experimental results in which it appears as a parameter - at least for mixtures. Professor Enderby (Bristol) posed the question as to whether the fluid is Newtonian near T_c and can one even define a viscosity ? Pr. Hohenberg pointed out that what one wants in the theory is the linear coefficient even if the fluid is not Newtonian. Professor de Gennes (Paris) pointed out that the non-linearity had been estimated and is small, as for most weakly divergent properties.

There followed a brief discussion on upper and lower critical consolution points in binary mixtures and how the closed two phase region arises.

Dr. Chieux (Grenoble) then gave a summary of the way in which neutron scattering can be used to study the critical region. He pointed out that most work so far was on static exponents but that dynamical properties were just beginning to be studied. He pointed out that for neutrons the fluid does not have to be transparent and corrosive fluids can be more readily contained. The k range available is higher than for light scattering and extends from 0.2 Å^{-1} to at least 0.001 Å^{-1} so that it is possible to be in the critical situation $k > \kappa$ much further from the critical point, and even so it is still possible to have $k < 1/a$. Consequently one can investigate ξ in the range of hundreds of Å or less. However the large samples required might give rise to gravity problems. He then considered binary mixtures more particularly. It was noted that the incoherent scattering, if appreciable, was a great nuisance. It was pointed out that for some binary mixtures it may be possible to choose the isotopes so that in S(0) only the contribution due to concentration fluctuations, $S_{cc}(0)$, survives and this is the quantity of most interest in binary mixtures. Typically even in the general case the S_{cc} term dominates so that one can get the 'γ' exponent in this way. There was then a discussion as to whether one can actually determine the exponent η

directly because of the dependence of S(k) on $k^{-2+\eta}$, as described above, since this appears to be too difficult to do by light scattering because k is too small. Some preliminary measurements of η have been done by this method using neutrons and X-rays. Pr. Hohenberg explained that η values in the literature are usually obtained from the scaling relations between exponents so that it is very desirable to have a direct measurement. Dr. Keyes (Yale) pointed out the desirability of knowing the form of the pair distribution function not only in the Ornstein-Zernicke limit but also for intermediate separations. Pr. Hohenberg pointed out that this information came out indirectly from the renormalization group theory. However Professor Powles thought that it was very desirable to actually measure g(r) over the whole range directly.

Dr. Goulon (Nancy) then turned the discussion to dielectric effects near the critical point. He pointed out that for binary mixtures the anomaly in the static dielectric constant is small but that, as shown by Piekara in the 1030's, there is a large anomaly in the non-linear dielectric behaviour in high electric fields. He presented results for a mixture of a strongly polar liquid, benzonitrile, in a non-polar one, isooctane. It is found that $\Delta\varepsilon/E^2$ is negative for non-critical systems but near the critical consolution point there is a large positive anomaly. A similar effect is observed in the Kerr-effect experiment. It was also mentioned that Debye had observed in 1965 that a strong electric field may shift T_c by about 0.05° but it only happens for non-linear $\varepsilon(c)$ in binary mixtures. Mistura showed in 1973 that this implies the divergence mentioned above. Dr. Goulon gave an explanation of this effect in terms of the distortion of transient 'drops' of higher concentration of one component in the electric field using classical electrostatics. In collaboration with Professor Oxtoby (Paris and Chicago) a semi-quantitative analysis has been made which explains the experimental results quite well.

In the discussion which followed,several old experiments were described in which one tried to observe a change in the critical point by doing various forms of violence to the fluid. For instance, Professor Alder (Livermore) described an old experiment in a centrifugal field (i.e. the sample was rotated rapidly !) which should shift the critical point by a second order effect on the fluctuations and they obtained a shift of order a degree for two liquids of different densities - however the interpretation is still not clear.

Dr. Quentrec (Nice) described some work he had done with Pr. Oppenheim in interpreting an effect found in the VV light scattering spectrum of liquid Xenon near T_c, which had remained unexplained. He reported that, contrary to the accepted explanation, the Brillouin peaks do not broaden (as η diverges) or move towards the centre (as $C_s \rightarrow 0$) and eventually disappear as one approaches T_c. Also Benedek observed an additional central Lorentzian which cannot be a Mountain line. Kawasaki's theory does not predict either effect. He offered an explanation by a generalisation of hydrodynamics which explicitly takes account of the large inhomogeneities when near T_c which was reminiscent of the Goulon-Oxtoby theory discussed above. There is a local broken symmetry which can be allowed for by introducing a new tensor mechanical dynamical variable which is not conserved since its lifetime is proportional to k^2. The new variable generates a new mode which leads to the extra central Lorentzian and also a generalised frequency-dependent shear viscosity which explains the behaviour of the Brillouin lines. There was some discussion of these results and the meeting then adjourned to a more congenial environment.

The following contributions can be found elsewhere:

P. Chieux - On the Use of Neutron Scattering for the Investigation of Critical Phenomena in Fluids
ILL report 77CH238S (1977).

J. Goulon, M. Hollecker, J.M. Thiebaut, and D. Oxtoby - Dielectric Properties of Binary Mixtures Near the Consolute Critical Point
to be published (1978).

P. Calmettes - Critical Behavior of the Mass Diffusivity and the Shear Viscosity of Binary Solutions
to be published (1978).

Informal Events

INFORMAL ROUND TABLE DISCUSSION ON SUPERCOOLED LIQUIDS

Convener : C.A. Angell

Purdue University, U.S.A.

No attempt has been made in this report to keep records on those participating in this discussion or on their contributions, since this round table has not been taped; hence no detailed statements of the discussion progress can be made. However, it is appropriate here to refer briefly to the subjects which were discussed and the extent of interest expressed by the conferees in them.

Considerable attention was given to the matter of ultimate crystallization of supercooling liquids. It was pointed out that at the equilibrium crystallization temperature the probability of crystallization in a pure liquid is very small and a distinction develops between the time scale on which the supercooled state is entered and the time scale on which it can be left. As the temperature decreases the probability of leaving the metastable state (i.e. of crystallization via homogeneous nucleation) increases while at the same time the time scale necessary for relaxation to lower energy but still fully amorphous state increases. The ultimate fate of the supercooling liquid depends on the way in which these time scales change with changing temperature. If the internal liquid state relaxation time becomes long while the probability of crystal nucleation is still small, then internal structure equilibrium becomes arrested and a glassy state results during normal cooling processes. In this case the nature of the glass transition becomes the phenomenon of interest.

Much attention in this discussion session was focussed on the case where the probability of escaping from the metastable state becomes high while the liquid relaxation times are still short. The phenomenon being studied under these circumstances is that of homogeneous nucleation, on which some interesting and very relevant neutron scattering experiments (dynamic structure factor studies) have recently been performed. The nature and implication of these experiments was given some discussion, as was also the observation of homogeneous nucleation in molecular dynamic studies.

Some time was devoted to a consideration of those properties or structural characteristics of the molecular constituents of molecular liquids which would make their nucleation rates relatively small and render them favorable candidates for use as solvents in supercooled liquid studies of various sorts.

Attention was also given briefly to the currently much researched problem of attaining amorphous states for liquid metal alloys by the use of very fast quenching schedules. The way in which the large temperature gradients which are imposed on splat cooled liquid samples can cause structural inhomogeneities in the resulting amorphous solid was pointed out. The discussion broke up after a final brief consideration of the Kauzmann paradox, i.e. the problem raised by the fact that in every case for which data are available, only a non-equilibrium phenomenon (the experimental glass transition at which $(\partial S/\partial T)_p$ changes) prevents the liquid entropy from falling to a value below that of the corresponding crystalline solid. This more-or-less untenable inversion of equilibrium properties would occur at temperatures only 20-50° below the glass transition according to entropy extrapolations based on crystal and supercooled liquid heat capacities.

TITLE OF POSTERS

Note: Names of participants of the School are underlined.

. Structural Analysis of Molten CuCl by Neutron Diffraction (S. Eisenberg[x], J.F. Jal[x+], W. Knoll[+], J. Dupuy[x]).

x Laboratoire de Physique des Matériaux, Lyon I, France.
+ Institut Laue-Langevin, Grenoble, France.

. Angular Correlation Functions for the Pure Dipolar Fluid in the Mean Spherical Model. (N. Quirke[x], J.W. Perram[x])

x Dept. of Math., University of Odense, DK 5230 Odense.

. Solutions of Alkali Metals in Molten Salts. Metal and Non Metal Transition and Concentration Fluctuations. (J.F. Jal[x+], P. Chieux[+], J. Dupuy[x]).

x Laboratoire de Physique des Matériaux, Lyon I, France.
+ Institut Laue-Langevin, Grenoble, France.

. Ionic Radii and Diffraction Patterns of Molten Alkali Halides. (C. Caccamo[x], M. Parrinello[x]).

x Istituto di Fisica, Messina, Italy.

. Temperature and Composition Dependence of the Nuclear Quadrupolar Relaxation Rate in Liquid Metals and Alloys. (W. Schirmaker[x])

x University of Bristol, U.K.

. A Lattice Model Showing Liquid–Gas and Metal Insulator Transitions. (S. Nara, T. Ogawa and T. Matsuhara)[x] – Progr. Theor. Phys. <u>57</u>, 1474 (1977).

x Physics Department, Kyoto University, Kyoto, Japan.

. Critical Proton Spin Relaxation in Liquid Chloroform. (Krynicki, J. Powles[x], Rigamonti) – Comm. in Phys. <u>1</u>, 183 (1976). – Mol. Phys. Dec. 1977.

x Physics Laboratory, The University, Canterbury, Kent, U.K.

. Comparative Study of an Internal Molecular Motion by I.R. Raman and Inelastic Neutron Scattering. (Application to the Ring-Puckering Motion of Cyclopentane in Condensed State). (M. Besnard[x], H. Jobic[+], J.C. Lassègues[+]).

x Institut Laue–Langevin, Grenoble, France.
+ Université de Bordeaux I, Talence, France.

. $S(Q,\omega)$ of Liquid Al at 1018 K. (O.J. Eder[x], Kunsch[x], J.-B. Suck[+]).

x Seibersdorf, Austria.
+ Institut Laue–Langevin, Grenoble, France.

. Deconvolution (M. Johnson[x]).

x N.B.R.U., Rutherford Laboratory, Chilton, Didcot, U.K.

. Diffusion in a Hard–Sphere Boltzmann Gas. (O.J. Eder[x])

x Seibersdorf, Austria.

. Mössbauer Scattering. (M. Soltwisch[x], M. Elivenspock[x], D. Quittman[x]).

x Freie Universität, Berlin, R.F.A.

. Low Frequency Raman Scattering in Amorphous Solid and Liquid KOH Water Solution (10 M). (M. Nardone, G. Signorelli and V. Mazzacurati)[x].

x Istituto di Fisica G. Marconi, Roma, Italy.

. Pressure Dependence of Dynamic Structure Factor of Liquid Argon. (P. Verkerk[x]).

x Interuniversitaire Reactor Institut Mekelweg 15, Delft, Holland.

. Computer Simulation in Biological Membrane. (J. Belle[x])

x Centre Paul Pascal, Talence, Bordeaux, France.

. AgI-type Solid Electrolytes. (K. Funke[x], R.E. Lechner[+]).

x Göttingen Universität, R.F.A.
+ Institut Laue-Langevin, Grenoble, France.

. OH^- Reorientation in Cubic NaOH (Neutron Quasielastic Scattering). (J.G. Smit[x], H. Dachs[x], R.E. Lechner[+]).

x Hahn-Meitner Institut, Berlin, Germany.
+ Institut Laue-Langevin, Grenoble, France.

. Coexistence Properties of the Lennard-Jones System. (A.C.J. Ladd[x], L.V. Woodcock).

x University Chemical Laboratories, Cambridge, U.K.

. Interatomic Potential of Liquid Rb. (W.S. Howells[x],J.R.D. Copley[x]).

x Institut Laue-Langevin, Grenoble, France.

. Diffusion in a Cosine Potential. The Neutron Incoherent Scattering Law. (A.J. Dianoux[x], F. Volino[x]).

x Institut Laue-Langevin, Grenoble, France.

. Crystallization from Supercooled Fluid Observed in Molecular Dynamics on Soft-Core Model. (M. Tanemura, Y. Hiwatari, H. Matsuda, T. Ogawa[x], N. Ogita and A. Veda).
(Geometrical Analysis of Crystallization of the Soft-Core Model) : To be published in Prog. Theor. Phys. 58, n° 4).

x Physics Department, Tokyo University, Tokyo, Japan.

. On the Theory of Surface Tension for Classical Fluids. (R. Evans[x]).

x H.H. Wills Physics Laboratory, Bristol, U.K.

. Structure and Temperature Dependence of the Resistivity Liquid Alkali Metals. (H. Minoo[x], C. Deutsch, J.P. Hansen[+]) - J. Phys. Lettres, 38, L.191, 1977.

x Physique Théorique et Hautes Energies, Paris XI, Orsay, France.
+ Physique Théorique des Liquides, Paris, France.

Closing Address

CLOSING ADDRESS

LIQUID STATE RESEARCH : WHY ?

P.G. de Gennes

Collège de France - Paris (France)

(edited from a tape recording taken at the final session of the Summer School on Liquids - Aleria, Corsica).

As you know I am here just as an "external observer" and I can only make very simple minded remarks. I have been educated by the books of Landau like many of us; you may remember a very hard sentence of Landau saying that "theories of liquids are neither convincing nor useful", therefore he refused to describe them. This severe statement refers to the state of affairs during the 1950's.

The merit of this school is to show us precisely to what extent things have changed. I am quite convinced that they did change. On the other hand, I am also convinced that now the art of liquids has reached its classical era and that in some instances it may soon become somewhat hellinistic; this will show up in my remarks.

In a certain sense the system of summer schools or Gordon conferences is auto-catalytic; a number of clever people being interested by the same problem, becoming really convinced that this problem is important. In some cases this self-catalytic process, which is highly non-linear, leads to a very strong focussing in a very small direction; this implies a danger that we have to keep in mind.

What justifications do we have for the existence of the school and for the enthusiasm that we saw here ? I will try and quote some partial answers below.

I. INTERATOMIC FORCES

We have heard very mixed statements about how much we can learn about forces between atoms or molecules by studies on the liquid state; there are some hopes; there are some difficulties. In the case of alkali metals, it is really gratifying to see how much a comparatively simple random phase picture can tell us about a physical system (except for certain wicked features like surface tension !). But that is an exceptional case, and in many cases I am amazed by the difficulties to be faced for more complex metals. But more generally the fact that the Friedel oscillations (which are very conspicious in the solid metals) become so inconspicuous in the liquid, as we heard from Dr. EVANS; this fact shows us how much the chaos of the liquid phase can smear out certain information.

In other simple systems like argon, the knowledge of the forces obtained from the liquid state appears very modest compared with that we get from molecular collisions. There are cases however where we can think of getting a little more. One branch where I see a hope is the kind of picosecond laser experiment which was described here by Dr. LAUBEREAU, because it gives us insight on very rapidly events which may be treated as a 2-body collision (or sometimes a 3-body collision). At that level we return to some of the simplifications of collision physics : I have good hopes for this particular technique.

II. LIQUID DYNAMICS

If we go to the other end - low frequency/long wave-length limit - we find a rather pleasant situation : we are dealing with a hydrodynamic limit which contains a lot of information on correlation functions. From what I heard here, we now assess the limits of that hydrodynamic picture rather well - and on this side I am happy.

On certain other sides I hoped for more novelty. There are many cases where it is interesting to see how a flow-field (not a uniform translation, but some shear-flow, for instance) couples to other physical phenomena; here I have not learnt exactly as much as wished. I am not clear whether this silence reflects the youth of the field, or if it is considered that these problems are just too complicated. Let me take a few examples. Flow birefringence was mentioned at one or two points but I am not at all clear about the microscopic interpretation of flow birefringence. To what extent can it be expressed in terms of a Kubo formula ? and to what extent does it bring specific information ? More generally, the effect of flows on many static or transport properties is a really fascinating subject. Coming from liquid crystals (where we have many more degrees of freedom) I was surprised not to hear more about this in simple liquids. For instance, does a shear flow change the electrical conductivity of an ionic solution. The effect should involve the product of a shear rate by a relaxation time of the liquid; that is a very small parameter, so probably it is hard to see(*).But there may be many more clever effects of this sort which would be interesting. Another effect which is connected to the microscopic dynamics, but which comes in macroscopic properties, and which is rather mysterious to me, is the SORET-DUFOUR effect. Let us take a liquid with one impurity in it, and we look at this impurity in a temperature gradient; it drifts in this gradient in a certain direction - either up or down - and with a certain speed : I would be very pleased to know what are the processes involved. This to me seems to be a very open question. It is also related now to a number of remarkable hydrodynamic instabilities which people begin to measure. To summarize liquid dynamics seems to be well understood in its relation to $S(q,\omega)$, but if we look at some more refined aspect like this DUFOUR coupling,I am still in the dark.

III. CRITICAL PHENOMENA

Another case of long ranges and low frequencies is critical phenomena. Here we have been educated in full detail by Professor HOHENBERG. Let us then ask the question : What is really the point of using liquid samples for critical phenomena (rather than, say, magnetic systems or superfluids, or superconductors or liquid crystals) ? There are a few advantages with the liquids which should be mentioned.

a) One is the fact that the liquid (although it is very disordered on a local scale) is on an average scale an ideally homogeneous system. There are no defects to pin a wall, an interface, as we meet them in crystals; all the complications that we have in crystals, related to small stray fields which are always present in the crystals and which complicate its transition are absent for fluids.

b) In liquids, we do not have the intricacies of coupling between order parameter and elastic variables; that is a very helpful simplification.

c) There is good coupling between the critical fluctuations and both the neutron and the light; this is very different from

(*) Note added in proof : we know now of one case where the times are long and the effect could be strong : suspensions of non spherical particles. See M. ADAM et al., Journal de Physique Lettres (to be published).

what we meet in other systems like superfluid helium where it is very difficult to couple to the order parameter, and with simpler systems like ferro-magnets. (It is very hard to couple linearly the electromagnetic field to the spin-up, spin-down variable of the ferro-magnet; in fact, the dominant coupling involves complicated terms which make it less interesting). Thus liquids are very favourable from the point of view of coupling with probing radiations.

d) Critical phenomena in liquids provide us with a very helpful time-scale. If you remember HOHENBERG's discussion this morning there is a characteristic time τ for a binary mixture, $\tau = \eta \xi^3/k_BT$ (η : viscosity, ξ : correlation length). This diverges strongly near the critical point T_c (essentially like $(T - T_c)^{-2}$. At $T - T_c \sim 0{,}1°$, one may achieve times that become really long; in particular you can realize a situation where a shear flow rate S is comparable to or larger than $1/\tau$. In the discussion last night, the question of a non-Newtonian viscosity came up; and that is one typical effect that you expect there. A more spectacular effect has just been measured recently by P. BERGÉ and coworkers. It is the change in the intensity of the light scattering pattern. Thus liquids are nice from the point of view of a time scale near the critical point.

There is an interesting difference in style or spirit between the people who come from phase transitions phenomena, and people who are primarily interested in liquids. When you come from the liquids side, you are really primarily concerned with local properties, be it something like a correlation function, be it refractive index (although the refractive index seems to me a terribly complicated animal). We know that most of these local properties reflect only very weakly the critical anomalies; most of them in fact are expected to scale very much like the energy $\Delta T^{1-\alpha}$, where α is a specific heat exponent. Take for instance the dielectric constant experiment which was described in a round table last night by Dr. GOULON. The first part, the simple experiment in low fields,showed a weak divergence of this type. We meet here a profound difficulty : the reaction of the universal critical properties on the local properties, which are of interest to this audience, is usually very small.

IV. MATERIALS RESEARCH

Materials research on liquids is fascinating.
There are real problems of our present day which are related to physical properties of unusual liquids. For the chemist, for instance, the importance of very unusual <u>polar solvent</u> has become really quite prominent; I have not seen yet the same interest among the liquid state physicists.

Another question, which I know from the liquid crystals (where it is a constant worry) is the problem of miscibility. You have synthetized molecule A and molecule B and you want to know if you will be able to make a mixture A + B. If you can, you will have some very interesting continually varying properties. The present state of the art seems to be very primitive. To guess what miscibility will be is hard. I realize the difficulties that would come if we would immediately address ourselves to complicated organic molecules. Note however that the question already arises in liquid metallic alloys : from what Dr. EVANS said the hope is that soon at least this random phase type of description will be available for such mixtures. But will it retain all the important features ?

Note that when I talk about miscibility I am not so much excited about the critical point aspect (which is reasonably under control), but more by the local properties which command that this particular mixture will be realizable. There are many other miscibility problems :

a) as I said, liquid crystals raise many questions of this sort;

b) miscibility effects are very striking in polymers. The physical reason is simple : when you have long flexible chains, the relevant interaction energy - to be compared with k_BT - is not the monomer-monomer interaction, but the sum of all interactions felt by one chain. This is 10^3 times bigger; for this reason you see very dramatic segregation effects; they command, in fact, many practical applications of polymers.

For example, if you use polystyrene to build a pot of yogourt, it is mechanically very poor; if you want to improve on that, you add rubber chains to the styrene before polymerising. But you immediately find that, at polymerisation, because of this segregation energy, the system separates into little drops; in fact this is helpful to resist fracture; fracture lines moving in this system enter the little blobs of rubber-like reagents and are stopped : you get what is called high-impact polystyrene. Because of cases like this, the question of segregation and phase-transitions in polymers are really important for many current applications.

I had the very lucky occasion here of listening to Professor BUCKINGHAM's ideas on the difference between deuterated species and hydrogenated species; there seems to be a very slight energy favouring segregation, in these mixtures. This, in a simple species like methane, is quite invisible but in a long molecule (of molecular weight few hundred thousand) it is probably enough to induce phase separation. This particular transition is interesting. I can show that the critical point should have classical exponents and a lot of interesting features.

I think that this materials research aspect of liquids is really nice. It showed up in the round table on solid state electrolytes : it might come up strongly in the future.

V. "NATURAL SLOPES" AND OTHER TRENDS

There is one thing that I tend to call the natural slopes of science. In this audience we clearly see some natural slopes for liquids. For instance, I am convinced that the study on polyatomic molecules will expand. Coming from liquid crystals, which are made of large molecules, I am to some extent tempted by this trend. However, I am not entirely sure of what really new concepts we shall learn from triatomic molecules. A similar remark holds for 3 body correlations.

Like most of us I often have the problem of advising graduate students : is it really good instruction for a student to work on a project in neutron experiments or in Monte Carlo calculations in liquids ? I am hesitant. I find that many of the recent contributions have been extremely careful but have not brought very novel concepts; I am worried to channel students through that system, I would perhaps prefer them to come to this field after having had an education in something different, be it polymers, metals, hydrodynamics, etc. I stand open to correction. But the question has some importance in view of the planning of future summer schools.

After stating worries, I should state hopes. I think that for the future, and possibly for future schools, the methodology that you people have built will really be extremely efficient in many directions. We see three generations :

a) the early schools like Haïfa and Menton - mainly concerned by the methods

b) this school (under the pressure of the active groups at Grenoble and Lyon) has been more in a sense centred on systems (but simple systems)

c) the future schools will have to move towards more complicated questions. Going from diatomic to triatomic molecules, etc. cannot be the whole answer. Let me now seek other possibilities.

One session which I found very stimulating here was the round table discussion about supercooling and glasses. The impact of the methods which you have on our knowledge of glasses will be very important. At present there are some very low temperature properties of glasses which have been rather well classified but the link between these and the microscopic bonding features is all to be built. I am convinced that you are the people who can do this in the future.

Another example is liquid crystals, and here the situation may become favourable in the future : I remember talking with L. VERLET a few years ago about the possibility of investigating liquid crystals by numerical methods on a local, molecular, basis; the conclusion was very pessimistic, because we are dealing here with very large and complex molecules (at least 6 degrees of freedom, free orientation, and free translation). However liquid crystals give us a very dense spectrum of intermediate possibilities between the solid state and the liquid state. The nematic phase (which is very close to a conventional liquid) is difficult. I am more tempted by studies on relatively ordered phases - for instance, the smectics systems where molecules are grouped in well defined layers, each layer being a two dimensional liquid, I am convinced that with the present tools one can study smectics rather well. Also liquid crystals have beautiful phase transitions. From the point of view of symmetry, one can find analogs to the λ point of helium, to liquid/gas, and to superconducting transitions, etc. And many of the techniques which you have set up have not been used yet for these materials.

In another direction, we have now data in various 2-dimensional fluids. I mentioned at one round table the problems of rare gas adsorbed on graphite. These are 2-dimensional liquids under a periodic potential which are very fascinating. Another family for which we just begin to have some really accurate data is the traditional Langmuir films : films at the surface of water of something like a fatty acid. These films show a number of phase transitions and the question of their equation of state is important. (See the poster by J. BELLE at this school). These films are related to certain applications, such as detergents, cosmetics, food processing and a number of things of this sort. [They are also claimed to be important for biological purposes (membranes); I am not sure that this is a good flag to hold, but it may be]. In any case, the equation of state of physical 2-dimensional liquids is a natural objective for the future schools.

Another branch which I may mention briefly is connected with flexible polymers; there are many polymer problems, where a reflection on the nature of the microscopic interactions would help enormously. In polymer dynamics, we begin to have a good vision of what happens at large scales, comparable to the size of one coil. But we know not enough on what happens on a local scale : for the future there will be a lot to explore here.

Still another question : we heard here from Professor ENDERBY and others that the hydration of ions is now quantitatively known for simple systems. It is a little bit more wicked for transition metal ions.

There may be a third step : hydration properties of small organic molecules like amino-acids. Once the amino-acids properties in water are understood, the properties of proteins in water could be attacked much better. In fact I would hope for some closer interaction between the liquid state experts and the people who are concerned with protein structure and function.

In such a list of open questions, I would like not to forget the science of colloid and interfaces. Here I am thinking, for instance, in terms of a micelle of 80 Å diameter. The talk we heard this morning by WEIS, on hard spheres carrying a permanent dipole, was meant, in his mind, to describe some features of simple diatomic fluids. To me, it is also interesting in view of the possible applications to magnetic colloïds where we have cobalt grains (typical diameter 80 Å) covered by a thin polymer sheet to prevent coagulation (i.e. : maintain a certain distance between magnetic poles of opposite sign). These cobalt grains interact by magnetic dipolar forces and their properties are quite unusual. For instance you can put in a field H such a suspension (a "ferro-fluid") and measure the correlation functions under H (something that WEIS did not mention explicitly). These correlation functions under H are interesting at moderate concentrations because dipolar systems tend to make long chains in the direction of H. More generally, there is a wealth of interesting problems in colloïdal physics. (And in fact, next week in Aussois, we are having a meeting similar to this one, concerning colloïds). At some point it will be interesting to have the two communities (liquids and colloïds) mixing a little bit (if they do not phase separate of course).

Other aspects of interface science where the present expertise on liquids might help very much are the questions of electrochemistry at liquid-solid interfaces, and of the various forms of chromatography, etc.

To summarize : I have some naive hopes and worries concerning the science of liquids. My own dream for the future summer schools is to see them occurring as joint projects :

Liquid experts + solution chemists
" " biochemists
" " liquid crystal scientists
" " colloïd and interface scientists
" " electrochemists

and so on. In many of these cases we may have some difficulties in establishing a common language, but this can be overcome with patience and good will - the same patience and good will which the present organisers have used so successfully during the last two weeks.

PARTICIPANTS

<u>Director and co-Director</u>

J. Dupuy	Département de Physique des Matériaux, Université Claude Bernard, Lyon, France. (LA 172).
A.J. Dianoux	Institut Laue-Langevin, Grenoble, France.

<u>Scientific Committee</u>

B. Alder	Lawrence Livermore Laboratory, University of California, U.S.A.
S. Bratos	Université Pierre et Marie-Curie, Paris, France.
P.G. de Gennes	Collège de France, Paris, France.
J.G. Powles	Physics Laboratory, The University, Canterbury, Kent, U.K.
T. Springer	Institut Laue-Langevin, Grenoble, France.
M.P. Tosi	Istituto di Fisica dell'Universita i CNSM del CNR, Roma, Italy.

<u>Organizing Committee</u>

C. Brot	Laboratoire de Physique de la Matière Condensée Université de Nice, France.
G. Chassagne	Département de Physique des Matériaux, Université Claude Bernard, Lyon, France.

P. Chieux	Institut Laue-Langevin, Grenoble, France.
P.G. de Gennes	Collège de France, Paris, France.
C.H.S. Dupuy	Département de Physique des Matériaux, Université Claude Bernard, Lyon, France.
G. Tourand	DPH/G Laboratoire Léon Brillouin, Gif-sur-Yvette, France.

Lecturers

J.P. Hansen	Laboratoire de Physique Théorique des Liquides Université Paris VI, Paris, France.
I.R. McDonald	Royal Holloway College, Egham, Surrey, U.K.
A.D. Buckingham	Chemical Laboratory University, Cambridge, U.K.
R. Evans	H.H. Wills Physics Laboratory, Royal Fort, Bristol, U.K.
F. Volino	Institut Laue-Langevin, Grenoble, France.
J.E. Enderby	H.H. Wills Physics Laboratory, Royal Fort, Bristol, U.K.
P. Hohenberg	TU München, Garching, Germany and Bell Laboratories, Murray Hill, U.S.A.
A. Laubereau	Physics Department, Universität München,
M.D. Zeidler	Institut für Physikalische Chemie, Universität Karlsruhe, R.F.A.
J.J. Weis	Laboratoire de Physique Théorique et Hautes Energies, Orsay, France.

Participants to the Round Tables

K. Funke	Institut für Physikalische Chemie der Universität Göttingen, R.F.A.
L. Pietronero	Brown Bovery Company, Baden, Switzerland.
P. Fulde	M.P.I. für Festkorperforschung, Heilbronnerstr., Stuttgart, R.F.A.
R.E. Lechner	Institut Laue-Langevin, Grenoble, France.
J.R.D. Copley	Institut Laue-Langevin, Grenoble, France.

S. Parrinello — Universita di Messina, Istituto di Fisica, Italy.

H. Versmold — Institute für Physikalische Chemie, Universität Karlsruhe, R.F.A.

J. Searby — Laboratoire de Physique de la Matière Condensée, Université de Nice, France.

P. Rouch — Laboratoire de Spectroscopie, Université de Bordeaux, Talence, France.

J. Calmettes — DPH/SRM, C.E.A. Saclay, Gif-sur-Yvette, France.

P. Chieux — Institut Laue-Langevin, Grenoble, France.

J. Goulon — Laboratoire de Chimie Théorique, Université de Nancy, France.

Technical Assistance

F. Parisot — Institut Laue-Langevin, Grenoble, France.

F. Giraud — Institut Laue-Langevin, Grenoble, France.

G. Guiraud — Département de Physique des Matériaux, Université Claude Bernard, Lyon, France.

Participants

M. Adam — DPM/SRM, C.E.A. Saclay, Gif-sur-Yvette, France.

M. Allen — Physical Chemistry Laboratory, University of California, U.S.A.

M. Alves-Marques — Centro de Fisica Materia Condensata, Lisboa, Portugal.

G. Andoloro — Istituto di Fisica, Palermo, Italy.

A. Angell — Department of Chemistry, Purdue University, West Lafayette, Indiana, U.S.A.

M. Battaglia — University Chemical Laboratory, Cambridge, U.K.

H. Beck — Institute of Theoretical Physics, University of Basel, Switzerland.

J. Belle — Centre Paul Pascal, Talence, France.

J. Bellissent	DPHG/G, Laboratoire Léon Brillouin, Saclay, Gif-sur-Yvette, France.
N. Bolt	Universiteit von Amsterdam JHV' T Hoff Institut, Holland.
J.P. Boon	Université Libre de Bruxelles, Belgique.
C. Caccamo	Istituto di Fisica, Messina, Italy.
D. Ceperley	Present address : Laboratoire de Physique Théorique - Paris XI, Orsay, France.
S. Cummings	H.H. Wills Physics Laboratory, Royal Fort, Bristol, U.K.
C. Da Fano	Present address : CECAM, Orsay, France.
P. Dore	Istituto di Fisica "G. Marconi", Universita di Roma, Italy.
I. Ebbsjö	AB Atomenergi Studsvik Fack, Niköping, Sweden.
C. Flytzanis	Laboratoire d'Optique Quantique, Ecole Polytechnique, Palaiseau, France.
W. Freyland	Institute of Physical Chemistry, University of Marburg, Germany.
I. Gibson	Physics Laboratory, University of Kent, U.K.
G. Giuliani	Istituto di Fisica "G. Marconi", Universita di Roma, Italy.
B. Guillot	Laboratoire de Physique Théorique des Liquides Université Paris VI, Paris, France.
A. Heidemann	Institut Laue-Langevin, Grenoble, France.
J.E. Hirsch	James Franck Institute, University of Chicago, U.S.A.
S. Howells	Institut Laue-Langevin, Grenoble, France.
A.Johnson Torbjön	Institut of Reactor Physics,Drottning Kristinasvög, Stockholm, Sweden.
T. Keyes	Yale University, Department of Chemistry, New Haven, U.S.A.

A. Khuen	Institut für Physikalische Chemie Hans Sommer Braunschweig, R.F.A.
J. Kleim	Faculté des Sciences, Ile du Saulcy, Metz, France.
M. Klein	Institut für Physikalische Chemie, Karlsruhe, R.F.A.
A. Ladd	University Chemical Laboratories, Cambridge, U.K.
P. Lamparter	Max Planck Institut für Metallforschung, Stuttgart, R.F.A.
J.C. Lassegues	Laboratoire de Spectroscopie Infrarouge, Université de Bordeaux, Talence, France.
H. Lekkerkerker	Faculteit Wetenschappen Vrije Universiteit Brussel , Belgium.
R. Lévy	Rensselaer Polytechnic Institute, Physics Dept., Troy, N.Y., U.S.A.
J. M'Halla	Laboratoire d'Electrochimie, Université Pierre et Marie Curie, Paris, France.
H. Minoo	Physique Théorique et Hautes Energies - Paris XI, Orsay, France.
M. Nardone	Istituto di Fisica, Universita di Roma, Italy.
D. Noreus	Institute of Reactor Physics Drottning Kristinasvög, Stockholm, Sweden.
K. Ogawa	Physics Department, Kyoto University, Kyoto, Japan.
D. Oxtoby	James Franck Institute, University of Chicago, U.S.A.
B. Quentrec	Laboratoire de Physique de la Matière Condensée, Université de Nice, France.
N. Quirke	Mathematics Institute, Odense Universiteit, Odense, Denmark.
E. Rapoport	Nuclear Research Centre, Yavne, Israël.
D. Ricard	Laboratoire d'Optique Quantique, Ecole Polytechnique, Palaiseau, France.

D. Rocca	Istituto di Fisica "G. Marconi", Università di Roma, Italy.
M. Rovere	Istituto di Fisica "G. Marconi", Università di Roma, Italy.
W. Schirmacher	H.H. Wills Laboratory, Royal Fort, Bristol, U.K.
K. Shou Pedersen	Institute of Physical Chemistry, Lingby, Denmark.
P. Sixou	Laboratoire de Physique de la Matière Condensée, Université de Nice, France.
J. Smit	Hahn-Meitner Institut für Kernforschung, Berlin, R.F.A.
M. Soltwisch	Freie Universität, Berlin, R.F.A.
J.-B. Suck	Institut Laue-Langevin, Grenoble, France.
K. Szumilin	Institute of Physics, Warsaw Technical University, Poland.
F. Van der Graaf	Universiteit van Amsterdam JHV 'T Hoff Institut La. Elehkoch, Amsterdam, Holland.
G. Venzl	Physics Department der TU München, Garching, R.F.A.
P. Verkerk	Interuniversitaire Reactor Institut, Delft Holland.
E. Weihreter	Institut für Atom- und Festkörperphysik, Freie Universität Berlin, R.F.A.
H. Weingärtner	Institut für Physikalische Chemie, Universität Karlsruhe, R.F.A.
J. Weis	Laboratoire de Physique Théorique et Hautes Energies, Orsay, France.
I. Yokoyama	School of Mathematics and Physics University of East Anglia, Norwich, U.K.

INDEX